Nuclear Ambitions

A Carnegie Endowment Book

Nuclear Ambitions

The Spread of Nuclear Weapons 1989–1990

Leonard S. Spector
with Jacqueline R. Smith

Routledge
Taylor & Francis Group
LONDON AND NEW YORK

First published 1990 by Westview Press

Published 2018 by Routledge
52 Vanderbilt Avenue, New York, NY 10017
2 Park Square, Milton Park, Abingdon, Oxon OX14 4RN

Routledge is an imprint of the Taylor & Francis Group, an informa business

Library of Congress Cataloging-in-Publication Data
Spector, Leonard S.
Nuclear ambitions : the spread of nuclear weapons 1989-1990 /
Leonard S. Spector, with Jacqueline R. Smith.
p. cm.
Includes index.
ISBN 0-8133-8075-8 – ISBN 0-8133-8074-X
1. Nuclear weapons. 2. World politics – 1985-1995. 3. Nuclear nonproliferation. I. Smith, Jacqueline R. II. Title.
U264.S629
355.8'25119-dc20 90-12835
CIP

ISBN 13: 978-0-367-01576-3 (hbk)
ISBN 13: 978-0-367-16563-5 (pbk)

Contents

Foreword

This is the fifth in the Endowment's series on the spread of nuclear weapons. Through these reports, the Endowment seeks to increase public awareness of the fact and the danger of nuclear proliferation and to stimulate greater attention to this vital issue by policymakers, the media, and the scholarly community.

The series was initiated with the publication of *Nuclear Proliferation Today* (covering developments in 1984) and continued with the issuance of *The New Nuclear Nations* (covering 1985), *Going Nuclear* (covering 1986), and *The Undeclared Bomb* (covering 1987 and 1988). *Nuclear Ambitions* updates these prior volumes with an examination of proliferation trends as of the summer of 1990.

Much has changed in the field of nuclear proliferation since the Endowment began its surveys. In South Asia, Pakistan has emerged as a *de facto* nuclear power, ending the regional nuclear monopoly India had enjoyed since its 1974 test. Today, as U.S.-Soviet relations enter a new era of cooperation, India and Pakistan are near the brink of war. Indeed it is not far-fetched to suggest that in 1990 the Subcontinent replaced Europe as the most likely arena of conflict between nuclear-armed foes.

In the Middle East, Mordechai Vanunu's revelations in 1986 confirmed long-standing speculation that Israel possessed a sizeable nuclear arsenal. Though it remains the region's only nuclear power, the growth of its adversaries' chemical weapon and ballistic missile capabilities has made Israel's cities increasingly vulnerable and has dramatically reshaped the strategic balance in the area. As Iraq's bellicosity has increased in recent months, the threat of a war that could escalate to the nuclear level is becoming a palpable danger in this region, as well.

Since 1984, North Korea's pursuit of a nuclear weapons capability has also become a source of growing concern. If the North's nuclear advances continue, Seoul's now-dormant interest in nuclear arms will inevitably be rekindled.

Pressures would also mount in Japan for increased defense expenditures, and conceivably for a reexamination of the country's non-nuclear stance. Such developments could have consequences of global significance.

As unsettling as these trends may be, other developments suggest that the threat of nuclear proliferation can be reduced under the right circumstances. If South Africa soon joins the Non-Proliferation Treaty, as is widely anticipated, its decision will be the result of a combination of domestic political change, regional accommodation, and non-proliferation diplomacy and regime-building by interested outside parties. Similarly, in Latin America, a confluence of positive developments in the political and security arenas led Argentina and Brazil to initiate a series of reciprocal visits to nuclear installations in 1987, a confidence-building measure which has greatly eased the two countries' long-standing nuclear rivalry. In a related area, American efforts through the Missile Technology Control Regime led Egypt and Argentina to withdraw from the Condor II missile program, which they were pursuing with Iraq, a result that again illustrates that in some instances, the trend toward further proliferation can be arrested.

These accomplishments may prove impossible to replicate in some other settings. Nonetheless, they demonstrate the importance of active non-proliferation diplomacy and the value of international norms, during a period when extraordinary changes on the international scene may offer new opportunities for slowing the spread of nuclear arms and related capabilities.

Leonard S. Spector, the Director of the Endowment's Nuclear Non-Proliferation Project, has been the author of the previous volumes in this series. He prepared *Nuclear Ambitions* with the assistance of Jacqueline R. Smith, the project's Assistant Director. A Senior Associate at the Endowment, Mr. Spector has worked in the field of nuclear non-proliferation for more than ten years, first at the Nuclear Regulatory Commission, and then as Chief Counsel of the Senate Energy and Nuclear Proliferation Subcommittee. His writings on non-proliferation issues have appeared widely. Before joining the Endowment, Ms. Smith was a project manager at Science Applications International Corporation, where she supported the U.S. Commander-in-Chief, Pacific, on a number of projects related to nuclear issues in the region. She has also served as a congressional staff aide in both the House and Senate.

As always, Endowment sponsorship of this report implies a belief only in the importance of the subject and the credentials of the authors. The views expressed are theirs. Comments or inquiries are welcome and may be addressed to the Carnegie Endowment for International Peace, 2400 N. St., N.W., Washington, D.C. 20037-1153.

Thomas L. Hughes
President
Carnegie Endowment
for International Peace

Preface

Like its predecessors in the Endowment's series on nuclear proliferation, *Nuclear Ambitions* surveys the status of nuclear developments in each of the emerging nuclear states, in a country-by-country format.

Every volume after the first—*Nuclear Proliferation Today* (New York: Vintage Books, 1984)—has also included an in-depth look at a particular facet of the nuclear proliferation problem. In the second volume of the series, *The New Nuclear Nations* (New York: Vintage Books, 1985), the thematic chapter described the many facets of clandestine nuclear trade under the title, "The Nuclear Netherworld." The lead chapter in the third volume in the series, *Going Nuclear* (Cambridge, Mass.: Ballinger Publishing Company, 1986), examined the risks of nuclear weapons or related nuclear assets being suddenly transferred from one regime to another as the result of war, revolution, or coup d'état. It used the rubric "nuclear inheritance" to describe this phenomenon, and reviewed relevant historical episodes in French Algeria, China, Vietnam, and Iran. The thematic chapter of the fourth volume, *The Undeclared Bomb* (Cambridge, Mass.: Ballinger Publishing Company, 1988), was devoted to the growing nuclear delivery capabilities of the emerging nuclear states. In the current survey, several of these themes are revisited in a new section entitled, "Global Trends," which also examines a number of additional topics, including the impact on nuclear proliferation of the dramatic changes in U.S.-Soviet relations.

Over time, the country-by-country studies have been expanded to give greater emphasis to domestic political developments in the emerging nuclear states and to the security concerns that give rise to their nuclear ambitions. In addition, the list of countries surveyed has grown. In *Going Nuclear*, a chapter on Iran was added after research disclosed that elements of Iran's Revolutionary Government had demonstrated an interest in acquiring nuclear arms. In the current volume, a chapter on North Korea has been included for the first time, in view of that country's growing nuclear capabilities.

Some revision of the historical analyses contained in earlier volumes has also been necessary, as relevant new evidence has surfaced. The notes indicate instances in which new information has led me to alter conclusions reached in a prior volume.

A word concerning the methodology I have used in preparing these surveys is also in order. The nuclear capabilities and intentions of the emerging nuclear states are shrouded in considerable secrecy. To build up a picture of the nuclear status of many of the countries at issue has therefore required a considerable amount of interpretation on my part. Through footnotes and explanations in the text, I have attempted to be as clear as possible about the basis for my conclusions, which, broadly speaking, rely on press reports, official statements and publications, and interviews—both on- and off-the-record—with U.S. and foreign officials.

In several cases, I have given particular weight to investigative reports by well-known journalists. Before doing so, however, I have attempted insofar as possible to confirm the substance of the report with knowledgeable official sources. When I have not been able to obtain such confirmation, I have treated the report as more speculative; suggested it represents a minority view; or, in some instances, chosen not to rely on it at all.

In preparing *Nuclear Ambitions*, I have had the good fortune of sharing authorship with Jacqueline R. Smith, the Assistant Director of the Endowment's Nuclear Non-Proliferation Project. This volume has been greatly enriched by her insights, analysis, and independent research. While her contribution with respect to North Korea, the emerging nuclear suppliers, and the spread of nuclear submarine programs are particularly noteworthy, the entire manuscript has benefitted greatly, in both substance and style, from her critical judgment and deft editorial hand.

Nuclear Ambitions has been prepared under the auspices of the Carnegie Endowment for International Peace. The project has been sustained by grants from the Carnegie Corporation of New York and the Rockefeller Brothers Fund. I greatly appreciate their generous support for this endeavor. I would also like to thank the Endowment's President, Thomas L. Hughes, for his continuing commitment to this project.

I am also greatly indebted to Ms. Betsy Hamilton, who has served as administrative assistant to this project for the past three years and has endured the hardships of helping bring forth two volumes in this series. In addition to her organizational and editorial contribution, her generosity and good spirits sustained the rest of us working on this project through many a difficult moment.

The work of four successive Carnegie Endowment interns, Felicia Wong, John Shields, Geoffrey Ritts, and Joel Kuester, was also essential to this effort. During their respective six-month stints, all four suffered the trials of maintaining the Non-Proliferation Project's ever-growing information base—the raw material for this book—and undertook research on a number of important topics. In addition, Messrs. Ritts and Kuester provided invaluable editorial assistance during the drafting and production phases. All of them have my sincere thanks.

I would also like to express my appreciation to Larry Fabian, Secretary of the Endowment, and to Pauline Baker, Mario del Carril, Gillian Gunn, Selig Harrison, Geoffrey Kemp, and Paul Kreisberg—all Carnegie Endowment Senior Associates—for their thoughtful comments on the manuscript and concerning the political and security questions confronting many of the countries discussed in this volume. And, once again, Jane Lowenthal and her assistants Lynn Meininger and Chris Henley at the Carnegie Endowment library proved that no research material, however obscure, could long escape their well-honed tracking skills. Thanks too to Mike O'Hare, the Endowment's chief administrative officer, for his help with staffing and for his infectious good cheer.

I would also like to express my thanks to Dr. Barry M. Blechman, who served as senior editorial advisor to the project and who first identified the need for a periodic survey of developments in this field.

In addition, *Nuclear Ambitions* has been greatly benefitted by an informal group of advisors that assisted in reviewing the manuscript. In addition to Dr. Blechman, the group included Dr. Joseph Nye, Director, Center for International Affairs, Harvard University; Dr. Albert Carnesale, academic dean, John F. Kennedy School of Government; and Mr. Charles Van Doren, former assistant director of the Arms Control and Disarmament Agency. I greatly appreciate their contributions in enhancing the text's balance and comprehensiveness. The judgments and views contained in this report, however, are my own and the participation of these individuals does not necessarily constitute an endorsement on their part of specific statements or conclusions.

In addition, a number of specialists provided important assistance that I would like to acknowledge, including David Albright, Joseph Bermudez, W. Seth Carus, Gary Milhollin, Harald Mueller, Randy Rydell, Warren Donnelly, and Peter Zimmerman.

My thanks also to the many investigative journalists whose work is cited throughout this volume. Without their contributions, this work could not have been written.

Finally, I wish to thank William Jackson for his fine copy-editing of the manuscript; Holly Horner, for her care in setting the manuscript into type; Diana Reganthal for her preparation of the index; and Brad Wye and Rick Clark, who prepared the maps used in the text.

Leonard S. Spector
Senior Associate
Carnegie Endowment for
International Peace
July 1990

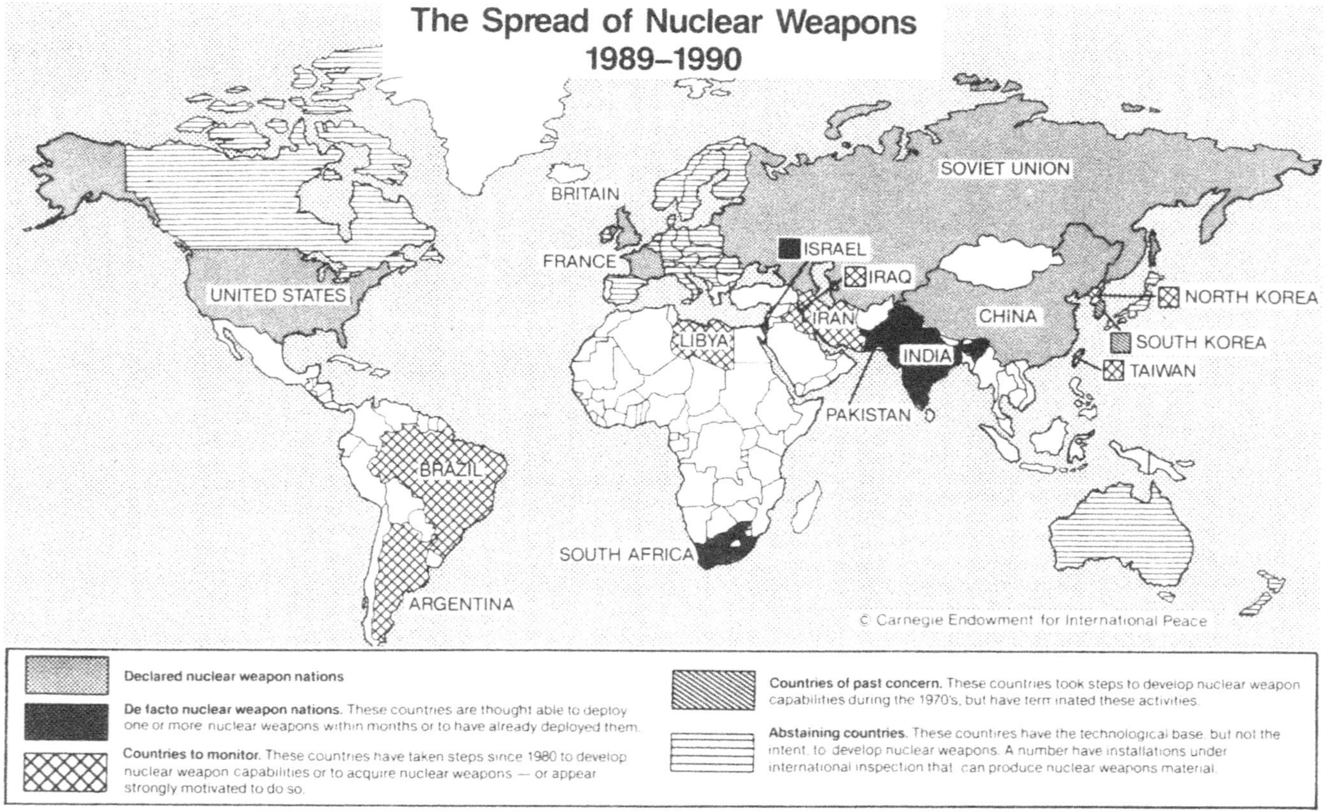

The Spread of Nuclear Weapons
1989–1990
SOVIET UNION
BRITAIN
FRANCE
ISRAEL
IRAQ
UNITED STATES
NORTH KOREA
CHINA
IRAN
LIBYA
SOUTH KOREA
INDIA
TAIWAN
PAKISTAN
BRAZIL
SOUTH AFRICA
ARGENTINA
© Carnegie Endowment for International Peace
Declared nuclear weapon nations
De facto nuclear weapon nations. These countries are thought able to deploy one or more nuclear weapons within months or to have already deployed them.
Countries to monitor. These countries have taken steps since 1980 to develop nuclear weapon capabilities or to acquire nuclear weapons — or appear strongly motivated to do so.
Countries of past concern. These countries took steps to develop nuclear weapon capabilities during the 1970's, but have terminated these activities.
Abstaining countries. These countires have the technological base, but not the intent, to develop nuclear weapons. A number have installations under international inspection that can produce nuclear weapons material.

Part I
Global Trends

As the 1990s began, dramatic changes were taking place within the international system, particularly in the area of U.S.-Soviet relations. A number of these developments will have an important impact on nuclear proliferation. At the same time, the spread of nuclear weapons and related capabilities has, itself, become an important engine of change on the international scene, recasting the strategic balance between rival states in several conflict-prone areas and redefining relations between the major powers and a number of their regional counterparts.

The country-by-country analyses that make up the bulk of this volume describe many of these developments from the narrow perspective of individual countries with demonstrated nuclear ambitions. In addition, however, certain common themes have emerged that warrant examination with a wider, more generic focus. This chapter adopts such a comprehensive global perspective to examine five important trends:

- The steady, though largely concealed, advances of undeclared nuclear weapon efforts in the developing world—efforts which increasingly involve countries hostile to the United States and through which new nuclear nations are now striving to progress well beyond rudimentary nuclear forces;
- The expected impact of changing U.S.-Soviet relations, in such areas as arms control and regional competition, on the spread of nuclear arms;
- Changing patterns in the proliferation of ballistic missiles and their impact on the spread of nuclear weapons;
- The critical dependence of undeclared nuclear weapon programs on clandestine nuclear commerce involving the industrialized nuclear supplier countries, and the growing potential of "second-tier" nuclear supplier countries to assist such programs; and

- The increasing interest of non-nuclear-weapon states in nuclear-powered submarines, a development likely to have an important long-term impact on the spread of nuclear weapon capabilities.

These trends are analysed separately in the five chapters that follow.

*

(Newcomers to the field of nuclear proliferation are encouraged to read Appendices A, B, and C at this point for a brief introduction to some of the technical concepts and institutions discussed throughout this volume.)

1
The Silent Spread of Nuclear Arms 1989-1990

Nuclear proliferation poses a grave danger to the international community, but rarely receives concerted high-level governmental attention in the United States or in other concerned countries. In large measure, this inattention has persisted because for many years there have been few publicly visible and unambiguous indications that nuclear arms are, in fact, spreading to additional states. No new country, for example, has openly declared that it has acquired nuclear arms since 1964, when the People's Republic of China joined the United States, the Soviet Union, Great Britain, and France, as the fifth country to announce that it possessed such weapons.

Similarly, since India's single nuclear detonation in 1974, no country is known to have tested a nuclear device for the first time, nor has India, itself, conducted further tests.* Moreover, those countries that have sought to develop nuclear arms since the mid-1960s have attempted to shield their activities from external scrutiny by denying any interest in such weapons and shrouding their activities in secrecy. Indeed, even when exposed and well publicized, individual steps taken by these countries in their quest for nuclear arms, such as the start-up of a sensitive new nuclear plant or the pursuit of a nuclear smuggling operation, can sometimes appear inconsequential in themselves, and rarely trigger sustained international concern.

If nuclear proliferation has been unannounced, it is real nonetheless. Indeed, the years since 1964 have seen the advent of Israel, India, Pakistan, and South Africa as *de facto* nuclear weapon states, *i.e.*, countries that have deployed nuclear weapons or could do so rapidly in a crisis. The other countries

---

* As discussed in Chapter 9, Israel may have conducted a clandestine nuclear test in the South Atlantic on September 22, 1979, with the assistance of South Africa.

Emerging Nuclear Weapon Nations 1989–1990

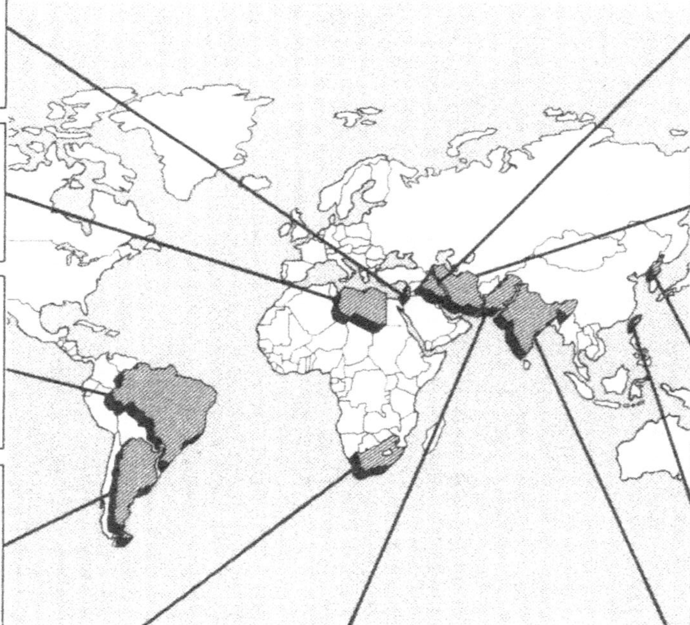

ISRAEL
- Thought to have obtained first nuclear weapons in late 1960's
- Probably has 50–100 undeclared A-bombs
- Since 1982 allegedly built "boosted" weapons that rely on H-bomb principle
- Thought to have deployed short-range nuclear-capable missiles, testing IRBM since 1987
- Not party to NPT

LIBYA
- Many years away from possibly building nuclear weapons indigenously
- Attempted to purchase atomic bomb in early 1970's, 1981
- Party to NPT

BRAZIL
- Exchanging n-plant visits with Argentina
- Civilian government opposed to nuclear arming, but military would keep option
- Has built facilities necessary for nuclear weapons capability as part of nuclear energy and research program, but not under international inspection
- Not party to NPT

ARGENTINA
- Exchanging n-plant visits with Brazil
- Civilian government opposed to nuclear arming
- Has built facilities necessary for nuclear weapons capability as part of nuclear energy program, but not under international inspection
- Not party to NPT

SOUTH AFRICA
- Able to build nuclear weapons since 1980–81
- Possible undeclared arsenal of 15 to 25 n-weapons
- Developing IRBM
- Not party to NPT

PAKISTAN
- Apparently obtained material for first atomic weapon in 1986
- Probably has essentials for 5–10 undeclared A-bombs
- Attempting to develop "boosted" n-weapons
- Tested nuclear-capable short-range missile in 1989
- Not party to NPT

IRAQ
- 5–10 years away from possibly building nuclear weapons indigenously
- Attempting to build plant to produce n-weapons material
- Earlier n-weapon effort thwarted in 1981, when Israel destroyed Osiraq reactor
- Party to NPT

IRAN
- 5–10 years away from possibly building nuclear weapons indigenously
- Some nuclear installations and weapons research inherited from Shah, reactivating n-research program
- Party to NPT

NORTH KOREA
- Has built large research reactor and may be building plutonium plant that could soon allow n-weapon production
- Party to NPT (1985), but above plants not under IAEA inspection since safeguards agreement with IAEA not yet signed

TAIWAN
- Has sizeable nuclear power program, but lacks facilities to produce material for n-weapons
- Built secret lab to extract plutonium in 1987, but dismantled unit under US pressure before plutonium obtained; (made similar attempt in mid-1970s, also thwarted by US)
- Party to NPT

INDIA
- Tested nuclear device in 1974
- Has essentials for 40–60 A-bombs and may be building undeclared nuclear arsenal
- Has greatly expanded nuclear weapons production capability in recent years, reportedly designing H-bomb
- Tested nuclear-capable short-range missile in 1988, tested IRBM in 1989
- Not party to NPT

NPT—The Nuclear Non-Proliferation Treaty. Requires all nuclear installations in a signatory country to be placed under International Atomic Energy Agency inspection.

Carnegie Endowment for International Peace

discussed in this book have also taken important steps toward nuclear arms, though none has yet reached this level. (Map 1.1 summarizes the current status of these programs.)

During 1989 and the first half of 1990, the pattern of slow, steady expansion of nuclear weapon capabilities in the developing world continued. First, based on what is known about its nuclear program, Israel must be presumed to have added a significant quantity of nuclear weapons material free from international non-proliferation controls to its pre-existing stockpile; in 1989, for example, it may have added enough new material for 5 to 10 nuclear weapons.

India, similarly, must be presumed to have augmented its stocks of such material at a roughly comparable rate. In addition, in 1989, it expanded its ability to produce such weapon-grade material, with the opening of a new nuclear power reactor not subject to International Atomic Energy Agency (IAEA) inspection.

Pakistan may have temporarily ceased production of weapons-grade nuclear material in early 1989, but even if it did so, it is thought to be adding to a stockpile of material that could be rapidly improved to the level needed for nuclear arms. Pakistan has also been expanding its overall weapons-material production capabilities and is thought to have taken steps toward the production of complete nuclear devices using its existing stocks of weapons-grade material.

South Africa, too, presumably added to its stockpile of unrestricted weapons-grade material during 1989—perhaps enough for 2 or 3 devices—and in mid-1988, opened a new non-IAEA-inspected uranium enrichment plant that could greatly increase its output of such material. In early 1990, however, under its new state president, Frederik W. de Klerk, Pretoria shut down a key production installation and may be preparing to relinquish its nuclear weapons option by joining the Nuclear Non-Proliferation Treaty.

Evidence also emerged during 1989 and the first half of 1990 that three additional countries—Iraq, Iran, and North Korea—had launched nuclear weapon programs. Iraq has begun a program to build a uranium enrichment plant and to acquire components for nuclear arms, steps which could permit it to manufacture nuclear weapons within the next five to ten years. Iran is also "actively pursuing" a nuclear weapons program, a senior U.S. intelligence official has declared,[1] and leaders of Iran's revolutionary government have openly called for the development of nuclear arms. North Korea, finally, is believed to be building a plutonium extraction plant, which, used in conjunction with other existing facilities, could permit it to manufacture its first nuclear device by the mid-1990s.[2]

In addition, in Latin America, Agentina and Brazil announced that they had produced 20-percent enriched uranium free from international non-proliferation controls, thereby demonstrating a mastery of the enrichment process that would enable them to produce weapons-grade uranium relatively quickly, if they chose to do so. The advent of civilian rule in the two countries and their implemention of a series of reciprocal visits to each other's nuclear installations, however, have substantially reduced the risk of nuclear proliferation in Latin American.

Thus, though they rarely made the headlines, during the recent past, nine regional powers expanded existing nuclear weapon capabilities or made important advances towards acquiring them. This constitutes by far the most important trend in the field of nuclear proliferation. While certain positive steps in South Africa, Argentina, and Brazil may have reduced the dangers of otherwise worrisome technological advances in these states, these hopeful signs have been greatly overshadowed by the disturbing developments elsewhere.

Hostile Proliferators

Within this broad context, a critically important new trend is emerging: the advent of potential nuclear powers with deeply ingrained hostilities toward the United States.

The history of nuclear proliferation from the mid-1960s through the mid-1980s involved almost exclusively nations that, if not always friends of the United States, were at least not openly antagonistic to American interests. The *de facto* nuclearization of Israel in the late 1960s, India in the early 1970s, South Africa in 1980-1981, and Pakistan in 1986 all fit this pattern. So, too, do the abortive nuclear weapon programs of Taiwan and South Korea in the 1970s, the secret nuclear weapons research by the Shah of Iran during this same period, and the gradual movement of Argentina and Brazil during the 1980s toward the nuclear threshold. Only Libya's unsuccessful attempt to purchase nuclear weapons from China in the early 1970s and Iraq's bid for a nuclear capability in the late 1970s—which was thwarted by Israel's destruction in 1981 of the Osiraq reactor—might have led to the emergence of nuclear powers antagonistic to the United States.

The 1990s, however, could witness the arrival of three nuclear powers that have shown themselves repeatedly to be hostile to American interests: Iraq, Iran, and North Korea.

Although Baghdad's moderating foreign policy during the late 1980s led Washington to reopen full diplomatic relations, Iraq remains a bitter foe of Israel, a close U.S. security partner. Iraq is also perceived as an increasing threat by Turkey, a member of the North Atlantic Treaty Organization (NATO) alliance, and by Kuwait and several other Arab Persian Gulf states, on whom the West is highly dependent for oil. Moreover, Iraq's use in 1988 of chemical weapons against civilians in Iraqi Kurdistan and the on-going expansion of its missile, chemical weapon, and other military capabilities raise questions about its long-term commitment to moderation in its external affairs. As its nuclear potential grows, Baghdad's relations with its Arab neighbors and the United States could easily sour, and a wide range of American interests in the region could well be threatened.

Similarly, over the past decade, relations between Iran's Muslim fundamentalist Revolutionary Government and the United States have been marked by profound animosity, which has been fanned by periodic diplomatic and military crises.[3] Iranian acquisition of nuclear weapons would increase Tehran's ability to intimidate the Arab Persian Gulf states, whose oil resources are critically important to the West, and could facilitate the spread of destabilizing Muslim fundamentalism in the region. By posing a potent new threat to U.S. forces, Iranian nuclear arms could significantly raise the stakes of any U.S. effort to project military power into this critically important area.

North Korean nuclearization, finally, would pose grave risks for South Korea, the United States' long-standing ally, and might neutralize the security benefits to South Korea of the U.S. nuclear presence in the region. Moreover, it would seriously heighten the vulnerability of U.S. forces in South Korea should a future conflict erupt with the North. Pyongyang's acquisition of nuclear arms would also place strong pressures on South Korea to develop a nuclear capability of its own, possibly negating more than a decade of U.S. non-proliferation efforts, and could ultimately lead Japan to re-examine its renunciation of nuclear arms—a step that could radically change the security environment in the region and beyond.

Compounding these various threats to American interests from the spread of nuclear arms to Iraq, Iran, and North Korea, is that all three have a history of supporting international terrorism (although Iraq is thought to have renounced this of late), opening still other avenues for mischief.

For a host of reasons discussed elsewhere in this book, the United States has limited the diplomatic pressure it has applied to halt the spread of nuclear weapons in the developing world. Undoubtedly one factor that has influenced American policy-makers has been the recognition that the countries at issue in the vast majority of cases were extremely unlikely to pose a direct nuclear threat to the continental United States, to U.S. allies, or to U.S. forces—even if the risk remained that a conflict involving one of the new nuclear states might lead to U.S. involvement.

This picture of relatively benign proliferation, however ill-founded, is beginning to change dramatically. It remains to be seen how Washington will meet this new challenge.

Beyond the Nuclear Threshold

A further important trend, which became increasingly manifest during 1989 and the first half of 1990, is that none of the emerging nuclear states that has acquired the ability to manufacture nuclear weapons has been content to limit itself to a rudimentary nuclear arsenal consisting of a small number of aircraft-deliverable atomic (*i.e.*, fission) bombs. (See Appendix B.) Not only have Israel, India, Pakistan, and South Africa all taken steps to expand the size of their respective potential nuclear weapons stockpiles, but all are also attempting to develop missile-based delivery systems, and the first three of this group are also moving to acquire advanced nuclear weapons. Simply stated, none of these countries has chosen to "hover" just above the nuclear threshold; all aspire to more advanced capabilities.

Israel is by far the most accomplished in the group. Israel has deployed surface-to-surface missiles with a range of 400 miles and is testing a system able to travel at least 900 miles.[4] The fact that Israel already has nuclear warheads for its deployed missiles has also been recently confirmed.[5] Moreover, Israel's orbiting of space satellites in September 1988 and April 1990 demonstrated its ability to produce nuclear-capable missiles with an intercontinental range, should it choose to do so.

Israel has also manufactured materials, *viz.* lithium deuteride and tritium, for advanced nuclear weapons that rely partly on nuclear fusion, the principle of the hydrogen bomb. British and American specialists have concluded that Israel has used these materials to produce advanced weapons, some of which may have yields as high as 100 kilotons (the equivalent of 100,000 tons of TNT). Such yields are comparable to those of individual warheads on some U.S. strategic missiles. (Israel's advanced nuclear weapons are discussed in detail in Chapter 9.)

India is also pursuing advanced delivery systems and the capability to manufacture sophisticated weapons, indicating its desire for a full-fledged nuclear force. It has announced two tests of a short-range, nuclear-capable missile, the Prithvi, and in 1989 tested the intermediate-range Agni for the first time. Although India may ultimately arm the Prithvi with a nuclear warhead, it will undoubtedly deploy it first as a conventional system.[6] The Agni, however, presents a different picture, since no nation has ever undertaken the enormously complex and expensive task of developing a missile with the 1,500-mile range of the Agni unless it intended to deploy the missile with a nuclear warhead. Officially, India claims the Agni is merely a "technology demonstrator," but there can be little doubt that it represents a bid by New Delhi to move toward an advanced nuclear capability.

India is also known to be producing precursor materials specifically needed to make tritium, a material used to manufacture certain types of advanced nuclear weapons.[7] The work is being undertaken at the Bhabha Atomic Research Center, the presumed site of India's nuclear weapons research program, adding to suspicions as to its purpose.[8]

Pakistan, long fearful of India's superior strategic capabilities, tested two short-range surface-to-surface missiles in 1989, the 50-mile range Hatf-I and the 185-mile range Hatf-II. The latter can carry a half-ton payload, sufficient for a nuclear warhead. Like India, Pakistan has also pursued the development of tritium-boosted weapons. Although it still appears to lack certain essential equipment to manufacture tritium, Pakistan has acquired quantities of tritium gas, a tritium purification and storage plant, tritium precursor materials, the design for a reactor that could be used for tritium production, and material for fabricating fuel for that reactor.[9]

South Africa, too, appears to be progressing toward more advanced nuclear systems. In mid-1989, the country tested a "booster" apparently in collaboration with Israel. A number of reports suggest that the missile was a version of the 900-mile Israeli system mentioned above, a range that implies that the missile, known as the Arniston,[10] is intended to carry a nuclear warhead. (Although South Africa's new State President Frederik W. de Klerk may be on the verge of relinquishing the country's nuclear weapons option, the Arniston was developed by his predecessor, apparently as part of an effort to develop an advanced nuclear weapons capability.)

These developments belie the popular notion that emerging nuclear weapon countries slacken their efforts after obtaining a few "bombs in the basement." In terms of the nuclear hardware they are developing, these countries are emulating the example set by the declared nuclear powers, even if they are not acknowledging their activities. Like the major powers, they are looking to missile delivery systems to ensure the penetration of enemy air defenses, to reach targets that may be beyond the range of their strike aircraft, and to diversify their nuclear forces in order to enhance survivability. And, like the declared nuclear weapon states, the new nuclear nations are apparently pursuing advanced nuclear weapons to obtain unquestionable "city-busting" capabilities.

The longer term implications of this trend are disturbing. In the years ahead, the greater military threat posed by advances in delivery systems and weapon sophistication will broaden and intensify nuclear arms races in the developing world. In South Asia, this pattern is already apparent. Moreover, as countries test longer-range missiles that have little utility without nuclear warheads, their claims that their nuclear programs are strictly peaceful will ring increasingly hollow. In effect, the deployment of long-range missiles by an emerging nuclear state could be perceived as tantamount to a declaration that it possesses nuclear arms.

In addition, as these countries seek to develop increasingly sophisticated nuclear weapons, internal pressures for nuclear testing will grow. Israel appears to have progressed far without them, but it is undoubtedly the most scientifically advanced of the *de facto* nuclear states, and the others are unlikely to be able to match its performance. Moreover, as described in Chapter 9, Israel may have relied on weapons test data obtained from France and the United States and, such assistance notwithstanding, may have found it necessary to conduct a clandestine test to prove out its advanced weapons designs. India has conducted one test of a simple nuclear device, and South Africa and Pakistan may have obtained weapons design data from Israel and China, respectively, which might have given each of these emerging nuclear states confidence, without testing, that its first-generation nuclear weapons will work. Nevertheless, pressures for nuclear testing may grow in all of these countries as they seek to develop advanced nuclear arms. If a breakdown of the *de facto* moratorium on nuclear testing by the nuclear threshold states occurs, it would mean the end of one of the few remaining restraints on their nuclear activities.

One consequence of the developments outlined above is that U.S. non-proliferation policy is changing its orientation with respect to Israel, India, and Pakistan, each of which faces continuing threats to its security and is already able to deploy nuclear arms in a crisis. In these cases, the United States is pursuing efforts to reduce the risk of such weapons being used, along with more traditional policies aimed at preventing the expansion of nuclear capabilities. (Only in the case of South Africa does U.S. policy appear to be still actively focused on the traditional goal of gaining the country's adherence to the NPT.)

In South Asia, the new thrust of U.S. policy is to urge India and Pakistan to expand confidence-building arrangements, such as the two countries' agreement not to attack each other's nuclear installations, and to continue to refrain from nuclear testing and from declarations that they possess nuclear weapons. In the case of Israel, Washington has long encouraged a political solution to the country's security dilemma through the Middle East peace process, but is now also helping Israel to develop defensive, anti-missile systems to reduce the risk of a catastrophic attack on that country, and thereby reduce the risk of nuclear retaliation by Israel.

Some analysts have also suggested that the United States may need to consider sharing technology with Third World countries to improve the safety of their existing nuclear systems and to reduce the risk of unauthorized use, but others fear this will only legitimize the spread of nuclear arms. Less controversial have been suggestions that the United States share its experience with the *de facto* nuclear powers on how to reduce the risk of unintended nuclear escalation during crisis situations.

*

Taken as a whole, the developments discussed above indicate that horizontal nuclear proliferation, *i.e.*, the spread of nuclear arms to additional countries, is continuing, with new efforts under way in Iraq, Iran, and apparently North Korea. At the same time, however, nuclear proliferation in the developing world is also acquiring a "vertical" component, as the *de facto* nuclear weapon states augment and advance their nuclear capabilities.

2 Changing American-Soviet Relations

Dramatically improving relations between Washington and Moscow and the rapid decline of Soviet dominance in Eastern Europe have influenced virtually every aspect of international affairs. These developments are likely to affect nuclear proliferation from a number of quarters.

Improved U.S.-Soviet relations have greatly improved prospects for superpower arms control, which can, in turn, be expected to strengthen the non-proliferation regime. At the same time, reduced East-West competition in regions of proliferation concern could open the door to coordinated U.S.-Soviet non-proliferation diplomacy. The diminishing influence of the superpowers in global affairs, however, may reduce the effectiveness of their efforts in this regard while simultaneously leaving politico-military vacuums that could encourage the emergence of new nuclear powers. These possible repercussions of the changing international environment are discussed below.

Superpower Arms Control and the Nuclear Non-Proliferation Treaty (NPT)

The new direction of superpower relations has created expectations that a significant reduction in the size of American and Soviet strategic arsenals may be possible through the conclusion of a strategic arms reduction treaty (START) during 1990. Progress towards such an agreement can be expected to bolster support for the Nuclear Non-Proliferation Treaty (NPT), the centerpiece of the non-proliferation regime.

As described in detail in Chapter 16, Article VI of the NPT requires the superpowers to "negotiate in good faith" to end their nuclear arms race—an obligation that was part of the bargain in the treaty under which non-nuclear-weapon states agreed to renounce nuclear arms and to accept International

Atomic Energy Agency (IAEA) inspections on all of their nuclear activities. The failure of the United States and the Soviet Union to live up to the spirit, if not the letter, of Article VI has proven controversial in the past and has weakened support for the NPT among some of its non-nuclear-weapon state parties, particularly those in the developing world.

Progress toward a START Treaty would be an important demonstration that the United States and the Soviet Union were meeting their Article VI obligations. This would help to shore up overall support for the treaty and would reduce the possibility of dissension over this issue at the Fourth Non-Proliferation Treaty Review Conference, to be held in late August and early September 1990. Progress on other U.S.-Soviet arms control agreements, such as a treaty to reduce conventional forces in Europe, would enhance the sense that the risk of nuclear war between the superpowers was diminishing and would undoubtedly be seen as further supporting the goals of Article VI. Prospects remain dim, however, for further progress towards a comprehensive nuclear test ban. Many states in the developing world have seen such a ban as the single most important step the superpowers could take toward fulfilling their Article VI obligations.

The NPT not only binds its 137 non-nuclear-weapon state parties to renounce nuclear arms and to accept comprehensive nuclear inspections; it also casts a penumbra that influences the nuclear activities of non-parties. Although India, Israel, Pakistan, and South Africa—the four *de facto* nuclear weapon states—remain outside the pact, the NPT has helped to establish an international norm against the spread of nuclear weapons. This norm has supported the non-proliferation diplomacy of the United States and other interested countries and has also contributed to decisions by all of the *de facto* nuclear states to avoid nuclear testing and refrain from declaring that they possess nuclear arms. Thus, to the extent that superpower arms control agreements bolster the NPT, such agreements should have a positive indirect influence on non-parties, as well.

Actual reductions of superpower nuclear arsenals will also improve the atmosphere for American and Soviet non-proliferation efforts by attenuating the element of hypocrisy inherent in non-proliferation initiatives advanced by countries that not only rely on nuclear armaments, but continue to modernize such forces. India, for example, has included substantial reductions in American and Soviet nuclear arsenals among its conditions for accepting formal nuclear restraints. While India's position may be cynically motivated, the failure of the great powers to slow the expansion of their nuclear forces during the 1980s has lent credibility to its posture. As discussed below, however, the new U.S.-Soviet relationship appears unlikely to generate substantial new superpower non-proliferation efforts for some time to come.

Multilateral Non-Proliferation Diplomacy

American and Soviet cooperation at the International Atomic Energy Agency and at past NPT Review Conferences has been excellent for many years. Improving bilateral relations, therefore, will likely have little visible impact in these fora.

The Missile Technology Control Regime (MTCR) presented a different picture, however, until early in 1990. The U.S.-led regime, formally established in April 1987 after operating informally for two years, consists of a set of parallel export controls adopted by seven Western governments—the United States, Canada, France, Great Britain, Japan, Italy, and West Germany. Belgium, Luxembourg, the Netherlands, and Spain have since joined the regime, and Sweden and Switzerland have agreed to implement comparable controls. The Soviet Union has not been a participant. Indeed, during the mid-1980s, when efforts to establish the MTCR were underway, Moscow transferred hundreds of Scud-B missiles, whose export would have been banned under the regime, to Iraq.[1] After the MTCR was established, moreover, the Soviet Union again transferred hundreds of Scud-Bs, this time to the Afghan government.

A joint communique issued by U.S. Secretary of State James Baker and Soviet Minister of Foreign Affairs Eduard Shevardnadze following their February 7-9, 1990 meeting in Moscow, however, announced that the Soviet Union had agreed to abide by the limits in the MTCR on future missile exports. According to U.S. government sources, the Soviet Union agreed specifically not to make further transfers of the Scud-B, with one major exception. Moscow refused to end exports of the system to Afghanistan, the one government to which Moscow was still providing the weapon in early 1990, except as part of an understanding with Washington that U.S. support for Afghan rebel forces would also be curtailed. Nonetheless, the Soviet decision to abide by the rules of the MTCR in all other respects is a useful step in building a regime to slow the spread of ballistic missiles and, as far as it goes, reflects the improved spirit of cooperation between the superpowers. (Missile transfers by China and North Korea, both outside the MTCR, remain a continuing problem.)

Regional Antagonism and Joint Non-Proliferation Efforts

One important element of growing U.S.-Soviet accommodation during the late 1980s was Moscow's decision to reduce its military commitments in three regions of proliferation concern: South Asia, the Middle East, and Southern Africa. Of particular significance were the withdrawal of Soviet troops from Afghanistan during the latter half of 1988 and early 1989; Soviet support for the Middle East peace process and its warnings to Syria, in late 1989, to cease its threatening bid for strategic parity with Israel; and Soviet encouragement of the phased withdrawal of Cuban forces from Angola, which began in 1989 and is scheduled to be completed in 1991.

As important as these developments have been in the Middle East and

South Asia, they have done little to reduce the basic security concerns that have motivated Israel, Pakistan, and India to acquire *de facto* nuclear weapon capabilities. Thus they are unlikely to affect the nuclear posture of any of these states. Conceivably, South Africa could be an exception to this pattern; the improved security climate in its region, associated in part with American-Soviet accommodation, could prove very significant. Indeed, this new climate may be the development that ultimately tips the balance in Pretoria in favor of joining the Non-Proliferation Treaty.[2]

In South Asia, some observers had anticipated that the Soviet withdrawal from Afghanistan might open the door to intensified nuclear non-proliferation diplomacy by one or both of the superpowers aimed at dampening the growing nuclear rivalry between India and Pakistan. This has not occurred, in part because the Soviet-backed Afghan government has not collapsed as American policy-makers had anticipated. This has led to continued U.S. reliance on Pakistan as a conduit for aid to anti-Soviet Afghan guerrilla forces. Moreover, the United States has been reluctant to impose sanctions against Pakistan for fear of undermining Pakistani Prime Minister Benazir Bhutto, the country's first democratically elected leader in over a decade.

Nor has the Soviet Union taken advantage of the easing of tensions over Afghanistan to pressure India, its own ally in South Asia, to accept added nuclear restraints. Indeed, in January 1988, while it was negotiating the withdrawal of its forces from the region, the Soviet Union leased a nuclear-powered submarine to India—the first-ever transfer of such a system. Later that year, moreover, Moscow's agreement to sell India two nuclear power reactors broke an eight-year *de facto* embargo by the principal nuclear suppliers on major new nuclear transfers to countries that, like India, had refused to place all of their nuclear activities under IAEA inspections. Coming at a time when India was significantly expanding its capacity to produce nuclear weapons material, the Soviet transfers signalled that Moscow was unlikely to support tough non-proliferation measures against New Delhi.

Opportunities may be better for superpower cooperation in addressing North Korea's possible efforts to build a nuclear weapons capability. Although North Korea ratified the Non-Proliferation Treaty in 1985, four years later it had not signed an implementing inspection agreement with the International Atomic Energy Agency, whose inspectors were thus unable to visit some of the country's nuclear plants. Washington and Moscow have coordinated efforts at the IAEA to prod North Korea to conclude the agreement, and Moscow appears to have responded to U.S. concerns by pressing Pyongyang directly to take this step. By June 1990, however, North Korea had not completed arrangements to permit comprehensive IAEA inspections.

Joint superpower efforts also could be useful in slowing Iraq's efforts to revitalize its nuclear weapons effort. Baghdad is dependent on Western credits and on Western and Soviet sales of advanced conventional arms. Despite clear indications of Iraqi interest in acquiring nuclear arms, by June 1990, a collaborative U.S.-Soviet effort to contain Iraq's nuclear ambitions was not in evidence.

Changing International Role

Improving U.S.-Soviet relations have been accompanied by a separate, but related phenomenon: the diminution of superpower dominance in global security affairs. Among other factors that have contributed to this decline of influence have been the relatively poor performance of the U.S. and, especially, the Soviet economies at a time of growing West European and Japanese economic strength; the disintegration of the Warsaw Pact alliance and the seemingly diminished need for American leadership through the North Atlantic Treaty Organization (NATO); the retrenchment of the superpowers' overseas military commitments; and the rise of potent new regional military powers, including Iraq and India.

As superpower influence wanes, the impact will undoubtedly be felt in the area of non-proliferation diplomacy, with Washington and Moscow finding it increasingly difficult to exercise the leverage needed to prevent the spread of nuclear arms to additional states. At the same time, regional states that have relied on superpower security guarantees, such as North and South Korea, and India, may begin to look to nuclear arms as an alternative to weakening commitments from Moscow or Washington.

On a separate front, as the military capabilities of the new regional powers grow and begin to rival those of some of the major powers—Iraq, for example, now possesses more main battle tanks than Great Britain and France combined—the global stratification of military might that has implicitly legitimized the division between the nuclear "haves" and "have-nots" could erode. This could further weaken the norm against the spread of nuclear armaments.

An additional proliferation danger stemming from declining superpower influence is the reorientation of European threat perceptions that may accompany the dismantling of the Warsaw Pact and the parallel scaling-down of NATO's military capabilities. The emergence of a unified Germany in Central Europe, in particular, promises to alter drastically the strategic landscape of the region, opening the possibility of a future German bid for Continental leadership and rekindling pre-World War II fears in Eastern and Western Europe of German hegemony. In NATO's southern tier, Italy and Turkey may become increasingly concerned about the threat posed by their Middle Eastern neighbors, namely Libya and Iraq, respectively, both of which are seeking nuclear arms. In this changing environment, additional European states or Turkey might seek to develop nuclear weapons, a possibility that has been largely dismissed as long as these nations were tied into the basic East-West alliance system.

For the foreseeable future, the emergence of a new nuclear power in this region appears unlikely, given domestic political opposition to nuclear arms in many of these countries and the reduced threat of general war in Europe. Moreover, all of the former Soviet Bloc states that have broken away from Moscow and all of the non-nuclear members of NATO are parties to the Non-Proliferation Treaty. In addition, West Germany is not only, like East Germany, a party to that pact but also gave assurances to the other Western European states at the time it gained its sovereignty in 1954 that it would not manufacture nuclear, biological, or chemical weapons within its territory.[3]

Domestic Turmoil

A final, if more tangential, proliferation concern relates to the pressures for autonomy by national groups within the Soviet Union that have been stimulated by Moscow's diminished authority in Eastern Europe. This is the risk that a militant separatist faction within the Soviet Union might attempt to gain control of Soviet nuclear weapons as a bargaining chip in its bid for independence.[4] In fact, in February 1990, a group of armed Azeri nationalists reportedly attacked a Soviet nuclear weapons storage area in Baku; they were repelled by Soviet forces without causing damage at the site. There is also evidence suggesting that during periods of turmoil in France and the People's Republic of China in the 1960s, elements opposed to the central governments of these states attempted to gain control of nuclear assets to advance their causes.[5] Moscow is said to have taken additional precautions to protect its nuclear weapons, and has removed smaller weapons from areas of turmoil within the Soviet Union.[6]

*

In sum, changing U.S.-Soviet relations can be expected to have a variety of impacts on the proliferation of nuclear arms. For the most part these will tend to support and strengthen current norms against the spread of these weapons and may facilitate joint diplomatic efforts to blunt the nuclear ambitions of some would-be nuclear states. The overall effect of the new U.S.-Soviet amity is likely to be limited, however, and some adverse consequences can be expected as the superpowers' role in world affairs diminishes.

3 Ballistic Missile Proliferation

The proliferation of ballistic missiles has become a fact of life in many Third World strategic environments. Hundreds of conventionally-armed missiles, with ranges of 200 to 400 miles, were launched by Iran and Iraq against each other's cities during the Gulf War, for example, and, during 1989, the Afghan government is said to have fired at least 300 comparable missiles against U.S.-backed rebel forces in that country. In early 1988, similarly, Israel threatened a pre-emptive attack against Saudi Arabia's CSS-2 intermediate-range missiles, recently acquired from China, and expressed increasing alarm about the growing missile threat from Syria. In South Asia, the Indian Army has ordered the domestically developed Prithvi short-range missile for deployment, while Pakistan, with help from China, conducted its first test-firings of two short-range systems in 1989. These are but a few examples that illustrate how surface-to-surface missiles have become a pervasive military factor in some of the globe's most conflict-prone regions.

Table 3.1 summarizes the missile capabilities of states in the developing world.

Ballistic missiles provide three major advantages in comparison to ground-attack aircraft. First, they can penetrate enemy air defenses with far greater certainty than aircraft, providing confidence that payloads will reach their intended targets. Indeed, despite on-going advances in anti-ballistic missile research, no country has effective defenses against a ballistic missile attack. Secondly, as countries graduate to intermediate- and long-range systems able to travel many hundreds of miles or more, ballistic missiles enable military planners to reach targets that are beyond the range of their strike aircraft (or which such aircraft can reach only with the use of vulnerable aerial refueling). Finally, ballistic missiles—especially if deployed in hardened shelters

Table 3.1
Ballistic Missiles in the Third World

Middle East & Africa

	Payload (lbs)	Range (miles)	Accuracy (CEP)	Propulsion	Source	Status
EGYPT						
FROG-7	1,000	40	440 yds	solid	USSR	Deployed
Scud-B	1,100	190	980 yds	1-stage liquid	USSR	Deployed
Vector	1,000	500-600	N/A	solid	Egypt/Argentina	R&D
Sakr 80	450	50	N/A	2-stage solid	Egypt/Iraq/North Korea?	Deployed
Scud-B	2,200	190	980 yds	1-stage storable liquid	Egypt/North Korea?	R&D
IRAN						
Scud-B	1,100	190	980 yds	1-stage solid [sic]	USSR or North Korea	Deployed
Oghab (Eagle)	N/A	25	N/A	N/A	Iran/North Korea	Deployed
Iran-130	N/A	80	N/A	solid	Iran/China	Deployed
IRAQ						
Scud-B	1,100	190	980 yds	1-stage liquid	USSR	Deployed
Scud-B (Al-Husayn)	N/A	375	1-2 mi	1-stage liquid	?	Deployed
Scud-B (Al-Abbas)	N/A	560	2-3 mi	1-stage liquid	?	R&D
Condor II	1,000	500-600	N/A	solid	Argentina/Egypt/Iraq	R&D[sic]
ISRAEL						
Jericho	N/A	400	N/A	2-stage solid	Israel	Deployed
Jericho Follow-ons	N/A	900	N/A	2-stage solid	Israel	Tested
Lance	600	80	400 yds	Liquid	USA	Deployed
LIBYA						
FROG-7	1,000	40	440 yds	solid	USSR	Deployed
Scud-B	1,100	190	980 yds	1-stage liquid	USSR	Deployed
SS-21 (?)	1,000	75	330 yds	solid	USSR	N/A
N/A	N/A	300	N/A	N/A	West German design	R&D
SAUDI ARABIA						
CSS-2 (DF-3A modified)	4,500	1600-1860	1.5 mi	1-stage cooled liquid	China	Deployed
SYRIA						
FROG-7	1,000	40	440 yds	solid	USSR	Deployed
Scud-B	1,100	190	980 yds	1-stage liquid	USSR	Deployed
SS-21	1,000	75	330 yds	solid	USSR	Deployed
NORTH YEMEN						
SS-21	1,000	75	330 yds	solid	USSR	Deployed
SOUTH YEMEN						
FROG-7	1,000	40	440 yds	solid	USSR	Deployed
Scud-B	1,100	190	980 yds	1-stage liquid	USSR	Deployed
SS-21	1,000	75	330 yds	solid	USSR	Deployed

SOURCE: U.S. Arms Control and Disarmament Agency, *World Military Expenditures and Arms Transfers, 1988* (Washington, D.C.: U. S. Government Printing Office, 1989).

Table 3.1
Ballistic Missiles in the Third World (continued)

Far East & Southwest Asia

	Payload (lbs)	Range (miles)	Accuracy (CEP)	Propulsion	Source	Status
INDIA						
Prithvi	2,200	150	N/A	liquid	India	Tested
Agni	1,500-2,000	1,550	N/A	2-stage liquid/solid	India	R&D
KOREA, NORTH						
FROG-7	1,000	40	440 yds	solid	USSR	Deployed
Scud-B	1,100	190	980 yds	1-stage liquid	North Korea	Deployed
KOREA, SOUTH						
Honest John	N/A	25	Unguided	solid	US and South Korea	Deployed
Korean SSM	N/A	110-160	N/A	2-stage solid	South Korea	Deployed
PAKISTAN						
Hatf I - King Hawk	N/A	50	N/A	1or 2 stage	Pakistan w/ technology from China & W.Europe	R&D
Hatf I I - King Hawk	N/A	187	N/A	1or 2 stage		R&D
TAIWAN						
Green Bee (Ching Feng)	N/A	60	N/A	solid	Taiwan	Terminated
Sky Horse	N/A	620	N/A	N/A	Taiwan	Terminated

South America

	Payload (lbs)	Range (miles)	Accuracy (CEP)	Propulsion	Source	Status
ARGENTINA						
Condor I	880	60	N/A	1-stage solid	Argentina	R&D
Condor II	1,100	500-600	N/A	2-stage solid	Argentina/Egypt/Iraq/-West European Consortium	R&D
BRAZIL						
MB/EE-150 (Orbita)	1,100	90	N/A	solid	Brazil	R&D
MB/EE-350 (Orbita)	N/A	215	N/A	solid	Brazil	R&D
MB/EE-600 (Orbita)	N/A	370	N/A	solid	Brazil	Concept
MB/EE-1000 (Orbita)	N/A	620	N/A	solid	Brazil	Concept
SS-300 (Avibras)	2,200	185	N/A	solid	Brazil	R&D
SS-1000 (Avibras)	N/A	740	N/A	solid	Brazil	R&D

SOURCE: U.S. Arms Control and Disarmament Agency, *World Military Expenditures and Arms Transfers, 1988* (Washington, D.C.: U. S. Government Printing Office, 1989).

or on mobile launchers—permit countries to increase the survivability of their retaliatory forces and to diversify them, a concept that has been central to the maintenance of East-West deterrence and that also may be applicable to military theaters in the developing world. Missiles have also become a source of prestige for developing states, symbolizing technological prowess, and longer-range systems, like nuclear weapons, have become indicators of major power status.

In economic terms, modern ground-attack aircraft, which are able to fly repeated sorties with multi-ton payloads, can deliver payloads to a target at a far smaller cost per ton than ballistic missiles, which can be used only once. Moreover, since most missiles in developing country arsenals are relatively inaccurate, only aircraft offer the precision necessary to destroy certain enemy military assets, such as command centers and runways. Nonetheless, because modern warplanes are extremely costly, difficult to replace, and needed in most Third World air forces for a wide variety of combat assignments, including air defense and close ground support, emerging regional powers have turned increasingly to missiles for rear-area strikes against cities and other sizeable defended targets rather than risk their aircraft on such missions.

The attributes that make surface-to-surface missiles attractive to the states acquiring them also make them highly threatening to potential adversaries. Regional powers, states such as Israel, for example, which may have effective defenses against enemy aircraft, have seen their cities become newly vulnerable to attack as the missile capabilities of neighboring countries have grown. Similarly, fears that an adversary might use its missiles to launch a surprise attack during a gathering regional crisis can lead the potential target country to attack enemy missiles first. This highly destabilizing possibility could greatly increase the risk of conflict and is one of the most serious threats posed by the spread of ballistic missiles.

Greatly exacerbating these dangers is the risk that several countries may soon be able to equip their surface-to-surface missiles with chemical warheads. Israel has apparently already taken this step[1] and Egypt, Iran, Iraq, Libya, North Korea and Syria, all of which are thought to possess chemical weapons, might also be able to do so. (India and Pakistan also have a chemical weapons potential, although they are not believed to have manufactured chemical arms.) Chemical warheads can greatly increase the effectiveness of ballistic missiles as terror weapons and can significantly enhance their potential utility against unprotected military targets.

Still more disturbing is the prospect of mating ballistic missiles with nuclear warheads. According to U.S. government documents, Israel has already taken this step.[2] India could follow suit in the next several years, and, as its missile capabilities improve, Pakistan may be able to do so soon thereafter. Argentina, Brazil, Iraq, Iran, and North Korea—all still some years away from acquiring nuclear arms but with growing missile capabilities—could also move toward nuclear-armed missiles by the end of the 1990s.

It is sometimes argued that by enhancing retaliatory capabilities, the proliferation of ballistic missiles improves deterrence and contributes to stability in regions of potential conflict. However, developments during the Gulf War tend to discount this view. In particular, the use of conventionally armed missiles by both Iran and Iraq against the cities of the other clearly demonstrated that both were willing to employ these weapons to gain advantage in combat, rather than merely to deter an adversary's use of such systems. Both states, moreover, employed chemical arms repeatedly during the war, another breakdown of deterrence. Whether this pattern would apply in the nuclear realm if, for example, one of Israel's adversaries were to acquire nuclear weapons, cannot be predicted, but these precedents are not encouraging.

As detailed in the sections of this book analyzing developments in individual emerging nuclear countries, the spread of ballistic missiles is likely to have a profound effect on nuclear proliferation in a number of settings.

- First, the intensifying threat posed by the missile (and chemical weapon) capabilities of Israel's adversaries has made Israeli cities increasingly vulnerable. In response, Israel has begun to hint that it may respond with nuclear weapons to such an attack, raising the possibility that a future Arab-Israeli conflict could escalate to the nuclear level, even if Israel were not threatened with annihilation.
- Second, if Iraq and North Korea continue to advance toward nuclear weapons, the growing missile capabilities of the two states will enhance their ability to retaliate against possible preventive attacks on their nuclear installations. A missile-based Iraqi or North Korean counter-strike—against an enemy city, for example—could lead to a further round of retaliation or even, in some circumstances, to war. On the other hand, if the missile capabilities of these two states successfully deter attacks on their nuclear facilities, they could facilitate the emergence of two new nuclear powers.
- Third, the growing ability of states in the developing world to deploy longer-range nuclear missiles is opening the possibility of dangerous confrontations with the declared nuclear powers. Since 1987, for example, Israel has been testing a follow-on version of its Jericho missile that will be able to reach targets in the southern Soviet Union, and perhaps beyond. Moscow has declared the system a threat and has issued a number of warnings to Israel not to deploy it. Similarly, India is developing the Agni intermediate-range missile, capable of reaching important targets in China. As it moves towards deploying the missile, which would certainly carry a nuclear warhead, nuclear tensions with Beijing will inevitably intensify. In the future, if North Korea acquires nuclear arms, it could deploy nuclear-tipped missiles that would pose a grave threat to U.S. forces in South Korea, while Iraq and Iran, if they cross the nuclear threshold, could menace the Soviet Union with missile-based nuclear systems.

Recent developments. A number of disturbing developments in the area of missile proliferation took place during 1989 and the first half of 1990. These are summarized in Table 3.2, at the end of this chapter. These developments illustrate four related trends in this area. First, and perhaps most troubling, through the actual use of missiles, reported deployments, and orders for missile systems, it appears that a number of developing countries are fully integrating such systems into their military forces.

Second, regional powers have continued their efforts to extend the reach of their missiles. The increasing range of regional missile systems—as reflected in tests of missiles currently under development and announced plans for the development of new systems—threaten both to intensify the nature of traditional regional conflict between more distant adversaries, such as Israel and Libya, and, as suggested above, to increase the risk that extra-regional parties may become involved because of the direct threat posed by the missile capabilities of some regional states.

Third, missile transfers—involving complete systems, as well as components and technology—have continued apace, despite the efforts of a number of countries to stem this flow through the 1987 Missile Technology Control Regime (MTCR). As noted in Chapter 2, in early 1990 the Soviet Union agreed to limit its exports of missiles and missile technology pursuant to the standards of the MTCR, all of whose formal members are Western states.

The MTCR has made selected contributions to easing the pace of missile proliferation. The United States, for example, credits the regime with slowing the Condor II program, involving Argentina, Iraq and, until September 1989, Egypt.[3] Nevertheless, the MTCR remains an insufficient constraint on the global spread of ballistic missiles, in part because the Soviet Union, notwithstanding its agreement to adhere to its limits in other respects, continues to supply Scud-B missiles to Afghanistan.

As of early 1990, China had not joined the MTCR. In 1988, Beijing sold Saudi Arabia the intermediate-range CSS-2, the first-ever transfer of such a system to a developing nation. At about this time, China also began to market a new class of relatively accurate missiles, with ranges up to 375 miles, to Iran and Syria, among other countries. Responding to U.S. demarches, Beijing subsequently assured Washington in late 1988 that it would cease offering the missiles, known as the "M" series, for sale. In mid-1989, however, stories appeared that Syria had signed an agreement to purchase "M" series missiles. In his December 1989 visit to Beijing (which proved controversial for appearing to condone China's harsh repression of dissenters which began in Tiananmen Square the previous June), U.S. National Security Advisor Brent Scowcroft was again given assurances by senior Chinese officials that the country would not sell medium-range missiles in the Middle East. No subsequent Chinese sales of such systems are known to have taken place, but in March

1990, press reports indicated that China was readying a shipment of short-range missiles to the region.[4]

The fourth trend in missile proliferation evident from Table 3.2 is the growing ability of the Third World states who are the object of the MTCR to cooperate on the development and production of surface-to-surface missiles. Iraqi, Argentine, and Egyptian collaboration on the Condor II project; joint Israeli-South African efforts; and North Korean ties to Egypt and Iran with respect to the production of Scud-Bs are all cases in point. This trend suggests the need for expanded international cooperation on efforts to stem missile proliferation possibly to include negotiations on this subject as part of wider discussions on easing regional tensions.[5]

*

The 1987 U.S.-Soviet Intermediate-Range Nuclear Forces (INF) Treaty may provide a useful model for limiting the global spread of ballistic missiles. For regions such as the Middle East where missiles have already been deployed, the INF Treaty stands as proof that the mutual elimination of such systems is indeed possible.

While the INF Treaty eliminated an entire class of US-Soviet weapons (missiles with ranges of 300-3,400 miles), significant activity continues in other countries with nuclear ambitions to produce or deploy missiles in these ranges. Thus, some have advocated the globalization of the INF Treaty.[6] One difficulty with this approach, however, is that it would entail the elimination of classes of French and Chinese nuclear missiles that these countries consider part of their strategic forces, a factor that would greatly complicate negotiations. Nevertheless, regional agreements to proscribe missiles within particular range bands may be a promising approach for reducing the dangers of missile proliferation.[7]

Whatever the merits of these and other prescriptions for halting the spread of ballistic missiles, recent developments, while illustrating the weaknesses of existing control mechanisms, also demonstrate that increasing international cooperation is possible and may prove more effective in the years ahead.

Table 3.2

Missile Proliferation in the Third World: Major Events of 1989-1990

Month	Development	Specifications
February 1989	***Afghanistan*** begins to take receipt of Soviet Scud-Bs from the USSR, and will use over 1,000 by year's end	190-mi. range, 1100-lb. payload
	Indigenous ***North Korean*** ability to produce Scud-Bs confirmed by U.S. government	190-mi. range
	Pakistan announces test firing of Hatf-1 SSM	50-mi. range
	Pakistan announces test firing of Hatf-2 SSM	190-mi. range
March 1989	***Argentina*** reportedly conducts test firing of Condor II in Patagonia	500-600-mi. range (Test: 315 mi.); 990-lb. payload
April 1989	***Brazil*** and the ***PRC*** announce a cooperative venture to produce & market launch vehicles (PRC Long March 2 & Brazilian VLS), launch facilities and services, & satellite-related equipment to Third World countries	N/A
	Brazil conducts successful test launch of the Sonda IV rocket	600 mi.
	Iran claims to have produced a new SSM, but may be a modified FROG or Scud	N/A
	Article states that ***North Korean*** Scud-Bs likely are deployed in hardened bunkers or mountain fortifications; reports collaboration between ***N. Korea*** and ***Egypt*** on designing an improved Scud[1]	N/A
May 1989	***Brazil*** conducts test launch of prototype satellite launching vehicle (VLS-2)	VLS-2 will launch 350-lb. satellite into 410-mi. orbit
	India conducts successful test launch of Agni IRBM	1500-mi. range (Test: 625 mi.); 1-ton payload
	Indian official asserts government is considering ICBM development	4365-5000 mi.
	Press allegation that ***Romanian*** plant to produce intermediate-range missiles is under construction and is based on plans of similar plants in ***Egypt*** & ***Iraq***, drawn up with the assistance of FRG firms[2]	N/A

Table 3.2

Missile Proliferation in the Third World: Major Events of 1989-1990

(continued)

Month	Development	Specifications
May/June 1989	***Indian*** officials indicate likely decision to integrate ballistic missiles into defense strategy	N/A
"Mid-1989"	U.S. government documents confirm ***Israel*** developing Jericho Follow-ons, & has deployed nuclear-armed Jericho I[3]	Jericho I: 400-mi. range Follow-ons: 900-mi. range
June 1989	Report that ***Israeli*** MAR-350 SSM could be complete by end of 1989; intended for sale to IDF and for export	65-mi. range; 735-lb. payload
July 1989	Test launch of ***South African*** "test booster rocket," possibly a Jericho Follow-on, or "Arniston," developed jointly with Israel	900-mi. range
August 1989	Unconfirmed report says ***Syria*** purchases M-9 SSMs from China (later denied by the PRC)	125-375-mi. range
September 1989	***Egypt*** withdraws from Condor II program	N/A
	Indian officials state Army has placed orders for Prithvi SSM	150-mi. range
	India conducts second Prithvi test	150-mi. range
	Syria reportedly conducts discussions with ***PRC*** on joint production of M-9 missiles	125-375-mi. range
	USSR asserts that ***Israel*** conducted SSM launch. Possible test of Jericho Follow-on or failed launch of Offeq-2 satellite from Shavit II launcher	Missile flew 810 mi.
October 1989	***India*** conducts successful test of the first stage of four-stage Polar Satellite Launch Vehicle (PSLV)	PSLV will lift satellites 420-600 mi. into polar orbit
	Extensive ***Israeli-South African*** missile collaboration reported[4]	N/A
	Libyan Al-Fatah ballistic missile reportedly under development with the aid of FRG technicians	300-450-mi. range
	Pakistan announces new missile development program	375-mi. range
November 1989	Article details essential role of Western technology, transferred to ***India*** since 1963, in Agni missile program[5]	N/A

Table 3.2

Missile Proliferation in the Third World: Major Events of 1989-1990

(continued)

Month	Development	Specifications
December 1989	Aside from CSS-2 sale to Saudi Arabia, ***Chinese*** government states "China has never sold nor is it planning to sell missiles to any Middle Eastern country," reaffirming pledge to U.S. not to sell M-9s to ***Syria*** or ***Iran***	N/A
	Iraq launches al-Abid 3-stage rocket (2nd and 3rd stages later reported to be dummies)	Under 1500-mi. range, liquid-fueled, 3 stages
	Iraq announces that two different SSMs, one called the Tammuz I, have been developed	1,250-mi. range
	Iraq announces SSM launch	1,250-mi. range
	Syria reportedly seeking ***North Korean*** help in acquiring SSMs, after cancellation of PRC sales	N/A
"Late 1989"	Possible ***North Korean*** deployment of improved Scud	280-375-mi. range
March 1990	***India*** conducts successful ground test of second-stage PSLV engine	N/A
April 1990	***Argentina*** suspends work on the Condor II	N/A
	Israel launches Offeq-2 satellite, presumably on a Shavit launcher	2,200-mi. range; 1,100-lb. payload

1. Joseph S. Bermudez, Jr. and W. Seth Carus, "The North Korean 'Scud B' Programme," *Jane's Soviet Intelligence Review*, April 1989, pp. 177-181.
2. "The Same Plant is Under Construction in Romania," *Der Spiegel*, May 8, 1989, pp. 166-68, translated in *JPRS-TND*, May 23, 1989, pp. 42-44.
3. Defense Intelligence Agency, *Surface-to-Surface Missile Systems Handbook—Free World (U)*, July 1989 (excerpt released to the Natural Resources Defense Council through the Freedom of Information Act); The United States Arms Control and Disarmament Agency (ACDA), *World Military Expenditures and Arms Transfers, 1988* (Washington, D.C.: ACDA, 1989).
4. Fred Francis, "NBC Nightly News," October 25 & 26, 1989.
5. Gary Milhollin, "India's Missiles—With a Little Help from Our Friends," *Bulletin of the Atomic Scientists*, November 1989, pp. 31-35.

4
New Patterns of Nuclear Trade

Background

As the 1990s begin, two overlapping patterns of nuclear trade are seriously undermining global efforts to slow the spread of nuclear arms.* The first is the growing ability of emerging nuclear states to obtain weapons-relevant nuclear commodities clandestinely from the advanced industrialized nations, the traditional nuclear supplier countries. These transfers have played a central role in advancing the nuclear weapon capabilities of Pakistan, India, Argentina, Brazil, and Iraq.

The second threatening trend has been the expanding commerce in weapons-relevant nuclear goods—both overt and covert—involving a group of less developed states with growing domestic nuclear industries. Only a minority of these "second-tier" or "emerging" nuclear supplier countries have formally accepted the strict nuclear export standards adopted by the advanced nuclear supplier states. Commerce involving the second-tier suppliers has already contributed importantly to nuclear weapon capabilities in India and Pakistan and is likely to grow in significance in the years ahead.

These two patterns of nuclear commerce have been observable for a number of years. However, a stream of disclosures since late 1988 have revealed that both were far more pervasive than previously thought.

During the 1970s, U.S. and international proliferation concerns regarding nuclear commerce focussed increasingly on the threat posed by anticipated

---

* Portions of this chapter were previously published as part of a special report. See Leonard S. Spector, with the assistance of Jacqueline R. Smith, *Nuclear Exports: The Challenge of Control*, (Washington, D.C.: Carnegie Endowment Nuclear Non-Proliferation Project, April 1990).

government-to-government sales of plutonium extraction and uranium enrichment facilities by several advanced countries—facilities that could allow their recipients to produce nuclear-weapons usable plutonium or highly enriched uranium. At the time, nuclear energy advocates in both the advanced and developing world viewed such facilities as necessary elements of full-fledged domestic nuclear energy programs, to which many regional powers aspired.

By the mid-1970s, however, U.S. policy-makers had become concerned that, even with the application of International Atomic Energy Agency (IAEA) safeguards to such facilities, these installations would enable the recipients to come dangerously close to nuclear weapon capabilities by permitting them to accumulate weapons-usable nuclear materials. If the recipient government were prepared to abrogate its safeguards commitments and had taken appropriate preparatory steps, it could rapidly transform such stocks into weapons during a crisis. Citing these dangers, the Ford and Carter Administrations successfully pressed the other advanced suppliers to halt prospective sales of such plants.[1]

Even as the advanced suppliers terminated these sales agreements, however, it became clear that the emerging nuclear weapon states had shifted their strategies for advancing the weapons-related elements of their respective nuclear programs. Rather than building nuclear weapon capabilities around IAEA-safeguarded facilities that had been acquired through government-to-government cooperation, the emerging nuclear states increasingly pursued the construction of unsafeguarded uranium enrichment, plutonium extraction, and related facilities on their own, while taking maximum advantage of any available outside assistance. Today, the nuclear facilities and activities of greatest proliferation concern in Argentina, Brazil, India, Iraq, North Korea, Pakistan, and South Africa all fall into this category. Israel's nuclear program, which was launched in the 1950s with secret assistance from France and has since been expanded with indigenously built additions, also largely fits this pattern. Table 4.1 lists these installations by type and country.

Assistance from more advanced states outside of government-to-government channels has been essential for most of these projects. To obtain such assistance, however, the emerging nuclear states have had to adopt strategies to circumvent the nuclear export control systems of the advanced countries, in part to maintain as much secrecy as possible about their nuclear activities. Moreover, the rules adopted by the advanced nuclear supplier countries through the nuclear Suppliers Group and the Non-Proliferation Treaty (NPT) Exporters Committee require that for all significant nuclear transfers, the recipient country must place the nuclear facility where the export is to be used under IAEA safeguards. Since the strategy of the emerging nuclear states has been to build facilities that are free from such restrictions and therefore potentially available to support nuclear weapons efforts, accepting nuclear transfers under these conditions would be self-defeating. Thus circumvention of these export controls has become imperative—

Table 4.1

Principal Indigenous Unsafeguarded Nuclear-Weapons Relevant Facilities in Countries of Proliferation Concern

Country	Enrichment Plant	Reactor*	Plutonium Plant
Argentina	Pilcaniyeu		
Brazil	Ipero		
India		Cirus (r) Dhruva (r) Madras I (p) Madras II (p)	Trombay Tarapur
Iraq	Believed to be developing clandestine plant		
Israel		Dimona (r)	Dimona
N. Korea		Yongbyon (r)	Yongbyon?
Pakistan	Kahuta	Possible clandestine reactor (r)	Rawalpindi ("New Labs")
S. Africa	Valindaba**		

* r = research reactor; p = electric power reactor

** Pilot-scale & semi-commercial-scale units. In early 1990, South Africa announced plans to shut down the former, believed to be its exclusive source of weapons-grade uranium.

Countries in italics are party to the Non-Proliferation Treaty and are obligated to place all of their nuclear installations under International Atomic Energy Agency (IAEA) safeguards once nuclear materials have been introduced there.

either by means of clandestine activities involving the advanced supplier countries, themselves, or by resort to less punctilious second-tier supplier states.

Clandestine Nuclear Transfers and the Advanced Nuclear Suppliers

During 1988 and 1989, investigations by a parliamentary committee and various prosecutors in West Germany, by private and congressional researchers in the United States, and by investigative journalists around the globe established that covert nuclear transactions originating in Western Europe and North America, or relying on middlemen based there, had played a central role in advancing the nuclear weapon capabilities of a number of developing countries. The breadth and diversity of the activities brought to light is extensive, and efforts to assess their full significance are complicated by the fact that many details have yet to be revealed and many of the persons charged with wrongdoing have not had the opportunity to respond in full. Nonetheless, it is useful to take stock of what has been disclosed in the public record, recognizing that accounts of these transactions may require revision as additional information becomes available.

Regulatory framework. The regulation of nuclear transfers is a highly complex process. In general terms, the nations that are members of the Nuclear Supplier Group (NSG) or that have adopted the NPT Exporters Committee Guidelines have promulgated domestic laws and regulations requiring would-be exporters to obtain export licenses for items specified on "trigger lists" that have been agreed to by these organizations; the term "trigger list" derives from the fact that the export of items on them "triggers" the requirement that the recipient place such items under IAEA safeguards (see Appendix C for a description of safeguards). The trigger lists (which are largely overlapping) include major nuclear facilities, such as nuclear power or research reactors, enrichment plants, plutonium extraction plants, and heavy water plants, as well as key components that have been "specially designed or prepared" for such facilities.

As the lists have been refined over the years, many such components have been identified and added to them. Nonetheless, ambiguities can arise as to whether a particular export falls within the trigger list criteria, a matter which must be resolved through the exercise of discretion by national export licensing authorities. The outcome can be all the more uncertain if, in making a license application, an exporter mischaracterizes the equipment or material involved and if the authorities lack the resources or the inclination to verify these declarations.

Even after an export has been properly authorized under the trigger list procedures and placed under the IAEA safeguards system, additional problems can arise. IAEA regulations, for example, do not require physical inspections to verify the transfer and receipt of certain less sensitive nuclear materials, such as unenriched uranium concentrate ("yellowcake"), or, if the transfer is between NPT parties, of heavy water. This can open opportunities, which have been exploited in the past, for the diversion of materials or equipment to end-users or purposes different from those listed in export documents.

Apart from trigger list items, a separate category of goods, known as "dual-use" items, are also subject to regulation by virtue of more general export licensing controls in advanced industrial states. Dual-use items, such as supercomputers, electronic components, and vacuum pumps, can contribute importantly to nuclear weapon programs but are not on the formal nuclear trigger lists because they are widely used in non-nuclear settings. For this reason, these items often can be freely exported, sometimes without specific export licenses, if intended for a non-nuclear purpose in a friendly country, but are subject to more stringent controls if a nuclear end-use is suspected or if the supplier has less cordial relations with the recipient country. Here, too, export licensing authorities exercise considerable discretion, both in determining whether a declaration of a non-nuclear end-use is specious—which can sometimes be inferred from the identity of the proposed recipient or from the specific characteristics of the item involved, such as its purity—and, if a legitimate nuclear end-use has been declared, in deciding whether to grant a license. Once again, there are obvious opportunities for unscrupulous exporters to attempt to defeat controls by falsifying licensing applications and shipping documents.

Finally, the export of certain non-nuclear items directly useful in the manufacture or delivery of nuclear weapons is controlled under regulations covering transfers of armaments and other military goods. In the United States, such items are enumerated on the Department of State's "Munitions List" and are licensed by that department under the Arms Export Control Act. (In contrast, non-military "dual use" items that might contribute to nuclear weapon programs are set forth on the COCOM list, established by the Western Coordinating Committee on Multilateral Export Controls, and are licensed by the Commerce Department, under the Export Administration Act.) The U.S.-made military-grade capacitors that Iraqi agents attempted to re-export illegally from Great Britain in March 1990 were on the U.S. Munitions List. As that episode demonstrates, unscrupulous exporters have employed various stratagems, including the falsification of end-use certificates and shipping documents, to circumvent these controls.

Clandestinity. In advancing their nuclear weapon capabilities, the emerging nuclear states have exploited these and other weaknesses in the nuclear export control systems of the advanced supplier states. In some instances, their agents have engaged in simple smuggling—for example, by attempting to export commodities without required export licenses or by making blatantly false declarations on export documents—and several individuals have been convicted of such deeds in criminal prosecutions. In other cases, however, the emerging nuclear states have seemingly complied with export licensing requirements, obtaining necessary export approvals by artfully minimizing any suspect characteristics of the commodities at issue, taking advantage of the *laissez-faire* attitude of licensing officials in some countries, or exploiting gaps in export regulations.

Common to every episode, however, has been an element of deception and an effort to shield these transactions from public view for fear that their true purpose might be exposed and efforts mounted to halt them. In this sense, virtually all of the attempts of the emerging nuclear states to obtain assistance in the advanced supplier countries for their own unsafeguarded nuclear installations have been, at bottom, clandestine in nature. Weapons-relevant activities in Pakistan, India, Argentina, Brazil, and Iraq have been critically dependent on such clandestine transfers over the past decade. (As described in Chapter 9, Israel, too, turned repeatedly to clandestine nuclear trade to advance its nuclear weapons program during the 1950s and 1960s but was less dependent on such sources during the 1980s, the period covered by this analysis.)

Pakistan. Since the late 1970s, Pakistan has relied heavily on covertly obtained technology and equipment to build its uranium enrichment facility at Kahuta, which apparently began producing weapons-grade uranium in 1986. As of mid-1988, information available publicly indicated that Pakistan had obtained clandestine assistance for the facility—and for an expansion of its enrichment capabilities either at Kahuta or at a second enrichment plant possibly under construction in the town of Golra[2]—from firms in Canada, Great Britain, the Netherlands, Switzerland, and the United States. The assistance included: design technology for the gas centrifuges that are the heart of the Kahuta plant; essential electronic components and measuring equipment; special "maraging" steel for the construction of the centrifuges; and vacuum pumps and other equipment for handling uranium hexafluoride gas within the facility, some of which hardware was specifically designed to handle weapons-grade material. In addition, between 1977 and 1980, with the help of a West German businessman, who was subsequently convicted for the smuggling operation, Pakistan obtained an entire installation for converting natural uranium into uranium hexafluoride, the form of uranium processed in uranium enrichment plants. The facility was built near Dera Ghazi Khan.

Pakistan is also thought to have received assistance in completing the "New Labs" plutonium extraction plant, near Rawalpindi. The unsafeguarded plant is not known to have begun operating. (All of Pakistan's plutonium-bearing spent nuclear fuel is subject to IAEA inspection and could not be processed at the New Labs facility unless the latter were also placed under the IAEA system, at least temporarily, a step Pakistan has not taken.[3])

It had also been reported by the mid-1980s that Pakistan had obtained from Sweden and the Netherlands equipment needed in the manufacture of nuclear weapons themselves, such as specialized X-ray cameras, known as flash X-ray machines, and specialized oscilloscopes. In addition, it attempted unsuccessfully to obtain from the United States high-speed electronic switches, or krytrons, used as nuclear weapon triggers, and reflector material for nuclear devices.[4]

During 1989, investigations in West Germany uncovered yet another major Pakistani smuggling network, involving two firms in the Federal Republic, Neue Technologien GmbH (NTG) and a subsidiary it established to mask its participation, Physikalische-Technische Beratung (PTB). Between 1983 and 1987, these firms reportedly assisted Pakistan's uranium enrichment efforts by providing specialized containers for the transport and storage of uranium hexafluoride and 130 metric tons of aluminum for use as centrifuge housings.[5] Apparently, these were to be used in conjunction with Pakistan's expansion of its uranium enrichment capability.

In addition, NTG and PTB provided critically important assistance to a new phase of the Pakistani nuclear weapons effort, the production of tritium for advanced nuclear weapons. According to various press accounts, NTG or PTB provided Pakistan with test quantities of tritium gas, a tritium purification and storage plant, tritium precursor materials, the design for a reactor that could be used for tritium production, and material and equipment for fabricating fuel for that reactor, including special American-made welding lasers.[6] Some of these items, like the uranium hexafluoride containers, were apparently exported from West Germany without appropriate licenses, and in early 1990, several NTG/PTB employees were indicted for these acts. In other cases, however, involving principally "dual-use" items, the firms were able to obtain the necessary approvals from export-oriented West German licensing authorities by glossing over the strategic importance of the items involved and using dummy corporations as recipients—a practice that repeatedly succeeded despite frequent warnings by the United States to Bonn of the true purpose behind these transfers.[7]

Through still other channels, Islamabad also obtained special West German milling machines and "hot isostatic presses," useful in machining weapons-grade uranium into nuclear-weapons components, and German-made mass spectrometers, needed to verify the enrichment level of the output of the Kahuta plant. Pakistan advised West German export authorities that these various dual-use items were intended for non-nuclear purposes and, again, despite U.S. protests, export licenses were granted.[8]

In sum, six ongoing Pakistani nuclear projects depended critically upon matériel or design information clandestinely obtained from the advanced nuclear supplier countries: the original Kahuta enrichment plant, the Dera Ghazi Khan uranium hexafluoride plant, a major expansion of Pakistan's enrichment capability, the possible construction of a secret reactor for the production of tritium, a tritium purification plant, and various material and equipment needed for the manufacture of nuclear weapons.

India. The new revelations about Pakistan's covert nuclear purchasing operations revealed that an already well recognized pattern of behavior was more pervasive than previously appreciated. Perhaps more troubling were the findings in 1989 by West German investigators that India's unsafeguarded nuclear program was equally dependent on external assistance—and that a West German middleman had played a central role in providing this aid.

Four unsafeguarded nuclear reactors—the Madras I and II nuclear power plants and the Cirus and Dhruva nuclear research reactors—together with two plutonium extraction plants at Tarapur and the Bhabha Atomic Research Center, form the backbone of India's nuclear weapons capability by permitting the production of plutonium not subject to international non-proliferation controls. All of the reactors, however, require heavy water to operate.

The countries exporting heavy water (with one important exception noted below) have long required that states, like India, that are not party to the NPT, place transferred heavy water under the IAEA system and accept IAEA safeguards on any plutonium produced in a reactor using the transferred material. To keep its reactors free from IAEA safeguards, India has attempted to produce this material domestically. However, India has experienced severe shortfalls in its domestic heavy water production, which hampered operation of the Madras I and II and Dhruva reactors.[9]

Despite this shortage, India was able to place these units in service between 1983 and 1985. How India obtained the necessary heavy water for the three reactors—which require some 500 metric tons in all—has remained shrouded in mystery. In late 1988 and 1989, however, West German investigators determined that India obtained more than 250 metric tons of the material with the assistance of a West German nuclear materials broker, Alfred Hempel. Ironically, the bulk of the heavy water—which most observers believe will ultimately be used to support India's development of a nuclear deterrent against China—originated in the latter country.[10] (A metric ton equals 1,000 kilograms or 1.1 short tons.)

In addition, in April 1990, Romanian President Ion Iliescu announced that the country's previous government, led by Nicolai Ceasescu, had diverted 12.5 tons of Norwegian-origin heavy water in its possession to India.[11] Hempel is thought to have master-minded this transaction, as well.

German investigators have unearthed an August 1, 1983 agreement between Hempel and the Indian Department of Atomic Energy's "Directorate of Purchase and Stores" for the long-term delivery of heavy water "from Europe to Bombay," although shipments of the material apparently began at least a year earlier. Hempel's usual *modus operandi* was to obtain the material under false pretenses from supplier countries, which also included Norway and the Soviet Union, by claiming that it was destined for end-users acceptable to the seller, typically in Western Europe. The Chinese material, however, would be shipped first to Sharja or Dubai in the Persian Gulf and then quietly rerouted to India. For Norwegian and Soviet heavy water, Hempel's declared end-users were usually West German, but he typically rerouted the material first to Basel or Zurich and then, to disguise its ultimate destination further, transshipped it to Sharja or Dubai, before sending it on to Bombay.[12] Norway has traced one shipment of its material to an Indian Department of Atomic Energy office in that city.[13]

Many specifics concerning Hempel's deals have now come to light. In the early 1980s, Hempel purchased substantial quantities of heavy water from two Chinese concerns, the Chinese National Chemicals Importing and Exporting Corporation and Sinochem. Given the sensitive nature of their product, the concerns were presumably closely tied to the Chinese government. However, neither they, nor Chinese authorities apparently took steps to verify the ultimate destination of the heavy water exports, nor was the requirement for safeguards applied, since this was not Chinese government policy at the time.[14]

By 1984, however, U.S. and IAEA representations to Beijing led China to begin restricting these exports by insisting on end-user documentation, according to West German press accounts. Hempel, however, was able to continue obtaining heavy water in lesser quantities from his Chinese sources through 1987, probably by using falsified end-user certificates. At this point, Chinese authorities apparently realized the shipments were going to a potential adversary, India, and put an end to them.

In the meantime, Hempel had developed Norway and the Soviet Union as additional sources of the material for India, albeit in considerably smaller quantities. Using false end-user certificates and fraudulently obtained West German import licenses, in December 1983, he assembled 15 metric tons of Norwegian and 4.7 metric tons of Soviet heavy water at Basel airport, which were shipped to Bombay via Dubai. Thereafter, Hempel repeatedly took advantage of a loophole in IAEA regulations to obtain numerous small quantities of the material from the Soviet Union for transshipment to India. The regulations stipulate that heavy water shipments of less than one metric ton between NPT parties do not need to be reported to the agency. Again, presumably using false end-user certificates indicating the material was going to West Germany or another NPT party, he arranged for numerous exports of Soviet heavy water in quantities of 970 or 990 kilograms—just below the IAEA limit. The total number of such transfers has not been revealed, but Hempel used the ploy so frequently between 1983 and 1987 that questions have been raised as to whether the Soviet Union was not a knowing participant in the scam, covertly supporting the nuclear aspirations of its Indian ally.[15]

To avoid obvious violations of West German nuclear export laws, Hempel pursued his activities outside of the country, often working through a subsidiary based in Switzerland, ORDA AG. West German export authorities repeatedly rebuffed U.S. demands that he be prosecuted for his activities on the grounds that he was operating beyond their jurisdiction. The West German parliamentary investigation determined, however, that at least occasionally, some of Hempel's heavy water shipments were transported through West Germany, which could have provided a basis for prosecution. Gary Milhollin, a professor of law at the University of Wisconsin, has argued that Hempel's misuse of German import licenses also offered a basis for legal action against him. Hempel, who consistently denied any wrongdoing, died in 1989 before any legal action had been initiated against him.[16]

While a number of details remain to be revealed, the overall pattern of these transactions is clear. Moreover, the very large quantities of heavy water involved leave little doubt as to the critical importance of this clandestine nuclear trade to India's ability to operate the three largest of the unsafeguarded reactors, all brought on line during the mid-1980s.[17] While highlighting the growing importance in global nuclear commerce of a key emerging supplier country, namely, China, the heavy water sales to India would not have been possible without the participation of a centrally placed middleman in the industrialized West. These sales thus again illustrate the crucial role that clandestine nuclear trade involving the traditional supplier countries has recently played in advancing emerging nuclear weapon capabilities around the globe.[18]

Argentina. In November 1983, Argentina's outgoing military government announced that for five years it had been secretly building a uranium enrichment plant near the town of Pilcaniyeu. Although Argentine nuclear aides asserted at the time that the plant had been built to produce only reactor fuel that would not be usable in nuclear weapons, there were strong grounds for believing that the secret project, which has never been placed under IAEA safeguards and was begun at a time of growing Argentine militarism, was initiated to provide the country with a nuclear weapons option. (See Chapter 13.)

Little has been revealed to date as to how Argentina built the Pilcaniyeu facility. Since the installation uses the long-established gaseous diffusion method for uranium enrichment, many observers tentatively accepted the assertion of Argentine nuclear officials that the plant was built largely with indigenous resources. The same officials, however, acknowledged that fifteen percent of the equipment in the facility had been obtained abroad. Given the secrecy surrounding the plant, this admission was an implicit acknowledgement that Argentina had engaged in clandestine nuclear purchasing activities.[19]

One key mystery remained: where Argentina had obtained the uranium hexafluoride feedstock to be processed in the facility. The country was not known to have a plant for converting natural uranium into this material and, until 1988, had never claimed to possess one.[20] West German investigations into the activities of Alfred Hempel have now provided the answer, highlighting once again the centrality of covert nuclear trade to current-day nuclear proliferation.

Like India, Argentina obtained key nuclear commodities from China with Hempel's help during the early 1980s. These commodities included sizable quantities of unsafeguarded heavy water and unspecified quantities of 20-percent enriched uranium. The former was presumably for use in a planned (but since cancelled) unsafeguarded research reactor, and the latter was intended both for use in Argentina's existing safeguarded research reactors and, apparently, for re-export, as fuel for several small research reactors that Argentina was providing to other developing states. These transactions had been revealed by the mid-1980s, but Hempel's role was not apparent at that time.[21]

The German investigation exposed not only Hempel's participation, but also uncovered that Argentina had imported some thirty to forty-five metric tons of Chinese uranium hexafluoride in 1981.[22] According to a British intelligence report obtained by the German parliamentary committee investigating Hempel's activities, an air charter company, Cargolux, was asked to transport the material from Hong Kong to Argentina. U.S. officials persuaded Cargolux not to go through with the deal because of the "failure of Argentina to give assurances that the material would not be used for military purposes."[23] Ultimately, Air France agreed to transport the material.

At the time, Argentina had no known use for uranium hexafluoride, since it was building the Pilcaniyeu enrichment plant in total secrecy.[24] Absent a clandestine source of the material, however, the Pilcaniyeu facility probably would have been delayed for a number of years. Argentina's reluctance to give China assurances that the material would not be used for military purposes provides added evidence that the underlying reason for building the enrichment facility was the desire of Argentina's then military government to acquire a nuclear weapons capability.

Brazil. Brazil, too, has relied on the clandestine acquisition of nuclear goods to develop such a capability. The vehicle for this effort has been Brazil's so-called "parallel" nuclear program, *i.e.*, a classified, unsafeguarded, supposedly indigenous nuclear R&D program, initiated in the late 1970s, which has operated largely under military control alongside the country's safeguarded civil nuclear energy program. The safeguarded program is based on imported facilities and technology, especially those transferred under a massive 1975 nuclear deal with West Germany.[25]

Since the early days of the parallel program, the Brazilian Navy had been attempting, with little apparent success, to develop a uranium enrichment capability using gas centrifuges. In September 1987, however, then Brazilian President José Sarney announced that Brazil had succeeded in using this technology to enrich uranium on a laboratory scale, and in April 1988, Brazil commissioned a pilot-scale enrichment facility at Ipero. The plant is believed to be producing uranium enriched to only 20 percent, not usable for weapons. However, a West German intelligence report, disclosed during the on-going nuclear export investigations in that country, stated that as early as 1987, Brazil had sufficient mastery of the uranium enrichment process to produce 70-percent enriched material, virtually weapons-grade.[26]

Few details are known about the development of the Ipero facility. Nonetheless, two senior Brazilian officials have acknowledged that clandestine assistance from the advanced supplier countries has played a key role in its construction. In June 1988, Rear Admiral Othon Luis Pinheiro da Silva, president of the Brazilian Navy's Special Projects Coordination Office, revealed in an interview that he had personally obtained under false pretenses a special West German lathe for machining cylinders for the Ipero centrifuges. He misrepresented the intended use of the lathe to its German vendors, claiming that it was for the manufacture of rocket shells.[27] Similarly, Brazilian National Nuclear Energy Commission Chairman Rex Nazareth, responding to allegations that Brazil had obtained a wide variety of nuclear equipment by smuggling it out of foreign countries as "tractor parts," agreed that his country had illicitly obtained essential matériel abroad. This, he stated, was probably known to the United States, but, he continued:

The catch is that it [the United States] has no way to prove it. I will not say a word about it, even under torture! Perhaps some day, when I write my memoirs, I will reveal just how it was done.[28]

The West German investigation has, nonetheless, unearthed one ploy Brazil has used—the transfer of technology from the safeguarded civil program to the unsafeguarded parallel program. The 1975 Brazil-West German nuclear trade agreement provided for the construction of a uranium enrichment plant in Brazil under IAEA inspection. Built in the town of Resende, the facility uses the jet nozzle technique, which differs considerably from the gas centrifuge technology used at Ipero. However, in both plants highly corrosive uranium hexafluoride gas must be processed in "cascades" of thousands of individual enrichment units. Brazil's unsafeguarded enrichment facility undoubtedly has benefitted from know-how acquired from the safeguarded, German-assisted program.

Under the terms of the agreement between the two countries, however, the leakage of technology between the two programs was prohibited. As a verification measure, the IAEA was to be notified each time technology, in the form of blueprints or equipment, was transferred to Brazil under the accord. Unfortunately, this system apparently has not worked as intended. A January 1987 internal German Foreign Ministry report obtained by the parliamentary committee investigating the country's nuclear export activities states that the IAEA had not, in fact, been informed by either party of the transfer to Brazil of many pieces of equipment used in the German-assisted enrichment plant. This would have left Brazil free to transfer them quietly to its unsafeguarded enrichment program or to duplicate them for use there. (The report was written eight months before Brazil's September 1987 announcement that it had mastered the enrichment process.) A second 1987 Foreign Ministry report stated that a number of the Brazilian nuclear specialists trained in Germany had been transferred from positions in the safeguarded, civilian nuclear program to the unsafeguarded parallel effort.[29]

Iraq. Iraq's program to develop a nuclear weapon capability remains far behind those of the countries discussed above. Nevertheless, as recent events have vividly revealed, it, too, is turning to clandestine nuclear transfers from the traditional nuclear suppliers to support this effort.

Iraq's reliance on this strategy has a long history. During the late 1970s, for example, it obtained from Italy radiologically shielded hot cells and related equipment suitable for producing plutonium from uranium specimens irradiated in the Osiraq reactor; the hot-cell deal was pursued in secret and revealed only in 1980.[30] Iraq also unsuccessfully attempted to obtain from suppliers in West Germany, Canada, and the United States, two commodities that could have been irradiated in the Osiraq reactor to produce plutonium: bulk depleted uranium and fuel pins containing this material.[31]

According to evidence obtained in a 1984 Italian prosecution, moreover, senior Iraqi military figures expressed interest in obtaining 74.6 pounds (33.9 kilograms) of plutonium—enough for several weapons—from an Italian arms smuggling ring purporting to have such material for sale.[32] One member of the ring, interviewed in 1985, claimed that he met first in Baghdad and then in Rome with members of the Iraqi military to discuss the sale; 1982 telexes between members of the smuggling ring contained in the prosecution file are consistent with this claim.[33] The deal fell through when, after a third meeting in Baghdad, the smugglers were unable to produce samples of the nuclear material. Although it is virtually certain that the plutonium offer was a hoax, possibly intended as a come-on to pave the way for sales of conventional arms, the episode suggests that through mid-1982, when negotiations on the matter ended, at least some in the Iraqi government remained interested in nuclear arming and were looking to Western sources for assistance.

More recently, following the route taken by Pakistan and Brazil, Baghdad appears to have launched a program to develop a gas centrifuge uranium enrichment capability, again relying heavily on imported technology. To date, Iraq has attempted to obtain various essential pieces of equipment from Western sources. These have included special vacuum pumps for an enrichment plant from a vendor in the United States, "maraging steel" used primarily in the manufacture of centrifuges, and rolling machines or lathes, also for manufacturing enrichment centrifuges, which Iraq allegedly received from a West German firm. The latter items apparently are similar to those obtained by Brazil. (China, whose role as an emerging nuclear supplier is discussed further in the next section, is reported to have supplied Iraq with special magnets needed to hold the rotating centrifuges in place as they operate.)[34]

These events form the background of the March 1990 arrest in London of five persons for attempting to export to Iraq military-grade capacitors used to trigger nuclear weapons.[35] The arrests were the result of an 18-month "sting" operation conducted by U.S. Customs Service officials, initiated when the agency was alerted by the U.S. manufacturer of the capacitors that Iraq was attempting to purchase the items through a London-based front company, Euromac. On March 29, 1990, Euromac, a related company, two Euromac officials, and three employees of Iraq's Al-Qaaqaa military research and development institute were indicted in San Diego for attempting to export the capacitors without obtaining the necessary export license from the State Department's Office of Munitions Control.[36]

Iraqi officials have not denied that Euromac was attempting to acquire the capacitors for use in Iraq, but they have claimed the units were intended for use in laboratory-based laser research.[37] The "ruggedized" military specifications of the capacitors belie this claim, however. Moreover, the U.S. customs officials, apparently relying on statements to U.S. undercover agents by various defendants in the case and related evidence, have insisted that the units were intended for use in nuclear weapons.[38]

Given the early stage of Iraq's efforts to build the facilities required to produce weapons-grade nuclear materials, it appears that the country remains five to ten years away from the possible manufacture of nuclear arms. The capacitor affair, however, highlights once again the importance to the country's nuclear weapons effort of matériel obtained illicitly from the advance supplier countries. Iraq's clandestine purchasing activities in the West make it the fifth emerging nuclear state that is known to have employed this route to pursue a nuclear weapons capability during the 1980s.[40]

*

The foregoing analysis leaves little question that clandestine nuclear transfers involving the advanced suppliers have become a central factor in the spread of nuclear arms. The role of West German citizens and West German firms has been particularly troubling, and there is good reason to believe that the laxity of West German export control authorities contributed seriously to the problem. However, other countries, including the United States, have also been the object of smuggling efforts by the emerging nuclear weapon states.

In early 1989, the West German government, whose liberal export practices have also facilitated the spread of chemical weapons and ballistic missile capabilities in the developing world, significantly tightened its export licensing procedures and introduced new legislation to increase the penalties for export control violations. The legislation would also subject German nationals operating outside of the Federal Republic to criminal penalties for assisting foreign nuclear weapon, chemical weapon, or ballistic missile programs, as Alfred Hempel did. The legislation was passed by the German *Bundestag* in a somewhat weakened form in June 1990 and was expected to be made more stringent during consideration by the upper chamber of the German parliament, the *Bundesrat*. Final action was expected by October.[41]

Emerging Nuclear Suppliers

Capabilities. China's export activities, as described above, have substantiated long-standing fears in the United States and in other concerned countries that the advent of new nuclear supplier states could facilitate the spread of nuclear arms.[42] Through the actions of a West German broker, Chinese exports of various nuclear materials appear to have contributed substantially to nuclear weapons capabilities in India and Argentina.

China is also believed, however, to have provided essential weapons-related nuclear aid directly to Pakistan, including the design of the nuclear device detonated in China's fourth nuclear test, uranium enrichment know-how, and—according to stories in the British press—quantities of weapons-usable highly enriched uranium sufficient for Pakistan to build two nuclear devices.[43] Although China pledged publicly in 1984 that it would not contribute to the spread of nuclear weapons and agreed to require IAEA safeguards on its nuclear exports, there is evidence that China has continued providing weapons-related aid to Pakistan and exported unsafeguarded nuclear matériel to other countries as late as 1987.[44] Similarly, Beijing is reported to be assisting Iraq to manufacture special magnets for a uranium enrichment plant essential to that country's apparent bid for nuclear arms.[45]

With respect to the other emerging nuclear supplier countries, the most serious anxiety for U.S. policy-makers is that the *de facto* nuclear powers—Israel, India, Pakistan, and South Africa—might similarly transfer nuclear weapons know-how, weapons-usable materials, or even nuclear weapons to others. There have been repeated reports, for example, that Israel has provided such technology to South Africa, and there has also been widespread speculation that Pakistan might be prepared to share its knowledge of nuclear weaponry—or nuclear weapons themselves—with other Islamic states. Though Pakistan is not known to have done this, it has reportedly begun training Iranian nuclear specialists and may have provided additional nuclear aid to that country.[46]

Apart from such direct aid for nascent nuclear weapon efforts, each of the *de facto* nuclear weapon states and China is also in a position to offer ostensibly non-military nuclear aid in such fields as uranium enrichment, plutonium extraction, research reactor design and construction, fuel fabrication, and uranium production. Comparable assistance is also available from many other countries that have not pursued nuclear weapons capabilities but possess growing civilian nuclear industries. Table 4.2 identifies the emerging nuclear supplier states and their respective export capabilities.

Since the early 1980s, as second-tier supplier capabilities have grown, the United States, with some success, has pressed these newcomers to the international nuclear marketplace to adopt the same controls on their nuclear exports as the advanced nuclear suppliers have implemented. Some new suppliers, such as South Korea and Spain, who are parties to the NPT, are obligated by the treaty to require recipients to place any exported matériel under IAEA safeguards. Other emerging suppliers, including Argentina, Brazil, China, and the *de facto* nuclear weapon states, are not bound by the treaty, but several have adopted the IAEA safeguards export requirement voluntarily. Indeed, the requirement is gradually becoming a norm for international nuclear commerce, to the point that unsafeguarded transactions are generally regarded as suspect. A number of the emerging supplier states have also agreed to exercise restraint in the sale of sensitive technologies, *i.e.*, uranium enrichment and plutonium extraction capabilities, a second principle adopted by the advanced suppliers through the Nuclear Suppliers Group guidelines.

While some of the emerging nuclear suppliers have thus adopted export policies consistent with those of the traditional supplier states, none has yet agreed to the strict standard adopted by the United States, Canada, Sweden,

Table 4.2 Export Potential of the Emerging Nuclear Suppliers

Country	Declared Nuclear Export Control Policy	Major Nuclear Export Capabilities: Alleged Nuclear Weapon-Related Exports to:	Nuclear Weapons, Technology (e.g., Design Data, Test Info) & Materials	Research Reactors	Power Reactors
ARGENTINA	• Requires written certification from importer of peaceful application. • *De facto* requirement that all nuclear exports be covered by IAEA safeguards, being codified.			■	◩
BRAZIL	• No announced general policy. • In specific bilateral agreements, has required safeguards on exports and retransfer of specified nuclear materials and technologies.			■	◩
China	• Pre-1975, supported third world acquisition of nuclear weapons to break superpower monopoly. • In 1984, agreed to require safeguards on all nuclear exports, but has not done so in all cases.	Pakistan Iraq	■	■	◩
INDIA	• No stated policy.		■	■	■
ISRAEL	• No stated policy.	S. Africa	■		
N. KOREA	• Status as NPT party creates requirement to submit all nuclear exports to IAEA safeguards. (NPT is not fully implemented in N.K. because no IAEA safeguard agreement has been concluded.)			■	
S. Korea	• Status as NPT party creates requirement to submit all nuclear exports to IAEA safeguards.				□
PAKISTAN	• No stated policy.	China	■	? ◩	□
S. AFRICA	• In 1987, agreed to comply with spirit and objectives of NPT and Nuclear Supplier Group Guidelines, i.e., to require safeguards on nuclear exports and to exercise restraint in exporting enrichment and plutonium extraction technology.	Israel	■	■	
Spain	• Status as NPT party creates requirement to submit all nuclear exports to IAEA safeguards.				
Taiwan	• As *de facto* NPT party, presumably requires IAEA safeguards on nuclear exports.				

Countries in italics are not parties to the NPT, which creates a requirement to submit all nuclear exports to IAEA safeguards.

Countries listed in all capital letters are potential proliferants.

Table 4.2
(Continued)

	Major Nuclear Export Capabilities							
	Major Reactor Components and/or Services	Fuel Fabrication	Reprocessing/Plutonium Recovery	Uranium Refining/Conversion	Uranium Enrichment/Production of H.E.U.	Heavy Water	Tritium	Zirconium
—	■	■	◩	■	■	◩		■ Tubing Metal
—	■	■	◩	■	■			◩ Metal
—	◩	■	■	■	■	■	■	■?
—	■	■	■	■	■	■	◩	■ Metal
—		■	■	■	■	■	■	
—		■	◩	■				
—	■	■	▨	■				
—		■	■	■	■	■	◩	□?
—		■	■	■	■			■
—	■	■						
—			▨					

LEGEND

- ■ Existing Capability
- ◩ Developing Capability
- □ Planned
- ▨ Evidence of prior intent to acquire capability, but no known capacity to date.

SOURCES: *World Nuclear Industry Handbook, 1989* (Sutton, Surrey, U.K.: Nuclear Engineering International, 1988); William C. Potter, ed., *International Nuclear Trade and Nonproliferation: The Challenge of the Emerging Suppliers* (Lexington), Mass.: Lexington Books, 1990); Leonard S. Spector, *The Undeclared Bomb* (Cambridge, Mass.: Ballinger Publishing Company, 1988); country chapters in this volume.

Australia, Japan, and, reportedly, West Germany that major nuclear exports be made only to countries that have accepted safeguards on *all* of their nuclear activities. One obvious reason for this reluctance is that many of the new supplier states have themselves refused to accept such comprehensive or "full-scope" safeguards on their own nuclear activities. Table 4.2 also summarizes the declared export policies of the emerging nuclear supplier states.[47]

Supplier Relationships. The increasing availability of technologies useful for the development of nuclear weapons is being accompanied by a growing trend towards nuclear cooperation within the developing world—and the prospect that additional countries could gradually join the roster of emerging suppliers.

Many countries of proliferation concern already have nuclear supply relationships with other industrializing countries that could benefit the former's nascent nuclear weapons programs. The scope of these relationships—ranging from broad agreements for nuclear cooperation, to specific civil and weapons-related transactions, to negotiations for future cooperation—is displayed in Table 4.3. In addition to nuclear supplier relationships, Table 4.3 also indicates major military trade relations through which the emerging nuclear suppliers have transferred aircraft, ships, tanks, artillery, and various missile systems to states with nuclear ambitions.[48] Such conventional military sales may indicate a predisposition to engage in nuclear cooperation with military overtones.

A number of troubling developments during the 1980s, in addition to China's unsafeguarded exports noted above, have highlighted the potential proliferation dangers of these various links involving the emerging suppliers. In 1987, Argentina, for example, agreed to help Iran refurbish its only research reactor and to provide low-enriched uranium fuel, under IAEA safeguards, for the unit. Though the reactor is too small to contribute to an Iranian nuclear weapons capability, it will provide useful training for Iranian technicians who might subsequently work on such a project. Similarly, both Brazil and Argentina have established nuclear ties to Iraq (which has apparently renewed its bid to acquire nuclear arms), and China is reportedly assisting Iraq produce specially designed magnets for the latter's nascent uranium enrichment program.[49] In another episode, through the machinations of the German NTG/PTB operation discussed earlier, Pakistan apparently obtained substantial quantities of zirconium metal from its nuclear rival, India. The material likely will be fabricated into cladding for uranium fuel that will be used to run a secret reactor Pakistan is said to be building; it is thought that the reactor will be used to produce tritium for advanced nuclear weapons.[50]

For the future, an additional concern is that the increasing independence from the Soviet Union of several countries in Eastern Europe with substantial nuclear export capabilities—including, Czechoslovakia, Hungary, Poland, and Romania—will add to the list of independent nuclear supplier states. Although all are parties to the NPT and most are also members of the Nuclear Suppliers Group, each of these countries will be seeking aggressively to in-

Table 4.3
Supplier Relationships [51]

Recipients	Suppliers: ARGENTINA	BRAZIL	CHINA	INDIA	ISRAEL	KOREA, N.	KOREA, S.	PAKISTAN	S. AFRICA	SPAIN	TAIWAN
Afghanistan				●							
Albania	◑		◐								
Algeria	●◐◑	○		●							
ARGENTINA	——	●◐	○◐	●○	▒ ○		▒ ●			▒ ◐	
Bolivia	●	●									
BRAZIL	● ◐	——	●◐	●○	▒ ●				◐	●	
Chile	●	●	●◐								
China	○	●○	——					● ▲		●	
Colombia	●	●								●	
Cuba	● ◑										
Ecuador	●○	●								● ◑	
Egypt	●◐	● ◑	●	●◑			●			○	
Gabon							◐				◐
Guatemala	●○										
Hong Kong			●								
INDIA	● ○◐	●○	▲	——						◐	
Indonesia	○		●								
IRAN	▒ ●◐◑		▒ ●○	○	▒	▒ ●?		▒ ◐?		◑	
IRAQ	●	▒ ●◐	▒ ▲	● ◐		▒			▒	▒ ◐	
ISRAEL	◐?	●			——				●?◯▲	○	▒
Jordan			○								
N. KOREA			▒ ●? ▲?			——					
S. Korea	●						——			●	
LIBYA	○◯△	▒ ◐		●		▒			△?		
Malaysia								●			
Mexico	●	◑								◐	
Morocco	○									○	
Niger								●			
Nigeria	●	○									
PAKISTAN	○		▒ ●◑▲	▲				——		▒ ◑	
Paraguay	●◑	●									
Peru	●◐◑			○							
Philippines											○
Romania	○◐		●	◐				●			
Saudi Arabia	○										
Somalia		◐									
S. AFRICA			◐		▒ ●?◐▲				——	▒	▒
Spain		●	●				●			——	
Sri Lanka				◑							
Syria	○			●							
Taiwan							● ◐		◐		——
Thailand											◐
Turkey	●○	○					○	●			
Uruguay	●○										
Venezuela	●	●○									
Vietnam				○							
Yugoslavia	●		●	●						◐	
Zaire	○										

LEGEND

- ● Civil cooperation agreement
- ○ Negotiations, unsigned or unratified agreement
- ◐ Civil transfer; training foreign personnel
- ◑ Civil contract
- ◯ Alleged civil transfer
- ▲ Reported transfer to nuclear weapons program or related facility
- △ Unsuccessful weapons-related transfer
- ▒ Arms sales to, and/or defense industrial cooperation with potential proliferants[5]

SOURCES: William C. Potter's Emerging Nuclear Suppliers Project, Associated with the Monterey Institute of International Studies; William C. Potter, ed., *International Nuclear Trade and Nonproliferation: The Challenge of the Emerging Suppliers* (Lexington, Mass.: Lexington Books, 1990). Note: Recipient countries listed in all capital letters are potential proliferants.

crease its exports, while loosening centralized controls over its industrial concerns. This could increase the risk of improperly regulated nuclear transfers, either as the result of deliberate *laissez-faire* national trade policies or as the result of unscrupulous middlemen preying on export-hungry nuclear firms unfamiliar with the complexities of international nuclear commerce.

On the more positive side, as noted earlier and detailed in column 2 of Table 4.2, some emerging supplier states are beginning to demonstrate a willingness to accept, unilaterally, established norms governing the transfer of nuclear materials and technology. Moreover, the expansion in nuclear power generating capacity worldwide has fallen far short of the plans and projections of the early 1970s. This slowdown has resulted in a much smaller international market for nuclear technology than originally anticipated, and this, in turn, has weakened the nuclear industrial base of a number of second-tier supplier states.

Nonetheless, the emerging nuclear supplier states that have adopted responsible export policies have often failed to institutionalize them by means of international agreements or other formal declarations, and, as a result, such policies could shift all too easily with changes in national leadership. Similarly, the scaled-down international demand for civilian nuclear technology could have the less desirable consequence of increasing competition for limited existing markets, a factor that could induce emerging suppliers to pursue irresponsible export policies, for example, by agreeing to transfer materials or technology without safeguards in order to "sweeten" a deal in the eyes of a prospective importer. Moreover, current emerging suppliers may attempt to stimulate world demand by directing their marketing efforts toward developing countries that have yet to build up a substantial nuclear infrastructure, with a view to reversing that situation.[52]

*

Without more uniform acceptance of the norms governing the transfer of nuclear materials and technology among the emerging nuclear supplier countries, there is a continuing danger that the contribution of these states to nuclear weapons programs in the developing world will grow. The 1980s have already seen the emergence of troubling nuclear ties among developing states, and new links can be expected in the years ahead. Despite the expanding capabilities of second-tier suppliers, many of these states remain dependent upon infusions from the nuclear industry of the advanced nuclear supplier states. Industrialized states should recognize this relationship and increase their vigilance in making nuclear sales to the second-tier suppliers to ensure that the reexport of nuclear technology and equipment is effectively controlled.

5
Nuclear Submarines

Events during the late 1980s demonstrated that the acquisition of nuclear-powered naval vessels by non-nuclear weapon states is likely to pose a growing long-term challenge to international efforts to control the spread of nuclear arms. Although these systems are not expected to carry nuclear weapons, their acquisition creates new opportunities for the diversion of nuclear materials to explosive purposes and lends legitimacy to the development of indigenous uranium enrichment capabilities in countries of proliferation concern that could also be used to support nuclear weapon programs. India, which acquired a nuclear-powered submarine from the Soviet Union under a short-term lease in 1988, and Canada, which announced a program to purchase similar vessels from France or Great Britain the same year, appear to have cancelled their plans to acquire nuclear submarines from foreign suppliers. Nevertheless, given their potential impact on the spread of nuclear arms, developments related to naval nuclear propulsion systems in these and other countries deserve to be monitored closely.

Background

Nuclear-propelled submarines, which are able to operate underwater for considerably longer periods than conventionally powered vessels, run on one of three kinds of fuel: high-, medium- or low-enriched uranium, with levels of enrichment of approximately 97.3 percent, 20 percent, or 7 percent respectively. Most nuclear submarines today are believed to use the first, which is also directly usable for nuclear weapons.[1] Reactor fuel consisting of 20-percent or 7-percent enriched uranium, though not usable for this purpose, contains relatively higher levels of plutonium after it is used. This material could be recovered through reprocessing and potentially employed for nuclear weapons.

Until recently, the spread of nuclear-powered submarines had been limited to the United States, the Soviet Union, Great Britain, France, and China—the five declared nuclear weapon states. This situation changed dramatically in 1988, when the Soviet Union leased a nuclear-powered 1960s-vintage Charlie I class submarine to India, reportedly for a three-year period. The same year, the norm against the spread of these systems was again challenged when Canada announced plans to purchase 10-12 nuclear-powered attack submarines (SSNs) from either Britain, whose vessels would use U.S. technology, or France; Washington had never before agreed to the sale of its nuclear submarine technology to a non-nuclear-weapon state. Brazilian and Argentine programs to develop nuclear submarines indigenously were also a growing source of concern at the time.

A major anxiety raised by the developments in 1988 was that they could weaken the Non-Proliferation Treaty by exploiting for the first time a key gap in its coverage. The treaty requires all parties to ensure that their nuclear exports to non-nuclear-weapon states for peaceful purposes be subject to International Atomic Energy Agency (IAEA) safeguards in the recipient country and, secondly, that all peaceful nuclear activities within a non-nuclear-weapon state that is party to the treaty be under the IAEA inspection system. Language in the treaty, however, provides for some exceptions. It allows parties to transfer nuclear material to a non-nuclear-weapon state without IAEA safeguards if the material is to be used by the recipient for "non-proscribed military purposes," *i.e.*, for non-explosive military purposes, such as naval propulsion. The treaty also permits non-nuclear-weapon state parties, such as Canada, to remove nuclear material from safeguards for this purpose.[2]

Heretofore, since no party had taken advantage of these exceptions, IAEA safeguards comprehensively covered all exports by NPT parties to non-weapon states and, if the non-weapon state was itself a party to the treaty, comprehensively covered all of its nuclear activities. The Soviet Union's submarine transfer to India and Canada's planned acquisition of British or French SSNs, however, created a separate category of transfers and domestic activities not subject to IAEA supervision. These developments, it was feared, would greatly complicate IAEA accounting and could open the door to the diversion to explosive purposes of nuclear materials that would otherwise be safeguarded. The possibility was all the more disturbing because of the widespread use of weapon-grade uranium as submarine reactor fuel.

Employing the NPT loophole, a nuclear-weapon state, for example, could surreptitiously provide a non-nuclear ally with a nuclear weapons capability by supplying it excessive quantities of weapons-usable uranium in the form of submarine fuel. Or a non-weapon state NPT party could attempt to remove from safeguards more material for submarine fuel than it really needed and use the excess for nuclear arms.

As a practical matter, the problem was minimized in the Soviet transfer to India, because Moscow had specified that it would periodically inspect the fuel that it had leased to India and would take the fuel back after it was used. Similarly, Canada, legally bound by the NPT not to manufacture nuclear explosives, had pledged to operate any submarines it acquired under a strict IAEA regimen that would (1) allow it to remove nuclear material from safeguards only by placing it in a submarine reactor and (2) require Canada to account for the fuel after use and to place it once again under IAEA oversight. Nonetheless, it was feared that the precedent of this potential transfer, as well as the transfer to India, would inevitably encourage additional sales of these systems, and in these instances, controls might be far less stringent.

A separate concern triggered by the events of 1988 was that the willingness of NPT states to transfer or use enriched uranium outside of safeguards for naval propulsion would legitimize the production in non-NPT states of unsafeguarded enriched uranium for this purpose. Unfortunately, once a non-NPT country produces the material for naval propulsion, the country would remain free to use it for nuclear weapons. Brazil, for example, has justified its uranium enrichment plant at Ipero as a necessary part of its nuclear submarine development program. Though it is currently producing only non-weapons-usable 20-percent enriched material, it could move to higher levels at a later date, claiming that it was merely producing submarine reactor fuel—but also acquiring a ready nuclear weapons option.[3]

Developments

As suggested above, recent activities concerning the spread of nuclear-powered submarines fall into two broad categories: international transfers of such submarines and efforts in the emerging nuclear states to build them domestically. In reality, however, these two categories are not as distinct as they may first appear, since existing nuclear submarine programs in Third World countries are heavily dependent on transfers of foreign technology, particularly that relating to conventional submarine hull construction. As detailed below, two West German shipyards have provided, or will soon provide, this technology to India, Brazil, Argentina, South Korea, and, it appears, South Africa, while Israel has ordered conventional submarines from one of them. The deals concluded with India and South Korea have been criticized domestically because of the excessive costs involved, and the government of India is currently investigating allegations that Indian officials accepted kickbacks from the German company.[4]

India. As originally contemplated, India's lease of the Soviet Charlie I Class nuclear sub was to be for training purposes, and it was envisioned that by the time the lease expired, India would possess an "indigenous alternative,"[5] using technology that it is now developing domestically.

Past uncertainty surrounding the future of Indo-Soviet leasing arrangements appears to have been laid to rest. Indian press accounts initially stated that the government of former Prime Minister Rajiv Gandhi had contemplated leasing additional Soviet nuclear submarines until India's own vessels were completed.[6] Some Soviet officials, however, noting the criticism surrounding the initial submarine lease, cast doubt on Moscow's willingness to expand the arrangement. A number of unsubstantiated reports surfaced in 1989 and early 1990 that India had taken delivery of a second Soviet SSN, possibly as a replacement for the first vessel, which was reported to have experienced serious technical difficulties. Although the Indian Government denied in 1989 that these problems stemmed from the sub's nuclear reactor, *Navy News and Undersea Technology* quoted Indian officials in early 1990 as stating that India had cancelled plans to acquire additional Charlie-class submarines because of severe radiation problems experienced with the first vessel.[7] The same report indicated that the Charlie I currently under lease will be returned when the lease expires. (The radiation exposure issue could create domestic political opposition to the Indian program, although none had emerged by early 1990.)

The indigenous program appears to be based in part on an agreement with the West German shipyard Howaldtswerke Deutsche Werft (HDW), under which India was to acquire three conventional type-1500 submarines. The first two, the *INS Shishumar* and the *INS Shankush*, were built in West Germany but the third, the *INS Shalki*, was constructed in India with HDW assistance. As the first "largely indigenous submarine," the experience that has been gained through production of the *Shalki* will provide much needed know-how for the construction of a hull suitable for a nuclear-powered vessel, a primary obstacle to India's efforts to develop its own nuclear-powered submarine.[8] A senior Indian Navy official has stated that India intends to have a number of its own domestically developed nuclear-propelled vessels by the year 2010. A stated rationale for this objective is to protect India from the possibility of dwindling oil supplies.[9]

Pakistan. Undoubtedly in response to these activities in India, Pakistan was reported in January 1990 to have begun negotiations with the People's Republic of China to obtain a nuclear-powered submarine of its own.[10] China has previously transferred conventional Romeo-class submarines to Pakistan. A transfer of a nuclear-powered boat would be another serious breach of the norm that has long prevailed in this area.

Canada. Canada's plan to acquire nuclear submarines from Britain or France was reversed early in 1989, when Ottawa announced that it had decided against the purchase of such vessels, primarily for budgetary reasons. Nevertheless, there have been reports that Canada may still be considering the acquisition of a more economical nuclear hybrid submarine, dubbed the SS_n, that is still under development. The SS_n is expected to be powered by a small nuclear power plant that will be integrated into conventional diesel-electric attack submarine (SSK) technology.[11] The nuclear plant will be used to recharge the submarine's batteries while the vessel is still submerged; existing SSK technology requires the submarine to surface for recharging.

The acquisition of the hybrid vessels would have implications for IAEA safeguards under the NPT similar to the acquisition of more traditional nuclear-powered subs with larger power plants. Moreover, the anticipated greater affordability of the new hybrid submarine could make it attractive to existing and aspiring mid-level powers, thereby increasing the risk of proliferating nuclear propulsion technology and related nuclear materials.[12] Interestingly, West German shipyards are reportedly focussing their own attentions on the development of a vessel propelled by such a hybrid.[13] These same shipyards also are providing submarines and submarine technology to countries with nuclear ambitions, some of which have expressed an interest in developing indigenous SSNs.

Brazil. As noted above, Brazil and Argentina each have launched a program to develop an indigenous nuclear-powered submarine. The Brazilian program has advanced the furthest. It has developed and is operating an indigenous nuclear research reactor, IPEN-MB-01, as a first step toward developing a nuclear propulsion system. It has also completed the design of a prototype submarine reactor, RENAP-1; the reactor is scheduled to be installed at the Aramar Experimental Center in 1991 and due to become operational in 1995.[14]

Brazil's submarine technology will be based on the West German type 209-1400 submarine. Brazil is scheduled to receive such a submarine built in West Germany in June 1990 and will then build three similar vessels at the navy yard in Rio de Janeiro. Its nuclear-powered submarine will be based on a design of the Brazilian Navy Engineering office that is expected to rely heavily on German engineering.[15] Some reports estimate that a Brazilian-produced SSN could be operational by 2010.[16] In the interim, the Brazilians could pursue the SS_n option, a technology in which it appears to have expressed an interest.[17]

Argentina. Argentina's program remains in the feasibility study stage. The fact that the program is still alive, however, indicates that the country remains interested in the possibility of acquiring nuclear-powered submarines—an interest that is unlikely to wane if Brazil's submarine program continues to make significant progress.

Argentina's navy, like that of India and Brazil, relies on West German conventional submarines, which might well serve as the starting point for Argentine efforts in the field of nuclear-powered subs. Argentina possesses two Howaldtswerke Deutsche Werft (HDW) 209-1200 submarines, ordered in

1968 and commissioned in 1974, and two TR 1700-type submarines, also of German origin. The acquisition of the latter two vessels was part of a larger contract with Thyssen Nordseewerke which stipulated that, after the initial purchase of two German-constructed subs, Asterillos Domecq Garcia in Argentina would build four additional vessels at a facility to be acquired from the German firm under license. The first two TR-1700s—purchased outright from Germany—were commissioned in 1984 and 1985. Argentina began construction of the second pair in 1983 and 1985. It became clear early in the effort, however, that the program would face serious financial constraints. Thus, the West German government gave Argentina permission to export the first two vessels to emerge from its shipyard, in order to generate additional operating revenues. Nevertheless, by 1988, the program had come to a standstill, and no vessel has yet been completed in Argentina.[18]

Other emerging nuclear-weapon states. Although South Korea, Israel, and South Africa are not known to be developing nuclear submarines, all three have substantial nuclear programs and burgeoning submarine acquisition or production efforts. And, all have concluded submarine purchase agreements with the same West German shipyard as India and Brazil, Howaldtswerke Deutsche Werft (HDW), some under surprisingly similar terms.

In 1987, the government of South Korea placed a secret order—that has since been publicly revealed—for three HDW submarines. The first is scheduled to be delivered in 1991. The remaining two vessels will be built in South Korea. Consistent with South Korean practice in acquiring major defense equipment abroad, these procurements were likely contingent on HDW's agreement to transfer know-how in the area of submarine construction. Total orders are expected to climb to nine, although one estimate was as high as eighteen.[19] U.S. security assistance reportedly has been extended to the ROK for the purchase of two diesel subs.[20] Since the U.S. has not produced diesel submarines for many years, and the ROK currently does not possess the necessary technology to build its own, it appears fairly likely that two of the vessels purchased from West Germany were acquired with U.S. funds.

Israel, too, is slated to acquire West German submarines—two HDW 209s. This procurement also will be financed with U.S. security assistance, and the work will be contracted through an American company so as not to violate "Buy American" regulations. The arrangement also avoids violation of German export laws proscribing the provision of war materials to areas of tension, since the submarines—to be built by HDW and Thyssen Nordseewerke—will be treated as if supplied to the United States. There is no indication at this time that Israel will be involved in production.[21]

South Africa, finally, was poised to purchase HDW 209-Class submarines in the early 1980s, but the procurement was scuttled when West Germany agreed to observe the United Nations arms embargo on South Africa, imposed in 1977. However, the South African company ARMSCOR received submarine blueprints from HDW, for an initial payment of forty million South African Rand.[22] These plans could serve as the basis for future South African submarine construction. Indeed, one report indicates that the information provided would have been sufficient to permit South Africa to build its own vessel.[23] Moreover, it appears that these plans were similar to those provided to India earlier in the decade. If this is the case, then the terms of the Indo-German agreement, whereby Germany reportedly agreed to keep details of the Indian submarine secret, may have been violated by HDW's transfer of the blueprints to ARMSCOR.[24] While Pretoria announced in July 1988 that it had shelved indefinitely its plans to build such vessels, reports surfaced in West Germany during 1989 that the program was continuing covertly.[25]

*

None of the states discussed in this chapter is likely to be able to launch its own nuclear-powered submarines for many years to come. Nonetheless, just as ballistic missiles and nuclear weapons themselves have slowly spread to a growing number of regional powers in difficult security environments, nuclear-propelled submarines can be expected to follow this pattern, and the field therefore deserves close monitoring. Moreover, by providing opportunities for the transfer of unsafeguarded nuclear weapons material and by legitimizing the production of such material in non-NPT states, the spread of nuclear-powered submarines can be expected to pose a growing challenge to global efforts to slow the proliferation of nuclear arms.

Part II
Asia

Three non-nuclear-weapon states in Asia are pursuing nuclear programs with military overtones: India, Pakistan, and North Korea. Their nuclear activities are examined in depth in the three following chapters.

India and Pakistan are currently capable of deploying small nuclear forces comprised of atomic bombs that could be delivered by advanced fighter-bombers, with India's capabilities being considerably greater than Pakistan's. Neither country is believed to have integrated nuclear weapons into its military forces, however, and it is possible that neither has manufactured complete nuclear devices. If growing tensions over Kashmir lead to major hostilities between these two powers, this situation could change rapidly, creating a serious risk of nuclear confrontation in South Asia. In the face of a looming conflict, moreover, it is possible that Pakistan might conduct a nuclear test, a step that would match India's single nuclear detonation in 1974 and serve as a warning to New Delhi of Pakistan's nuclear potential. Neither India nor Pakistan is a party to the Nuclear Non-Proliferation Treaty (NPT).

North Korea, which cannot yet manufacture nuclear weapons, is constructing the facilities that could give it this capability during the 1990s. It is a party to the NPT, but has not yet permitted International Atomic Energy Agency (IAEA) inspectors to visit all of its nuclear installations as the treaty requires. If the North continues its provocative nuclear activities, pressures will mount in South Korea to revitalize its nuclear weapons program, which has been dormant since the late 1970s. Ultimately, Japan might be led to reexamine its non-nuclear status as a result of nuclear developments in the Koreas. (South Korea's nuclear capabilities and the concerns that it and Japan have expressed over North Korea's nuclear program are discussed in Chapter 8.)

*

Taiwan is the fourth non-nuclear-weapon state in Asia that has demonstrated an interest in nuclear arming. Although it is not known to have an active nuclear weapons program at this time, relevant research may be continuing.[1]

In 1987, moreover, according to U.S. officials quoted in the press, Taiwan secretly began construction of a small-scale plutonium extraction unit. In March 1988, under U.S. pressure, it agreed to halt work on the facility and to dismantle a 40-megawatt research reactor, located at the Institute for Nuclear Energy Research at Lung Tan. The reactor was supplied by Canada in 1969. U.S. officials suspected that Taiwan intended to extract plutonium from the reactor's spent fuel clandestinely, as part of an effort either to develop nuclear weapons—if, indeed, a final decision to pursue this course had been taken by Taiwan's leaders—or at least to create the option to build such arms at a later time.[2]

Taiwan would probably have needed several years to acquire sufficient plutonium for a nuclear device and would also have had to circumvent IAEA safeguards covering the Lung Tan reactor's spent fuel.[3] U.S. officials stated that Taiwan had not violated these IAEA safeguards, however, indicating that no plutonium extraction had taken place.[4]

Washington viewed the clandestine construction of a plutonium separation laboratory as a violation of Taiwan's assurances that it would not pursue research in this field, which it had given the United States in 1976. The assurances had been offered after an almost identical episode that year, when the United States had discovered that work on a secret plutonium extraction unit was under way. The earlier episode had been preceded by numerous attempts by Taiwan dating back to 1969 to obtain a plutonium extraction capability.[5]

In the wake of the 1976 plutonium-separation incident, Taipei agreed to dismantle the laboratory unit involved in that controversy, but continued to operate the Lung Tan research reactor. By 1988, it had apparently accumulated enough plutonium-bearing spent fuel from the unit for more than ten weapons (if the plutonium were separated).[6] While this fuel had been subject to periodic inspection by the IAEA, as an added precaution, Washington arranged for the material to be brought to the United States for permanent storage. Now that the reactor is also shut down, there appears to be no potential source of weapons-grade nuclear material on the island other than the spent fuel from the country's nuclear power reactors, material which is closely monitoried by the IAEA.[7]

*

Finally, the activities of the People's Republic of China, a nuclear-weapon state since 1964, are also critically important to the risk of nuclear proliferation in the region. China's role is discussed at various points in the following three chapters.

China poses a potential nuclear threat to India, a factor which has contributed to New Delhi's rejection of bilateral non-proliferation arrangements with Pakistan and to its efforts to develop an intermediate-range ballistic missile. In addition, China is believed to have provided important assistance to Pakistan's nuclear weapons program during the mid-1980s and, as explained below, apparently, is continuing to do so. It is also thought to be assisting Pakistan's ballistic missile program.

In North Asia, Beijing's close ties to North Korea may place China in a position to urge Pyongyang to curtail its provocative nuclear activities. Although China is not known to have assisted the North Korean nuclear weapons effort, it is believed to have assisted its ballistic missile development program.

As detailed in Chapter 4, moreover, throughout much of the 1980s, China sold nuclear materials without IAEA safeguards to a number of emerging nuclear weapon states, including South Africa and Argentina. Thus Chinese actions appeared to contribute to the risk of proliferation not only in Asia, but in other regions, as well.

Through the 1970s and early 1980s, China's policy was not to oppose the spread of nuclear weapons to additional countries, a stance it adopted as a means of diminishing the power of the United States and the Soviet Union. This policy, together with growing concerns about Beijing's nuclear assistance to Pakistan and its unsafeguarded nuclear sales, led the Reagan Administration in 1982 to suspend talks it had initiated with Beijing on an agreement for cooperation in the field of nuclear energy.[8]

In mid-1983, China began an apparent shift in its position, and the U.S.-China talks were resumed. In January 1984, China joined the IAEA and subsequently advised the United States that in the future it would require IAEA safeguards on its nuclear exports to non-nuclear-weapon states. In a series of statements beginning that month, top Chinese leaders declared China's firm commitment to non-proliferation. Premier Zhao Ziyang stated in an address to the Sixth National People's Congress in May 1984, for example, that China does "not engage in nuclear proliferation by helping other countries to develop nuclear weapons."[9] Based on these developments, a nuclear trade agreement was initialed in Beijing during President Ronald Reagan's April 1984 visit.

The Reagan Administration did not submit the accord for the required review by Congress until July 1985, over a year later, reportedly because of concerns that Chinese nuclear aid to Pakistan was continuing.[10] After further talks, however, China agreed to implement its non-proliferation policy, according to U.S. officials, "in a manner consistent with basic non-proliferation practices common to the United States and other suppliers," a circumlocution apparently intended to signify that any Chinese assistance to Pakistan had been halted.[11] China has repeatedly denied that anything beyond normal scientific exchanges ever took place.[12] In the fall of 1985, China also announced it would voluntarily place some of its civilian nuclear installations under IAEA safeguards, matching a step previously taken by all other nuclear-weapon states.

The U.S.-China agreement went into effect in December 1985 after a lengthy congressional review.[13] Through a joint resolution, however, Congress required that before a license could be issued for any nuclear exports to China under the agreement, the President would have to certify to Congress, after further negotiations with Beijing, that China was not assisting other states to develop nuclear weapons; that the provisions of the accord would be "effective" in ensuring that U.S. nuclear exports were used exclusively for peaceful purposes; and that the United States would retain certain controls over China's extraction of plutonium from spent fuel produced from U.S.-supplied fuel or in a U.S.-supplied reactor.[14] As of mid-1990, these certifications had not been made, principally because of U.S. concerns that China was continuing to provide assistance to Pakistan's nuclear weapons program.[15]

(China's sale in 1988 of intermediate-range ballistic missiles to Saudi Arabia and its subsequent pledges to the United States not to transfer additional medium-range missiles to the Middle East are discussed in Chapter 3. Beijing's 1989 agreement to sell a nuclear power reactor to Pakistan is addressed in Chapter 7.)

6
India

India's development of a nuclear weapons capability, which dates from the late 1960s, has been prompted by a range of motives. These have included concerns over the nuclear threat from the People's Republic of China, aspirations for regional pre-eminence and international recognition, and, during the 1980s, the desire to maintain a clear edge over Pakistan's growing nuclear potential. Over the past decade, India has greatly expanded its nuclear weapon production capabilities, constructing a range of nuclear facilities free from International Atomic Energy Agency (IAEA) safeguards or other non-proliferation controls. Indian leaders, however, have declared that the country has not manufactured nuclear arms and will not do so unless "compelled" to take this step because of Pakistan's increasing nuclear potential.

As of mid-1990, India was generally believed not to have deployed nuclear arms as part of its military forces. Its possession of a growing stockpile of nuclear weapons material, however, gave it the ability to take this step rapidly in a crisis, and some reports have indicated that India may have manufactured complete nuclear weapons and readied the aircraft that would be used for nuclear missions.[1] India's extensive efforts to enlarge its plutonium-production capability during the 1980s, in part by means of clandestine nuclear trade, its production of materials needed for advanced nuclear weapons, and its testing of a nuclear-capable intermediate-range ballistic missile all suggest that the country might become an overt nuclear power during the 1990s. In the more immediate future, a continuing crisis with Pakistan over Kashmir has raised the possibility that the first major conflict in history between two nuclear-armed belligerents might soon begin.

Background

Indian nuclear decision-making takes place against a backdrop of broader strategic, foreign policy, and domestic political considerations, and is directly influenced by specific developments in the nuclear programs of India's potential adversaries, particularly Pakistan. Here, centuries old Hindu-Muslim antagonisms remain a central factor, even among the governing elites in both countries, and the three wars fought between the two since gaining their independence from the British in 1947 have left a legacy of deep distrust.

Early years. In retrospect, it appears that the groundwork for an Indian nuclear weapons program was laid in the 1950s.[2] Nonetheless, until the mid-1960s, India's public nuclear stance was one of unequivocally rejecting the development of nuclear arms. In late 1964, shortly after China's first nuclear test and at a time when India's defeat in its 1962 border war with China was still vivid, Prime Minister Lal Bahadur Shastri is believed to have launched a program to reduce the lead time needed to build nuclear arms to six months. Publicly, Shastri continued to declare that he opposed nuclear weapons, but—in a major departure from past official statements—he stated that he supported the development of nuclear explosives, provided they were for peaceful purposes, such as tunnelling.[3] Since peaceful nuclear explosives are essentially indistinguishable from nuclear weapons in all important respects, Shastri had given official sanction to the development of an Indian nuclear weapons option.[4]

Indira Gandhi, who took over as Prime Minister in 1966, initially downplayed the nuclear weapon issue. In late 1969, however, she again made the development of peaceful nuclear explosives a declared objective of India's nuclear program. Mrs. Gandhi's bid for a nuclear explosive capability likely was a veiled response to continuing Chinese nuclear advances. In 1968, the year that India refused to sign the Non-Proliferation Treaty and thereby retained the option to build nuclear arms, China conducted its first thermonuclear, or "hydrogen," bomb test. (See Glossary). In 1969, moreover, China was involved in serious border clashes with the Soviet Union, which undoubtedly called up memories in New Delhi of the 1962 hostilities between China and India.

The 1971 Indo-Pakistani War left India by far the dominant power in the region, while Pakistan was dismembered, with the creation of the independent country of Bangladesh in the territory that had formerly been East Pakistan. Three years later, in May 1974, while Mrs. Gandhi was still prime minister, India conducted its first and only nuclear test, which it referred to as a peaceful nuclear explosion.[5] The prior failure of New Delhi to respond directly to the Chinese nuclear weapons program and the fact that after the 1974 test Mrs. Gandhi did not proceed to build a nuclear arsenal suggest that the test was intended as a symbolic, rather than military, counter to China's nuclear capabilities and as a political gesture, aimed at enhancing India's international stature. In addition, India may have been responding indirectly to the intervention of the United States in the 1971 Indo-Pakistani War, when Washington had moved the aircraft carrier *USS Enterprise* (presumed by India to be nuclear armed) toward the Bay of Bengal,[6] in an effort to constrain India from invading West Pakistan. The 1974 test and India's subsequent conventional military build-up were no doubt partly an effort to establish India's credentials as a major regional power that would be less vulnerable to such external pressures in the future.

Growing nuclear threat from Pakistan. Former Pakistani Prime Minister Zulfikar Ali Bhutto is believed to have launched Pakistan on the path to nuclear arms in 1972, in the wake of the country's 1971 defeat.[7] This was two years prior to India's nuclear test, and although India was known to have the potential to manufacture nuclear arms, Pakistan's overriding objective at the time was probably to offset its rival's conventional military superiority.

As Pakistan's quest for nuclear arms became known in the late 1970s, New Delhi began to reexamine its nuclear posture. Although Moraji Desai, who succeeded Mrs. Gandhi as prime minister in 1977, was committed to a non-nuclear policy, official attitudes began to shift during the tenure of his successor, Charan Singh, who held the office from July 1979 until January 1980. In August 1979, Singh appeared to open a new era in Indian nuclear thinking when he declared, "We do not want to join the race to make a bomb, but if Pakistan sticks to its plans to assemble a bomb, we will perhaps have to reconsider the whole question."[8]

Indira Gandhi, who was returned to office at the beginning of 1980, tended to minimize this emerging threat in her public utterances. However, she soon declared that she would "not hesitate from carrying out nuclear explosions . . . or whatever is necessary in the national interest," and in 1981 and again in 1983 it was reported that India was maintaining a nuclear test site in a state of readiness.[9] In addition, several reports suggested that Mrs. Gandhi had authorized work on improved nuclear weapon designs and, possibly, the fabrication of nuclear weapon components.[10]

Moreover, Mrs. Gandhi quietly began to enlarge India's actual capability to produce nuclear arms and, despite strong pressure from the United States, refused to place the key installations involved under International Atomic Energy Agency safeguards, thus keeping them free of legal constraints that might limit their use for nuclear weapons. The facilities, which remain outside of the IAEA system and now form the backbone of India's nuclear weapons potential,[11] include:

- two nuclear power reactors, Madras I and II;[12]
- the Dhruva research reactor at the Bhabha Atomic Research Center (BARC), near Bombay;[13]
- the refurbished plutonium extraction plant at BARC;[14] and
- the Tarapur plutonium extraction plant.[15]

The capabilities represented by these facililities are discussed in detail below; together they have greatly increased India's nuclear potential in comparison to its ability to produce one bomb per year in 1974.

In the summer of 1984, nuclear tensions with Pakistan increased, spurred by new revelations of Pakistani nuclear smuggling and of reports that China had given it nuclear-weapon-design information.[16] These anxieties were heightened by press accounts stating that Mrs. Gandhi was being urged to launch a pre-emptive attack against Pakistan's Kahuta uranium enrichment plant.[17] Overall Indo-Pakistani relations also deteriorated seriously, as India accused Pakistan of fomenting unrest by Sikh nationalist groups in India's Punjab region and of making incursions into the Siachen Glacier region of Kashmir, which led to a series of border clashes. Against this background, in mid-October 1984, Mrs. Gandhi departed from her earlier muted statements, and in an address to a group of army commanders, declared that the Pakistani nuclear program was "a qualitatively new phenomenon in our security environment," which must add a "new dimension" to India's defense planning.[18]

Pakistani advances; increased Indian concerns. On October 31, 1984 Indira Gandhi was assassinated by Sikh members of her bodyguard, and her son, Rajiv Gandhi, became prime minister. By April 1985, he had taken his mother's expressions of concern about the Pakistani nuclear program a step further. In a series of interviews and statements, which established the pattern of Indian pronouncements on the subject that has continued ever since, Gandhi openly accused Pakistan of seeking to build nuclear weapons and declared that this development was causing India to reconsider its position against nuclear arming.[19] Indeed, in a June 1985 interview in Paris, he hinted that India had already made the components for nuclear arms and, if it chose to, could assemble such weapons rapidly.[20]

In July 1985, a U.S. television report stated that Pakistan had detonated the non-nuclear portion of a nuclear device, a significant milestone toward nuclear arming.[21] In response, on August 8, following a series of parliamentary calls for action, Indian Atomic Energy Commission Chairman Raja Ramanna, who had led the team that had developed and tested India's nuclear device in 1974, announced the start-up of the Dhruva research reactor. Rather than merely extolling India's achievement in commissioning one of the world's largest civilian non-power reactors, the announcement carefully underscored the fact that the plant would give India the ability to produce plutonium from domestic technology and fuel—a none-too-subtle declaration that India had a new source of nuclear weapons material free from all non-proliferation controls.[22] The announcement, widely publicized in Pakistan, conveyed a clear warning that India intended to maintain its superiority in any regional nuclear arms race.[23] Moreover, taken with Rajiv Gandhi's comments, the timing and substance of Ramanna's remarks also made clear that concerns over nuclear developments in Pakistan had become by far the dominant factor in Indian nuclear decision-making. In the wake of these events, in late September 1985, rumors again surfaced in Pakistan that India was about to attack the Kahuta enrichment plant, a plan supposedly conceived by Indira Gandhi but postponed after her death.[24]

Indian nuclear milestone. In the second half of 1985, India began to separate plutonium at the Tarapur reprocessing plant from unsafeguarded Madras-I spent fuel;[25] thus India obtained for the first time plutonium that was indisputably free from any external non-proliferation controls.[26] At the time, India appeared to possess ample stocks of plutonium for its peaceful nuclear program, but this material was subject to non-proliferation restrictions of one type or another.[27] This suggests that New Delhi wished to use the new material for the one purpose to which India's other plutonium could not be put—the manufacture of nuclear arms. While the extraction of plutonium from spent fuel does not necessarily mean that the Gandhi government had decided to build nuclear weapons, it does provide strong evidence that, at a minimum, Gandhi was taking steps to ensure that India would have the option to do so rapidly if circumstances warranted.

Momentary hopes for peace; threat of war. Despite these growing tensions on nuclear issues, a slow thaw in overall Indo-Pakistani relations was evident throughout much of 1985. The warming trend reached its peak on December 17, when Pakistani President Zia ul-Haq visited New Delhi and reached agreement with Prime Minister Gandhi on a step-by-step process for normalizing relations, including new discussions to end border clashes over the disputed Siachen Glacier region in Kashmir, the expansion of trade links, and the resumption of talks on draft peace treaties.[28] On the nuclear front, the two leaders announced that they had reached an agreement to refrain from attacking each other's nuclear installations.[29] The no-attack understanding did not restrict Indian or Pakistani nuclear activities as such. Nonetheless, it appeared to be a cautious initial confidence-building measure, raising hopes that more comprehensive initiatives might follow.

The anticipated process of overall normalization launched in December 1985 soon bogged down, however, largely because of domestic political difficulties on both sides—including Sikh unrest in the Indian state of Punjab and the tumultuous return to Pakistan in April 1986 of Benazir Bhutto, daughter of Pakistan's former prime minister. Each country accused the other of being the unseen hand behind these and other domestic challenges, and the efforts to normalize relations stalled.

By the end of 1986, relations had reached a new low point. Indian military exercises near the Pakistani border, dubbed Operation Brass Tacks, led to a series of mobilizations and counter-mobilizations in December 1986 and January 1987 that some feared might lead to the outbreak of major hostilities. This particular episode was resolved after talks between President Zia and Prime Minister Gandhi led to an agreement on mutual troop withdrawals.[30] Nevertheless, overall tensions between the two countries continued over the Siachen

Glacier; allegations of Pakistani support for extremists in India's Punjab state; and Pakistani accusations that India was contributing to sporadic rioting and terror bombings throughout the country.[31]

Continuing nuclear competition. In parallel with these developments, both states continued the build-up of their nuclear capabilities. In November 1986, their nuclear rivalry intensified, following a U.S. press report that Pakistan had produced weapons-grade uranium for the first time at its Kahuta uranium enrichment plant, in effect, giving it the ability to manufacture nuclear arms.[32] Less than twenty-four hours later, the chairman of the Indian Atomic Energy Commission, Raja Ramanna, declared that India had mastered the same process and could enrich uranium to "any level," although subsequent reports indicated that India had achieved a relatively modest level of only two percent.[33] (Uranium used in nuclear weapons is normally enriched to 90 percent or more.) Later reports also revealed that India had built an unsafeguarded experimental enrichment facility, using the gas centrifuge method, at the Bhabha Atomic Research Center. Ramanna's announcement demonstrated India's considerable technical capabilities, but, like his 1985 announcement of the start-up of the Dhruva reactor, the timing of his statement on the enrichment program revealed New Delhi's acute sensitivity to nuclear developments next door.[34]

Five months later, in March 1987, nuclear frictions were again exacerbated when interviews with the head of Pakistan's classified uranium enrichment program, Dr. A. Q. Khan, and Pakistani President Zia appeared to confirm that the country had acquired the ability to manufacture nuclear weapons.[35] Despite the strained environment which these statements created, Indian policy-makers reacted cautiously, reiterating past pronouncements that India was prepared to develop a nuclear deterrent in response to Pakistan, if necessary, but never openly declaring that this step was being taken.[36]

India's U.S. gambit; test-ban rebuff. New Delhi did, however, use the Khan and Zia statements as part of a continuing campaign during 1987 to press the United States to halt Pakistan's nuclear advances. In particular, the Gandhi government publicized Pakistan's progress toward nuclear arms in order to obtain a cut-off of U.S. military aid to Pakistan, which had increased markedly in the wake of the 1979 Soviet invasion of Afghanistan.[37]

In late September 1987, as the tide in Congress seemed to be turning against further assistance in response to news of Pakistani efforts to smuggle nuclear commodities out of the United States, Pakistan's Prime Minister Mohammad Khan Junejo attempted to seize the initiative by proposing that the two South Asian rivals agree to a ban on nuclear tests.[38] Unlike earlier Pakistani nuclear arms control proposals, this one did not require the comprehensive restrictions on India's nuclear activities that New Delhi was known to find objectionable,[39] and it appeared consistent with the actual nuclear testing policies of both countries.[40] India, however, promptly rejected Junejo's initiative, labeling it "not serious" because it reflected a regional approach to nuclear issues that failed to include China.[41] Gandhi also may have maintained the hope that American pressure on Pakistan would obviate the need for India to make parallel concessions through bilateral negotiations with Islamabad. Gandhi pressed this point during his visit to Washington in October 1987, rejecting President Ronald Reagan's appeal that India enter into a dialogue with Pakistan on the nuclear question.[42]

The Indian gambit proved unsuccessful, however, and, in late December, Congress adopted legislation authorizing $480 million in economic and military aid to Pakistan.[43] Despite concerns over Pakistan's nuclear activities, U.S. lawmakers were unwilling to risk a breach with Pakistan at a time when negotiations over the withdrawal of Soviet troops from Afghanistan were reaching a critical point.[44]

In early 1988, although the Reagan Administration had recently certified to Congress that Pakistan did not "possess" an operable nuclear explosive device,[45] administration officials publicly acknowledged for first time that Pakistan was developing nuclear weapons and that the country's enrichment program was being pursued for that purpose.[46] In response, New Delhi invoked its well-rehearsed formulation that Pakistan's nuclear activities would force India to reconsider its policy against nuclear arming, but that India was still not manufacturing nuclear weapons itself.[47]

Growing Indian capabilities. Gandhi's attempt to influence Washington and his numerous statements on the nuclear question at home again highlighted that while India often invoked the nuclear threat from China as a means of parrying unwelcome proposals for limiting Indian nuclear activities, Pakistan had, in fact, become the central factor in Indian nuclear decision-making. Throughout 1987 and early 1988, New Delhi appeared to be working actively to meet this growing challenge by adding to its stocks of unsafeguarded plutonium, in particular by bringing the Dhruva reactor up to its full 100-megawatt capacity.[48] In a May 1988 interview, M. R. Srinivasan, who had succeeded Ramanna in early 1987 as the head of the Department of Atomic Energy and chairman of the Indian Atomic Energy Commission, declared that India planned to acquire tons of plutonium during the next decade, ostensibly for India's nuclear power and research program.[49] This amount also would be sufficient to produce hundreds of nuclear weapons.

Additional, albeit conflicting, information came to light in 1988 concerning the status of the Indian nuclear program. While there appeared to be little question that India was stockpiling plutonium, a U.S. State Department official testified to Congress in February 1988, that India had not "actually taken the final step to acquire nuclear weapons" and, in off-the-record interviews, other administration and congressional sources stated that they believed that India was not developing an undeclared stockpile of nuclear devices or their components.[50] A few months later, however, a press report cited U.S. intelligence sources as stating that India began producing nuclear weapons at a rate of twenty weapons per year beginning in late 1986.[51] It also quoted a State Department aide as stating that India had manufactured a number of nuclear devices between 1974 and 1977 (when Moraji Desai became prime minister and halted the process), but that the State Department did not believe that India had an on-going program to manufacture nuclear weapons.

Strategic environment. By mid-1988, a number of developments had begun to redraw South Asia's strategic landscape.

First, although India was undisputably the region's preeminent military power, it had become clear by this time that Pakistan had attained the status of *de facto* nuclear-weapon state, ending the nuclear monopoly India had enjoyed for more than a decade and significantly altering the local security balance. The development not only intensified the threat that India had historically perceived to its west, but also undercut hopes India might have harbored of regaining the unchallenged dominance it had enjoyed in the region following the 1971 Indo-Pakistani War. Pakistan's acquisition of a rudimentary nuclear deterrent also tended to depreciate the value of India's conventional military superiority in this theater. In response, New Delhi sought to maintain a significant edge over Pakistan in the nuclear sphere, by expanding its nuclear infrastructure and developing a variety of ballistic missiles. Having rejected the alternative of a regional nuclear non-proliferation regime of some kind, India appeared to have opted for an open-ended, if undeclared, nuclear arms race with its rival to the west.

Pakistan's strengthened ties to the United States between 1980 and mid-1988 created an additional source of frustration for New Delhi. Though American arms and political support for Islamabad did not pose a direct military threat to India, they inevitably intruded on Indian prerogatives, freeing Pakistan to play an independent role in regional affairs and permitting a distant power to pursue its own objectives in a sphere that would otherwise have come increasingly under Indian influence.[52]

During this period, moreover, Pakistani Prime Minister Zia placed increasing stress on his country's Islamic identity, providing extensive support for Muslim fundamentalist guerrilla groups fighting Soviet forces in Afghanistan, forging links with the revolutionary government of Ayatollah Khomeini in Iran, and expanding ties to the Arab Persian Gulf states.[53] In the view of Indian observers, in addition to bolstering Pakistan's autonomy, these developments raised the specter of a pan-Islamic anti-India front and heightened the danger of Pakistani-inspired unrest among India's sizeable Muslim population, particularly in predominantly Muslim Kashmir.

Notwithstanding these challenges, by this point, India was increasingly perceived as a regional superpower. Under Rajiv Gandhi, the country undertook a major expansion and modernization of its conventional forces, which left it unquestionably the region's strongest military power.[54] The most notable force improvements included the acquisition of the Soviet Union's most advanced fighter, the MiG-29; the deployment of a second aircraft carrier; and the lease of a nuclear-powered Charlie I-class submarine from the Soviet Union in January 1988.[55] The development of a short-range dual-capable ballistic missile, the Prithvi, which was first tested in February 1988, enhanced India's capabilities in both the conventional and the nuclear sphere.[56]

Gandhi also had demonstrated a readiness to intervene in neighboring states, committing tens of thousands of Indian troops to a peacekeeping effort in Sri Lanka beginning in 1987, aimed at ending an insurrection by radical elements of the country's Tamil minority. In 1988, similarly, Gandhi airlifted troops to the Maldives to thwart an attempted coup there, an action which earned an expression of appreciation from the United States.[57] Nevertheless, India's assertiveness cast it in the role of "regional bully" in the eyes of many of its neighbors and undoubtedly heightened Pakistani security concerns.

Finally, Gandhi pursued closer relations with the People's Republic of China in an attempt to reduce the threat from this quarter and to dampen Beijing's support for Pakistan. These efforts led to an historic summit in Beijing in December 1988. The meeting ushered in an era of relatively harmonious relations between the two countries, although a long-standing border dispute between them remained unresolved and Beijing's ties to Islamabad appeared to be as strong as ever.

Political and Technical Developments

Nuclear capabilities enlarged. As the 1990s opened, India was continuing to expand its ability to produce plutonium free from IAEA safeguards or other external non-proliferation restrictions. The most important new addition to its roster of unsafeguarded facilities was the 235-megawatt (electric) Narora I nuclear power reactor, which began operating in mid-1989. The plant, in theory, can produce approximately 132 pounds (60 kilograms) of plutonium annually in its spent fuel,[58] possibly enough for 12 weapons, assuming 11 pounds (5 kilograms) of plutonium per device.[59]

Each of India's two previously commissioned unsafeguarded power reactors, Madras I and II, could also produce enough unsafeguarded plutonium for 12 weapons annually.[60] Two Indian research reactors, located at the Bhabha Atomic Research Center (BARC), are additional important potential sources of plutonium: the Cirus reactor, which produced plutonium for the 1974 test device, and the Dhruva reactor. The former can produce about 20 pounds (9 kilograms) of plutonium annually, and the latter 55 pounds (25 kilograms), which together would be almost enough for 7 devices per year.[61]

Past statements by Indian officials, as well as Indian Department of Atomic Energy annual reports, have indicated that plutonium is being extracted from spent fuel removed from the Madras I, Cirus, and Dhruva reactors.[62] Moreover, as noted earlier, M. R. Srinivasan, the Chairman of the Indian Atomic Energy Commission from February 1987 until January 1990, repeatedly declared that by the next century India plans to acquire several tons of the material free from IAEA restrictions to serve as fuel for advanced "breeder" reactors.[63] The breeder program has suffered a number of major setbacks and delays, but the effort to obtain a plutonium stockpile—potentially sufficient for hundreds of nuclear weapons—has continued.[64]

Taken together, the capabilities of these installations could permit India to manufacture enough plutonium for more than 40 weapons each year. Actual production, however, appears to be far lower. Since they became operational in the mid-1980s, for example, the Madras I and II and Dhruva reactors have been plagued by problems that have reduced their output by at least a half, and plutonium extraction plants rarely perform at their design levels.[65] A more realistic and conservative estimate, which takes account of these factors, is that India is currently capable of producing 15 nuclear devices annually and that, during 1989, it may have produced enough plutonium for 5 to 10 devices.[66]

Estimating India's existing stocks of plutonium is more difficult because India has used some of its plutonium for the 1974 nuclear test, for its experimental breeder reactor, and for other research purposes. Nonetheless, it would be reasonable to estimate that, as of mid-1990, India probably had enough plutonium for 40 weapons and possibly enough for 60.[67]

India is also known to be undertaking projects relevant to the production of thermonuclear weapons at the Bhabha Atomic Research Center, the presumed site of India's nuclear weapons research program.[68] These activities include the production of precursor materials specifically needed to make tritium—a material used to manufacture certain types of advanced nuclear weapons, including thermonuclear, or "hydrogen," bombs[69]—and work on nuclear fusion.[70] It has been widely speculated that if Pakistan were to test an atomic bomb akin to India's 1974 device, India would respond with the test of a thermonuclear weapon to demonstrate its continuing lead in the field.

Heavy water smuggling. The Madras I and II, Dhruva, and Narora reactors, all of which began operation between 1983 and 1989, require heavy water to function. As detailed in Chapter 4, convincing evidence has emerged that from the early 1980s through 1987, India engaged in extensive clandestine purchases of heavy water to enable it to operate these facilities without having to place them under IAEA safeguards.

Since India was first accused of smuggling the material in 1986,[71] senior officials in New Delhi have insisted that, except for heavy water openly obtained from the Soviet Union for the safeguarded Rajasthan nuclear power reactors, the country imported no heavy water *whatsoever* during the 1980s.[72] Yet by mid-1990, a number of facts had surfaced which belie this claim:

- In late 1988, Norway announced that it had traced an illicit 1983 transfer of 15 tons of Norwegian-origin heavy water to India;[73]
- in April 1990, the government of Romanian President Ion Iliescu announced that in 1986 the country's former government, led by Nicolai Ceausescu, had diverted 12.5 tons of Norwegian heavy water in its possession to India;[74]
- by late 1989, an on-going West German parliamentary investigation had exposed numerous similar operations orchestrated by a West German businessman, Alfred Hempel, involving the clandestine transfer of hundreds of tons of the material to India;[75]
- a 1988 report by the Indian Comptroller and Auditor General had concluded that between 1978 and 1986, India's indigenous heavy water facilities had produced less than a third of the heavy water needed to start the Dhruva reactor and the two Madras reactors;[76]
- the 1988-1989 Annual Report of the Indian Department of Atomic Energy had acknowledged that major problems continued to plague a number of operating heavy water plants and that the start-up of two new plants would be delayed; and
- the Indian press had disclosed that the Department of Atomic Energy had $1.4 billion for heavy water imports during 1989-1990, more than twice the allocation for domestic production, another indicator that the country's indigenous capabilities in this area have been far from sufficient to meet its needs.[77]

In the face of this evidence, New Delhi's claims that during the 1980s India did not receive a single shipment of heavy water through clandestine trafficking deserve to be treated with considerable skepticism.

As noted in Chapter 4, in 1984, India also imported 210 pounds of high-purity beryllium, used in nuclear weapons and nuclear reactors under questionable circumstances.[78] India apparently could not produce beryllium of this quality at the time.[79]

India has for years portrayed its own nuclear program as entirely open and "transparent," and thus presumably peaceful. In contrast, it has cast Pakistan, whose nuclear smuggling operations have been well publicized, as a nuclear outlaw bent on obtaining nuclear arms "by hook or by crook."[80] Now that India also has been shown persuasively to have relied on clandestine nuclear purchases to expand its nuclear weapons capability, New Delhi has lost its claim to transparency and has created grounds for deep suspicions about its intentions. Indeed, it is fair to ask whether a country would engage in such extensive clandestine nuclear dealings *except* for the purpose of supporting a nuclear weapons program. It is difficult to imagine a nation risking exposure and international censure simply to produce electricity or manufacture radioisotopes for medical treatment free from international monitoring.

The government of Prime Minister Vishwanath Pratap Singh, who succeeded Gandhi following parliamentary elections in November 1989, has pursued investigations of other illicit international trading activities that took place during Gandhi's tenure, such as the possible bribery of Indian officials during the early 1980s in connection with the purchase of artillery from the Swedish firm, Bofors, and potential kickbacks to Indian officials on the purchase of submarines from West Germany's Howaldtswerke Deutsche Werft (HDW). Conceivably, India's heavy water purchases will also be scrutinized, but as of mid-1990, no inquiry into the heavy water matter had been initiated. Indeed, there was reason to assume that Singh would not pursue the matter, since as discussed below, he appointed Raja Ramanna, the man who chaired the Indian Atomic Energy Commission throughout most of this period, to a prominent position in the Indian Defense Ministry.

The Agni missile. Adding to concerns about India's long-term nuclear ambitions has been its development of the Agni, an intermediate-range ballistic missile which can deliver a 2,200 pound (1,000 kilogram) payload to a range of 1,550 miles (2,500 kilometers).[81] India successfully tested the system for the first time on May 22, 1989. The missile's substantial payload would enable it to carry a nuclear warhead easily.

India appears to be developing the system partly for prestige and partly for strategic reasons. In terms of prestige, Indian officials see the missile as symbolizing India's arrival as a major power and as demonstrating the country's credentials as a technologically advanced state.

Strategically, given the Agni's range, it appears that India is developing the system as a counter to China's nuclear arsenal, although the missile would also have utility against Pakistan. If based in Assam, the portion of India that is east of Bangladesh, the Agni could reach a number of strategic targets and key cities in eastern China, including Beijing, Nanking, and Canton. Even if the Agni were deployed somewhat more centrally in eastern Bengal province, the missile would remain capable of striking important targets in central China, such as Chungking, although the large cities near the Chinese coast would be out of range. Without the Agni, China's coastal cities would remain out of range of India's manned bombers, and India could reach targets in central China only with such aircraft flying vulnerable, high-altitude missions.

Conversely, the Agni does not appear to be necessary to deal with the threat from Pakistan. India's 150-mile range nuclear-capable Prithvi, first tested in 1988 and already ordered by the Indian army for use with conventional warheads, can reach key targets there, including Lahore, Islamabad, and Karachi.[82] Many of India's advanced fighter-bombers could also reach these targets with nuclear payloads, even using low-altitude flight paths, which consume fuel at high rates, in order to avoid Pakistani air defenses. Thus, an entirely new, 1,500-mile range missile would not be needed for this purpose, although the Agni's range would have the advantage of allowing India to base it beyond the reach of Pakistani missiles and aircraft.

India has long perceived itself to be threatened by China's nuclear forces and consistently has alluded to this threat in justifying its unwillingness to enter into bilateral nuclear arms control talks with Pakistan. Nonetheless, it is unlikely that Beijing has deployed nuclear forces against India. Rather, it is widely assumed that China has focussed its limited nuclear deterrent on meeting the threat from the Soviet Union's vastly superior nuclear arsenal; indeed, in January 1989, Rajiv Gandhi was reported by the Indian press to have "categorically stated that Indian intelligence believes there are no [Chinese] nuclear missiles targeted on Indian territory."[83] Moreover, India has never deployed even a rudimentary, aircraft-based nuclear deterrent against China, suggesting that, rhetoric aside, it has never considered itself under anything more than a potential nuclear threat from this quarter.

China's posture could change dramatically, however, if India deployed the Agni. Such a step would inevitably increase tensions with Beijing and could well lead either to a significant expansion of Chinese nuclear forces or to a reorientation of a portion of China's existing arsenal, particularly if Sino-Soviet tensions continue to ease. Either eventuality could stimulate a nuclear arms race between the two Asian giants. China has already expressed concerns about the Agni, as has Pakistan.[84]

Seeking to allay such concerns and deflect pressure from the United States to halt development of the system, Indian officials have stressed that for the time being, the Agni is only a "technology demonstrator" and that no decision has been made to deploy it. They also have emphasized that the Agni is being developed as a highly accurate conventionally-armed system, insisting that, even if ultimately deployed, it would not carry nuclear warheads. In commenting on the test of the Agni, for example, then-Prime Minister Rajiv Gandhi declared that the missile was only "an R&D vehicle, not a weapon system." He continued:

> Agni is not a nuclear weapons system. What Agni does is to afford us the option of developing the ability to deliver non-nuclear weapons with high precision at long ranges. This provides us with a viable non-nuclear option of the greatest relevance to contemporary strategic doctrines.[85]

In March 1990, it was reported that a second test of the missile would be conducted later in the year and that full-scale production of the system was also being planned; a second test had not taken place, as of mid-June, however.[86]

Given India's oft-repeated concerns about the nuclear threat posed by China and Pakistan, its disclaimers that the Agni is not intended as a nuclear system have been greeted with disbelief. Indeed, there are strong grounds for fearing that if India deploys the Agni the event will signal its emergence as an overt nuclear power.[87]

No state has ever undertaken the enormously costly and complex task of developing intermediate-range missiles without arming them with nuclear warheads.[88] It is improbable that India would deviate from this pattern, particularly when the principal adversary it hopes to deter through deployment of the Agni possesses nuclear-armed missiles of sufficient range to reach targets throughout India. Indeed, if India did not possess at least some nuclear-armed Agnis, it might be effectively forestalled from using conventionally armed versions of the missile for fear that an attack on high-value Chinese military targets would lead Beijing to respond with nuclear weapons, against which India would have no countervailing capability.

If India kept its Agni force entirely conventional, moreover, it would also have to worry about Chinese miscalculation in a crisis. Since India has not ratified the Non-Proliferation Treaty or taken other steps to permit verification that it is not manufacturing nuclear arms, China would inevitably fear that at least a portion of India's Agni force carried nuclear warheads. In a crisis, this would give Beijing a strong incentive to attack India's missiles preemptively rather than risk nuclear destruction. Similarly, if India launched some of its Agnis first and China detected this, the latter would have to assume that it was under attack by high-accuracy Indian *nuclear* missiles that might be attempting to destroy China's own nuclear systems, including those in facilities that were hardened against conventional attack. Under these circumstances, Beijing would have strong incentives to launch its nuclear forces after it received warning of the Indian launch rather than ride out the attack. Again, India could be at risk of a nuclear strike with no means of responding.

Although some well-respected U.S. strategists have argued that the United States should rely more heavily on highly accurate conventionally armed missiles and reduce the role of nuclear systems, their reasoning is of questionable relevance to India's situation.[89] Unlike India, for example, under the strategy these analysts propose, the United States would retain a powerful missile-based nuclear deterrent to inhibit escalation by its adversaries to the nuclear level. Moreover, the proposed U.S. strategy would be based on the acquisition of hundreds, or even thousands, of inexpensive short- and medium-range "stand-off" missiles, to be fired from aircraft or naval platforms, rather than on costly surface-to-surface missiles like the Agni, of which India would be likely to deploy only 50 to 100.[90]

The dangers of deploying a conventionally armed Agni force are surely appreciated by Indian defense planners. Under the circumstances, it seems highly improbable that, in fact, they intend to deploy such a costly and provocative system without nuclear warheads. Indeed, for most observers, deployment will be perceived as tantamount to a declaration that India is a nuclear weapon state.

No-attack pledge; nuclear stance. India has long insisted that it will not enter into bilateral nuclear arms control arrangements with Pakistan, arguing that such regional arrangements do not address the continuing nuclear threat the country faces from China nor the broader global issues of superpower nuclear disarmament. Nonetheless, as described earlier, at a December 1985 summit meeting in New Delhi, Rajiv Gandhi and then-Pakistani President Mohammed Zia ul-Haq orally agreed that their two countries would refrain from attacking each other's nuclear installations.

In December 1988, Pakistan's newly elected prime minister, Benazir Bhutto, and Gandhi signed a written agreement codifying the verbal no-attack pledge in Islamabad.[91] Both states then proceeded to ratify the agreement, a process that was completed by the end of 1989. This left only two steps to bring the agreement into force: the exchange of instruments of ratification and the exchange of lists of facilities covered. The latter is of particular importance, inasmuch as Pakistan is thought to have built several nuclear facililties that presumably would be covered by the accord but which it has never publicly acknowledged.[92] Because of domestic political concerns, Benazir Bhutto may refuse to list these installations for fear of seeming to give away Pakistani nuclear secrets. If she fails to list them, however, India could accuse Pakistan of bad faith, which would largely vitiate the benefits of the agreement as a confidence-building measure.

Unfortunately, by the time the two states had completed their respective ratifications of the accord, relations between them had deteriorated badly, making it politically infeasible to proceed.[93] As of mid-June 1990, the agreement was still in limbo.

The improvement in bilateral relations during the first half of 1989, permitted progress on a range of divisive issues in addition to the no-attack agreement.[94] In July, Gandhi made a state visit to Islamabad, the first such visit by an Indian leader in several decades.

Gandhi lost his post as Prime Minister in India's November 1989 elections. During his remaining time in office in 1989, India's declared nuclear policy remained largely unchanged from that of recent years. In particular, Gandhi continued to rebuff Pakistani proposals for bilateral non-proliferation agreements, such as a formal ban on nuclear testing in the region, despite U.S. efforts to press New Delhi to begin a dialogue with Islamabad on these issues.[95] In addition, in the wake of Soviet force withdrawals from Afghanistan in 1988 and early 1989, Gandhi once again pressed the United States unsuccessfully to restrict economic and military aid to Pakistan because of that country's nuclear activities.[96] New Delhi hoped that the threat of U.S. aid reductions would be accompanied by intensified U.S. pressure on Islamabad to halt its nuclear weapons program.

Gandhi's continuing effort to project India as a regional superpower and as an important player on the international stage strongly influenced regional nuclear affairs during his last year in office. Gandhi maintained the momentum of the country's military build-up, despite budgetary pressures,[97] and, in a dispute over a trade pact, imposed a harsh economic blockade against Nepal, after it sought to strengthen ties to China.[98] Such Indian "muscle-flexing" undoubtedly reinforced Pakistan's interest in expanding its nuclear capabilities as a counterweight to India's apparent bid for hegemony in South Asia.

Nuclear policies continued under Singh. V. P. Singh's November 1989 election victory brought little change in Indian nuclear policy, despite the fact that Singh's Janata Dal Party coalition included the militant Hindu BJP Party, the only faction that during the election had called for India to abandon Gandhi's ambiguous nuclear stance and become an overt nuclear power.[99] Although BJP Party members were not included in the Singh government, the party's support is essential to Singh's parliamentary majority, a factor that gives it a degree of influence over Singh's decision-making on nuclear affairs.

During his first weeks in office, Singh named moderates to key foreign policy posts. In the defense and nuclear fields, however, he appointed two senior officials widely believed to support India's acquisition of nuclear weapons: Raja Ramanna, who was appointed Minister of State for Defense, and P. K. Iyengar, who was named head of the Indian Atomic Energy Commission.[100] The two had played leading roles in India's 1974 nuclear test and almost certainly were involved in India's apparently illicit heavy water acquisition activities. Ramanna was Chairman of the Indian Atomic Energy Commission from August 1983 until February 1987, during the height of these activities, and, during most of this period, Iyengar headed the Bhabha Atomic Research Center, the site of the then-newly completed Dhruva reactor, which urgently needed heavy water. Although Iyengar succeeded to the Atomic Energy Commission post in 1990 on the basis of seniority, Singh undoubtedly considered Ramanna's hawkish nuclear credentials when appointing him to his political position at the Defense Ministry.

Immediately after taking office, Singh sought to ease tensions with India's neighbors and, at one point, offered to hold a dialogue with Pakistan on nuclear issues—talks his predecessor had long rejected.[101] Singh rapidly backed away from this gesture, however, and subsequently embraced both Gandhi's opposition to regional arms control discussions and his ambiguous posture on India's own nuclear course.[102]

Kashmir and the threat of war. In late 1989, relations with Pakistan seriously deteriorated as violence by Muslim separatists erupted in Kashmir, a Muslim-majority state in northern India. The disputed area has long been a source of friction between the two countries and was an important cause of the Indo-Pakistani wars in 1948 and 1965. India accused Pakistan of fomenting the new round of unrest, which continued to escalate during the first half of 1990 despite increasingly harsh efforts by New Delhi to suppress it.[103]

As tensions grew, strident rhetoric by Indian and Pakistani leaders inflamed passions on both sides of the border, and there were mounting concerns that war might break out.[104] In April, for example, Prime Minister Singh stated that India should be "psychologically prepared" for an armed conflict with Pakistan over Kashmir and warned that Pakistan would "pay a very high price" for fomenting unrest in the region.[105] A leader of the BJP Party went further and suggested that Pakistan's very survival might be at stake, declaring that if war comes, "Pakistan will cease to exist."[106] In Pakistan, Prime Minister Bhutto declared in March that her country was prepared to fight a war of "1,000 years" over the region and would donate $4.8 million to Kashmiri Muslim fundamentalists, while her Army Chief of Staff stated that "Pakistan stands committed to support the just cause" of the people of Kashmir.[107] Discussions between the two countries' foreign ministers in New York in late April reportedly did little to calm the situation, nor was a visit to the region in May by U.S. Deputy National Security Advisor Robert Gates successful in persuading Indian and Pakistani leaders to reduce tensions.[108]

Nuclear dimension. If India and Pakistan go to war over Kashmir, it would be the first major military conflict in history between two states with ready access to nuclear arms. Such a conflict would hold the gravest dangers for the region and could have a profound impact on global affairs. By comparison, the hostilities between Chinese and Soviet forces in 1969—the only time two nuclear states have actually clashed—were limited, involving skirmishes for control of a relatively small and sparsely populated disputed border area.[109]

If significant hostilities appeared imminent, there is good reason to fear that Pakistani and Indian leaders would ready their nuclear forces. India's potential arsenal would be far superior. It would consist of 40 to 60 weapons which could be delivered on any of some 200 advanced aircraft in the Indian Air Force, including the British-French Jaguar bomber, various Soviet MiGs, and the French Mirage 2000. Pakistan's nuclear force, by contrast, would probably comprise only five to ten bombs, and the country possesses only 40 to 50 first-class strike aircraft, all U.S.-supplied F-16s.

Fearing the worst from India and with only a handful of weapons and top aircraft potentially available, the first concern of Pakistani military strategists would be protecting the country's nuclear "ace-in-the-hole" from pre-emptive attack. In such a situation, they inevitably would demand that the country's nuclear devices be assembled, dispersed to several air bases, and mounted on aircraft that could be kept on alert. Preparations might also be made for a nuclear test, to be set off as a warning if India decided to launch an invasion.

India would surely get wind of any Pakistani nuclear preparations, either directly or through the intelligence capabilities of the Soviet Union, its longstanding military patron. Alternatively, Indian planners might simply assume that Pakistan was taking these steps.

To avoid being placed at a disadvantage, India would also have to ready its weapons and would probably do so in a way that would be easily observed. It might, for example, send a large day-time convoy from an atomic research center to a major air base, assuming that it would be spotted by a U.S. intelligence satellite and that the information would be relayed to Pakistan.

Whatever the details, both sides would surely take their not-quite-bombs out of the "basement" and make them militarily usable. Even if the crisis were to subside without a shot being fired—and undoubtedly both the United States and the Soviet Union would bring considerable pressure to bear to achieve this result—the nuclear history of the Subcontinent would be forever changed. Moreover, as word leaked out about the deployments and about the importance that both sides had attached to nuclear arms during the crisis, nuclear ambitions elsewhere in the developing world would be greatly encouraged.

If major hostilities do break out, it is possible that the threat of nuclear reprisal will deter the use of nuclear arms. Simply because nuclear deterrence appears to have prevented conflict between the superpowers over a span of forty-five years, however, such an outcome cannot be taken for granted elsewhere. In the Iran-Iraq War, for example, where both sides had potent, albeit non-nuclear, weapons, deterrence appears to have failed. Thus, each side attacked the other's cities with ballistic missiles even when it knew that the opposing side possessed the capability to retaliate in kind. Iraq, moreover, continued to use chemical weapons against Iranian forces even after Iran demonstrated its relaliatory capability by using them against Iraqi troops.

In addition, the unusually weak political position of leaders in both Pakistan and India could make nuclear restraint all the more difficult. It is unlikely that Benazir Bhutto would have the authority to restrain the country's military from conducting a nuclear test or firing a nuclear warning shot, for example, if Indian forces gained a foothold in Pakistani territory. Nor is there reason to assume that Prime Minister Singh would be prepared to resist calls for nuclear retaliation if Pakistan used nuclear arms first.

At least one senior Indian strategist has already commented on the potential nuclear dimension of the Kashmir crisis. In an April 1990 interview, former Indian Army Chief of Staff General Krishnaswami Sundarji stressed that:

> any sensible planner on this side of the border is going to assume Pakistan does indeed have [sic] nuclear weapon capability. And by the same token, I rather suspect the view from the other side is going to look very similar. . . .

He then went on to give the most direct indication to date by any senior Indian official of India's readiness to deploy nuclear weapons:

> On the other side there may be the odd person who has kidded himself into believing that they have the nuclear weapon and we don't. . . . If it is such a hypothesis that some people are working on on the other side, then it's stupid. The sooner they wake up to this reality the better.[110]

Sunderji went on to state, however, that he believed the risk of war was low.

Prospects

Assuming that the threat of conflict subsides, the antagonism of the recent past make it very unlikely that Singh will reduce India's commitment to the slow but steady build-up of its nuclear weapon capabilities, including development of the Agni, which the country has pursued for years and which appears to have wide popular support. At the same time, Singh will probably refrain from nuclear tests or overt declarations that the country possesses nuclear arms, despite the prodding of the BJP, because of the likelihood of international criticism, which could include the imposition of sanctions, and the risk of triggering a wide-open nuclear arms race with Pakistan. The presence of the BJP in Singh's coalition makes it all the more unlikely, however, that he would reconsider India's long-standing rejection of bilateral nuclear arms control measures with Pakistan, although if relations were to improve somewhat, efforts would probably be made to effectuate the two states' agreement not to attack each other's nuclear facilities. Additional confidence-building measures in related security areas, such as improved communication between the two countries' military establishments, would probably be pursued as well, while an undeclared nuclear deterrence relationship evolved.

Although this would preserve the nuclear status quo in the region for the time being, as suggested above, there is reason to be concerned that over the next decade India's posture will gradually shift and that the country will emerge as a recognized nuclear power.

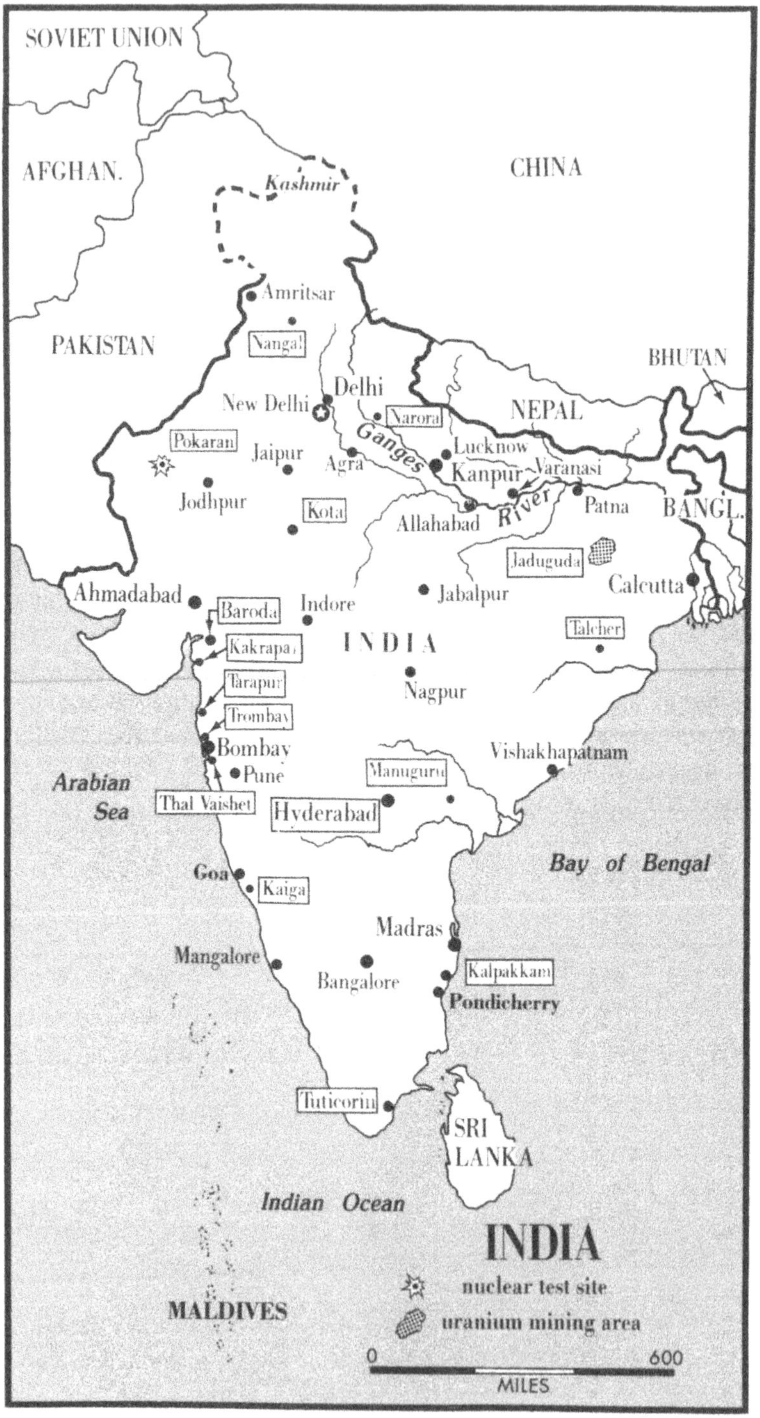
SOVIET UNION
AFGHAN.
CHINA
Kashmir
Amritsar
Nangal
PAKISTAN
BHUTAN
Delhi
New Delhi
Narora
NEPAL
Pokaran
Jaipur
Agra
Ganges
Lucknow
Kanpur
Varanasi
Jodhpur
Kota
Allahabad
River
Patna
BANGL.
Jaduguda
Calcutta
Ahmadabad
Baroda
Indore
Jabalpur
Kakrapar
INDIA
Talcher
Tarapur
Nagpur
Trombay
Bombay
Vishakhapatnam
Pune
Manuguru
Arabian Sea
Thal Vaishet
Hyderabad
Bay of Bengal
Goa
Kaiga
Madras
Mangalore
Bangalore
Kalpakkam
Pondicherry
Tuticorin
SRI LANKA
Indian Ocean
INDIA
nuclear test site
uranium mining area
MALDIVES
0
600
MILES

India

Power Reactors/Operating or Under Construction

Tarapur I (light-water/low-enriched uranium, 150 MWe)[a]
- supplier: General Electric (U.S.)
- start up: 1969
- fuel source: U.S.; France after 1982
- safeguards: yes

Tarapur II (light-water/low-enriched uranium, 150 MWe)[a]
- supplier: General Electric (U.S.)
- start up: 1969
- fuel source: U.S.; France after 1982
- safeguards: yes

Rajasthan I, Kota (heavy-water/natural-uranium, 207 MWe)[a]
- supplier: Canadian General Electric (Canada)
- start up: 1972
- fuel source: Initial load, half Canadian Westinghouse, half Indian; subsequently all Indian through 1989, then some French fuel[b,c]
- heavy water: 130 metric tons from Canada and U.S.; 80 metric tons from USSR in 1973.[a]
- safeguards: yes

Rajasthan II, Kota (heavy-water/natural-uranium, 207 MWe)[a]
- supplier: Larsen and Toubro (India), following termination of Canadian assistance in 1976.
- start up: 1980[g]
- fuel source: India through 1989, then some French fuel[b,c]
- heavy water: USSR,[b] India
- safeguards: yes

Madras I, Kalpakkam (heavy-water/natural-uranium, 220 MWe)
- supplier: Larsen and Toubro (India)
- start up: 1983
- fuel source: India
- heavy water: India, PRC, USSR, Norway, Romania[e]
- safeguards: no

Madras II, Kalpakkam (heavy-water/natural-uranium, 220 MWe)
- supplier: Larsen and Toubro (India)
- start up: 1985[e]
- fuel source: India, PRC, USSR, Norway, Romania[e]
- heavy water: India (?)[c,j]
- safeguards: no

Narora I (heavy-water/natural-uranium, 220 MWe)
- supplier: Larsen and Toubro/Walchandnagar Industries, Ltd. (India)
- start up: 1989[g]
- fuel source: India
- heavy water: India and other sources (?)[e]
- safeguards: no

Narora II (heavy-water/natural-uranium, 220 MWe)
- supplier: Larsen and Toubro/Walchandnagar Industries Ltd. (India)
- start up: 1990 (est.)[a]
- fuel source: India
- heavy water: India
- safeguards: no

Kakrapar I (heavy-water/natural-uranium, 220 MWe)
- supplier: Larsen and Toubro/Walchandnagar Industries, Ltd. (India)
- start up: 1991[a]
- fuel source: India
- heavy water: India
- safeguards: no

Kakrapar II (heavy-water/natural-uranium, 220 MWe)
- supplier: Larsen and Toubro/Walchandnagar Industries, Ltd. (India)
- start up: 1992[a]
- fuel source: India
- heavy water: India
- safeguards: no

Kaiga I, Karnataka (heavy-water/natural-uranium, 220 MWe)
- supplier: Larsen and Toubro/Walchandnagar Industries, Ltd. (India)
- start up: 1995[g]
- fuel source: India
- heavy water: India
- safeguards: no

Kaiga II, Karnataka (heavy-water/natural-uranium, 220 MWe)
- supplier: Larsen and Toubro/Walchandnagar Industries, Ltd. (India)
- start up: 1995[g]
- fuel source: India
- heavy water: India
- safeguards: no

Rajasthan III, Kota (heavy-water/natural-uranium, 220 MWe)[a]
- supplier: India
- start up: 1995[d]
- fuel source: India[a]
- heavy water: India (?)
- safeguards: no

Rajasthan IV, Kota (heavy-water/natural-uranium, 220 MWe)[a]
- supplier: India
- start up: 1995[d]
- fuel source: India[a]
- heavy water: India (?)
- safeguards: no

Uranium Resources/Active Mining Sites/Uranium Mills

- reasonably assured reserves: 46,090 metric tons[h]
- currently active site: Jaduguda[a]
- mills in operation: Jaduguda[a]

Uranium Purification (UO_2)[e]

Hyderabad
- capacity: 50 tons per year[a]
- supplier: India
- start up: 1984[a]
- safeguards: no

Uranium Conversion (UF_6)

Trombay/Bhabha Atomic Research Center (B.A.R.C.) (?)

- capacity: sufficient for experimental enrichment program at B.A.R.C., presumed
- supplier: India, presumed
- start up: 1984?
- safeguards: no

Enrichment

Trombay/B.A.R.C.

- type: ultracentrifuge
- capacity: pilot scale, 100 (?) centrifuges, uranium reportedly enriched to "less than two percent"[i]
- supplier: India
- start up: 1984?
- safeguards: no

Heavy Water

Nangal

- capacity: 14 metric tons per year[b] (actual output lower in recent years)[e]
- supplier: Linde Gmbolt (West Germany)[b]
- start up: 1962
- safeguards: no

Baroda

- capacity: 45 metric tons per year[b]
- supplier: GELPRA (Swiss-French)
- start up: 1977 (closed 1977-80; intermittent operation thereafter)[e]
- safeguards: no

Tuticorin

- capacity: 49 metric tons per year[b]
- supplier: GELPRA (Swiss-French)
- start up: 1978 (has operated at only 20% capacity)[e]
- safeguards: no

Talcher

- capacity: very little production[b,e]
- supplier: Friedrich Unde Gmbolt (West Germany)[b]
- start up: 1979
- safeguards: no

Kota

- capacity: 85 metric tons per year (actual production substantially lower)[b,e]
- supplier: India, Canada (Canadian aid terminated 1974)[j]
- start up: 1984[k]
- safeguards: no

Thal-Vaishet

- capacity: 110 metric tons per year (actual output substantially lower)[k,g,e]
- supplier: Rashtriya Chemicals and Fertilizers (India)
- start up: 1987
- safeguards: no

Manuguru

- capacity: 185 metric tons[f]
- supplier: India
- start up: 1990 (est.)[g]
- safeguards: no

Hazira
- capacity: 110 metric tons per year[f]
- supplier: KRIBHCO and Projects Development India Ltd. (India)
- start up: 1990 (est.)[g]
- safeguards: no

Fuel Fabrication

Hyderabad
- capacity: 80 metric tons/yr. for Rajasthan, Madras, Tarapur, Narora, and FTBR blanket assemblies; being expanded to 225 metric tons/yr.[a,f,l]
- supplier: India
- start up: 1971[r]
- safeguards: partial[r]

Trombay/B.A.R.C.
- capacity: 135 tons per year[a]; sufficient for Cirus, Dhruva, FBTR fuel pins
- supplier: India
- start up: late 1960s (Cirus), 1981-1982 (Dhruva), 1983-1984 (FTBR)
- safeguards: no

Tarapur (Mixed Uranium, Plutonium Oxide - MOX)
- capacity 20 tons per year of MOX fuel, using 60-70 kg plutonium[m]
- supplier: India
- start up: 1990 (est.)[g]
- safeguards: partial? (Safeguards would apply when plant is processing plutonium produced in safeguarded reactors.)

Reprocessing (Plutonium Extraction)

Trombay/B.A.R.C.
- capacity: 30 metric tons of spent fuel per year (CIRUS, Dhruva fuel); has been enlarged[a,b]
- supplier: India (with U.S. technology in early 1960s)
- start up: 1966; shut down, 1974; restarted late 1983 or early 1984[n]
- safeguards: no

Tarapur
- capacity: 100 metric tons of spent fuel per year; 135-150 kg. plutonium per year when at full capacity[a]
- supplier: India
- start up: trial runs, 1979[b]; full scale operations, 1982[a]
- safeguards: partial (only when plant is processing spent fuel from safeguarded reactors)

Kalpakkam
- capacity: laboratory scale
- supplier: India
- start up: 1985
- safeguards: no

Kalpakkam
- capacity: 125 metric tons per year (for Madras I, II, and FTBR)[f]
- supplier: India
- start up: 1991 (est.)
- safeguards: no

Research Reactors

Apsara, Trombay/B.A.R.C. (light-water/medium-enriched-uranium, 1 MWt)[o]
- supplier: India
- start up: 1956
- fuel source: United Kingdom
- safeguards: no

Cirus, Trombay/B.A.R.C. (heavy-water/natural-uranium, 40 MWt)[o]
- supplier: Canada
- start up: 1960; fully operational, 1963
- fuel source: Canada, then India
- heavy water: United States
- safeguards: no

Zerlina, Trombay/B.A.R.C. (heavy-water/variable-fuel, 400 Wt)[o]
- supplier: India
- start up: 1961; decommissioned 1983[f]
- fuel source: India?
- safeguards: no

Purnima II, Trombay/B.A.R.C. (uranium-233)[f]
- supplier: India
- start up: 1984, dismantled 1986? 1987?[f]
- fuel source: India
- safeguards: no

Purnima III, Trombay/B.A.R.C. (uranium-233)
- supplier: India
- start up: 1989 (?)[p]
- fuel source: India
- safeguards: no

Kamini, Kalpakkam (uranium-233, 30 KWt)[f]
- supplier: India
- start up: 1988[q]
- fuel source: India
- safeguards: no

Dhruva (formerly R-5) Trombay/B.A.R.C. (heavy-water/natural-uranium, 100 MWt)[d]
- supplier: India
- start up: 1985[f]; fully operational, 1988
- fuel source: India
- heavy water: India, PRC, USSR, Norway, Romania[e]
- safeguards: no

FBTR, Kalpakkam (fast breeder/plutonium & natural uranium, 42 MWt, 15MWe)
- supplier: India, with French assistance[b]
- start up: 1985
- fuel source: India
- safeguards: no

Sources and Notes

a. *World Nuclear Industry Handbook, 1990* (Sutton, Surrey, U.K.: Nuclear Engineering International, 1990).

b. David Hart, *Nuclear Power in India, A Comparative Analysis* (London: George Allen and Unwin, 1983).

c. "IAEA Safeguards to Apply to French Nuclear Fuel," Delhi Domestic Service, 1230 GMT, October 12, 1989, reprinted in *Joint Publications Research Service (JPRS)-Nuclear Development and Proliferation (TND)*, October 26, 1989, p. 33.

d. *The Annual Report of the Department of Atomic Energy, 1988-1989* (Bombay: Indian Department of Atomic Energy, 1989).

e. Concerning the poor performance of Indian heavy water plants and the use of foreign heavy water in the Dhruva and Madras I and II reactors, see discussion in this chapter accompanying notes 71-77 and the section on India in Chapter 4. GELPRA is a consortium of Sulzer Brothers, Cie de Construction Mecaniques (a French licensee of Sulzer Brothers) and Société Air Liquide (France). See "India and France," *Nucleonics Week*, July 4, 1985.

f. *The Annual Report of the Department of Atomic Energy, 1986-1987* (Bombay: Indian Department of Atomic Energy, 1987).

g. *The Annual Report of the Department of Atomic Energy, 1988-1989* (Bombay: Indian Department of Atomic Energy, 1989).

h. Reasonably Assured Reserves at less than $130 per kilogram. OECD Nuclear Energy Agency and the International Atomic Energy Agency, *Uranium, Resources, Production and Demand* (Paris: OECD, 1986), pp. 223-225. Reference in note g. states that reasonably assured reserves amount to only 29,000 metric tons. India also has two plants for recovering uranium from copper tailings at Surda and Kakha.

i. According to one report, the pilot scale enrichment facility consists of one hundred centrifuges and had "achieved enrichment of less than 2%." See Ivan Fera and Kannan Srinivasan, "Keeping the Nuclear Option Open: What It Really Means," *Economic and Political Weekly*, December 6, 1986, p. 2119. See also "DAE Chief Denies Industrial HEU Production at BARC," *Nuclear Fuel*, June 29, 1987.

j. Jairam Ramesh, "Problems in Production of Heavy Water Discussed," *The Times of India*, reprinted in *JPRS-TND*, August 18, 1982, p. 11.

k. *The Annual Report of the Department of Atomic Energy, 1984-1985* (New Dehli: 1985).

l. "Promotion of Nuclear Power Plants Emphasized," *Delhi General Overseas Service*, 1010 GMT, March 15, 1989, printed in *Forign Broadcast Information Service-Near East and South Asia*, March 16, 1989, p. 53.

m. "Plutonium to Be Used as Fuel for Rajasthan Plant," *The Times of India*, December 15, 1988, reprinted in *JPRS-TND*, March 20, 1989, p. 21.

n. "Trombay Is Ready for Work," *Nuclear Engineering International*, (June 1983): 6; "Interview, Nuclear Energy," *India International Center Quarterly*, Vol. 10, No. 4, 1983, p. 379. Some sources state plant was shut down in 1972, not 1974.

o. Albert Wohlstetter, ed., *Swords From Plowshares* (Chicago: University of Chicago Press, 1979), p. 206.

p. "More on Kamini Reactor," *The Hindu*, August 28, 1989, reprinted in *JPRS-TND*, October 26, 1989, pp. 29-30. Purnima III is a full-scale zero-power mock up of the Kamini reactor. The Purnima II reactor system was dismantled to make way for the Purnima III experiments.

q. " 'Third Stage' Reactor to Go Critical," *Nuclear Engineering International* (February 1990): 16.

r. *Annual Report of the Department of Atomic Energy, 1982-1983* (Bombay: Indian Department of Atomic Energy, 1983).

7
Pakistan

Driven by fears of Indian domination and a desire for prominence in the Islamic world, Pakistan launched a program to acquire nuclear arms in 1972. Today, some four years after it first achieved the ability to produce such weapons, Pakistan probably could deploy five to ten nuclear bombs for delivery by aircraft. Under Benazir Bhutto who, in November 1988, became the country's first democratically elected leader in nearly two decades, Pakistan has continued to expand its ability to produce weapons-grade nuclear material free from International Atomic Energy Agency (IAEA) safeguards and has continued to resort to clandestine nuclear trade to advance its nuclear program. It is also actively pursuing the development of nuclear-capable ballistic missiles. After years of insisting that Pakistan's nuclear program was entirely peaceful, moreover, senior Pakistani officials have begun to invoke the military value of the country's nuclear capability in their public statements.

President George Bush certified in late 1989 that Pakistan still did not "possess a nuclear explosive device," and there is evidence that Bhutto may have slowed certain narrow aspects of the country's nuclear program. Pakistan has also refrained from nuclear testing, leaving time for a stabilization of the incipient nuclear competition on the Subcontinent. The continuing crisis with India over Kashmir has seriously increased the risk of war, however. This threat could lead Islamabad to take the remaining steps needed to ready nuclear arms for military deployment.

Background

Pakistani leaders have repeatedly denied that their nation is developing nuclear arms. Nevertheless, the available evidence leaves no doubt that Pakistan is indeed pursuing such a course, a conclusion shared by the United States, the Soviet Union, India, and other nations.

Although Pakistan, like India, refused to join the Non-Proliferation Treaty (NPT) in 1968, thereby leaving itself free to develop nuclear arms, Pakistan's nuclear weapons program was not initiated until shortly after the country's devastating defeat in the 1971 Indo-Pakistani War. The following year, according to eye-witnesses, then-Prime Minister Zulfikar Ali Bhutto announced his plan to develop nuclear arms at a secret meeting of Pakistan's top scientists and nuclear aides in Multan.[1] The program appears to have been aimed at countering India's substantial conventional military superiority and its significant, but then still undemonstrated, nuclear capability. India's nuclear test in May 1974 gave added impetus to the Pakistani program.[2]

Plutonium route. Initially, it appears that Bhutto intended to pursue nuclear weapons by acquiring a large "reprocessing" plant from France for extracting plutonium from spent power-reactor fuel. Although France insisted that this facility and any others based on its technology be placed under International Atomic Energy Agency safeguards, the installation would have nonetheless permitted Pakistan to accumulate plutonium. Pakistan had no apparent need for this material in its civilian nuclear power program, which then, as now, has only one small, natural-uranium-fueled nuclear power reactor.[3]

By 1976, American concerns over Pakistan's possible misuse of the plant had intensified to the point that President Gerald Ford dispatched Secretary of State Henry Kissinger to Islamabad and then to Paris in an unsuccessful bid to head off the reprocessing plant sale. In August 1977, however, France reversed its stance and suspended deliveries for the plant after U.S. nuclear aides showed their French counterparts intelligence data revealing Pakistan's intentions. The following month, to underscore its concern, Washington cut off economic and military aid to Islamabad. It was restored in August 1978, however, after France stopped all performance under the plutonium plant contracts.[4]

Indeed, Bhutto himself acknowledged the veiled purpose behind the French deal, after he was ousted by Pakistan's military in July 1977. In a testament written in 1978 shortly before his execution at the hands of his successor, General Zia ul-Haq, he declared that Pakistan was "on the threshold of a full nuclear capability." "All we needed," he wrote, "was the nuclear reprocessing plant."[5]

Enrichment route. Plutonium extraction, however, was not the only route Pakistan was pursuing to obtain nuclear arms. In 1975, Pakistan began acquiring hardware and technology for a uranium enrichment plant using ultra-high-speed centrifuges, a plant potentially capable of producing highly enriched uranium, the alternative nuclear weapons material to plutonium.[6] (See Glossary.) A key figure in the effort was Dr. Abdul Qadeer Khan, a German-trained metallurgist, who worked intermittently at a classified centrifuge enrichment plant at Almelo, the Netherlands, where he is believed to have gained access to key plans and listings of component suppliers.

Official and press investigations have confirmed Dr. Khan's role in developing Pakistan's enrichment capability.[7] Apparently, returning to Pakistan for a visit following India's 1974 test, Khan encouraged his government to master the uranium enrichment process, and advised them as to the necessary facilities. When the program seemed slow in getting under way, Khan agreed to head the operation himself and quit his post in Holland to return to his native country in 1976.

The project was made autonomous from the Pakistani Atomic Energy Commission the same year, possibly to protect its secrecy, and it was pursued with considerable urgency, as evidenced by the fact that various elements of the program were undertaken simultaneously. In 1978, construction began on a pilot-plant at Sihala that was placed in operation the next year, while construction also was started on a full-scale facility at Kahuta.[8]

To support the enrichment project, Pakistan relied increasingly on smuggling and black market transactions to obtain hardware for the Kahuta plant and several related installations from a number of Western nations, often in violation of their export control laws. The strategy would become a hallmark of the Pakistani nuclear program. (See Chapter 4 for additional data on Pakistani smuggling activities.)

Perhaps the most egregious case related to Kahuta involved the smuggling from West Germany between 1977 and 1980 of an entire plant for converting uranium powder into uranium hexafluoride, the easily gasified material that is the feed for the Kahuta enrichment facility. In March 1985, a West German court convicted Albrecht Migule for the deed.[9] The plant, located in the town of Dera Ghazi Khan, continues to operate, producing material that is essential for the Pakistani nuclear weapons effort. Pakistan, it may be noted, has refused to place the Kahuta facility or the Dera Ghazi Khan plant under IAEA inspections and, indeed, has never publicly acknowledged the existence of the latter installation.

U.S. response. In May 1979, Washington made public its own concerns over the Pakistani enrichment program by terminating aid to Islamabad for a second time. In announcing the decision, Assistant Secretary of State Thomas Pickering reasoned in congressional testimony that Pakistan's enrichment program was not justified by its nuclear energy needs. "We are concerned, therefore," he concluded, "that the Pakistani program is not peaceful but related to an effort to develop a nuclear-explosive capability."[10]

The fall of the Shah of Iran in January 1979 and the subsequent turmoil of the Iranian Revolution resulted in the loss of U.S. access to posts in Iran used to gather intelligence about the Soviet Union. Hoping to relocate these listening stations in Pakistan, in late 1979, the Carter Administration decided to reverse course and offer Pakistan economic and military aid at increased levels, a decision which was given further impetus when Soviet forces began their occupation of Afghanistan on Christmas Day 1979.[11] Although Pakistani president General Zia ul-Haq rejected the Carter Administration's initial offer of $400 million, he ultimately accepted a six-year, $3.2 billion aid package proffered by President Ronald Reagan in 1981. The package included an agreement to sell Pakistan forty advanced F-16 fighter-bombers.

In approving the aid in 1981—by which time, U.S. listening posts had been relocated in China and American concerns were focused on meeting the Soviet challenge in Afghanistan—Congress granted Pakistan a six-year exemption from a U.S. non-proliferation law, known as the "Symington Amendment," which had triggered the April 1979 aid cutoff. This provision had prohibited such assistance because of Pakistan's importation of enrichment equipment and its simultaneous refusal to place all of its nuclear facilities under International Atomic Energy Agency inspection. At the same time as it granted the six-year exemption, Congress strengthened a portion of the law prohibiting U.S. aid to any non-nuclear-weapon state that subsequently detonated a nuclear explosive device.[12]

In addition to its hopes that American assistance would strengthen Pakistan as an anti-Soviet bulwark in the region, the Reagan Administration argued that the aid package would check the Pakistani nuclear weapons program by providing an alternate means, namely advanced conventional arms, for that country to protect its national security. Although this position was not vindicated by events, U.S. officials subsequently stated that American assistance helped to restrain Pakistani nuclear behavior, contributing, for example, to the country's decision not to conduct a nuclear test or actually deploy nuclear arms. Many disagreed. Indeed, the debate in the United States over whether to extend aid to Pakistan in the face of mounting evidence of its nuclear advances continued throughout the 1980s.

1984: Progress toward weapons capability; krytron affair; Chinese assistance. In early 1984, Dr. A. Q. Khan announced that the Kahuta plant had succeeded in producing enriched uranium.[13] President Zia subsequently confirmed the point but stressed that only low-enriched, non-weapons-grade material had been produced.[14] A June 1984 speech by Senator Alan Cranston, meanwhile, declared that the Pakistani nuclear weapons program was continuing, and Reagan Administration officials at the time privately confirmed that this effort included nuclear-weapon design work and the acquisition of hardware from abroad.[15]

Also in June, still further evidence of Islamabad's pursuit of nuclear arms surfaced when three Pakistani nationals were indicted for attempting to smuggle fifty krytrons—high-speed electronic switches used in nuclear weapons—out of the United States. Two of the three were released after turning state's evidence and the third, Nazir Vaid, ultimately pleaded guilty to a lesser charge, serving a total of three months in jail before being deported to Pakistan.[16] Official Pakistani denials that the defendants were linked to the Pakistani government only made matters look worse when cables, taken from the defendants at the time of their arrest, revealed that the parts had been ordered by S. A. Butt, a director of supply and procurement for the Pakistani Atomic Energy Commission.[17] The episode left little doubt about Islamabad's nuclear intentions.

The krytron affair was soon followed by press accounts that quoted Reagan Administration officials as saying that the People's Republic of China was assisting Pakistan, its long-time ally, in operating the Kahuta plant and had given Pakistan the design of a nuclear weapon.[18] Such data might allow Pakistan to manufacture reliable nuclear arms without the need for a nuclear test. Although China denied the reports that it was aiding Pakistan, the Reagan Administration took the extraordinary step of halting for nearly a year the formal approval of a highly publicized nuclear trade pact with China, suggesting that the evidence available to the U.S. government was persuasive.[19]

The September 1984 Reagan letter. Apparently in response to Khan's enrichment breakthrough, the krytron affair, and China's nuclear assistance, in September 1984 President Reagan sent a letter to President Zia expressing strong U.S. concern over Pakistan's continuing nuclear activities and threatening, but not specifying, "grave consequences" if Zia used the Kahuta enrichment plant to produce uranium enriched to more than the relatively innocuous five-percent level.[20] (Uranium enriched to ninety percent or more is used for nuclear weapons.) Such a limitation on the Pakistani nuclear program would keep it an important step short of nuclear arms. In this respect the Reagan letter went considerably further than existing U.S. legislation, which made the detonation of a nuclear device the trigger for U.S. sanctions, but which would not have penalized Pakistan's acquisition of an untested (though potentially reliable) nuclear arsenal.[21]

According to unnamed Reagan Administration officials quoted in the press, Pakistan subsequently gave Washington written assurances that the five-percent level would not be exceeded, and then-Pakistani Prime Minister Muhammad Khan Junejo confirmed in a 1986 interview that Pakistan had given such pledges.[22] Despite these apparent confirmations, in 1988, Pakistan's ambassador to the United States, Jamsheed K. A. Marker, denied that Pakistan had ever given assurances on the precise level of enrichment that it would undertake at Kahuta.[23]

1985: Non-nuclear tests, new U.S. laws. Pakistan's seeming acquiescence in the American restriction came at a time of its increasing dependence on the United States. In the fall of 1984, some 40,000 additional Soviet troops were deployed in Afghanistan, many close to the Pakistani border, bringing the total to 120,000. At the same time, Soviet planes began repeatedly violating Pakistani air space in order to bomb Afghan refugee camps in Pakistan. As these incursions continued, in March of 1985, the Reagan Administration agreed to supply Pakistan with sophisticated air-to-air missiles, and key congressional committees demonstrated their support by approving another installment of the $3.2 billion in U.S. aid begun in 1981.[24]

Congress authorized this assistance in late July, barely three weeks after a report that Pakistan had passed yet another major milestone in its quest for nuclear arms: the successful test of the non-nuclear triggering package for a nuclear weapon.[25] In approving continued U.S. assistance, however, Congress tightened existing U.S. non-proliferation restrictions by enacting the "Pressler Amendment." The provision stipulates that at the beginning of each fiscal year, the President must certify that Pakistan does not "possess a nuclear explosive device" before aid can be disbursed or arms sold to that country.[26]

In addition, as a result of the Vaid krytron affair (and a similar episode disclosed in 1985, involving the smuggling of krytrons to Israel in the early 1980s), Congress enacted another proviso, known as the "Solarz Amendment," prohibiting aid to any non-nuclear state found to have smuggled items from the United States for use in a nuclear explosive device.[27] (Interestingly, Pakistan's successful triggering-package test suggests that it may have obtained some krytrons through yet-undiscovered channels, although they may not have been those designed to military specifications of the kind Vaid and his associates were seeking.)

Turning Point. It now appears that late 1985 was a crucial turning point in U.S. efforts to slow Pakistan's nuclear weapons program. Sometime prior to October of that year—despite what American officials claim were explicit pledges given to Washington—Pakistan crossed the "red line" that had been set out in the September 1984 Reagan letter and produced uranium enriched to more than five percent. Whether this was widely understood by members of Congress when they authorized aid to Islamabad in July 1985, is not known, but according to a subsequent news story, the Reagan Administration was aware of the development by October of that year. When President Reagan met with General Zia that month at the United Nations, however, he did not challenge the Pakistani leader about the violation, undoubtedly in order to avoid a rift with a key ally assisting U.S. efforts to oust the Soviet Union from Afghanistan.[28] Thus, the United States acquiesced in Pakistan's decision to move toward production of weapons-grade uranium, the key remaining step in its development of a *de facto* nuclear weapons capability. The Zia government had taken a gamble in crossing the five-percent "red line," but, fully aware of administration and congressional actions since 1981, had correctly read American priorities.

In March 1986, the Reagan Administration announced that it would provide Pakistan a second six-year aid package, to begin in October 1987, in the amount of $4.02 billion.[29] From this point onward, there could be no doubt in Islamabad that the Reagan Administration was aware of its advances at Kahuta and was prepared to look the other way.[30]

Threshold Crossed. A series of press stories in July 1986, citing Reagan Administration sources, stated that Pakistan was considered to have the capacity to build nuclear arms or to be on the verge of having that capability, with one account stating that all that remained was for Pakistan to assemble the components.[31] By the fall of 1986, a "Special National Intelligence Estimate," circulated to senior U.S. officials, had concluded that Pakistan had produced weapons-grade material.[32] According to the report, Pakistan had also conducted two additional tests during the year of the non-nuclear portions of the device.[33] U.S. aid remained unaffected, however, and on October 27, President Reagan certified—as he had late in 1985—that Pakistan did not "possess" a nuclear explosive device, a finding necessary under the 1985 Pressler Amendment to permit the disbursement of funds and continued arms sales to Islamabad.[34] According to administration sources, the 1986 non-possession finding was based on the view that despite Pakistan's advances in enrichment, it had still not fabricated certain key weapon components.

Nonetheless, India and other observers would henceforth have to assume that Pakistan could rapidly assemble at least one nuclear device in the event of a future war. Thus Pakistan, it appears, had become a *de facto* nuclear weapon state, albeit at this juncture only in a very limited sense.

Relations with India. Pakistan has sought nuclear weapons principally to meet the threat from India's conventional military superiority and substantial nuclear potential, as well as to counter more subtle forms of Indian dominance in regional affairs. In addition, Pakistani leaders undoubtedly perceived that nuclear arming would enhance Pakistan's image as the most technically advanced state in the Muslim world and that nuclear status would cement Islamabad's military ties with the Persian Gulf states. Nuclear arms were also no doubt valued for reasons of domestic politics. As a symbol of nationhood, they could help unite a society that was yet to be fully integrated into a modern state—and, given the popular support for the nuclear program, could help Zia consolidate his political base.[35]

These additional factors notwithstanding, any normalization of Indo-Pakistani relations that might occur could advance the prospects for mutual nuclear restraint, an objective that Pakistan has pursued as a second track of its nuclear policy for more than a decade.[36] At the other extreme, the outbreak of bilateral hostilities could lead to increasingly serious consequences as the nuclear capabilities of both countries grow.

After the 1971 Indo-Pakistani war, the two countries experienced a period of "cold peace," marked by non-interference in each other's internal affairs. This era—which nonetheless saw both nations pursue nuclear-weapons development programs—came to an end in 1983, when bilateral relations turned increasingly hostile. An uprising of separatists in Pakistan's Sind Province in August 1983 sparked accusations by Islamabad that India was aiding and encouraging the rebels. Over the following year, New Delhi, in turn, accused

Islamabad of fomenting unrest by Sikh nationalists in India's Punjab region and of making incursions into Kashmir, which led to a series of border clashes. These issues have troubled relations repeatedly in subsequent years and, despite periods of harmony, mutual antagonisms became so intense in the early fall of 1984 that general war was considered a distinct possiblity.

After the assassination of Indian Prime Minister Indira Gandhi in late October 1984, President Zia launched a "peace offensive" aimed at improving ties with India's new leader, Rajiv Gandhi. Though a brief warming trend ensued, by mid-1985 a decline was again evident. On the nuclear front, in June and July Zia and his top aides accused India of provoking a nuclear arms race by building up its nuclear capabilties and by refusing Pakistan's repeated offers of mutual inspections and bilateral non-proliferation pledges. The effectiveness of this campaign was severely undercut, however, by the July 1985 report, noted earlier, that Pakistan had tested the non-nuclear triggering package for a nuclear device.

Despite continued advances in both the Pakistani and Indian nuclear programs, overall relations with New Delhi managed to improve in the latter half of 1985, leading to a watershed Zia-Gandhi meeting in New Delhi in December 1985 at which the two leaders orally agreed not to attack each other's nuclear installations and took steps toward normalizing relations in a number of areas.[37]

The warming trend was short-lived, however, in part because of domestic political difficulties on both sides, including serious anti-government disturbances in the two countries, which each accused the other of secretly abetting.[38] The first published report that Pakistan had enriched uranium to 30 percent, it may be added, appeared in March 1986; this may have contributed to increased tensions between the two countries.[39]

1987: Hints of weapons program; expanding uranium enrichment capabilities. As described in the preceding chapter on India, at the end of 1986—by which point Pakistan had apparently acquired its first stocks of weapons-grade uranium—Indo-Pakistani relations deteriorated seriously, and major Indian military exercises near the Pakistani border, known as Operation Brass Tacks, led to a series of mobilizations and counter-mobilizations that some feared might lead to war. The crisis eased after agreement was reached on a mutual pull-back of forces, but bilateral frictions continued, as each country accused the other of interfering in its internal affairs and as skirmishing on the disputed Siachen Glacier in Kashmir continued. In this uncomfortable security environment—made all the more stressful by the continued Soviet presence in Afghanistan—Pakistan continued to advance its nuclear capabilities.

On March 1, 1987, a controversial interview with the head of Pakistan's classified uranium enrichment program, A. Q. Khan, was simultaneously published in Islamabad, New Delhi, and London. In the interview—written by the well-known Indian journalist Kuldip Nayar—Khan was quoted as declaring that Pakistan had built nuclear arms.[40] Khan was also reported to have explicitly stated that Pakistan had produced weapons-grade uranium, declaring, "We have upgraded it to ninety percent to achieve the desired result."

Immediately after the interview appeared, the Pakistani government issued a release containing a denial by Khan in which he claimed that the piece had attributed to him "false and concocted statements" and insisted that Pakistan's enrichment program was entirely peaceful.[41] The ensuing tangle of denial,[42] confirmation by an independent source,[43] and questions arising from Nayar's extensive reliance on an account previously published by Khan,[44] blunted the article's political impact—particularly in the United States, where congressional consideration of continued U.S. aid to Pakistan was getting under way—and prevented it from being used as an authorative acknowledgement of Pakistan's nuclear ambitions. Nonetheless, the interview was generally taken seriously and sparked an outcry in India.[45]

The alarm caused by Khan's remarks was exacerbated in late March when President Zia, in an interview with *Time* magazine, stated that "Pakistan has the capability of building the Bomb whenever it wishes." "Once you have acquired the technology which Pakistan has," he continued, "you can do whatever you like." Here too, however, there was no direct confirmation of Pakistan's weapon-related activities, as Zia stressed that Pakistan did not intend to manufacture nuclear arms and asserted that it had not yet enriched uranium to weapons grade.[46] Zia later declared that he had been speaking only of Pakistan's broad technological capabilities, and had not meant to imply that his country was pursuing an on-going nuclear weapons program.[47]

Whatever the ambiguities in the Khan and Zia statements concerning the achievements of the Pakistani uranium enrichment program, it soon became apparent that Pakistan had undertaken a serious effort to expand its capabilities in this regard. In April 1987, the West German weekly, *Stern*, published a lengthy article detailing a West German and Swiss investigation of an extraordinary Pakistani nuclear smuggling operation. The operation involved the custom-manufacture in Switzerland of major pieces of equipment designed specifically for the production of highly enriched uranium and the secret transfer of closely held blueprints of parts of West Germany's own uranium enrichment facility at Gronau.[48] The effort was orchestrated by a German materials processing and electronics firm, Leybold-Hereaus GmbH, which contracted with a Swiss industrial equipment manufacturer, Metallwerke Buchs, to build the equipment in question. Significantly, according to Western officials quoted by the press, the operation appeared to be part of a Pakistani effort to acquire the hardware for a second uranium enrichment plant.[49] Pakistani officials denied Pakistan was involved in the affair, which received little attention in Washington, calling it a "smear."[50]

A second Pakistani smuggling operation, this one in the United States, did attract American attention, however. On July 11, 1987, Arshad Pervez, a

Pakistani-born Canadian, was arrested in Philadelphia for attempting to export illegally to Pakistan "maraging" steel, an especially strong form of the metal used in uranium enrichment centrifuges, and beryllium of a quality used in nuclear weapons themselves to increase their yield.

Pervez's activities appeared to violate the 1985 Solarz Amendment, described earlier, which prohibits U.S. assistance to non-nuclear-weapon states that illegally attempt to export nuclear commodities from the United States to develop nuclear weapons. The incident led several prominent U.S. legislators, including Congressman Stephen Solarz, to oppose continued U.S. security assistance to Pakistan unless it accepted verifiable restraints on the production of weapons-grade uranium.

As discussed at length in Chapter 4, during this period, Pakistan was also engaged in a series of major smuggling operations in West Germany aimed at acquiring equipment and know-how needed to produce tritium, a material used in advanced nuclear weapons. The existence of these operations was not revealed publicly until 1989, but the Reagan Administration was aware of them at this earlier point and unsuccessfully sought German assistance to block them.

In August 1987, Under Secretary of State for Political Affairs Michael Armacost visited Islamabad to express U.S. concerns over these developments and to seek verifiable restraints on Pakistan's nuclear activities. Not unexpectedly President Zia rebuffed these entreaties.[51] Members of Pakistan's parliament, both in the ruling Pakistani Muslim League Party and in the opposition, forcefully echoed Zia's stance in early September.[52]

Pakistani arms control initiative. In September 1987, Pakistani Prime Minister Mohammad Khan Junejo announced in a speech to the United Nations General Assembly that in June he had proposed a bilateral ban on nuclear testing to Indian Prime Minister Gandhi. Although Junejo's public disclosure of the initiative appeared timed to placate American critics of Pakistan's nuclear program, the fact that the ban had been proposed privately in June suggests that it was aimed more at India and represented a new element in Pakistan's longstanding efforts to engage New Delhi in bilateral nuclear-arms-control arrangements. These efforts appear to represent a second track of Pakistani nuclear policy aimed at placing limits on India's nuclear potential, even as Pakistan has sought to achieve a nuclear weapons capability of its own.

Pakistan has over the years offered to enter into a wide range of comprehensive bilateral nuclear arrangements with India, including reciprocal ratification of the NPT, mutual acceptance of IAEA safeguards on all nuclear installations, comprehensive bilateral nuclear inspections, the establishment of a nuclear-weapon-free zone in South Asia, and formal mutual pledges not to manufacture nuclear arms. India has rejected these proposals in large part because they tend to treat India and Pakistan as equals and because they would restrict India's right to develop nuclear arms at some future time to deter China, a nuclear-weapon state since 1964. Junejo's test ban offer, in addition to avoiding the comprehensive controls that would constrain India's nuclear options, also appeared more practical than prior proposals because it was apparently consistent with the actual nuclear testing policies of both countries, *i.e.*, it recognized, and sought to institutionalize, the restraint both were exercising in this sphere. India rejected the initiative, however, as "not serious." [53]

Official U.S. acknowledgement. On October 1, 1987, delays in passage of U.S. foreign aid and other appropriations legislation led to a temporary suspension of new U.S. assistance to Pakistan.[54] As Congress debated funding measures for the new fiscal year, on October 22, U.S. Ambassador-at-Large for Non-Proliferation Richard T. Kennedy acknowledged in congressional testimony that Pakistan had enriched uranium to more than five percent, a significant departure from previous Reagan Administration statements.[55] In late December, despite this admission, revelations that Pakistan was constructing a second enrichment plant near the town of Golra, and the conviction of Arshad Pervez, Congress voted $480 million in aid for Pakistan.[56] At the time, negotiations on the withdrawal of Soviet forces from Afghanistan had brought a Soviet pull-out within reach, and the lawmakers were unwilling to withhold assistance for America's principal ally in the region, which continued to play a crucial role in channeling U.S. military assistance to anti-Soviet Afghan guerrilla groups.

1988: A Pakistani deterrent; death of Zia. At the end of December 1987, President Reagan again certified that Pakistan did not "possess a nuclear explosive device" pursuant to the Pressler Amendment, enabling assistance for the new fiscal year to be disbursed to Islamabad. The next month, Reagan waived the Solarz Amendment, prohibiting assistance to non-nuclear countries that engage in nuclear smuggling in the United States, as it applied to Pakistan. The latter decision amounted to an admission that the Pakistani government had been behind the Pervez smuggling operation. In congressional testimony on this matter, the administration for the first time publicly acknowledged that Pakistan was developing nuclear weapons, that the country's enrichment program was being pursued for this purpose, and that Pakistani smuggling activities appeared to be ongoing.[57]

Moreover, on March 6, 1988, the *New York Times Magazine* published a detailed report by Hedrick Smith on the Pakistani nuclear program based on information from U.S. government sources. The account stated that Pakistan had accumulated enough highly enriched uranium for four to six nuclear weapons; Smith said in a subsequent interview that he had been told that Pakistan had accumulated 220 pounds (100 kilograms) of the material. Sources quoted in the piece also confirmed earlier accounts that Pakistan's weapons are based on a Chinese design and provided the new information that the weapons weighed only 400 pounds (180 kilograms) when completed,[58] indicating that they were considerably more advanced than the first U.S. weapons, which weighed approximately 10,000 pounds (4,500 kilograms). The article went on to state that Pakistan had fabricated virtually all of the components for these devices and made it clear that Pakistan could rapidly deploy nuclear weapons in any future conflict, even if technically speaking, it did not yet "possess" a fully assembled nuclear device, as President Reagan had certified in December.[59] Reagan Administration sources subsequently suggested in off-the-record interviews that, while not all of the details in the piece were correct, its overall characterization of the status of the Pakistani nuclear program was accurate.[60]

Thus, by early 1988, it became evident that Pakistan had joined the club of *de facto* nuclear powers. In terms of turning materials and components into an arsenal of nuclear weapons, however, Indian capabilities would surpass those of Pakistan for the foreseeable future, owing to its much larger stockpiles of weapon-grade material. Nevertheless, at this point, a rough-hewn form of nuclear deterrence had emerged on the Subcontinent.[61] Indeed, in a July 1988 interview with a member of a delegation sponsored by the Carnegie Endowment for International Peace, which was examining nuclear proliferation in South Asia, President Zia declared:

> With respect to their [India's] nuclear capabilities, if they create ambiguity, that ambiguity is the essence of deterrence. The present programs of India and Pakistan have a lot of ambiguities, and therefore in the eyes of each other, they have reached a particular level, and that level is good enough to create an impression of deterrence.

Zia went further in a second interview the same month by stating that the region had achieved "a stable nuclear deterrent relationship based upon ambiguity as to whether India or Pakistan had nuclear weapons, and if they did, how many they possessed."[62] With the tacit acknowledgement that Pakistan may possess a *de facto* nuclear weapons capability, these statements marked a major departure from Zia's past insistence that Pakistan's nuclear program was devoted exclusively to peaceful purposes.

Political and Technical Developments

On August 17, 1988, Zia was killed in a suspicious plane crash.[63] In December, following Pakistan's first free elections for a national leader in nearly twenty years, Benazir Bhutto, the daughter of former Prime Minister Zulfikar Ali Bhutto and Zia's longtime antagonist, became prime minister.[64] Prior to her election, Ms. Bhutto had suggested on a number of occasions that she was prepared to curb the country's nuclear build-up,[65] and her close ties to the United States appeared to make her more susceptible than Zia to U.S. pressure on the issue. At the same time, the withdrawal of Soviet forces from Afghanistan, which had begun in May 1988 and would be completed in February 1989, seemingly freed Washington to apply such pressure more forcefully than had been possible in the past. These events thus raised the possibility that Pakistan might slow its nuclear weapons effort.

By mid-1990, however, there seemed to be little to distinguish Pakistan's nuclear posture from that of the Zia years, although there was evidence that some limited aspects of the country's nuclear-weapons related activities had been curbed, at least temporarily.

Zia's nuclear legacy; Reagan certification. On November 18, 1988, two days after the parliamentary elections that would bring Ms. Bhutto to power, President Ronald Reagan again certified to Congress that Pakistan did not "possess a nuclear explosive device," pursuant to the Pressler Amendment.[66] By this time, however, it was widely believed that Pakistan possessed sufficient quantities of weapons-grade uranium for a number of weapons and had acquired most, if not all, of the other necessary components. Apparently reflecting his recognition of the advanced stage of the Pakistani nuclear weapons effort, President Reagan qualified his conclusion in a letter accompanying his certification of "non-possession," stating:

> I have . . . taken into account the fact that the statutory standard as legislated by Congress is whether Pakistan possesses a nuclear explosive device, not whether Pakistan is attempting to develop or has developed various relevant capacities. Based on the evidence available, and on the statutory standard, I have concluded that Pakistan does not possess a nuclear explosive device.
>
> ***
>
> The Congress should be aware that as Pakistan's nuclear capabilities grow, and if evidence about its activities continues to accumulate, this process of annual certification will require the president to reach judgments about the status of Pakistani nuclear activities that may be difficult or impossible to make with any degree of certainty.[67]

Thus, when Ms. Bhutto took office, she inherited not only a *de facto* nuclear weapons capability, but also the increasing risk that within a year American aid would be terminated unless modifications were made in the direction of Pakistan's nuclear program.

By December 1988, Pakistan was thought to possess the ability to produce enough weapons-grade nuclear material for two or three weapons annually at its Kahuta uranium enrichment plant, a capability it apparently had acquired in 1986. By the time Bhutto took office, Pakistan thus probably had enough nuclear material for roughly five to ten nuclear devices.[68] In December 1989, Zalmay Khalilzad, a former U.S. official who served in the Reagan Administration until the end of 1988 and had intimate knowledge of the Pakistani nuclear program, confirmed in a scholarly article that Pakistan had indeed been accumulating weapons-grade material, although he did not estimate the quantities that might be on hand.[69]

As noted earlier, it was also believed by late 1988 that Pakistan had obtained the design of a nuclear device from China,[70] had tested the non-nuclear components of such devices,[71] and had taken many of the steps needed to fabricate complete devices.[72] However, it appeared that Pakistan had not stockpiled *all* of the components that would be needed to build a nuclear device. Under a 1987 State Department interpretation of the Pressler Amendment, if Pakistan had manufactured all of the necessary components and needed only to assemble them, this would have constituted "possession of a nuclear explosive device" and would have precluded President Reagan's November 1988 certification.[73] Nonetheless, Reagan's finding suggested that whatever weapon parts remained to be fabricated could be manufactured quickly and perhaps without detection by the United States. It has been speculated that among the remaining steps Pakistan may have needed to complete as of late 1988 were: the precise machining of highly enriched uranium into weapons parts; the design and fabrication of an aerodynamic casing for the device, suitable for suspending it beneath an aircraft; and the fusing for the device, to ensure it would detonate at a predetermined altitude when released.[74]

Expanding capabilities. In addition to operating the Kahuta facility and taking a number of steps towards the manufacture of nuclear devices, by late 1988, Pakistan was also attempting to expand its ability to produce weapons-grade material. As detailed earlier in this chapter and in Chapter 4, using equipment and know-how clandestinely obtained in Western Europe, Pakistan was thought to be enlarging the Kahuta plant and, reportedly, to be building a second enrichment facility in the town of Golra.[75]

Pakistan also had taken important steps to acquire the ability to produce tritium for advanced nuclear weapons,[76] and had apparently begun construction of an unsafeguarded research reactor as part of this effort.[77] That facility could also be used for the production of plutonium. By the mid-1980s, Pakistan had completed the "New Labs" plutonium extraction plant, near Rawalpindi, where spent fuel from the new reactor could be processed. The plutonium facility was not known to have begun operations, however.[78]

The reactor has been referred to as a "large, pool-type" unit, with a possible rating of 50 megawatts, which would make it suitable for both tritium and plutonium production.[79] Little else has come to light about the facility's size or design. Pakistan is said to have sought unsuccessfully certain materials needed for construction of the plant from West German sources in 1986 but was able to acquire other items needed to manufacture fuel for it from these sources in 1987.[80] Construction may have begun by the time Ms. Bhutto took office; even if this were true, the reactor probably was a number of years from completion. It is not known to be operating today.[81]

New Missile. On February 11, 1989, Pakistani Army Chief of Staff General Mirza Aslam Beg, announced that Pakistan had successfully tested two new surface-to-surface ballistic missiles, the Hatf I and the Hatf II. According to

Beg, the former had a range of 50 miles (80 kilometers) and the latter of 186 miles (300 kilometers), making them roughly comparable in terms of range to the Soviet FROG and Scud-B missiles, respectively.[82] Both weapons reportedly had payloads of 1,100 pounds (500 kilograms), in effect making them powerful enough to carry a relatively crude nuclear warhead.[83] The larger missile could reach major targets in northern and western India, and, although New Delhi is at the margins of its range, the Hatf II could reach the Indian capital if its payload were reduced.[84] According to the U.S. Arms Control and Disarmament Agency, Pakistan received assistance in developing the missiles from China and from West European concerns.[85] As of mid-June 1990, neither missile was known to have been deployed or to have undergone additional flight tests. Later in 1989, Beg announced that Pakistan was working on a missile with a range of 375 miles.[86]

Control over nuclear, foreign policy. Though Bhutto had become prime minister, her authority over sensitive foreign policy and defense issues was considerably constrained. Despite her status as the head of the Pakistani government, there were serious doubts as to whether she had acquired political control over the country's nuclear program, inasmuch as her authority over the Pakistani Army and the country's powerful nuclear bureaucracy, which jointly run the program, was uncertain.[87]

As a practical matter, Bhutto's ability to govern required the support of the Pakistani military, which had been the dominant force in Pakistani politics ever since Zia's seizure of power in 1977.[88] Indeed, the fact that Army Chief of Staff Beg, rather than Ms. Bhutto, announced the testing of the Hatf I and Hatf II in February appears to have been an attempt by Beg to enhance the Army's prestige in its continuing effort to jockey for position vis-à-vis the country's elected leadership. Moreover, given the widespread popularity of the Pakistani nuclear weapons effort, the country's senior nuclear aides could easily turn public opinion against Ms. Bhutto if she sought to constrain the program against their wishes. These factors appeared to give the Army and the country's nuclear establishment an effective veto over Bhutto's policymaking in the nuclear sphere.

Bhutto's authority was further limited by her weak political position. Within parliament, Bhutto's Pakistani People's Party (PPP) commanded only a plurality of seats, requiring her to govern through a coalition, and Bhutto was unable to consolidate her position in the country's January 1989 bi-elections. In addition, Bhutto was obliged to do political battle with two former Zia supporters: President Ghulam Ishaq Khan, with whom Bhutto was constitutionally required to share power, and Nawaz Sharif, a leader of the opposition Islamic Democratic Alliance, who governed Punjab, Pakistan's most populous and wealthy state. Bhutto had a series of debilitating confrontations with both figures in succeeding months.[89]

In addition, Bhutto's control of the secretive Inter-Services Intelligence Directorate (ISID) was in doubt. Under Zia, the ISID had been responsible for one of the most sensitive and vexing areas of Pakistani foreign policy, covert support of the anti-Soviet Afghan resistance and the channelling of millions of dollars in U.S. military assistance to them. The withdrawal of Soviet forces from Afghanistan promised to end the guerrilla conflict there and to prompt the exodus of the three million Afghan refugees living in Pakistan. Under Zia, however, the ISID had become committed to installing a Muslim fundamentalist government, sympathetic to Pakistani interests though not necessarily popular in Afghanistan, as a successor to Afghanistan's Soviet-backed regime. The ISID was also said to be responsible for alleged Pakistani support for militant separatist groups within India, including Sikh extremists active in India's Punjab region and Muslim fundamentalists in Kashmir. Bhutto's control over the ISID's clandestine operations and her ability to set policy on these crucial issues also remained in question.

Possible nuclear restructuring. To reassure Washington that the country was not acquiring a nuclear explosive device, Bhutto needed to constrain certain aspects of Pakistan's nuclear program. While her political situation greatly limited her freedom of action with respect to Pakistani security policy, it appears that during the spring of 1989 she found a tentative solution to these conflicting demands.

At this time, despite the Reagan certification, it was not clear that Congress would support continued military assistance to Pakistan at the high levels of the recent past.[90] Of particular interest to the Pakistani military was a proposed sale of 60 advanced F-16 fighter-bombers over eight years, to complement the 40 such aircraft Pakistan had received between 1981 and 1987. Such assistance had been deemed necessary since 1981 to bolster the country's defenses against possible Soviet pressure from neighboring Afghanistan. After Moscow completed the withdrawal of its forces from the region in February 1989, this rationale no longer applied, and it became clear that future U.S. military aid would be used by Islamabad exclusively to counter the traditional threat from India. Although U.S. policy-makers were aware that past military assistance had been used in part for this purpose, Washington had never explicitly recognized the Indian threat as a legitimate basis, in itself, for U.S. support.[91]

The fragility of Bhutto's political base offered a new rationale for U.S. aid at a time when the Soviet-Afghan threat was abating. During the late winter and spring of 1989, U.S. analysts increasingly argued that high levels of security assistance were needed to ensure continued support by the Pakistani military and other hawkish factions for the country's fledgling democracy under Bhutto.[92] The argument gained considerable backing in large part because of Bhutto's personal popularity in Washington. In effect, it began to appear that Bhutto would be able to obtain the U.S. assistance sought by Pakistan's armed forces, assistance which might not be available if it undermined her authority or forced her from office.

By the spring of 1989, it appeared that this new balance had given Bhutto the leverage she needed to work toward a compromise on the nuclear issue with other elements of Pakistan's power structure. On May 18, U.S. Central Intelligence Agency Director William Webster, testifying before a Senate committee, stated:

> Clearly, Pakistan is engaged in developing a nuclear capability. What creates problems for the United States is whether that capability has reached a point that it implicates various amendments that apply to other assistance and relief for Pakistan.[93]

Responding to a reporter's question about Webster's comments, Congressman Stephen J. Solarz, the chairman of the House Subcommittee on Asian and Pacific Affairs, replied that a senior Bush administration official had informed him that the Pakistanis were responding to U.S. insistence that they redirect their nuclear program but that "the implication of this testimony is that they haven't."[94] Though qualified by the second remark, the congressman's initial comment implied that by May an accommodation of some type regarding the Pakistani nuclear program was being implemented in Islamabad.

In June, Bhutto visited Washington. In an address to a joint session of Congress she stated, with obvious reference to the standard of the Pressler Amendment, "Speaking for Pakistan, I can declare that we do not possess, nor do we intend to make, a nuclear device."[95] At the time of her visit, Bush Administration officials were quoted in a number of press stories as indicating that Bhutto had slowed a key element of Pakistan's nuclear program, the production of weapons-grade uranium.[96]

Though few details were revealed, it appeared that under the new *modus vivendi* Pakistan would keep its existing stocks of weapons-grade material and the other components that comprised its five to ten almost-ready-to-assemble nuclear devices. In addition, Pakistan would continue to enrich uranium to an intermediate level, producing and stockpiling material which could not be used directly in nuclear arms but which could be upgraded, perhaps in a matter of weeks, to that level. The U.S. objective of limiting Pakistani enrichment to the production of 5-percent enriched uranium (also not suitable for nuclear arms and more difficult to upgrade) appeared to have been abandoned, despite the importance that had been attached to this benchmark in President Reagan's 1984 letter to General Zia. Moreover, Pakistan's clandestine efforts to obtain equipment and materials for nuclear arms would apparently continue unabated.[97]

These reports suggest that Bhutto was able to strike a bargain with the various elements of Pakistan's security establishment under which the country's nuclear bureaucrats and other supporters of nuclear arms were persuaded to accept a modest restriction on the country's nuclear activities in order to ensure that the U.S. assistance sought by the Pakistani military could be obtained.[98]

An arrangement of this kind would leave the Pakistani nuclear-weapons effort largely intact and would do little to stem its expansion, thus falling far short of U.S. non-proliferation objectives. On the other hand, it would place a cap on the number of nuclear devices that the country might assemble quickly. This could open the door to possible negotiations between Pakistan and India for additional restraints. Indeed, shortly after Ms. Bhutto's departure from Washington, President George Bush is believed to have telephoned Indian Prime Minister Rajiv Gandhi to propose Indo-Pakistani negotiations on the issue, albeit without success.[99]

By late spring it was also increasingly clear that despite the withdrawal of Soviet troops from Afghanistan, neither the Bush Administration nor Congress was prepared to terminate U.S. aid or apply other sanctions in an attempt to force Pakistan to halt its weapons-related nuclear activities. Continued U.S. aid to Islamabad had wide support in Washington, not only to sustain Pakistani democracy, but also to ensure continued Pakistani assistance for anti-Communist Afghan guerrillas, whose efforts to topple the Soviet-backed government in Kabul were faltering unexpectedly. Under the circumstances, U.S. policy-makers appear to have settled for only a marginal limitation of Pakistan's nuclear status.

Improved relations with India. As detailed in Chapter 6, from the time Ms. Bhutto took office through the middle of 1989, Pakistan's relations with India improved considerably. On December 31, 1988 the two countries signed a written codification of the 1985 oral understanding between Zia and Gandhi under which both countries agreed not to attack each other's nuclear installations, and, in July, Rajiv Gandhi made a state visit to Islamabad, the first by an Indian leader in several decades. (The no-attack agreement is not yet in force, however. See Chapter 6.) Ms. Bhutto also renewed the offers made by her precedessor for a variety of bilateral non-proliferation and nuclear arms control measures, but India continued to reject these proposals.[100]

The momentum of the improving ties between the two countries was lost in the fall of 1989, however, as both leaders faced domestic political crises. On November 1, Bhutto narrowly survived a no-confidence vote in parliament; three weeks later she disclosed that, had her opponents prevailed, the opposition had planned to arrest her and other PPP leaders for treason. Also in November, Gandhi was succeeded as Prime Minister by V. P. Singh, after the Congress (I) party lost its parliamentary majority in national elections. By the time Singh took office, Indo-Pakistani relations had become seriously inflamed because of the eruption of Muslim separatist violence in Kashmir, which India accused Pakistan of fomenting.

Nuclear advances. In July 1989, the West German magazine, *Der Spiegel* quoted a report by German intelligence officials as stating that Pakistan was conducting wind-tunnel tests of the casing of a nuclear bomb designed to fit under the wing of an F-16—a possible violation of the U.S. sales contract covering the aircraft.[101] The report also noted that Pakistan had learned how to program the computers on the aircraft for nuclear bombing missions.[102] Clandestine Pakistani efforts to acquire matériel for the country's nuclear weapons effort also continued unabated through the fall of the year, according to U.S. officials quoted in the press.[103]

In August, moreover, pressures appeared to mount for Bhutto to manufacture nuclear arms in response to India's growing conventional and nuclear capabilities. The previous month, in a Pakistani defense journal, retired General Khalid Mahmud Arif, who had served as vice chief of the army staff under Zia and was reportedly responsible for managing the construction of the Kahuta plant,[104] declared:

> Since India is relentlessly pursuing her nuclear military effort, Pakistan needs to review her nuclear policy. Her effort in this field already enjoys bipartisan and popular support and is above political controversy. Her national interest demands that her nuclear capability be taken to its logical conclusion.[105]

These views received considerable attention and, in general, were widely applauded.[106]

Apparently seeking to respond to such calls for a more overt nuclear weapons effort, Bhutto declared in August that Pakistan had sufficient nuclear "knowledge" to deal with any threat, stating in an interview:

> We do have the knowledge but I think there is a difference between knowledge and capability. So we do have knowledge, if confronted with a threat, to use. But we do not, in the absence of any threat, intend to use that knowledge. In fact, as a matter of policy my government is firmly committed to non-proliferation [but] any one-sided action by any country in the subcontinent can trigger off a nuclear arms race.

Though still ambivalent, these remarks reflected an important shift from her statements in June in their clear invocation of the military and deterrent potential of the Pakistani nuclear program.[107]

In early October, Army Chief of Staff Beg offered a more direct statement of Pakistani nuclear policy, although it was by no means clear whether he was speaking on behalf of the country's civilian leadership. In an interview, he explicitly called attention to the military dimension of Pakistan's nuclear capability, stating:

> Both the nuclear option and the missiles [that Pakistan is developing] act as a deterrence and these in turn contribute to the total fighting ability of the army, which acts as a deterrent to the enemy.[108]

In effect, Beg resurrected the declaratory stance that Zia had begun to articulate prior to his death, abandoning the long-standing claim that the country's nuclear program had exclusively peaceful objectives.

F-16 sale; Bush certification; Symington Amendment. Washington's readiness to continue high levels of assistance to Pakistan appeared relatively unaffected by these developments. During July and August, Congress reviewed the proposed sale of 60 F-16 fighter-bombers to Pakistan, and raised no objections. A House subcommittee, whose members were familiar with the allegations that Pakistan was preparing some previously transferred F-16s for nuclear delivery, did, however, obtain assurances from the Bush Administration that the sale would be halted if, in accordance with the Pressler Amendment, the President subsequently determined that Pakistan possessed a nuclear explosive device. In addition, Executive Branch witnesses noted that none of the F-16s previously transferred to Pakistan was configured for nuclear delivery and that the contract for sale of the new aircraft would prohibit their use with nuclear arms and would also prohibit any alteration of the warplanes, including alterations that might be needed for the delivery of nuclear ordnance, without the express approval of the United States.[109]

On October 5, barely a week before Beg's remarks noted above were reported, President Bush again certified that Pakistan did not "possess a nuclear explosive device," under the terms of the Pressler Amendment. Since U.S. officials had indicated after President Reagan's 1988 certification that such a finding would not be possible in 1989 unless Pakistan modified its nuclear activities, the Bush certification implied that Bhutto had, indeed, altered the program, presumably in the manner discussed earlier, in order to meet the "non-possession" requirement. However, Bush's letter accompanying his finding was equivocal. Bush noted President Reagan's statement in 1988 that "as Pakistan's nuclear capabilities grow, and if evidence about its activities continues to accumulate," it might become impossible to reach judgments about the status of its nuclear activities with any degree of certainty. Bush then stated that, since the Reagan certification, "Pakistan has continued its efforts to develop its unsafeguarded nuclear program," implying that a finding of non-possession had become increasingly problematic in the intervening year. After repeating Prime Minister Bhutto's June statement renouncing nuclear weapons, Bush appeared to conclude that this assurance offset any ambiguities created by Pakistan's nuclear activities during 1989:

> Based on my own discussions with the Prime Minister and other extensive diplomatic contacts with Pakistan, I believe the government of Pakistan is fully aware of the requirements of U.S. law governing security assistance and of the serious consequences that would ensue, particularly with regard to regional security, should it prove impossible to make the certification required by law. The Administration will continue to make clear to Pakistan that our ability to certify depends on evidence that the assurances given by the Prime Minister are being implemented.[110]

Given this tortured analysis, it is not surprising that U.S. aides would later state that Bush had "great difficulty" in making the non-possession finding. Thus, while Bhutto may have checked certain narrow aspects of the Pakistani nuclear weapons program, there was little reason to believe that the pace of the country's overall nuclear effort had slowed.

The President's sidestepping on the matter of certification was mirrored by Congress' willingness to overlook another statutory non-proliferation measure. Since 1981, congressional debate on the question of Pakistan's nuclear program had centered on whether or not to exempt Pakistan from the 1976 Symington Amendment, which prohibits U.S. economic and military assistance to non-nuclear weapon states that import uranium enrichment technology but refuse to place all of their nuclear activities under IAEA safeguards. Pakistan's nuclear activities violate this standard, and in 1981, in order to permit renewed aid to Pakistan following the Soviet invasion of Afghanistan, Congress had enacted legislation suspending the application of the provision to Pakistan for six years. In late 1987, after a year-long debate, Congress again suspended the provision, this time for a two-and-a-half-year period ending in April 1990. Then, in November 1989, with no debate in either house, the suspension was extended once again for a period of one year, until April 1991, as Congress appropriated funds for foreign aid, including those sought by the Bush Administration for Pakistan.[111]

Given the history of controversy surrounding this subject, the lack of debate was surprising, indeed. It appeared to reflect the fact that Congress had come to view Pakistan's enrichment program as a *fait accompli* and that, along with the Bush Administration, had shifted its focus to the question of whether Pakistan was manufacturing nuclear weapons themselves. Nonetheless, the extension of the Symington Amendment waiver was considerably shorter than the three-and-a-half-year suspension originally sought by the Bush Administration in early 1989, suggesting that Congress remained concerned about Pakistan's nuclear behavior and wished to monitor developments closely.

Chinese, French reactor sales. On November 16, 1989, during a visit to Islamabad, Chinese Premier Li Peng announced that China would sell Pakistan a 300-megawatt nuclear power reactor, under a nuclear cooperation agreement the two states had signed in 1986. Although the reactor itself and any plutonium produced from its operation were to be under IAEA safeguards, Beijing did not require as a condition for the sale that Pakistan accept such safeguards on its entire nuclear program, a step that would have effectively required Islamabad to give up its nuclear weapons capability. Since the mid-1970s, when Pakistan's interest in nuclear arms became apparent, no other nuclear supplier has been prepared to make major nuclear sales to Pakistan unless it accepted this condition. Moreover, for much of the 1980s, as the result of a U.S. initiative, no supplier had made a major nuclear sale to *any* country that had not accepted such "full-scope" safeguards. In 1988, however, the Soviet Union broke this *de facto* embargo, by agreeing to sell two large nuclear power reactors to India, its longtime security partner in the region. This precedent undoubtedly paved the way for Beijing to make a comparable sale to its own regional ally.[112]

The reactor will be a replica of China's first indigenous nuclear power plant, which China now has under construction, a unit moderated with "light," or ordinary, water and fueled with low-enriched uranium. That Pakistan would purchase such a costly facility from so unproven a supplier leaves little doubt that political factors were dominant in the arrangement. The deal allowed Beijing to cement relations with its regional partner at a time when it was facing increasing international isolation because of its violent repression the previous June of the country's pro-democracy movement. Similarly, the agreement allowed Benazir Bhutto to boost her standing with her country's nuclear advocates after her apparent efforts to slow elements of Pakistan's nuclear weapons program earlier in the year.

In a second reactor deal, on February 21, 1990, French President François Mitterrand announced in Islamabad that he would authorize French industrial concerns to sell Pakistan a 900-megawatt nuclear power plant. As China had done, Mitterrand stated that France would not require as a condition for the sale that Pakistan accept IAEA safeguards on all of its nuclear activities, although any French reactor and material it produced would be subject to IAEA coverage. The reactor will be fueled with low-enriched uranium and use light water as a moderator. The announcement made France the first Western supplier to break the *de facto* embargo on major new nuclear sales to states refusing full-scope safeguards, but Mitterrand dismissed U.S. protests on the matter.[113]

Although it is not known how Pakistan will finance the two reactor purchases, the resulting transfer of nuclear equipment and material is unlikely to make a direct contribution to Pakistan's nuclear weapons efforts. Nonetheless, the nuclear power facilities will strengthen the country's overall nuclear infrastructure and train and employ numerous nuclear scientists, engineers, and skilled construction workers. This could contribute indirectly to Pakistan's military nuclear activities. Moreover, the fact that China and France have declared their readiness to assist Pakistan in this sensitive area gives the impression of condoning Pakistan's weapons-related nuclear activities, and thus goes against the grain of international non-proliferation efforts.

Pakistani nuclear transfers. With its growing nuclear mastery, Pakistan has aquired the ability to assist other states to develop nuclear weapons. Indeed, over the years, there has been widespread speculation that Pakistan might be prepared to share its knowledge of nuclear weaponry—or nuclear weapons themselves—with other Islamic states. Though Pakistan is not known to have done this, it has reportedly begun training Iranian nuclear specialists and may have provided additional nuclear aid to that country.[114] India and Israel, in addition, have expressed concern that Pakistan might cooperate with Saudi Arabia to help it arm its Chinese-supplied intermediate-range ballistic missiles with nuclear warheads. Similarly, Iraq appears to be developing a uranium enrichment capability based on the same technology as that used by Pakistan (*i.e.* high-speed centrifuges) and to have received assistance from China, as

Pakistan did; this could be an indicator that direct Pakistan-Iraq nuclear ties in this sensitive area are developing.[115]

Kashmir and the threat of war. As detailed in the previous chapter examining India, during the latter months of 1989 and the first half of 1990, Indo-Pakistani tensions mounted over Kashmir, and the risk that war might break out increased dramatically. Since such a war would be the first major conflict between two nuclear-armed states, the potential impact on the region and on international security more generally could be very grave, indeed. These important issues, including the likelihood that Pakistan would ready and disperse its nuclear weapons to avoid an Indian pre-emptive attack if hostilities appear imminent, are discussed in full in Chapter 6.[116]

Prospects

For the near term, the most decisive influence on Pakistani nuclear policy is likely to be the growing threat of war with India. Even if the current crisis subsides, it is all too likely that Pakistan will have taken additional steps to ready its nuclear arms so as to ensure that they could be rapidly deployed, if needed. India, too, is likely to take such actions, with the result that both states will be very close indeed to possessing deployed nuclear forces. As both countries move from "not-quite-bombs in the basement" to this new status, the *de facto* nuclear arms race on the Subcontinent would intensify.

For Pakistan, such developments—which Pakistani military planners may insist upon and which Benazir Bhutto may be largely powerless to prevent—could trigger a crisis with the United States because of the likely inability of President Bush to certify that Pakistan did not possess a nuclear explosive device. With Indian hawks calling for a war that would eliminate Pakistan as a national entity, however, nuclear advocates in Islamabad may be little concerned about the risks from this quarter.

Assuming that war is avoided and that any differences with the United States are minimized, it is unlikely, given the continuing threat from India and the influence of Pakistani military and nuclear establishments, that Islamabad will deviate from its nuclear policies of the recent past: the slow but steady expansion of its nuclear weapon production capabilities; reliance on clandestine purchases abroad of needed hardware and materials; and the manufacture of virtually complete nuclear devices. Presumably, if Kahuta switched in 1989 from the production of weapons-grade uranium to the production of uranium enriched to an intermediate level, this would continue. This would leave Pakistan with a static stockpile of about five to ten quickly completable nuclear devices, while its overall nuclear weapons potential slowly grew. In addition, to avoid an all-out nuclear arms race with India and the risk of irretrievably alienating the United States, Pakistan is likely to continue to refrain from nuclear testing.

Improved relations with New Delhi could lead to the implementation of the no-attack agreement and to other limited confidence-building measures, but India is likely to continue its opposition to the more ambitious non-proliferation and arms control measures that Pakistan has proposed in the past. Most probable is that both states will continue their respective *de facto* nuclear build-ups and their development of advanced delivery systems, as an undeclared nuclear deterrence relationship evolves between them.

SOVIET UNION
CHINA
Kashmir
Islamabad
Golra
Mardan
Kahuta
AFGHANISTAN
Peshawar
Sihala
Rawalpindi (PINSTECH)
Gujranwala
Chashma
Lahore
Jhang Maghiana
Faisalabad
Quetta
Multan
Dera Ghazi Khan
River
Chagai Hills
PAKISTAN
Bahawalpur
Sukkur
Indus
IRAN
Karachi
Hyderabad
INDIA
Arabian Sea
PAKISTAN
possible test site
0
400
MILES

Pakistan

Power Reactors/Operating or Under Construction

KANUPP, Karachi (heavy-water/natural-uranium, 125 MWe[a])

- supplier: Canadian General Electric (Canada)
- start up: 1972
- fuel source: Canada[b], also Pakistan after 1980
- heavy water: United States and Canada[c]
- safeguards: yes[d]

Uranium Resources/Active Mining Sites/Uranium Mills

- reasonably assured reserves: sufficient for Kahuta enrichment plant[e]
- currently active site: Dera Ghazi Khan
- mills in operation: Dera Ghazi Khan;[aa] Lahore (?)[f]

Uranium Conversion (UF_6)

Dera Ghazi Khan

- capacity: max. 198 metric tons of hexafluoride per year[g]
- supplier: CES Kalthof G.m.b.H. of Freiburg (West Germany)[h]
- start up: 1980[h]
- safeguards: no

Heavy Water

Multan

- capacity: 13 metric tons[c]
- supplier: Belgonucléaire (Belgium)[f](?)
- start up: 1980 (?)[f]
- safeguards: no

Karachi

- capacity: upgradation unit to serve KANUPP, quantity unknown[f]
- supplier: Canada (?)
- start up: 1976
- safeguards: (?)

Enrichment

Kahuta

- type: ultracentrifuge
- capacity: 110 pounds (50 kilograms) per year of weapons-grade uranium (est.)[i,p,v]
- supplier: Pakistan, with extensive use of equipment and technology clandestinely acquired from various Western countries.[r]
- start up: 1984 (partial); weapons-grade uranium first produced in 1986[j]; production of weapons-grade uranium possibly suspended in 1989[bb]
- safeguards: no

Sihala

- type: ultracentrifuge[f]
- capacity: experimental scale
- supplier: same as Kahuta (presumed)
- start up: 1979[t]
- safeguards: no

Golra (reported)[s]

- type: ultracentrifuge
- capacity: ?
- supplier: Pakistan with equipment clandestinely aquired from various Western countries
- start up: Reported to be under construction, December 1987[s]
- safeguards: no

Fuel Fabrication

Chashma/Kundian

- capacity: sufficient fuel for KANUPP[c]
- supplier: Pakistan, plans from Canada[k]
- start up: 1980[l]
- safeguards: no

Reprocessing (Plutonium Extraction)

Chashma

- capacity: 100 metric tons of spent fuel; 100 to 200 kg of plutonium per year[g]
- supplier: SGN (France)
- start up: France terminated this project in 1978; construction may be continuing.
- safeguards: uncertain; agreement between Pakistan, France, and IAEA provides for safeguards, but these provisions not yet in force.

New Labs, Rawalpindi

- capacity: capable of extracting 10 to 20 kg of plutonium per year[q]
- supplier: SGN (France), Belgonucléaire (Belgium)[g,m]
- start up: cold tests 1982; plant believed not to be operating.[n]
- safeguards: no (but might be subject to safeguards as a "replicated" plant under Pakistan-France-IAEA agreement covering Chashma plant, if safeguards under this agreement are implemented).

PINSTECH, Rawalpindi

- capacity: experimental scale[o]
- supplier: Pakistan (?); plans from Great Britain[o]
- start up: ?
- safeguards: no

Tritium

Tritium Purification Plant, located 150 km south of Rawalpindi[x]

- capacity: 100,000-240,000 curies of tritium gas; 0.5-1.0 grams of tritium per day[y]
- supplier: Neue Technologien GmbH (NTG) and Physikalisch Technische Beratung (PTB) (both West Germany)[y]
- start up: test operated 1987[x]
- safeguards: no

Research Reactor

PARR, Rawalpindi (light-water/highly enriched uranium, 5 MWt[a], expanded to 10 MWt by late 1989[u])

- supplier: United States[a] (through the IAEA)
- start up: 1965[a]
- fuel source: United States
- safeguards: yes[a]

Pakistan Atomic Research Reactor-II (PARR-II), Rawalpindi (pool-type; light-water/enriched uranium(?), 27 KWt)[u]

- supplier: Pakistan, China[u]
- start up: 1989[u]
- fuel source: Pakistan(?), China(?)
- safeguards: yes

Research Reactor, location unknown (pool/highly-enriched uranium, 50 MWt)[w]

- supplier: Pakistan, with equipment clandestinely acquired from various western countries[w]
- start up: construction thought to have begun in mid- to late-1980s[w]
- fuel source: Pakistan
- safeguards: no

Sources and Notes

a. International Atomic Energy Agency, *The Annual Report for 1981* (IAEA, 1982), pp. 73-76.

b. Canada terminated fuel supplies in 1976 because of Pakistan's unwillingness to sign the Non-Proliferation Treaty; since 1980, plant has run partly on Pakistani-produced fuel and on remaining Canadian-supplied material.

c. David Hart, *Nuclear Power in India: A Comparative Analysis* (London: George Allen and Unwin), p. 133; "Pakistan AEC Journal Says Nation Has Mastered Front End of Fuel Cycle," *Nuclear Fuel*, March 25, 1985, p.10.

d. From September 1980 until March 1983, IAEA was unable to certify that no diversion of spent fuel occurred. Milton Benjamin, "Pakistan Backs Atomic Safeguards," *Washington Post*, November 17, 1982.

e. "Scientist Affirms Pakistan Capable of Uranium Enrichment, Weapons Production," *Nawa-I-Waqt* (Lahore), February 10, 1984, translated in *Joint Publication Research Service-Nuclear Proliferation and Development*, March 5, 1984, p. 36; "Pakistan AEC Journal . . .," *Nuclear Fuel* (see note c).

According to press reports in November 1979, Pakistan had obtained as much as 100 metric tons of uranium concentrate, or "yellowcake," not subject to International Atomic Energy Agency monitoring from Libya. The material had originally been purchased by Libya from Niger and then re-exported to Pakistan, possibly along with material that Pakistan had itself purchased from Niger through normal, above-board channels. John J. Fialka, "West Concerned by Signs of Libyan-Pakistan A-Effort," *Washington Star*, November 25, 1979.

f. P. B. Sinha and R. R. Subramanian, *Nuclear Pakistan* (New Delhi: Vision Books 1980), pp. 35, 121.

g. Steve Weissman and Herbert Krosney, *The Islamic Bomb* (New York: Times Books, 1981), p. 219 (UF_6); pp. 80-84 (New Labs); p. 81, (experimental reprocessing unit); "Germans Fostered Climate for Bombs," *Der Stern*, July 2, 1981, pp. 96-99.

h. Weissman and Krosney, *The Islamic Bomb* (see note g), p. 219, John M. Geddes, "Bonn Says Firm Illegally Sent Pakistan Gear That Can Be Used for Atomic Bombs," *Wall Street Journal*, July 16, 1981.

i. Senator Alan Cranston, "Nuclear Proliferation and U.S. National Security Interests," *Congressional Record*, June 21, 1984, p. S 7901; "Inside Kahuta," *Foreign Report*, May 1, 1986.

j. "Scientist Affirms," *Nawa-I-Waqt*, pp. 43-44 (see note e); Hedrick Smith, "A Bomb Ticks in Pakistan," *New York Times Magazine*, March 6, 1988, p. 38.

k. "Nuclear Facilities in the Middle East," U. S. Department of State, submitted in *Hearings on the Israeli Air Strike*, June 18, 19, and 25, 1981, 97th Congress, 1st Session (Washington, D.C.: Government Printing Office, 1981), p. 40.

l. James Katz and Onkar Marwah, *Nuclear Power in Developing Countries* (Lexington, Mass.: Lexington Books, 1982), p. 268.

m. According to Weissman and Krosney, in *The Islamic Bomb*, France may not have been aware of SGN participation.

n. Cranston, "Nuclear Proliferation and U.S. National Security Interests" (see note i). Cranston states facility has operated with radioactive material, but U.S. officials deny this.

(Personal communication). Seymour M. Hersh, "Pakistani In U.S. Sought to Ship A-Bomb Trigger," *New York Times*, February 25, 1985.

o. Thomas W. Graham, "South Asian Nuclear Proliferation and National Security Chronology," Center for International Affairs, Massachusetts Institute of Technology, 1984, citing Weissman and Krosney, *The Islamic Bomb* (see note g), p. 81.

p. Hedrick Smith, "A Bomb Ticks in Pakistan," *New York Times Magazine*, March 6, 1988, p. 38.

q. Milton R. Benjamin, "Pakistan Building Secret Nuclear Plant," *Washington Post*, September 23, 1980.

r. See Leonard S. Spector, *The New Nuclear Nations* (New York: Vintage Books, 1985) Chapter II, "The Nuclear Netherworld," and accompanying notes, especially note 31. See also discussion in text above.

s. Simon Henderson, "Pakistan Builds Second Plant to Enrich Uranium," *Financial Times*, December 11, 1987. But see, "Reports of Pakistan Uranium Plant Weighed," *New York Times*, January 10, 1988; "Pakistan Denies New Enrichment Plant," *Nuclear Engineering International* (February 1988): 7.

t. Hafiz R. Khan, "The Kahuta Story," *Defence Journal*, (August 1986): 11-13.

u. "Pakistan: Research Reactor Inaugurated," *Nucleonics Week*, December 14, 1989, p.15.

v. See note 68 in this chapter.

w. Mark Hibbs, "U.S. Repeatedly Warned Germany on Nuclear Exports to Pakistan," *Nuclear Fuel*, March 6, 1989; Mark Hibbs, "German Firm's Exports Raise Concern About Pakistan's Nuclear Capabilities," Nuclear Fuel, March 6, 1989.

x. Mark Hibbs, "Pakistan Rebuts Proliferation Charge, But Germans Step Up Investigation," *Nuclear Fuel*, April 3, 1989, pp. 6-7.

y. Mark Hibbs, "German Firms Exported Tritium Purification Plant to Pakistan," *Nuclear Fuel*, February 6, 1989, pp. 6-7.

z. See notes 95 and 96 of this chapter and accompanying text.

aa. *World Nuclear Industry Handbook, 1990* (Sutton, Surrey, U.K.: Nuclear Engineering International, 1990).

bb. See note 96 for this chapter and accompanying text.

8
North Korea

Under the hard-line Communist rule of President Kim Il Sung, North Korea has emerged as one of the most heavily militarized and repressive states in Asia. For four decades the North has sought to achieve reunification with South Korea by violence—through direct military attack, terrorism, and political and social destabilization—thus posing a continuing threat to its neighbor on the Korean Peninsula. By the end of the 1980s, there was considerable evidence that, in furtherance of its unbridled political ambitions, the country had launched an indigenous nuclear weapon development program. While some U.S. government analysts believed that North Korea was many years away from acquiring all of the capabilities needed to manufacture nuclear arms, worst-case estimates by other U.S. government specialists suggested that Pyongyang could develop a rudimentary nuclear arsenal before 1995.

North Korea ratified the Non-Proliferation Treaty (NPT) in 1985. Nevertheless, a number of key North Korean nuclear installations remained outside International Atomic Energy Agency (IAEA) safeguards as of mid-1990 because Pyongyang had not yet concluded an agreement with the agency, as required by the treaty, for the comprehensive application of IAEA safeguards on all of the country's nuclear activities.

North Korea's increasing ties with other renegade states, including growing military collaboration with Iran, have compounded concerns over its nuclear program by raising the possibility that Kim Il Sung, or a successor, might one day transfer nuclear weapons or related technology to other countries.

Background

The possibility of nuclear proliferation in North Korea arises within a unique context: Kim Il Sung's repeated attempts to dominate the South by force have led the United States to guarantee South Korea's security with the promise to use nuclear weapons, if necessary, in its defense. As a result, North Korea is the sole nuclear aspirant in the developing world today that faces an overt nuclear threat from a superpower. The roots of this relationship, as well as the North's intense hostility toward South Korea, date from the close of the Second World War.

Division and conflict. Following the surrender of Japan on August 15, 1945, the two major Allied powers in the region, the United States and the Soviet Union, shared responsibility for disarming Japanese troops in Korea and for overseeing the subsequent trusteeship of the country. The Peninsula was divided at the 38th parallel to delineate provisional zones of U.S. and Soviet occupation. Deliberations on the creation of a Korean government were soon marked by controversy arising from the fundamentally different political systems and objectives of the two occupying powers. A 1947 United Nations resolution stipulated that elected Korean representatives should themselves determine the form of the future government. The Soviet Union rejected this formula, but free elections, observed by the U.N. Temporary Commission on Korea, proceeded in the southern half of the Peninsula in May 1948. The government of the Republic of Korea (ROK) was formally inaugurated in August of that year. The Democratic People's Republic of Korea (DPRK) was proclaimed in September, with Kim Il Sung as premier. In 1948, both North and South claimed sovereignty over the entire Peninsula, triggering a bitter and enduring competition.

Soviet troops had largely withdrawn from the Peninsula by the end of 1948, leaving behind a number of military advisers. The United States, too, began a process of military withdrawal that was completed in 1949, with the exception of a 500-man Korean Military Advisory Group.[1] On the heels of the U.S. withdrawal, and bolstered by Soviet-supplied matériel, North Korean armed forces crossed the 38th parallel to "liberate" the South in June 1950. Seoul, situated near the dividing line, fell in three days, and North Korean forces continued to push their way south, overpowering the ill-equipped Southern army.[2]

Unforeseen by Kim Il Sung, the United States—under the auspices of the United Nations, and with the active involvement of several other U.N. member states—committed forces to repel the North's invasion. Following a successful landing at Inchon on September 15, 1950, U.S.-led forces cut off and destroyed Kim Il Sung's invading army, regained territory to the 38th parallel, and continued to advance northward. In late October 1950, at a point in the hostilities when U.N. forces had driven nearly to the North Korea-China border, China entered the war to support its Communist ally, in part to prevent U.S. forces from gaining a permanent foothold so near Chinese territory. The Chinese overwhelmed U.N. troops and drove them southward.

Ultimately the war stalemated slightly north of the 38th parallel and negotiations led to a return to the territorial *status quo*, leaving the Peninsula still divided and economically devastated. An armistice was signed on July 27, 1953, establishing a demilitarized zone (DMZ) between North and South at the original dividing line. The armistice remains in effect today. After the conflict ended, moreover, the United States and the Republic of Korea concluded a Mutual Defense Treaty which provided for the stationing of U.S. troops on South Korean soil. Thus, while no treaty guarantees the peace between the two Koreas, U.S. troops have helped to prevent renewed hostilities by acting as a "trip wire" that would ensure American involvement in the event of a Northern attack.[3]

Nuclear threats. Although the United States never employed nuclear weapons during the war, their possible use was an ever-present element of the conflict. After China intervened, for example, U.N. Forces Commander, General Douglas MacArthur, was known to believe that air attacks against China and the use of nuclear weapons in the North were necessary to prevent total defeat.[4] While MacArthur never openly called for this step, his tendency to operate independently of Washington—which ultimately led to his recall by President Harry S. Truman—created early concerns even among U.S. allies that the General might decide on his own to use nuclear weapons. Similarly, a November 1950 statement by President Truman that the United States had given "active consideration" to the use of atomic weapons in Korea stirred further controversy. Finally, during the negotiations to end the conflict in 1953, the Eisenhower Administration moved nuclear delivery systems (bombers and artillery pieces) to U.S. bases in the area (possibly to Okinawa) and sent a number of indirect diplomatic signals to Beijing intended to suggest that the United States was prepared to use nuclear arms in the event that peace efforts broke down and large-scale fighting resumed.[5]

In early 1954, less than a year after the conclusion of both the Korean armistice and the Mutual Security Treaty with the Republic of Korea, the United States adopted the doctrine of massive retaliation as its basic nuclear strategy. This strategy, a response to increasing Soviet pressure around the globe, stated that the United States would be prepared to respond to armed attack "instantly, by means and at places of [its] choosing," effectively declaring a readiness to employ nuclear weapons on a massive scale in response to major conventional aggression.[6] As a corollary, Washington announced that it was prepared to use theater nuclear weapons, *i.e.*, nuclear weapons specifically designed to be used within a theater of military operations, to resolve regional conflicts. Later that year, the Chairman of the U.S. Joint Chiefs of Staff, Admiral Arthur W. Radford, disclosed that U.S. ground forces in the Far East had received atomic artillery, including cannons and rockets. Radford was more explicit during a trip to Seoul in early January 1955, when he stated that the United States "would be ready to use atomic weapons, if needed" in the event that the North renewed aggression on the Peninsula.[7]

Early nuclear activities in the North. North Korea's nuclear research program was launched against the backdrop of these historic events. In 1947, even as the United Nations was debating the political future of the Peninsula, the Soviet Union reportedly took the first steps to develop an atomic industry in its area of occupation. In support of its nascent nuclear weapons program, Moscow is believed to have conducted geological surveys of northern Korea for uranium-bearing mozanite sand.[8] Soon afterward, mining and exports of uranium to the USSR reportedly began in the North.

In 1952, in the midst of the Korean War, China also turned to North Korea as a source of "radioactive material." At the time Moscow and Beijing were cooperating on nuclear research, including that related to weapons.[9]

In July 1955, representatives of the North Korean Academy of Sciences, along with scientists and technicians from all the Socialist countries, attended a nuclear energy conference held in Moscow, one month before the U.S.-sponsored "Atoms for Peace" conference in Geneva. The following year, North Korea entered into an agreement with the Soviet Union on nuclear research under which a number of North Korean scientists were reportedly sent to the Dubna Nuclear Research Institute for training. In 1959, Pyongyang signed a second agreement with Moscow on nuclear cooperation and concluded a separate nuclear accord with the People's Republic of China.

In 1964—the year China became the world's fifth declared nuclear weapon state—the DPRK is reported to have discovered approximately 4 million tons of uranium deposits at Hamhung, Unggi, and Haegumgang; this report appears to be exaggerated, however, since reserves of this magnitude would be larger than those of Australia, South Africa, the United States, or Canada, countries thought to have some of the world's richest endowments of the ore.[10] In 1964, North Korea also established its first nuclear research center at Yongbyon, where a small Soviet-supplied research reactor was installed in the following year.

During the mid-1960s, allegedly, North Korean nuclear specialists also maintained close ties to China, which, following its rift with Moscow in 1960, had pursued its own nuclear weapons program. According to one account in the South Korean press, North Korean nuclear scientists have been present at Chinese nuclear tests since 1964, although the report does not explicitly describe the extent to which China may have shared nuclear-weapons technology with its ally.[11]

Continuing tensions; the South's nuclear bid. The 1960s were marked by sustained tension between the two Koreas, as Kim Il Sung pursued a policy of reunification though subversion and violence directed against the South. The most belligerent action occurred in January 1968, when the North sent a 31-man commando team to Seoul in an attempt to assassinate South Korean President Park Chung Hee; the team was stopped just 500 yards from President Park's residence. North Korea's seizure two days later of the U.S. intelligence-gathering ship *Pueblo* in international waters off the coast of North Korea, and the North's downing in 1969 of a U.S. EC-121 reconnaissance aircraft, also over international waters, reinforced American and South Korean perceptions of Kim Il Sung's hostile intentions.

Notwithstanding this pattern of aggression, in 1970 the Nixon Administration announced plans to withdraw American forces—then numbering 60,000—from South Korea, beginning with a cut of 20,000 men in 1971. The move was part of a broader withdrawal of U.S. forces from Asia, under a policy known as the Nixon Doctrine, which coincided with the scaling back of U.S. involvement in Vietnam. This U.S. action appears to have been an important factor in South Korea's decision in the early 1970s to launch a secret nuclear weapons program.[12] A second North Korean assassination attempt on the life of President Park in 1974 intensified Southern anxieties during this period; although the President was not harmed, his wife was killed in the incident.

The United States apparently learned of the South Korean nuclear weapons effort shortly after India's 1974 nuclear test, when Washington began to seek evidence of nuclear ambitions by other developing states.[13] U.S. pressure on Seoul led it to abandon the program and to sign the Non-Proliferation Treaty in 1975. Some nuclear-weapons relevant activities apparently continued through 1978, however, raising questions about the country's commitment to the treaty, at least during the latter years of President Park Chung Hee's tenure (1961-1979).[14]

U.S. nuclear presence. In conjunction with its effort to check the South Korean nuclear weapons program and to allay concerns over a rapidly growing military threat from the North in the wake of the fall of Vietnam, the United States in 1975 reiterated its readiness to defend South Korea. Indeed, Secretary of Defense James R. Schlesinger warned on June 20, 1975, that the United States would retain the option to use nuclear weapons or to introduce additional ground troops in response to a Northern attack on the South. Confirming the presence of nuclear weapons in the South, he stated:

> It is, I believe, known that we have deployed nuclear weapons in Europe and Korea along with our forces, and that those nuclear weapons are available as options for the President.[15]

The U.S. commitment to South Korea was called into question again, however, in early 1977, when shortly after his inauguration, President Jimmy Carter announced his intention to fulfill a campaign pledge to withdraw all American ground forces from the country—some 30,000 troops—over a period of four to five years. The U.S. decision was greeted with alarm in South Korea. This prompted Carter on May 30, 1977, to underscore that South Korea remained under the U.S. nuclear umbrella, stating that U.S. tactical nuclear weapons deployed in the Pacific region could be used to defend South Korea against an invasion from the North, if necessary.[16] The Carter troop withdrawal was only partially implemented; in 1979, Carter reversed course, in part because of opposition in Seoul, but also as a result of new, higher U.S.

intelligence estimates of North Korea's military capabilities that intensified opposition to the pullback in the United States.[17]

Initiation of the North's nuclear weapons effort. Although evidence remains sparse as to the exact timing and motivations for his decision, it appears that by the end of the 1970s, Kim Il Sung had initiated efforts to build the infrastructure needed for a nuclear weapons program. The cornerstone of the program is an indigenous, 30-megawatt research reactor that was built, apparently between 1980 and 1987, near Yongbyon, alongside the Soviet-supplied reactor mentioned above.[18]

The 30-megawatt reactor is thought to be of the "gas-graphite" design, a relatively rudimentary technology dating from the 1940s, which uses natural uranium as fuel. This simple design is well suited to the production of plutonium and has been cited as the logical choice for would-be nuclear states with limited industrial capabilities.[19] The Yongbyon reactor is also unusually large for a reactor intended to be used as part of a developing country's civilian nuclear research program. Indeed, the vast majority of research reactors in the developing world are in the 1- to 5-megawatt range. However, indigenous reactors used to produce plutonium in the early stages of the Indian and Israeli nuclear programs had power levels roughly comparable to that of the new Yongbyon reactor.[20]

The gas-graphite design also avoids the need for enriched uranium or heavy water—two commodities that are extremely difficult to produce and which North Korea would have had to import, presumably from China or the Soviet Union. Both states might have objected to North Korea's enlarging its nuclear infrastructure, and Moscow, as a party to the Non-Proliferation Treaty, would have been obliged to require the North to place any imported material (and the facilities where it was used) under International Atomic Energy Agency safeguards. In the late 1970s and early 1980s, China was apparently prepared to make nuclear transfers without IAEA safeguards, but North Korea might not have felt confident that it would continue this policy.

By pursuing an autonomous program, Pyongyang has been able to avoid these possible constraints on the larger Yongbyon reactor. Indeed, Kim's decision to pursue this route appears to support the conclusion that he wished to maintain the option to use the Yongbyon unit to produce material for nuclear arms.

One detailed report in the South Korean press states that construction of the Yongbyon reactor began in 1980, and the plant is believed to have begun operating in 1987.[21] This seven-year timeframe is consistent with more generalized estimates of the time a developing country would require to build such a facility.[22] Moreover, assuming that the North Korean program is, in fact, entirely indigenous, to operate the reactor in 1987, North Korea also would have needed to develop a substantial nuclear infrastructure, a task that could have been accomplished in the intervening years. Such an infrastructure would include an operating uranium mine; a mill for concentrating the uranium ore into "yellowcake;" a facility (known as a "conversion" plant) to purify that material; the ability to mine and purify graphite for the reactor; and a facility to fabricate fuel elements for insertion into the reactor.[23]

Taken alone, the construction of a 30-megawatt reactor, though suspicious, would not convincingly establish that a country was pursuing nuclear arms. In the mid-1980s, however, the North would begin work on a plant to separate plutonium produced in the Yongbyon facility. Since there is no legitimate non-military use for plutonium in a nuclear program as rudimentary as North Korea's, the decision to build the plutonium plant provides strong evidence that Pyongyang is currently seeking to acquire a nuclear weapon capability and that this was the purpose behind its earlier construction of the 30-megawatt reactor, as well.

Possible motivations. If these estimates are accurate, Kim's decision to develop a nuclear weapon capability probably was made sometime during 1979 or 1980. A number of additional factors support this hypothesis. First, the decision to acquire nuclear arms would have been consistent with the major build-up of the country's forces during the late 1970s, as revealed by the findings of the 1979 U.S. Army intelligence study mentioned earlier.

By the end of the 1970s, moreover, it had become clear that the nuclear-backed American military presence in South Korea was likely to remain indefinitely and that, even if American forces withdrew, South Korea might choose to develop nuclear weapons of its own. To pursue its policy of reunification through force, the North would need to neutralize this opposing nuclear capability. The ability of North Korea to threaten Southern cities with nuclear devastation could impose unacceptable costs on the use of nuclear weapons by the United States or South Korea, raising the prospect that the North could prevail in a future conflict by gaining a preponderance of conventional forces.

In addition, although the North is generally estimated to have possessed the military edge during this period, Kim Il Sung's apparent decision to acquire nuclear arms may have been influenced by concerns that the regional balance would soon shift because of the South's growing strength.[24] In a 1979 announcement, President Carter indicated that the withdrawal of American forces would be deferred until 1981. This delay not only meant that U.S. forces would continue to be deployed in the region, but was specifically intended to permit South Korea to build up its own military capabilities in the interim so that the U.S. pull-out could resume.[25] At approximately the same time, the South Korean economy began a dramatic upswing that caused it to outpace Northern economic performance by an increasing margin. Thus Kim may have seen nuclear arms as a necessary counter to the prospect of a strengthened South Korean military (with or without nuclear arms of its own) that would be backed by a superior industrial base.

Other factors related to Pyongyang's longer-term strategic objectives may have also contributed to Kim's decision to pursue a nuclear option. Kim, for example, may have believed that an independent DPRK nuclear arsenal would reduce his political and military dependence on Moscow and Beijing and provide greater freedom of action to pursue his designs on the Peninsula.

Similarly, Kim may have perceived that the domestic and foreign impacts of developing nuclear weapons would boost the legitimacy of his regime and help ensure its future survival under the leadership of his son and heir-designate, Kim Jong Il. The elder Kim, who had ruled North Korea for thirty-two years in 1980, had begun as early as 1973 to position his son to succeed him. At the Sixth Congress of the Korean Workers' Party in 1980, this succession formula became more apparent when Kim Jong Il was elected to positions that made him second only to his father in seniority within the party.[26]

With the succession in mind, Kim Il Sung may have reasoned that nuclear arms would provide tangible proof of the technological and military accomplishments of the Kim dynasty. He may also have believed that the North's possession of nuclear weapons would deter Seoul from attempting to emulate his own tactics and interfere in the country's succession process after his death or retirement. Finally, Kim may have concluded that in the event that reunification on Pyongyang's terms were impossible, the North's possession of nuclear weapons would tend to ensure, at the very least, the continued existence of the Democratic People's Republic on a divided Peninsula and rule out political absorption by the South.

North-South relations in the 1980s. Tightly restricted diplomatic interaction between North and South commenced in 1971, when statements were exchanged between Park Chung Hee and Kim Il Sung. The first meetings (outside of armistice-related talks) between representatives of the two Koreas took place later that year, as discussions on the reunification of divided families began under the auspices of the Red Cross. It was not until December 1989, however, that 300 visitors from each side of the DMZ were able to cross the dividing line for a brief visit.[27]

In the intervening years, relations between the two states suffered from continuing tensions, which were gravely exacerbated by Northern terrorist activity. On August 5, 1983, for example, North Korean commandos attempted to infiltrate the perimeter of the Wolsong nuclear power plant near Pusan with the apparent intention of sabotaging the facility.[28] Two months later, North Korean agents attempted to assassinate South Korean president Chun Doo Hwan during a visit to Burma (now Myanmar) at the beginning of an important Asia-Pacific tour. Chun was spared when the bomb intended to kill him was detonated prematurely, but four members of his Cabinet were killed in the blast. In a third incident, North Korean terrorists planted explosives on a Korean Airlines (KAL) flight in December 1987; the plane crashed off Burma following the explosion, killing all those on board. The action is generally believed to have been intended to undermine international confidence in the South's ability to maintain security during the 1988 Olympic games.[29]

These extraordinary displays of the North's disregard for the norms of international behavior coincided with the build-up of its nuclear potential and have spawned continuing suspicions about the objectives of the country's nuclear program.

Ties with Moscow and Beijing. The Soviet Union and the People's Republic of China have been Kim Il Sung's most valuable allies over the years, and the vicissitudes of Sino-Soviet relations have permitted him to play the two nations off one another. This strategy has benefitted the North by ensuring that at least one, and usually both, of these neighboring powers have provided Pyongyang with much needed security assistance and diplomatic support at all times.

The country's ties with the Soviet Union experienced a marked revitalization after Kim Il Sung visited Moscow in 1984. Following that visit, the two countries entered into a series of agreements and cooperative efforts, including a 1985 agreement under which the Soviet Union undertook to provide North Korea with four nuclear power reactors. Military ties also were strengthened considerably. In 1986, for example, Pyongyang expanded Soviet naval access to North Korean ports and Soviet rights to overfly North Korean territory. That same year, the two countries also began conducting joint naval exercises. The USSR has also provided the North with modern military equipment, MiG-23 fighter-bombers, and most recently, the advanced MiG-29 fighter-bomber.[30] (These transfers were completed by 1989, and it is unclear whether additional equipment of this quality will be provided under a follow-on program.)

The North's warming relations with the Soviet Union during the last half of the decade were accompanied by increasing distance from China. Beijing's liberalization at home, its increased ties with the West, and its suspicions concerning closer military ties between the DPRK and the USSR combined to chill relations with Pyongyang.

Ratification of the NPT. In December 1985, after construction had begun on the Yongbyon reactor, North Korea ratified the Non-Proliferation Treaty. The decision is believed to have been taken at the urging of the Soviet Union. Apparently the United States had pressed Moscow to seek North Korean ratification of the pact in 1984, after Washington had discovered the construction at Yongbyon and had determined that North Korea might be attempting to develop a nuclear weapons capability. Moscow's 1985 offer to sell the North four nuclear power reactors was undoubtedly intended to make Kim's acceptance of the treaty more attractive.

Under the NPT, North Korea would be obligated to place all of its nuclear installations under IAEA inspection to ensure that no nuclear material produced in them was used for nuclear explosives. Since the North apparently had only one new nuclear reactor under construction at the time and was not known to be planning a facility for extracting plutonium, it appeared that NPT adherence would largely eliminate the risk of the country's acquiring nuclear arms. Moreover, since the South had ratified the NPT in 1975 and did not possess any means to produce weapons-grade nuclear material, it appeared that nuclear developments would be stabilized at a non-provocative level.

Unfortunately, as of mid-1990—more than four years after ratifying the NPT—the North had not accepted the comprehensive, or "full-scope," IAEA inspections required under the treaty, by signing an implementing agreement with the agency. Moreover, as discussed below, the country subsequently began operating the reactor at Yongbyon without IAEA inspections and started construction of a plutonium extraction facility. This behavior suggests that Kim Il Sung's acceptance of the treaty in 1985 may have been only a ploy to avert international pressure to curtail his nuclear program and that he had no intention of accepting the treaty's operative elements.

Nuclear assistance from abroad. Apart from the aid obtained from the Soviet Union and China, North Korea has received foreign assistance from other sources for its nuclear activities, some of which has probably benefited the North's largely indigenous nuclear weapons effort. North Korea, for example, has received technical assistance from the International Atomic Energy Agency in such areas as uranium prospecting and the analysis of uranium ore and concentrates—capabilities that are necessary to enable a country to produce uranium indigenously.[31]

Kim also appears to have begun resorting to clandestine transactions. Sometime between 1984 and 1987, a West German chemicals and metals firm, Degussa AG, illegally re-exported a small quantity of U.S.-origin zirconium, valued at $700, to North Korea. The zirconium may be used to encase uranium rods in the 30-megawatt Yongbyon unit, or may have been imported for scientific analysis to benefit indigenous production efforts. Though seemingly insignificant in itself, the episode suggests the North is prepared to exploit weaknesses in the international nuclear control system to enhance its nuclear activities.[32]

Another disturbing development, while not directly involving collaboration in the nuclear realm, involves North Korean cooperation with Iran to produce missiles that could deliver a nuclear warhead. In 1987, these two countries reportedly concluded an agreement worth $500 million, which included Iran's purchase of 100 North Korean-produced Scud-B missiles. It is believed that the approximately $400 million that remains will be used, at least in part, to build a missile production facility in Iran, and possibly to support North Korean efforts to produce an improved Scud, a missile that would allow the North to reach all parts of South Korea.[33] Iranian-North Korean collaboration in this strategically important field may serve as a precursor to joint efforts in the development of nuclear arms, a capability which Tehran is also seeking.

Political and Technical Developments

Growing evidence of nuclear weapons program. Some time in late 1988 or early 1989, U.S. satellite photographs are reported to have revealed that North

Korea had begun construction of a plutonium extraction plant near the Yongbyon nuclear complex.[34] In March 1990, U.S. Department of Defense officials publicly confirmed that Washington believed such a project to be under way.[35] Since there is no legitimate use for plutonium in the North's peaceful nuclear program, these officials assume that the facility is intended to produce material for nuclear weapons.[36]

The Yongbyon reactor is thought to have been operating since 1987, although not necessarily at full power. It appears, moreover, that as of early 1990, it still contained its original charge of fuel. Although a 30-megawatt reactor, when operating at full power, can produce enough plutonium in its spent fuel each year for approximately one nuclear device, by mid-1990 the North may not have accumulated enough material, even in unseparated form, for its first nuclear weapon. However, as the operation at the Yongbyon reactor stabilizes, the North will begin to accumulate plutonium-bearing spent fuel. When the plutonium extraction plant near Yongbyon is completed, the North could begin processing its accumulations of spent fuel to obtain plutonium and, soon thereafter, could begin fabricating nuclear weapons, assuming it had also obtained other necessary elements, including the design for such arms and the non-nuclear triggering package components.

U.S. analysts are apparently divided over how quickly North Korea could achieve a nuclear weapons capability. Using worst case assumptions, the Department of Defense apparently fears that Pyongyang could reach this point by the mid-1990s, but other elements of the U.S. government believe that this estimate greatly overestimates the North's scientific and technical competence and that the country could not achieve a nuclear weapons capability before the end of the decade.

IAEA inspections under the NPT. Compounding concerns about North Korea's nuclear activities has been the North's refusal to sign a safeguards agreement covering all of its nuclear activities with the IAEA, as required under the Non-Proliferation Treaty. Such an agreement would grant the agency access to the Yongbyon reactor to verify that spent fuel from it was not being diverted to a nuclear weapons program. The safeguards agreement also would grant the IAEA authority to place the nearby plutonium plant under verification procedures once nuclear materials were introduced there and would entitle the agency to obtain design information about the plant during its construction to facilitate the subsequent application of such safeguards. Inspections would also be required for North Korea's uranium conversion and reactor fuel fabrication plants.[37] Thus the agreement would be a critical step toward constraining North Korea's nuclear program. (The IAEA already inspects the smaller Soviet-supplied reactor at Yongbyon under a 1977 bilateral agreement covering that installation exclusively.)[38]

Pyongyang's refusal to accept comprehensive IAEA safeguards had continued for some four-and-a-half years by mid-1990. At the end of the eighteen-month grace period permitted by the treaty,[39] North Korea claimed that the IAEA had presented an incorrect version of the proposed agreement, which was based on the IAEA's non-NPT safeguards arrangements and which would have imposed uniquely stringent controls on North Korea in comparison to other NPT parties. The IAEA acknowledged the mistake, sent Pyongyang a new draft agreement, and granted the DPRK a second eighteen-month grace period, which expired at the end of 1988.

Negotiations have continued, however. North Korea reportedly proposed significant changes in the second draft accord, including new language in the agreement's preamble objecting to the presence of U.S. nuclear forces in South Korea. The agency rejected these proposals, on the ground that the NPT safeguards agreement is a standardized document, applicable to all non-nuclear-weapon state NPT parties equally; negotiations are limited to questions concerning the application of safeguards at specific installations.

In early 1990, North Korea appeared to accept the point, but then reportedly conditioned its acceptance of the agreement on the withdrawal of U.S. nuclear forces from South Korea, or on the provision of specific assurances that the United States would not use nuclear arms against the North. The United States has rejected this position, insisting that North Korea is legally obligated to accept full-scope safeguards by virtue of the fact that it is a party to the NPT and that it has no right under the treaty to impose additional conditions before taking this step.[40] As of mid-June 1990, the impasse between the IAEA and Pyongyang continued. New talks between the IAEA and North Korea were scheduled for July 1990.

The North's emphasis on the removal of U.S. nuclear weapons from the South has also figured prominently in an arms control proposal that it put forward in July 1987 and elaborated more fully in November 1988 and again in May 1990. The revised proposal called for trilateral talks with the United States and South Korea on a multi-year process of mutual North-South troop reductions to a level of 100,000, together with parallel U.S. force withdrawals; the proposal calls for the United States to withdraw ground troops and any nuclear weapons on the Peninsula to the southernmost portion of South Korea in the first year, and to remove both entirely by the end of the second, as North-South troop cuts were implemented. The North's proposal has met a chilly reception in Seoul and Washington, where less ambitious North-South confidence-building measures are favored prior to discussions of major force cutbacks. The North's bid at the IAEA to obtain a nuclear *quid pro quo* from the United States in return for accepting full-scope IAEA safeguards is a new Northern approach to achieve a longstanding objective: the removal of the nuclear threat associated with the U.S. presence in the South.[41]

Growing concerns. The situation has created increasing unease on the part of South Korea, the United States, and Japan. In early July 1989, U.S. defense officials reportedly briefed their South Korean counterparts on their concerns about the North's nuclear program. Soon afterwards, the South Korean press carried a series of detailed stories on the subject, and a number of editorials decried the emerging nuclear threat to the country. In October, at a session with reporters in Washington following his meeting with President George Bush, South Korean President Roh Tae Woo objected to the North's failure to accept safeguards, a further indication of South Korean anxieties on the subject.

Senior U.S. officials have also expressed their concerns about the North's nuclear program with growing frequency since late 1989.[42] Under Secretary of Defense Paul Wolfowitz has been the most explicit. Referring to the North's 1983 attempt to assassinate then South Korean President Chun Doo Hwan in Rangoon (now Yangon) and to the North's bombing of the KAL flight in 1987, he stated in testimony before the Senate Armed Services Committee:

> It's very hard to know about [North Korea's nuclear] intentions. What we know is what they are doing. What they're doing is developing a nuclear program that includes, among other things, a reprocessing [*i.e.*, a plutonium extraction] capability. There is absolutely no reason whatsoever why a country, why any small country, needs reprocessing as part of a nuclear program. And moreover, we've encountered enormous North Korean resistance to accepting safeguards, international inspection, on their facilities, which is something that they are actually obliged to accept . . . under the Non-Proliferation Treaty. . . .
>
> So that's what they're doing, and given the track record that this is a country that makes it a practice of blowing up foreign leaders in foreign countries, blowing up airplanes, I think one has to assume the worst in terms of intentions.[43]

U.S. initiatives to gain North Korean acceptance of its IAEA safeguards obligations have included behind-the-scenes diplomatic pressure at the quarterly meetings of the agency's Board of Governors; encouraging supportive statements on the issue by concerned states at IAEA meetings; and efforts to have the Soviet Union press its ally more forcefully to take this step. The communiqué following the February 1990 meeting between U.S. Secretary of State James A. Baker, III, and Soviet Foreign Minister Eduard Shevardnadze reflected these last efforts. It declared:

> The Soviet side noted that the DPRK is close to concluding an agreement with the IAEA on nuclear safeguards. The U.S. side expressed its desire that the agreement be concluded swiftly and observed faithfully.[44]

Tokyo, too, has been increasingly disturbed by North Korea's nuclear activities. In February 1990, for example, Deputy Chief-of-Cabinet Noburo Ishihara told reporters that Japan was "very concerned" about the North's nuclear program, and other sources noted that the North's refusal to accept IAEA safeguards has "raised suspicions" about its intentions.[45]

Inspections insufficient. Even if Pyongyang were to agree promptly to IAEA oversight, its current nuclear program would remain highly provocative. Under IAEA rules, if the North were to complete a plutonium extraction plant, it would be legally entitled to stockpile plutonium, as long as the material were under IAEA inspection. Unfortunately, this would leave North Korea dangerously close to possessing nuclear weapons. It is this very possibility that has led Washington to pressure Seoul over the years not to build a plutonium facility, notwithstanding the fact that it would be under IAEA oversight because of the country's status as a party to the NPT.[46]

Growing missile, chemical weapon threat. According to Director of Naval Intelligence Admiral Thomas A. Brooks, North Korea has been able to manufacture the Scud-B missile for some time.[47] China has apparently assisted Pyongyang in manufacturing the missile, possibly by providing key components.[48] North Korea is believed by U.S. officials to have exported the system to Iran.

The 190-mile system, which carries a payload of 2,200 pounds (1,000 kilograms), can reach targets in much of South Korea, including Seoul, but excluding the southern third of the country, which remains out of range. U.S. officials have also concluded that North Korea has probably manufactured chemical weapons, which could be placed on its Scuds, adding to the threat posed by the missiles.[49] Eventually, as the North's nuclear capabilities grow, it could equip the missile with nuclear warheads.[50]

The North's missile capabilities, moreover, have direct implications for the nuclear environment on the Peninsula. First, the system would give Kim Il Sung the ability to retaliate against major cities in South Korea if that country were to launch a "preemptive" attack against a North Korean nuclear facility.[51] Although the South's ability to launch a counter-strike might lead the North to limit its initial retaliation, South Korea would have to assume that it would suffer a serious blow for attempting to prevent the emergence of a Northern nuclear capability by force, and, indeed, such an overt military action could well lead to war. Thus, in effect, the North's existing missile capabilities and chemical warfare potential may be providing an umbrella under which it can pursue its nuclear ambitions with relative impunity.

A second nuclear-related threat posed by the North's missiles is that they can be used to target South Korea's nuclear power reactors. A successful attack on a nuclear reactor could have catastrophic effects in a country the size of South Korea. A large-scale radiation release would result either from a direct missile hit on a reactor containment vessel or from the destruction of a reactor's cooling system or control room. The former would cause a (non-nuclear) steam explosion that would spew the reactor's radioactive contents high into the atmosphere, forming a plume that could contaminate tens of thousands of square miles and cause thousands of immediate deaths. The latter would cause a meltdown of the reactor's core, releasing large amounts of radiation into both the ground and the atmosphere. In addition to the death and environmental destruction that would ensue, a successful attack on a nuclear power plant could also cause widespread demoralization or even panic in the population of the South.[52]

At present, four South Korean power reactors are within range of the North's Scud-B missiles.[53] Although Scud-Bs are not accurate enough to target reliably a reactor containment vessel itself, they constitute a more realistic threat to larger nuclear installations with several reactors and extensive support facilities. Moreover, the DPRK is said to be enhancing its missile capabilities with the development of a new generation of Scuds, the Scud-PIP.[54] These more accurate Scuds, which may be ready for deployment in 1990, reportedly will have the range to strike any target on the southern half of the Peninsula. As the North augments its missile capabilities both qualitatively and quantitatively, and as new power reactors in the South increase the number of potential targets,[55] the possibility of nuclear sabotage and blackmail of the South by the North will loom ever larger. Given Kim Il Sung's demonstrated willingness to attack the South's nuclear installations as seen in the 1983 Wolsung incident, Seoul cannot overlook the existence of such a nuclear "weapon" in the hands of the North.[56]

Increasing isolation. In July 1988, South Korean President Roh Tae Woo called for additional North-South cooperation on humanitarian, trade, and economic issues, as a means of easing bilateral tensions. Under this policy of *nordpolitik,* Roh also indicated that he would pursue expanded ties with Socialist states and envisioned that North Korea, in its turn, would seek to improve relations with countries such as Japan and the United States.[57] The approach has been successful for the South, but has offered fewer dividends for the North.

Since Roh's July 1988 announcement, for example, Mongolia and all of the states of Eastern Europe (with the exception of Albania and East Germany) have established full diplomatic relations with Seoul. Soviet and Chinese ties to South Korea have also grown considerably. Trade between these two and the South far exceeds that with the North, and the Soviet Union, which opened a trade mission in Seoul in April 1988, announced in March 1990 that it had agreed in principle to normalize relations with the South.[58] Following his June 1990 summit with President George Bush, Soviet President Mikhail Gorbachev met with Roh in San Francisco, a step that for all practical purposes recognized the South and dealt a fatal blow to Kim's decades-long efforts to deny the legitimacy of the government in Seoul.

For its part, the North has participated in a limited dialogue with the United States, which commenced with talks in Beijing in December 1988, and it has garnered a number of modest political concessions from Seoul.[59] On balance, however, Pyongyang has been the loser in its ongoing competition with the South for international acceptance and status.

Kim Il Sung's relations with his principal allies have also become unsettled, a trend which became manifest in 1988, when both Moscow and Beijing refused to join with Pyongyang in boycotting the Olympics in Seoul. More recently, the disparity between Kim's Stalinist-style autocracy and Soviet President Mikhail Gorbachev's commitment to economic and political reforms at home and in Eastern Europe has become increasingly apparent. In September 1989, moreover, a top Soviet advisor to Mikhail Gorbachev, Georgi Arbatov, stated that his country would not support the North if it initiated another war with South Korea, a step plainly intended to dash any hopes Kim may still harbor of reuniting the country by force of arms.[60] Gorbachev's meeting with Roh Tae Woo was the most glaring manifestation of the growing rift between Moscow and Pyongyang. These increasing tensions led Kim to rebuild ties with Beijing in mid-1989, after his hard-line counterparts there crushed China's pro-democracy movement and reversed the trend toward political liberalization. China, too, has indicated that it is not prepared to support Kim in a war of reunification, however.

Domestic uncertainty. In parallel with its foreign policy achievements, South Korea's leadership has also been groping toward greater domestic political stability by institutionalizing a variety of democratic and other political reforms. Two principal opposition leaders joined with President Roh in organizing a new political party in 1990. One of these former opponents, Kim Young Sam, traveled to the USSR to discuss the future of ROK-Soviet relations. Popular political participation has also increased, as evidenced in the record voter turn-out for Roh's election. Nonetheless, politics in South Korea remains tightly tied to personalities rather than institutions, and the consolidation of democratic reforms, though progressing, remains to be achieved.

In North Korea, however, the threat to domestic instablility is growing. Despite Kim Il Sung's efforts, spanning nearly two decades, to ensure a smooth transfer of power to his son, Kim Jong Il, such a transition is far from assured. Although considerable authority is said to have been passed already to the younger Kim,[61] the latter's personal shortcomings, past domestic opposition to his leadership, and the example of the overthrow of Stalinist-style dictatorships elsewhere in the Communist world make it increasingly likely that Kim Il Sung's death will be followed by a bitter struggle for power.[62] In May 1990, the elder Kim, now 78, was given another four-year term in office.[63]

Links with other renegade states. During the late 1980s, as Pyongyang experienced greater frictions with Moscow over its domestic and foreign policy and as its international isolation increased, Kim Il Sung sought to build ties with three other states on the fringes of the international system that were led by autocratic, anti-Western figures like himself: Iran, under Ayatollah Ruhollah Khomeini; Cuba, under Fidel Castro; and Romania, under Nicolai Ceausescu.[64] Despite the changes of leadership in two of these countries, these links could lead to cooperation with one or more of these states in the nuclear field; indeed, there is evidence suggesting that such nuclear ties may have been established.

North Korea's military ties to Iran date to the Iran-Iraq War, during which Pyongyang was a significant supplier of conventional arms (including Scud-B missiles) to Tehran. Further, as previously noted, in 1987, the two countries are believed to have entered into a major agreement involving the purchase of additional North Korean Scuds and the joint development of an advanced version of the missile. Ties between the two countries were expanded with the signing of two new agreements for cooperation in January and December 1989.

The content of the first accord is not known, but it has been referred to in press accounts as an agreement for "economic, scientific, and technological cooperation."[65] North Korea's 1985 agreement with the Soviet Union, under which the latter agreed to transfer four nuclear power reactors to the North, was also termed an agreement for "economic, scientific and technical cooperation."[66] This suggests the possibility that the Iran-North Korea pact may also cover cooperation in the nuclear field. Supporting such speculation is that one of the negotiators for the agreement appears to have been Iran's Minister of Mines and Metals, the Iranian official most likely responsible for locating and mining the country's uranium deposits.[67] In the fall of 1989, Iran announced the discovery of major new uranium deposits and plans to mine and process the material, raising the possibility that North Korea may have shared its know-how in this field—obtained from its own experience and from the IAEA—with its Middle East trading partner.

The second Iran-North Korea agreement, a protocol "on exchanges of military exercises and operations experiences and on exchanges of military and science information," by its very nature confirmed the widely reported security ties between the two states. It also points to the likelihood of cooperation in some scientific endeavors in the military sphere, raising again the possibility of weapons-related nuclear collaboration. The agreement was signed in December 1989, when Mohsen Reza'i, commander of Iran's Revolutionary Guards Corps, visited North Korea with a sizeable entourage.[68]

North Korea has had long-standing ties to Cuba and joined the latter in sending troops and advisors to Angola during the 1980s to support the Luanda government in its civil war against U.S.-backed rebel forces. The countries appear to have drawn closer as Kim Il Sung and Fidel Castro have each sought to withstand the tide of reform sweeping the Communist world. Although the two states are not known to have engaged in nuclear cooperation, in February 1990, a delegation from the North Korean Nuclear Energy Ministry, led by Minister Choe Hak Gun, visited Cuba and met with Fidel Castro. During the meeting, Castro expressed Cuba's solidarity with North Korea by supporting Pyongyang's newest reunification plan.[69]

North Korea appears to have a more diversified nuclear infrastructure and to be less dependent on foreign nuclear assistance than Cuba. The latter, however, is completing construction of two Soviet-supplied nuclear power reactors and undoubtedly has a cadre of reasonably well-trained nuclear technicians. Although Cuba is not considered to have an interest in acquiring nuclear weapons and will place the reactors under IAEA safeguards, some portions of

North Korea's nuclear weapons efforts might benefit from technical exchanges with Havana. The North, moreover, has plans to build four Soviet-supplied nuclear power reactors comparable to those in Cuba, suggesting another, less controversial area for cooperation. The February 1990 visit by North Korean nuclear specialists suggests that at least some modest collaboration in the nuclear field is taking place.[70]

While ties between North Korea and Romania had been an important part of each country's diplomatic landscape in past years, the fall and execution of Nicolai Ceausescu and his wife, Elena, in December 1989, and the advent of more popularly backed leadership in that country, appear likely to bring a cooling of relations. During the 1980s, Ceausescu launched a program to develop ballistic missiles with the help of a number of West European companies and at one point threatened Hungary with the claim that Romania could make nuclear weapons if it chose to.[71] There has been only one report of North Korean-Romanian cooperation in either of these fields, an account suggesting that Romania (and East Germany) sold enriched uranium to Pyongyang.[72]

Support for international terrorism. Adding to concerns over North Korea's possible acquisition of nuclear weapons is the country's support for international terrorism. The North's abetting of terrorist activity has not been limited to that directed against the South, but has also included the provision of training and funding for non-Korean terrorist organizations. Although the North's support for external terrorist organizations was heaviest during the 1970s,[73] the country remains on the list of states identified by the U.S. State Department as supporting such renegade activities today. North Korea also has established ties with other countries known to support international terrorism, such as Libya and Iran. This dimension of North Korean behavior raises the concern that in the future, Pyongyang might transfer sensitive nuclear materials, or nuclear arms themselves, to terrorist groups, with potentially catastrophic consequences for international stability.

Prospects

The emergence of a North Korean nuclear weapon capability would raise a host of disturbing issues. Given its diplomatic isolation, history of hostile acts against South Korea, support for international terrorism, and unbridled arms transfers to Iran and others, North Korea could emerge as the quintessential nuclear renegade state. Its continued resistance to IAEA safeguards could be a signal of its intention to withdraw from or abrogate the NPT, and if it were to take this action, there would be little reason to expect it to refrain from nuclear testing. Either step would be a severe blow to current international non-proliferation norms.

Indeed, as noted earlier, even if Pyongyang were to accept safeguards, its program would remain highly provocative, since the North would be legally entitled to stockpile plutonium and, having built the necessary facilities, would presumably do so. By the time the North's plutonium plant is ready to operate, Pyongyang could have accumulated several years' worth of spent fuel. This could give North Korea the wherewithal for several nuclear weapons soon thereafter and could then permit the country to add to its stockpile at a rate of one or two weapons per year. North Korea would thus possess a ready nuclear option. As noted earlier, views within the U.S. government differ as to how long it might take the North to reach this stage, with some analysts believing it could acquire a nuclear weapons capability by the mid-1990s and others holding that it will take the North until the end of the decade.

U.S. officials believe that the risk of war on the Korean Peninsula is low today, assuming that Pyongyang does not exaggerate its strength or the degree of instability in the South.[74] However, Secretary of Defense Richard Cheney has noted that the risk of a no-warning attack against the United States and its allies is greater in Korea than in any other part of the globe.[75] The acquisition of nuclear arms, moreover, might give the North a sense of invincibility that could encourage its aggressive tendencies. Even if it refrained from open hostilities, it might attempt to use nuclear blackmail to topple the Seoul government or obtain the expulsion of U.S. forces and thereby create conditions conducive to reunification of the country under Northern control.

If Pyongyang were to use its nuclear capability more benignly—for example, as a deterrent against U.S. nuclear weapons—the development would still drastically alter the security equation on the Peninsula. Such a step could largely neutralize the security benefits to Seoul of U.S. nuclear guarantees and, by exposing U.S. forces to a new class of threat, could reduce South Korea's overall confidence in American defense commitments.

At a minimum, a North Korean nuclear option would trigger powerful pressures in the South for the development of a comparable capability. Despite its sizeable and advanced civil nuclear infrastructure, South Korea has not built the facilities necessary to produce nuclear weapon materials. This willingness to foreswear a nuclear option has been largely the result of American diplomatic pressure and continuing U.S. security guarantees. In response to budgetary constraints, however, Washington has begun a gradual reduction of its forces in South Korea,[76] a step that could increase security anxieties there and intensify pressures for a South Korean nuclear option. As the North approaches the nuclear weapons threshold, the possibility of a South Korean pre-emptive attack could also grow, a step that would gravely increase the risk of war.

Another disturbing consequence of the acquisition of nuclear arms by North Korea, whether or not matched by the South, is that this development could well prompt Japan to reevaluate its own security posture. Official policy and public opinion in Japan continue to oppose nuclear weapons. Nevertheless, possession of nuclear weapons by a government not formally allied to

Japan—*i.e.*, either North Korea or South Korea—could lead to unprecedented debate in Japan over the country's long-standing non-nuclear position.

Even if a reversal in Japanese nuclear policy did not ensue, a nuclear-armed North Korea could prompt Japan to increase its defense expenditures and military deployments dramatically, particularly as the U.S.—an important security partner for both Japan and South Korea, who has played the role of regional stabilizer for the past 45 years—begins to draw down its forces in the region. The reemergence of Japan as a major military power would trigger anxieties throughout Asia. It might also lead to a dangerous Japan-China arms race and could ultimately lead to a major alteration in the global balance of power.

In the immediate term, efforts to avert the risks of a nuclear-armed North Korea—risks that include the danger that Pyongyang might be prone to transfer sensitive nuclear technology or materials to others—have focused on gaining its acceptance of IAEA safeguards on the Yongbyon reactor and related facilities. Ultimately, however, it will be necessary to ensure that the North does not complete the plutonium facility.

The Soviet Union, which has demonstrated considerable commitment to non-proliferation over the years, would appear to be in the best position to dissuade North Korea from this path. Unfortunately, the fact that the North has resisted IAEA safeguards for over four years, despite presumed Soviet interventions, suggests that even Moscow may not have the necessary influence in this arena. The Soviet Union has apparently suspended further cooperation with the North on the construction of the four nuclear power plants it promised until the North fulfills its NPT obligations, and work on the units has apparently not yet begun.[77] So far, however, this sanction appears to have been insufficient to alter Pyongyang's stance.

One potential bright spot on the horizon is the possibility of a fundamental change in the government of North Korea during the next several years that could lead to a dramatic reversal of the country's foreign and nuclear policies. The fall of hard-line Communist regimes in Eastern Europe, Soviet pressures for reform, the model of the pro-democracy movement in China, and the South's increasing lead in economic performance and international recognition are inevitably causing pressures for change in Pyongyang. With Kim Il Sung nearing his eightieth birthday, a change of leadership in the North appears likely in the next several years. Though Kim Jong Il may prove even more erratic and aggressive than his father, his tenure in office could be brief. In sum, a combination of international pressure on the North to restrict its nuclear program and a shift to more liberal leadership in Pyongyang may avert the emergence of the DPRK as a nuclear-weapon state.

Source: Joseph S. Bermudez, Jr. and additional research.

North Korea

Research Reactors

Yongbyon IRT (pool/highly enriched uranium(?), 4 MWt)[a]
- supplier: USSR
- start up: 1965-early 1970s
- fuel source: USSR
- safeguards: yes

Yongbyon (gas-graphite/natural uranium, 30 MWt)[b]
- supplier: North Korea
- start up: 1987(?)
- fuel source: North Korea
- safeguards: no*

Reprocessing (Plutonium Extraction)

Yongbyon[c]
- capacity: size unknown
- supplier: North Korea, China(?)[d]
- start up: 1993(?)
- safeguards: no*

Uranium Resources/Active Mining Sites/Uranium Mills

- uranium reserves: "4 million tons"[e]
- currently active mining sites: Pyongsan,[f] Hamhung(?), Unggi(?), Hae Kumgang(?)[g]
- mills in operation: Kusong(?)[h]

Uranium Purification (UO_2)—presumed, location unknown

- capacity: unknown
- supplier: North Korea(?)
- start up: pre-1987
- safeguards: no*

Fuel Fabrication—presumed, location unknown

- capacity: unknown
- supplier: North Korea(?)
- start up: pre-1987
- safeguards: no*

* Although North Korea has signed the NPT, it has yet to conclude an implementing safeguards agreement with the IAEA. Presumably, after North Korea has done so, these facilities would be covered by IAEA safeguards.

Sources and Notes

a. International Atomic Energy Agency (IAEA), *The Annual Report for 1981* (Vienna: IAEA, July 1982), p. 71. The IAEA report also lists as safeguarded a 0.1 MWt critical assembly at Yongbyon, which North Korea apparently acquired from the Soviet Union at about the same time as the IRT reactor. "North's Nuclear Capability Assessed," *Hoguk*, July 1989, translated in *Foreign Broadcast Information Service (FBIS)-East Asia (EAS)*, August 3, 1989, p. 25.

b. "North Believed Closer to Producing Nuclear Arms," *Korea Times*, July 6, 1989, reprinted in *FBIS-EAS*, July 6, 1989, pp. 39-40; Joseph S. Bermudez, Jr., "N. Korea—Set To Join the 'Nuclear Club'?" *Jane's Defence Weekly*, September, 23, 1989, pp. 594, 597; "Military Cooperation With USSR Intensifies," *Naewoe Tongsin*, October 13, 1989, translated in *FBIS-EAS*, January 22, 1990, p. 15; "Nuclear Weapons Capability Discussed by ROK Paper," *Korea Herald*, June 4, 1989, reprinted in *JPRS-TND*, June 28, 1989, pp. 3-4.

c. Ann MacLachlan and Mark Hibbs, "North Korea Begins Negotiations With IAEA; South Korea Interested in Coprocessing," *Nuclear Fuel,* October 2, 1989, pp. 1, 11-12; John J. Fialka, "North Korea May Be Developing Ability To Build Nuclear Weapons," *Wall Street Journal,* July 19, 1989; Don Oberdorfer, "North Koreans Pursue Nuclear Arms," *Washington Post,* July 29, 1989.

d. MacLachlan and Hibbs, "North Korea Begins Negotiations . . ." (see note c).

e. "North's Nuclear Capability Assessed" (see note a). It is unclear whether this claim is a credible one, as uranium reserves of this magnitude would make North Korea one of the world's leading nations in uranium reserves, comparable to the U.S., Canada, and Australia.

f. "North Believed Closer to Producing Nuclear Arms" (see note a); Bermudez, "N. Korea . . ." (see note b).

g. "Military Cooperation With USSR Intensifies" (see note b); "North's Nuclear Capability Assessed" (see note a).

h. "Military Cooperation With USSR Intensifies" (see note b) (refers to a "uranium dressing station"); "North's Nuclear Capability Assessed" (see note a) (refers to a "new uranium processing plant" in Kusong).

Part III
The Middle East

For the next five to ten years, Israel is likely to remain the only Middle Eastern nation to possess nuclear weapons or the ability to manufacture them. Based on the revelations of former Israeli nuclear technician Mordechai Vanunu and additional information that has come to light over the years, there is compelling evidence that Israel possesses a substantial nuclear arsenal, consisting of 60 to 100 weapons, some of which are many times more powerful than the atomic bombs used against Hiroshima and Nagasaki. It has also been established authoritatively that Israel possesses 400-mile range missiles, for which both nuclear and chemical warheads are available. Israel is not a party to the Non-Proliferation Treaty (NPT).

The three other Middle East states treated in this volume, Iraq, Iran, and Libya, have demonstrated an interest in acquiring nuclear weapons during the past decade, but are unlikely to attain this objective for many years. While all three are parties to the Non-Proliferation Treaty, their past official statements and conduct raise doubts about their commitment to the treaty.

During 1989 and early 1990, evidence emerged that Iraq had revitalized its efforts to acquire a nuclear weapons capability in the wake of its eight-year war with Iran. The Iraqi nuclear program had been dormant since 1981, when Israel destroyed the Osiraq reactor outside of Baghdad. Iraq's renewed nuclear efforts, albeit unlikely to bear fruit until well into the 1990s, appear to be based around the construction of a uranium enrichment plant using high-speed gas centrifuges. Iraq has also developed substantial chemical weapons and ballistic missile capabilities, and it employed both extensively in the war with Iran.

Iraq's renewed nuclear activities have undoubtedly stimulated Iran to begin its own program to acquire nuclear arms, although little specific evidence of an

Iranian nuclear weapons program has been made public. Iran's efforts appear to be limited to research, and the country remains at least ten years away from the production of nuclear arms. Tehran is believed to be expanding its chemical weapon and missile capabilities actively, however, and like Iraq, used both types of weapons during the Gulf War.

The longstanding ambition of Libyan leader Muammar Khadafi to acquire nuclear weapons appears to be stymied. Nonetheless, Libya's military capabilities, which include the construction of a massive chemical weapons production plant and the acquisition of increasingly sophisticated delivery systems, could pose an increasing threat to a number of regional states and to Southern Europe.

Developments in these four states—Israel, Iraq, Iran, and Libya—are discussed in detail in the remainder of this section. Several additional introductory points concerning the Middle East nuclear environment, however, deserve comment.

In the late 1970s, Egypt announced a major nuclear power program but has yet to start construction of its first nuclear power plant. It is believed to have sought Chinese aid in developing nuclear arms during the 1960s[1] and is known to have used chemical weapons in 1967, when its troops were fighting in North Yemen.[2] Since concluding its peace treaty with Israel in 1979 and ratifying the NPT in 1981, however, Egypt's nuclear intentions have appeared to be entirely peaceful. Egypt has, however, continued to enhance its missile capabilities, reportedly working with North Korea to obtain improved Soviet-designed Scud-B missiles and participating, until mid-1989, with Iraq and Argentina in a project to develop the 600-mile Condor II missile.[3] Presumably, any new missiles Egypt might acquire will be armed with conventional or chemical warheads.

Syria, Israel's principal regional adversary, has never been considered a serious aspirant for nuclear weapons. It possesses both chemical weapons and short-range ballistic missiles, however, which Israel perceives to be an increasing threat.

Although it lies outside the region, Pakistan's nuclear capabilities (discussed in Chapter 7) also may have an impact on Middle Eastern security. An Islamic state with ties to a number of countries on the Persian Gulf, including Iran, Pakistan is believed to possess the essentials for five to ten nuclear devices. It is unlikely, even once Pakistan acquires longer-range delivery systems, that the country would pose a direct nuclear threat to Israel or that it would transfer nuclear weapons to Israel's regional adversaries. However, Pakistan might be prepared to provide technical assistance for the weapons-oriented nuclear programs of countries with which it is friendly, such as Iraq, which is attempting to build a uranium enrichment plant on the same design as Pakistan's Kahuta facility, or Iran, which reportedly has already received some nuclear aid from Islamabad.[4]

At a minimum, Pakistan's growing nuclear capabilities could serve as a symbolic counterweight in the Islamic world to that of Israel. The extent of Israeli concerns about the so-called "Islamic bomb" is not clear. In recent years there have been occasional rumors, however, of Israeli discussions with India concerning a possible joint attack on Pakistan's key uranium enrichment plant at Kahuta.[5]

*

Although Israel will continue to enjoy a nuclear monopoly in the Middle East for at least five to ten years, the dangers stemming from nuclear proliferation in the region are growing. In large measure, this is the result of the expanding ballistic missile and chemical weapon capabilities of the Middle Eastern states mentioned above.

For many years it was assumed that Israel would not consider resorting to nuclear weapons unless its very existence were threatened. Today, however, as the growing ballistic missile and chemical weapons capabilities of its adversaries make its civilian population increasingly vulnerable, Israel appears to have heightened its reliance on nuclear arms, raising the possibility that a future Arab-Israeli conflict could escalate to the nuclear level even if Israel were not threatened with annihilation.[6] The fact that nuclear-capable Israeli missiles will soon be able to reach parts of the Soviet Union—a development that Moscow has publicly decried—could introduce yet another dangerous dimension to the Middle East nuclear environment.

Iraq's apparent revitalization of its nuclear weapons program poses additional new risks for the region. With its potent missile and chemical weapon capabilities, Iraq could now retaliate forcefully against Iran, Syria, or Israel, should any of them attempt to thwart its nuclear ambitions by attacking its nuclear facilities; Iraq lacked this retaliatory potential in 1981, when the Osiraq reactor was destroyed.

Since an Iraqi counter-strike—against an Israeli or Iranian city, for example—could lead to a further act of retaliation or even, in some circumstances, to war, the emergence of a nascent Iraqi nuclear weapons effort could be the tinder that ignites a new round of hostilities in the region with a serious escalatory potential. On the other hand, if Iraq's non-nuclear capabilities, in combination with various defensive measures, successfully deter an attack on its nuclear facilities, they will have served as an umbrella that permitted the emergence of the region's second nuclear power.

Despite these disturbing trends, several more positive developments may be reducing the dangers stemming from nuclear proliferation in the Middle East. Most important is the new relationship between the United States and the Soviet Union, which, in conjunction with the developments in late 1989 in Eastern Europe, has sharply reduced the risk of conflict between the superpowers. In the Middle East, this can be expected to translate into a determination on the part of Washington and Moscow to limit the scope of any future hostilities in the region so as to avert the possibility of direct American or Soviet military involvement. In effect, the long-standing fear that escalation within the region might trigger a superpower confrontation with nuclear overtones appears to be diminishing.

One indication of the Soviet Union's interest in reducing tensions in the area has been its warnings to Syria that it would not support the latter's bid to achieve "strategic parity" with Israel—a posture that implies the acquisition of threatening offensive capabilities—but would press Damascus to adopt the new Soviet strategic outlook based on "reasonable defensive sufficiency."[7] The new line not only promises to reduce the threat to Israel from Syria itself, but also to lessen significantly the likelihood of Soviet support for possible Syrian aggression. Moreover, the new Soviet stance appears to undercut past claims by Damascus that Moscow had agreed to supply it with nuclear arms if Israel employed such weapons against Syria.[8] This, too, would help ease the risk of a regional crisis escalating into a superpower nuclear confrontation.

At the end of 1989, despite serious tensions in the region, prospects for major conflict appeared to be relatively low. Factors contributing to this situation included the exhaustion of Iran and Iraq following the eight-year Gulf War, initiatives by both Iraq and Libya to temper their foreign policies and strengthen ties to the more moderate Arab states, and the fact that Syria, for want of potential Arab allies who might help to balance the odds, appeared unlikely to open hostilities against far more powerful Israel. Nonetheless, many serious sources of tension in the region remained, including growing Syrian-Iraqi frictions in Lebanon, the continuing impasse over Palestinian aspirations for a homeland, an uprising of Palestinians in the Israeli-occupied West Bank and Gaza Strip territories, long-standing Syrian, Libyan, and Iraqi hostility toward Israel, the lack of a final settlement in the Iran-Iraq War, and the anxieties stemming from nuclear weapons programs in Israel and Iraq. Indeed, the spring of 1989 saw a serious resurgence of tensions between Israel and its neighbors that reawakened fears of possible hostilities in the region.

*

By 1990, the spread of chemical weapons and ballistic missiles in the Middle East had created for many observers a sense that the regional arms race was spiraling out of control. By this point, to be sure, virtually every militarily powerful country in the region was in possession of chemical arms and ballistic missiles with a range of approximately 200 miles, while Iraq, Israel, and Saudi Arabia had missiles of considerably greater range. (See Table 3.1.) Several of these countries were also believed to be developing biological weapons. International non-proliferation efforts aimed at curbing these capabilities, including export controls adopted by the advanced Western democracies, under the aegis of the Australia Group (for chemical weapons) and the Missile Technology Control Regime (for ballistic and cruise missiles), appeared to have achieved few results.

Such export controls have been imposed only since the mid-1980s, however, and can be expected to have an increasing impact on efforts by regional states to expand and improve existing chemical warfare and missile programs. Since few of the Arab states or Iran, for example, have fully integrated chemical weapons production or missile manufacturing capabilities, they will remain dependent to a greater or lesser degree on external assistance from more advanced countries. An international convention to ban the manufacture of chemical weapons is coming closer to fruition, and the United States and the Soviet Union are negotiating the details of plans to reduce their stockpiles of chemical weapons in the next several years. These developments are likely to strengthen the commitment of potential suppliers to stopping the spread of chemical weapons.

Moreover, the Soviet Union, as noted in Chapter 3, has agreed to restrict the sale of missiles covered by the MTCR, and has not made new sales of Scud-Bs in the region for several years. The People's Republic of China has also pledged not to sell medium-range missiles there, leaving North Korea as the only remaining supplier of such systems.

Whatever the difficulties of containing the spread of chemical armaments and of relatively unsophisticated missiles, preventing the proliferation of nuclear weapons presents a far more hopeful picture. The long-standing international consensus on nuclear non-proliferation and the diverse multilateral treaties and institutions that seek to address the problem mean that international pressures to constrain new nuclear weapon programs can be brought to bear more easily in this field. Moreover, such efforts can be mounted over a longer period and can be concentrated on fewer targets because of daunting technological challenges confronting countries seeking to develop nuclear arms indigenously.

As the 1990s unfold, Iraq has emerged as the second state in the Middle East, after Israel, aggressively seeking to achieve a nuclear weapons capability. Given Iraq's rudimentary nuclear infrastructure, it will be heavily dependent on foreign sources for nuclear materials, equipment, and know-how, and thus vulnerable to other countries' export controls. Moreover, given its need to rebuild after the devastation of the Gulf War and its desire to expand and improve its conventional military forces, it is likely to be highly dependent on Western financing and on Western and Soviet sales of advanced weaponry. Such dependence could permit foreign governments to gain sufficient leverage to dissuade Baghdad from pursuing its nuclear quest, notwithstanding Iraq's enduring concerns over the possibility of a resurgent Iran.

It is not possible to predict with accuracy whether non-proliferation diplomacy, however assertive, would be able to overcome the national security concerns that appear to be motivating Iraq's bid for nuclear arms. Nor is it possible to anticipate whether an Iraq-centered policy could succeed without commensurate efforts to constrain Israel's nuclear arsenal and to suppress Iran's reawakening interest in nuclear arms. Even so, powerful diplomatic tools are available for slowing Iraq's nuclear bid, and there are strong reasons for using them. It remains to be seen how the international community will address this challenge.

9 Israel

The revelations in late 1986 by former Israeli nuclear technician Mordechai Vanunu, now widely accepted as genuine, have confirmed the existence of Israel's nuclear arsenal, which probably consists of 60 to 100 devices. Some of these are of an advanced design that could make them many times more powerful than the atomic weapons used in World War II.[1] During 1989 and the first half of 1990, it must be presumed that Israel added several more nuclear weapons to its preexisting arsenal.

Also in 1989, British scientist Frank Barnaby—the only nuclear expert to speak with Mordechai Vanunu—disclosed that Vanunu had told him that, in addition to producing sizeable quantities of plutonium, Israel has been producing enriched uranium since 1979 or 1980, initially using centrifuges for this purpose, and then, beginning in 1981, also using advanced lasers.[2] Since highly enriched uranium—like plutonium—can be used as the core of a nuclear weapon, this capability may have allowed Israel to expand its nuclear arsenal beyond the size estimated above.

Israel has deployed medium-range (400-mile) surface-to-surface ballistic missiles intended for use with nuclear warheads.[3] In September 1988 and April 1990, it orbited space satellites using a rocket that could be modified to serve as a missile able to carry a 1,000 pound payload at least 2,000 miles.[4] Since May 1987, moreover, Israel has conducted several tests of a new missile, which has flown as far as 800 miles. Presumably, the missile uses a number of the same booster stages as Israel's satellite launcher. Accordingly, there is reason for concern that Israeli missiles will ultimately have a range considerably greater than 800 miles, bringing a substantial portion of the Soviet Union within reach.

Iraq's apparent revitalization of its nuclear weapons program will be of increasing concern to Israel in the years ahead. Today, Iraq has the ability to retaliate against an attack on its nuclear facilities with chemically armed missiles, which it lacked in 1981. If pressures for military action mount in Israel, the risk of conflict in the region could increase dramatically. Indeed, tensions over Iraq's nuclear ambitions and its growing non-nuclear potential, could replace Israeli-Syrian frictions as the most likely source of renewed hostilities between Israel and its Arab antagonists.

On a related front, as surface-to-surface missiles and chemical weapons capabilities proliferate more generally among Israel's adversaries, Israel's nuclear arms are likely to play an increasingly important role as a deterrent to attacks with such advanced weapons—a factor that could heighten the risks of nuclear escalation in any future Arab-Israeli conflict. These and other developments are discussed below.

Background

In October 1986, the London *Sunday Times* published a lengthy story disclosing many new details about Israel's undeclared nuclear weapons program. The exposé was based on information and photographs supplied by Mordechai Vanunu, who, from August 1977 to November 1985, had worked as a nuclear technician at Israel's classified Dimona nuclear complex in the Negev.

Vanunu's disclosures. Vanunu told the *Sunday Times* in great detail how Israel produced plutonium for nuclear weapons at an underground plutonium extraction plant where he worked and where, he stated, the material was fabricated into components for nuclear weapons. Thus he unambiguously confirmed the existence of an Israeli nuclear weapons program.

He also stated that Israel produced 88 pounds (40 kilograms) of plutonium annually, several times more than previously imagined. Vanunu's photos included pictures of a model of the internal components of an Israeli atomic device and of an actual weapon component. This led Dr. Theodore Taylor, a former U.S. nuclear weapons designer who was consulted by the *Sunday Times*, to conclude that Israel could build a nuclear weapon using only 8.8 pounds (4 kilograms) of plutonium—rather than the 11 to 17.6 pounds (5 to 8 kilograms) usually assumed necessary—and perhaps even less. On the basis of this information, Taylor, and Dr. Frank Barnaby, a nuclear physicist who had worked in the British nuclear weapons program, estimated that Israel's nuclear arsenal might comprise as many as 100 to 200 weapons, not the 20 to 25 usually cited in previous estimates.[5] (As discussed later, some U.S. government analysts disagree with this new estimate.)

Vanunu also stated that Israel had produced weapons that use nuclear fusion, the principle of the H-bomb. Vanunu backed up his claim with a photo of a weapon component said to be made of lithium deuteride, a fusion material. He also stated that Israel had produced tritium, another fusion material, which is used to "boost" fission weapons to enhance their efficiency. (See Glossary and Appendix B). On the basis of this evidence, Taylor and Barnaby concluded that Israel had indeed produced such advanced nuclear weapons. No new information has emerged to dispute their conclusions in this regard.

Vanunu's abduction. Vanunu disappeared from his London hotel several days before the *Sunday Times* story was published. Apparently, a female Israeli intelligence operative lured him to Rome, where he was abducted by other Israeli agents and taken to Israel.[6] He was subsequently tried by a secret tribunal and convicted of espionage and treason on March 24, 1988, receiving an 18-year prison term.[7] Israel's decision to detain and prosecute Vanunu indicates that his revelations were considered a major security leak and tends to confirm the authenticity of his disclosures. Indeed, since the *Sunday Times* story appeared, the authenticity of Vanunu's data and photographs has not been seriously challenged.[8]

Early Israeli program.[9] Vanunu's revelations and other information published in recent years, especially a highly detailed book by French journalist Pierre Péan,[10] help to fill some of the gaps in what is known about the history of Israeli nuclear affairs. Still, a number of important questions remain unanswered.

The Israeli nuclear weapons program was launched in the fall of 1956, in the wake of the Suez crisis. At the time, France's Socialist government, led by Guy Mollet, was deeply committed to Israel's survival, and the two states confronted threats stemming from Arab nationalism—Israel because of its isolated position in the Middle East and France because of growing unrest in French Algeria. France secretly pledged to assist Israel to develop nuclear arms and agreed to supply it with a sizable plutonium-producing reactor to be built at Dimona, in the Negev, some 40 miles from Beersheba.

In mid-1957, with French Atomic Energy Commission approval, Israel signed an agreement with the French firm of St. Gobain Techniques Nouvelles for the construction of several additional facilities at the Dimona site, including the key installation—where Vanunu would subsequently work—for extracting plutonium from the Dimona reactor's spent reactor fuel.[11] (During this period, France and Israel also collaborated on the development of a short-range nuclear-capable missile, which Israel deployed in the late 1960s.[12])

Dimona reactor enlarged. Excavation work for the Dimona reactor and for the plutonium extraction plant began early in 1958.[13] Israel was to receive a 24-megawatt research reactor, nominally based on France's EL-3 experimental reactor, built at Saclay between 1954 and 1957.[14] In reality, Péan writes, the Israeli reactor had at least a 40-megawatt capacity enabling it to produce enough plutonium for two or three weapons per year, rather than enough for the one to two devices that a 24-megawatt reactor would have been able to produce.[15] In light of Vanunu's revelations, the actual size of the reactor has become the key variable in assessing Israel's current nuclear weapons capabilities.

De Gaulle; weapon design assistance. After becoming President of France in June 1958, Charles De Gaulle authorized continued work on the reactor but sought to slow further construction of the sensitive plutonium extraction unit. According to Péan, however, his instructions on the extraction unit were thwarted by senior French bureaucrats sympathetic to Israel, and major construction proceeded through 1959 and much of 1960.

During this period, France also gave Israel important information on the design and manufacture of nuclear weapons themselves.[16] Israel may also have obtained data from France's first nuclear test, which took place in February of 1960.[17]

Confirmation. Francis Perrin, the scientific head of the French Atomic Energy Commission from 1951 to 1970, was intimately involved with the French-Israeli nuclear program. He confirmed several key points of Péan's account in a 1986 on-the-record interview with the *Sunday Times*, acknowledging that France supplied the Dimona reactor and the plutonium extraction plant and that for at least two years during the late 1950s, France and Israel collaborated on the design and development of nuclear weapons.[18]

U.S. concerns. In early 1960, De Gaulle reiterated his instructions that work should cease on the Israeli plutonium facility, but, according to French participants interviewed by Péan, work continued until late in the year.[19] It was suspended only when the United States discovered in December 1960 that Israel was building the reactor.[20] Under U.S. pressure, on December 21, Israeli Prime Minister David Ben-Gurion acknowledged that Israel was building the reactor, which he said was rated at 24 megawatts, and declared that the facility would be used exclusively for peaceful research and training.[21] To allay American concerns, Ben-Gurion agreed to permit American specialists to visit the reactor for verification.[22]

Plutonium plant secrecy maintained. When the story about the Dimona reactor broke in 1960, there was no indication that the United States knew of the existence of the plutonium processing plant, apparently because the facility was built underground. According to Vanunu, all of the plutonium processing equipment was housed in six subterranean floors of the building, which were topped with two stories of innocuous changing rooms and offices. The unit (which was not completed until at least 1966) would have required an eighty-foot deep excavation, but if this had been dug in 1958, as Péan suggests, before American suspicions were fully aroused, it is conceivable that this part of the Dimona complex could have remained undiscovered.

Uranium. As construction proceeded on the reactor and plutonium plant in the late 1950s, Israel was still missing two key items, uranium fuel for the reactor and heavy water, a commodity needed to permit the fissioning of the uranium within the reactor—the process by which some uranium atoms are transmuted into plutonium.[23] Israel obtained uranium on the world market from a number of sources, mainly in the West and in Africa.[24] The latter probably included French-controlled mines in what are today Gabon, the Central African Republic, and Niger; South Africa may have been another source.[25] Argentina also supplied 15 tons (13.6 metric tons) of uranium, and Israel may have produced some of the material itself as a by-product of its phosphate mining and processing operations.[26]

Heavy water. Israel obtained its initial supply of heavy water—some 22 tons (20 metric tons)—from Norway in 1959.[27] In 1986, Norway revealed that it had imposed what were at the time unusually strict controls over the material, obtaining Israel's pledge to use it exclusively for peaceful purposes and obtaining the right to inspect it to verify Israeli compliance. However, Norway conducted only one such inspection, in 1961. This was before the Dimona reactor was completed, and the heavy water was still in storage.[28]

Israel would have needed at least 20 tons (18 metric tons) of heavy water to operate even a 24-megawatt reactor. Since Israel is not known to have ever obtained substantial quantities of heavy water from any other source until possibly the late 1980s, it appears that Israel continued to use the Norwegian material in the Dimona reactor until this time, *i.e.*, throughout a period when the unit was being employed to produce plutonium for nuclear arms. This would have violated Israel's pledge to Norway.[29] In the wake of the Vanunu revelations, this possibility led Norway in late 1986 to attempt to exercise its inspection rights, triggering a major controversy with Israel, discussed below. Israel subsequently acknowledged that the Norwegian material was used in the Dimona unit but claimed that it did not violate the agreement covering the material.[30]

The plutonium plant: work continues. Following the public disclosure of the Dimona reactor in late 1960, De Gaulle's order to halt construction on the plutonium extraction facility was apparently enforced, but work was allowed to continue on the reactor itself. By this stage, however, Israel had acquired the plans for the plutonium facility, according to Péan, and was able to recruit the key French subcontractors that had worked on France's own plutonium plant at Marcoule to complete the job.[31]

U.S. visits. Péan states that the Dimona reactor was operating at full capacity at the end of 1962 and that the first plutonium extraction tests took place at the plutonium unit in 1965; by 1966 or 1967, he writes, Israel had apparently acquired enough plutonium for its first weapon.[32] Vanunu also states that the plutonium plant began operating in 1966.[33] From 1964 to 1969, however, U.S. nuclear specialists making annual visits to the Dimona complex did not detect evidence that Israel was operating a plutonium extraction plant.[34] Apparently, the visits—which never amounted to rigorous "inspections" —were so infrequent and limited in scope that the U.S. officials could not, in fact, be completely confident about what was occurring at the site. Israel also reportedly engaged in deliberate deception. Vanunu, relaying stories he heard from more senior co-workers, told the *Sunday Times* that the door to the stairs leading down to the subterranean plutonium plant, where he later worked, was bricked up for the annual U.S. visits, so that all the U.S. specialists saw were the innocuous two upper stories of the building. And Péan claims that even the operations of the reactor were disguised. The inspectors were taken not to the unit's actual control room, but to a mock-up, whose meters were connected to simulators that showed the reactor to be operating at a very low power level.

In 1969, the U.S. site-visit team complained in writing about Israel's restrictions on its activities, but Israel refused to alter them.[35] The Nixon Administration, for reasons that are still unclear, dropped the matter, and the visits ceased.

Whatever the details, by virtue of Vanunu's disclosures, the existence of the plutonium plant is known today; the U.S. site visits are known to have been seriously limited; Péan's account of the construction arrangements is persuasively documented and has been essentially confirmed by Francis Perrin; and Péan's and Vanunu's independent reports of deception have a similar ring that gives them an air of credibility. While questions remain, for now the evidence in the public record most strongly supports the conclusion that the plutonium plant was completed in the mid-1960s without detection by the U.S. site visit teams.[36]

Possible diversion of U.S. weapons-grade uranium. It is also widely believed that during the mid-1960s, Israel was able to obtain 220 pounds (100 kilograms) of U.S.-owned highly enriched uranium—material that, like plutonium, can be used as the core of a nuclear weapon. The material is thought to have been stolen from a uranium fabrication plant in Apollo, Pennsylvania, owned by the Nuclear Materials and Equipment Corporation (NUMEC). The material would be enough to manufacture four to six weapons, using the standard measure of 33 to 55 pounds (15 to 25 kilograms) of material per device.[37] The plant was to have processed the material, received between 1962 and 1965, for the U.S. Atomic Energy Commission, which discovered the inventory discrepancy in 1966.

According to a former senior official of the U.S. Central Intelligence Agency, Carl Duckett, by 1968 the agency had concluded that "the most likely case" was that the "NUMEC material had been diverted and had been used by the Israelis in fabricating weapons."[38] The episode has remained controversial, however, and conclusive evidence that Israel obtained the material is still lacking.[39] Neither Vanunu's disclosures nor Péan's book have shed any additional light on the matter.

The 1967 Arab-Israeli War and its aftermath. If Péan is correct that Israel obtained plutonium for its first nuclear weapons in 1966, or if Israel was able to divert highly enriched uranium from NUMEC in the mid-1960s, this would suggest that Israel possessed the ability to manufacture at least a handful of nuclear weapons at the time of the June 1967 Six-Day War. If so, these weapons are not known to have figured in that conflict.

Following the 1967 war, Israel's Soviet-backed Arab adversaries—principally Egypt, Iraq, Libya, and Syria—refused to negotiate a permanent peace settlement, and military tensions continued, eventually leading to the so-called "War of Attrition" with Egypt in March 1969. Against this background, several figures known to favor Israel's acquisition of nuclear weapons, including Moshe Dayan, returned to positions of authority in the Israeli government, and Israel apparently accelerated and expanded its nuclear weapons efforts. In November 1968, it is widely believed to have undertaken a secret mission to acquire large amounts of processed uranium ore, or "yellowcake." Using intermediaries, Israeli agents are believed to have arranged to have 200 tons of yellowcake shipped from Antwerp to Genoa and then to have orchestrated its "disappearance" while it was in transit on the Mediterranean Sea in a vessel that had been secretly purchased by an Israeli front company.[40] Converted to uranium metal, the yellowcake could make 168 tons of fuel for the Dimona reactor, enough for perhaps five to seven years' operation of a 40-megawatt unit (the size Péan estimated), assuming it was operating to produce plutonium and was therefore being refueled several times a year.

In addition, by 1972, Israel had built three phosphoric acid plants from which it was able to extract uranium as a by-product, using an indigenously developed technique. The plants theoretically could produce about 100 tons (90 metric tons) of uranium annually.[41] According to Vanunu's testimony, Israel is also able to extend its uranium supplies by recovering unfissioned uranium during the plutonium extraction process.[42]

The 1973 Arab-Israeli War. By the time of the 1973 Arab-Israeli conflict, it appears that Israel did indeed possess nuclear arms, and in 1976, *Time* magazine reported that in the early days of that war, when Israel was being threatened on two fronts by Egypt and Syria, Prime Minister Golda Meir took steps to deploy nuclear warheads on some of Israel's short-range (260-mile) Jericho missiles.[43] Use of these weapons proved unnecessary, however, once the tide of battle turned. Former and current U.S. officials have confirmed that Israel secretly readied nuclear weapons for possible use at this time, although not necessarily in conjunction with the missiles.[44]

In the aftermath of the 1973 war, a series of hints by senior Israeli officials and statements by authoritative sources in the United States lent further credence to the view that Israel possessed nuclear arms.[45] Virtually all of these sources, if they described Israel's nuclear capabilities, indicated that the country was probably able to produce only one or two weapons per year using the Dimona reactor and associated installations.[46] If these estimates were correct, and if Israel had continued to produce weapons at this rate, today it would have an arsenal of perhaps 30 to 50 weapons, or slightly more if the NUMEC material were included in the calculation. It was not until the 1980s that reports began to suggest that Israel's arsenal might be considerably larger.[47] Until Vanunu's disclosures, however, there seemed to be little basis for such larger estimates.

Iraqi reactor. On June 7, 1981, Israel destroyed Iraq's large "Osiraq" research reactor outside Baghdad. This episode and its aftermath are discussed in Chapter 11.

Further smuggling. During the 1980s, Israel is known to have engaged in at least one additional clandestine transaction that may have contributed to its nuclear capabilities. On May 16, 1985, federal prosecutors in Los Angeles indicted Richard K. Smyth for illegally exporting 810 high-speed electronic switches to Israel between 1980 and 1983. The switches, known as krytrons, are used in the triggering mechanisms of U.S. nuclear weapons.[48]

The Israeli Defense Ministry at first denied, but subsequently acknowledged, being in possession of the switches, which, it claimed, had been employed exclusively for research and development on conventional weapons, such as laser range-finders.[49] Israel later returned several hundred of the krytrons that had not been used.[50] Smyth disappeared shortly before his trial and is still a fugitive. Reportedly he is in Israel, but the United States has not sought to have him extradited.[51]

Smyth's smuggling activities on behalf of Israel may have involved far more than krytrons. According to two separate reports, in 1982 he illegally obtained special butyl rubber compounds, used as a binder in solid rocket fuel, for Israel's medium-range Improved Jericho (sometimes referred to as the "Jericho II") missile. Subsequently, according to these accounts, he smuggled specially hardened aluminum extrusions to be used as the rocket's body, as well as inertial gyroscopes for its advanced guidance system.[52] One of the reports quoted a former U.S. under secretary of state as saying, "What Smyth did was pass on to Israel virtually hundreds of millions of dollars of advanced U.S. technology."[53]

Nuclear tests. Israel is not known to have conducted a full-scale nuclear test. It has been widely assumed that the country's undeclared nuclear arsenal has been developed, in part, through the testing of non-nuclear components and computer simulations—and through the acquisition of weapons design and test information from abroad.

In addition, as noted earlier, there is strong evidence that Israel collaborated with France during the late 1950s on the development of atomic weapons and may also have obtained data from France's first nuclear test, which took place in the Sahara in February 1960.[54]

Israel may also have obtained data from U.S. nuclear tests at approximately this time. According to a May 1989 U.S. television documentary, Israel was able to gain access to information concerning U.S. tests from the 1950s and early 1960s and then shared this information with France in return for French assistance in constructing the Dimona nuclear complex. The test data could well have included the results of U.S. boosted and thermonuclear weapons that were being developed at the time.[55] If the documentary is correct, the data could have assisted Israel during the early 1980s, when it apparently developed weapons using these technologies, and possibly could have given Israel enough confidence in its weapon designs to obviate the need for full-scale nuclear tests.

It has also been alleged that Israel obtained assistance in the design of nuclear warheads for its medium-range Improved Jericho missiles from émigré U.S. scientists.[56] One report states that the system "carries a warhead derived from the U.S. XW-58 warhead."[57] More recently, the U.S. General Accounting Office—the investigative arm of Congress—found that during the 1980s, Israeli scientists had surprisingly open access to certain weapons-related research at U.S. nuclear weapons laboratories.[58]

Finally, there has been speculation that a signal detected on September 22, 1979, by a U.S. monitoring satellite, which was then over the South Atlantic, was, in fact, the flash from a low-yield Israeli nuclear test—possibly of a tactical nuclear weapon or of the fission trigger of a thermonuclear device. A panel of experts convened by President Carter's Office of Science and Technology Policy determined that the event "probably" was not a nuclear detonation, but the panel's conclusions were disputed by the Defense Intelligence Agency and the Naval Research Laboratory. As a result, the event remains shrouded in uncertainty, although in recent years, the weight of opinion within the U.S. government appears to have moved toward the conclusion that the event was an Israeli test using South African facilities.

Strategic environment. Throughout its history, Israel has confronted a series of intensely hostile regional adversaries who have refused to recognize its right to exist and against whom it has fought three major wars in which the country's very survival has sometimes been at risk. To ensure its security, Israel has sought to maintain a margin of conventional military superiority over its foes sufficient to discourage their resort to force and to ensure victory in the event of conflict—without the use of nuclear arms. Indeed, Israel's conventional military edge has long been so substantial that for many years it reasonably could be argued that the risk was quite small that Israel might resort to nuclear arms to ensure its survival. This situation may be changing as Israel's adversaries acquire the ability to threaten its population centers with ballistic missiles equipped with chemical warheads, against which Israel currently has no defense.

Although the combined strength of Israel's potential Arab adversaries has for some time far outstripped that of the Israeli Defense Forces, the latter's qualitative superiority in virtually all areas leaves little question that Israel could readily defeat any single opponent and many combinations of opponents.[59] Given Israel's peace treaty with Egypt, the deep animosities that divide Iraq and Syria, and the slowly moderating foreign policies of Jordan, Saudi Arabia, and the Arab states of the Persian Gulf, the possibility of a successful unified Arab attack on Israel has become increasingly remote in recent years.

Nonetheless, the end of the Iran-Iraq War found both of these states with formidable military forces. This is particularly true for Iraq, which remains hostile toward Israel and now possesses the capability to strike its territory with chemically armed ballistic missiles. Similarly, after Israel's crushing defeat of Syrian air and ground forces in the 1982 Lebanon War, Damascus aggressively rearmed with Soviet assistance. For much of the 1980s, its stated goal was to achieve "strategic parity" with Israel; to this end, it is believed to have developed a substantial chemical weapons capability and has acquired a variety of ballistic missiles, including the highly accurate, short-range Soviet SS-21. Libya, which has also acquired chemical weapons and is seeking a missile and air-strike capability sufficient to reach Israel, similarly poses a potentially serious threat to Israeli security. Israel also has expressed deep concerns about Saudi Arabia's acquisition in 1988 of conventionally armed Chinese CSS-2 intermediate-range missiles.

The uprising of Palestinians living in the Israeli occupied West Bank and Gaza Strip, which began in December 1987, has also confronted Israel with a serious security challenge. Although on a day-to-day basis, Israel's principal military concern has been maintaining internal law and order, the right-wing Likud Party, which has led Israel's coalition governments since 1986, considers continued control over the strategically located West Bank area (and to a lesser extent, the Gaza Strip) to be essential for Israel's security. It thus has resisted proposals for Palestinian autonomy that might change the status quo. Somewhat ironically, a number of more dovish Israeli analysts, who might be expected to oppose Israel's nuclear weapons program, have argued that the existence of this capability is sufficient in itself to ensure Israeli security and that ceding the West Bank as part of a settlement of the Palestinian issue would not, therefore, seriously affect Israel's ability to defend itself.[60]

Israeli nuclear policy in the wake of the Vanunu disclosures. In the aftermath of Vanunu's revelations, Israeli leaders continued to deny that Israel possessed nuclear arms and reiterated the longstanding position that Israel "will not be the first country to introduce nuclear weapons into the Middle East."[61] Nonetheless, a number of second-echelon Israeli officials began alluding to the country's nuclear capabilities with unusual openness.

Shortly after the *Sunday Times* ran the Vanunu story, for example, Major General Amnon Shalhaq, chief of Israeli Defense Force Intelligence, noted in a lecture that one reason the Arab states know they cannot annihilate Israel is because, "the Arabs believe Israel also has nonconventional power."[62] In January 1987, Gideon Raphael, Israel's ambassador to the United Nations, made a comparable allusion to Israel's nuclear might, declaring in an op-ed piece, that Syria's chemical warfare capabilities ". . . could mean that the next war between Syria and Israel will degenerate into a contest between chemical and radiation weapons. . . ."[63] Later that month, Amos Rubin, then recently installed as an economic advisor to Israeli Prime Minister Yitzhak Shamir, was still more explicit, stating in an interview that, "to enable Israel to abstain from dependence on nuclear arms calls for $2 to $3 billion in U.S. aid."[64]

Although there have been few such official references to Israel's nuclear might since early 1987, as noted below, Israel's missile tests and two well-publicized space launches have inevitably drawn attention to this capability. On a number of occasions, moreover, Israeli leaders, in emphasizing Israel's ability to retaliate massively to a chemical attack, have implicitly invoked the country's nuclear arsenal.

Political and Technical Developments

Current capabilities. Estimates of the size of the Israeli nuclear arsenal vary significantly. Based on the data supplied by Mordechai Vanunu, the *Sunday Times* projected that Israel might have as many as 200 devices. The consensus among U.S. officials who have attempted to harmonize Vanunu's testimony with other relevant information, however, is that Israel's nuclear armory probably contains fewer than 100 weapons, and perhaps no more than 50 or 60. Even such a smaller stockpile, however, would permit Israel to use a number of its nuclear weapons tactically, *i.e.*, against military targets, during any conflict with its neighbors, while keeping a number of weapons in reserve to threaten enemy cities and thereby deter actions that might imperil Israel's existence.

The key piece of information provided by Vanunu was that the plutonium extraction plant where he worked produced a total of about 88 pounds (40 kilograms) of plutonium per year—possibly enough for ten weapons, if former U.S. weapons designer Theodore Taylor is correct that Israel has the sophistication to build weapons using only 8.8 pounds (4 kilograms) of plutonium per device. During the eight years that Vanunu worked at the plutonium facility (1977-1985), enough plutonium for 80 weapons could have been produced.

In his book on the Israeli nuclear program, published in 1989, Frank Barnaby states that the plutonium plant began operating in 1966 and reached full capacity in 1972; Barnaby also assumes that the plant continued to operate at this rate after Vanunu left in 1985.[65] Thus by 1990, the facility would have operated at full capacity for eighteen years, producing ten weapons' worth of plutonium per year, or enough plutonium for about 180 weapons.

If the Dimona reactor was only 24 megawatts, the power level announced by Israel in 1960, it could irradiate only enough fuel to produce plutonium sufficient for two devices per year, however, not for the ten bombs suggested by Vanunu's data. This discrepancy led the *Sunday Times* to hypothesize that the reactor had been enlarged nearly six-fold, possibly to 150 megawatts.

Barnaby agrees that a 150-megawatt reactor would be required for the larger plutonium-output capacity. According to Barnaby, Vanunu told him that the reactor's output was increased twice before his arrival at Dimona, the first time to 70 megawatts. Barnaby concludes that the second upgrade "presumably" raised the unit's power to the needed 150 megawatts, and adds, "To disperse the extra heat produced, a new large cooling unit was added to the reactor."[66] Barnaby does not indicate whether this is his own speculation or a detail provided by Vanunu.

U.S. view. U.S. government specialists, however, doubt that the Dimona reactor was enlarged so dramatically because, they claim, there is no evidence to suggest that the needed additional cooling units have been built.[67] They believe that the reactor operated initially at 40 megawatts, because of the cooling efficiencies permitted by the desert climate, and, based on Vanunu's statements, concede that the reactor may have been enlarged to 70 megawatts before his arrival in 1977. If the reactor had been enlarged to this level in the early 1970s—perhaps, in 1972—and had operated through 1989 at the higher level, total plutonium output might amount to enough for about 90 weapons, altogether; if it was not enlarged, only enough plutonium for 50 to 60 weapons could have been produced.[68]

To explain why the throughput of the plutonium plant was so high during Vanunu's tenure, U.S. analysts speculate that Israel had trouble running the plutonium plant for a number of years, which would have created a large backlog of spent fuel from the Dimona reactor for processing. Vanunu was hired as part of a large group of technicians in 1977 and was part of a 1985 lay-off that involved 180 workers at the Dimona site. An increase and sudden reduction in staff would be consistent with the hypothesis that extra manpower was needed for a number of years while a backlog was being worked off.

For now, the matter cannot be resolved. Péan's assertions that the reactor has always been larger than publicly described and that the Americans who visited the complex in the 1960s were deceived on this point may be correct. Alternatively, the reactor could have been enlarged secretly, eluding U.S. intelligence analysts, as Vanunu claimed. Or, as U.S. analysts suggest, the reactor may never have operated at 150 megawatts, but only at an intermediate power level, while operating difficulties in the plutonium plant led to a backlog and the subsequent periods of high throughput described by Vanunu. All of these estimates would suggest, however, that Israel possesses a nuclear armory larger than previous U.S. estimates of 50 or so weapons, which assumed that the reactor was never enlarged, but still somewhat short of the highest alternative projections of 200 weapons. (Moreover, to these various estimates of plutonium-based weapons should be added any weapons that Israel may have produced with highly enriched uranium cores, using either material obtained from the U.S. NUMEC plant in the 1960s or from Israel's own enrichment facility.)

U.S. acknowledgement. In November 1989, the Natural Resources Defense Council, a private environmental group that monitors nuclear issues, released a U.S. Defense Intelligence Agency document entitled, "Surface-to-Surface Missile Handbook—Free World," which it had obtained under the Freedom of Information Act.[69] One page of the document is headed, "Jericho I," and under the column for type of warhead "nuclear, chemical, or HE [high-explosive]" are listed. The document also states that the system is "deployed on a wheeled TEL [transporter/erector/launcher] vehicle." The document is one of the few explicit indications of the U.S. view that Israel possesses nuclear arms and has deployed missiles to carry them. (It also represents the first published indication that the United States believes Israel has missiles equipped with chemical warheads.) Israel's missile program is discussed below.[70]

Enrichment. As mentioned earlier, Mordechai Vanunu revealed to Frank Barnaby that in addition to producing sizable quantities of plutonium for nuclear weapons, Israel has been producing enriched uranium with centrifuges since 1979 or 1980 and with advanced lasers since 1981.[71] This capability may have allowed Israel to expand its nuclear arsenal further. Barnaby does not, however, state whether Israel is enriching uranium to weapons grade or only to lower levels. Nor does Barnaby indicate how large the Israeli enrichment capability might be.[72] Finally, testifying before Congress in early 1989, CIA Director William Webster confirmed that Israel has been pursuing research on the use of lasers for enriching uranium, but he indicated that the program had only limited funding. He did not comment on whether Israel also has a centrifuge enrichment capability.[73]

Advanced nuclear weapons. Vanunu's claim that Israel had produced tritium and lithium deuteride suggests that the Israeli arsenal consists, at least in part, of advanced nuclear weapons. Presumably, the tritium is used by Israel in the manufacture of "boosted" nuclear weapons. These are nuclear devices that rely principally on the fissioning of plutonium or highly enriched uranium for their yield, but in which this reaction is enhanced by the presence of tritium. The tritium undergoes an atomic fusion reaction when the device is detonated, releasing a stream of neutrons that greatly improves the efficiency of the fission reaction. Boosting thus allows weapon designers to achieve a greater nuclear yield than could otherwise be obtained from a given quantity of plutonium or highly enriched uranium.[74] In practice, this means that some Israeli nuclear weapons may have yields several times greater than the 20-kilotons nominally assumed to be the size of developing country nuclear devices.

In addition, former U.S. nuclear weapons designer Theodore Taylor, after examining Vanunu's photo of an unusually configured nuclear weapon component made from lithium deuteride, suggested that Israel also may be building "super-boosted" weapons—a term he coined to characterize the Israeli design—with a yield of perhaps 100 kilotons, the size of some warheads on U.S. strategic missiles.[75] Presumably, the manufacture of full-fledged, multistage thermonuclear "hydrogen" bombs with megaton yields would require an extensive nuclear testing program and thus remains beyond Israeli abilities, since Israel is not known to have conducted the necessary tests.[76]

According to Frank Barnaby, in 1977, Israel apparently began producing small quantities of lithium-6, a key precursor for the production both of tritium and of lithium deuteride.[77] Full-scale production of lithium-6 began in 1984, and continued for two years; tritium production (and presumably lithium deuteride production) also began the same year. Altogether Israel produced approximately 375 pounds of lithium-6, which Barnaby estimated would have allowed Israel to produce enough lithium deuteride for 35 fusion weapons—presumably of the super-boosted type identified by Taylor.[78] Or, Israel could have used some of the lithium-6 to produce tritium for boosted weapons. The proportion of lithium-6 that may have been used for each purpose is not known.

The foregoing analysis suggests that Israel probably has at least 60 to 100 nuclear weapons. These weapons most likely use plutonium, but some may rely on highly enriched uranium. Of these 60 to 100 devices, a number probably are boosted with tritium to increase their yield to perhaps 40 or 50 kilotons, while other weapons have been "superboosted" using lithium deuteride to give them a yield in the 100-kiloton range.

Emerging long-range missile capability. Israel's capabilities for delivering nuclear weapons with ballistic missiles have also grown dramatically in recent years. The precise status of the country's missile program remains uncertain, inasmuch as Israel maintains considerable secrecy about its efforts in this sphere. Nonetheless, several elements of the program can be described.

First, during the 1960s, Israel developed with French assistance a 260-mile nuclear-capable ballistic missile, known in Western intelligence circles as the Jericho. The missile is noted explicitly in a 1974 CIA analysis of the Israeli nuclear program.[79]

Second, Israel has deployed an improved version of the missile. Apparently this took place by the early 1980s. The modernized Jericho was brought to public attention in a pair of May 1985 reports in *Aerospace Daily*, which described the system as having a 400-mile range and an inertial guidance system that greatly improved its accuracy.[80] The author of the reports subsequently stated in a television interview that the missile carries a nuclear warhead that is about 2 feet long, 20 to 22 inches in diameter, and 226 pounds (100 kilograms) in weight.[81] Further evidence of the system's existence emerged the following year in a *New York Times* report, which described the missile as having a 1650-pound (750-kilogram) payload and disclosed that during the late 1970s Iran had worked with Israel in developing the system.[82] Finally, any doubt as to the system's existence was removed in 1989, when the U.S. Arms Control and Disarmament Agency (ACDA), in its annual survey of world military expenditures, listed the Jericho as a deployed system and described it as a two-stage, solid-fuel missile with a 400-mile range.[83]

Third, since 1987, Israel has been developing one or more new missiles and has been testing a version with a 900-mile range; by the end of 1989, the system had been tested to a range of 800 miles.[84] ACDA's annual global military expenditure survey again provides confirmation, referring to "Jericho Follow-ons;" U.S. officials indicate that the plural was used because more than one new system may be under development. The report characterizes the Jericho Follow-ons as "tested" and lists them as two-stage, solid-fuel systems, with a range of 900 miles.[85] U.S. officials in off-the-record interviews elaborated by stating that the system tested since at least 1987 does have this range, but that another version of the Jericho Follow-on still under development may have a considerably greater range; moreover, it is not clear which system Israel may ultimately deploy.[86]

Satellite launches. On September 19, 1988, Israel publicly demonstrated its rocket capabilities for the first time by orbiting its first space satellite, using a powerful launcher, which it has named the "Shavit II."[87] Though Israel has released few details, the satellite launch appears to have been a technological *tour de force*. Weighing 340 pounds, the Israeli spacecraft, known as Offeq ("Horizon") 1, not only outsized Sputnik I and America's first satellite, Explorer I, but was nearly six times as large as the first satellite that Japan orbited in 1970 and more than four times the size of India's first, lofted in 1980. (Indeed, in March 1987 and again in July 1988, India failed in its first two attempts to orbit a 330-pound unit.)

The three-stage Shavit ("Comet") booster was unusually powerful. To avoid the risk of a launch mishap over Arab territory, Israel had to place Offeq 1 in a highly unusual east-to-west orbit and could not take advantage of the momentum from the Earth's west-to-east rotation as the other spacefaring countries do. Moreover, comparable U.S. and Indian rockets use four stages.[88]

Any rocket that can be used as a space launcher can be adapted for use as a surface-to-surface ballistic missile, if its payload and trajectory are suitably modified and a re-entry vehicle is added. One analysis, prepared by a researcher at the U.S. Lawrence Livermore National Laboratory, estimated that if reconfigured as a missile, the Shavit II could carry a one-ton (900-kilogram) payload just over 3,000 miles and could carry a 1,100-pound (500-kilogram) payload nearly 4,700 miles, in effect making the missile an intercontinental system.[89] The analysis assumed that a warhead and guidance package would be substituted for the Shavit II's third stage. A separate analysis, which was undertaken as part of a 1989 study for the Aspen Strategy Group and was based on somewhat different assumptions about the Shavit II, concluded that a missile based on the satellite launcher could carry a 2,200-pound (1,000-kilogram) payload over 2,000 miles and a 1,100-pound (500-kilogram) payload close to 3,000 miles, again giving the system a reach well beyond the Middle East.[90]

Israel launched a second satellite on April 3, 1990, again demonstrating its considerable rocket capabilities. Offeq II's size and orbit were roughly comparable to those of Offeq I.[91]

These analyses of the Offeq I launch indicate that if one version of the

Jericho Follow-on is essentially the same as the first two stages of the Shavit, the Follow-on would have a range that is far greater than indicated by the tests of the system that had been conducted by the end of 1989. This would mean that the system was intended, at least in part, to reach targets outside the Middle East—the most obvious assumed target being Soviet territory.[92] Possibly, this is the system that Israel has been testing since 1987, since its trajectory could be readily adjusted to limit its flight to the 800 or 900 miles in its reported tests.

Indeed, it is difficult to imagine that Israel, with its over-stretched defense budget, would develop two completely different sets of boosters for missile and space-launch purposes, when one set could be used for both missions. However, the ACDA report mentioned above, along with suggestions from a number of U.S. government sources, indicate that more than one version of the Follow-on is apparently being developed—presumably one with a 900-mile range, which is being tested, and one with a considerably longer range, which had not yet been tested by the end of 1989.[93] If it is assumed both of these systems use boosters already developed by Israel for other purposes, the best working assumption may be that the two-stage, 900-mile Jericho Follow-on uses one of the lower stages from the Shavit II and one from the Improved Jericho. This would give it a range in excess of the latter, but less than that of the former; a longer-range version of the Follow-on that may be under development could use both of the first two stages from the Shavit II.

Soviet warnings. The Soviet Union has repeatedly expressed concern over Israel's development of the Jericho Follow-on.[94] Even if the range of the system Israel ultimately deploys were only 900 miles, it could reach the Soviet cities of Baku, Tbilisi, and possibly, Odessa. In the wake of a test of the missile in September 1989, Soviet Foreign Ministry spokesman Gennady Gerasimov indicated that Moscow considered the missile to be a threat to Soviet security. Gerasimov declared:

> Israel is known to possess a technological basis necessary for the creation of nuclear weapons. The availability of delivery systems makes Israel a source of destabilization, far exceeding the boundaries of the Middle East region.[95]

Collaboration with South Africa. Israel has had close political ties to South Africa and the two states have cooperated extensively in the area of conventional arms. It also has been alleged for many years that Israel and South Africa have collaborated in the development of nuclear arms and that Israel may have used South African facilities to conduct a nuclear test in the South Atlantic in September 1979. Although a nuclear relationship has not been conclusively established, circumstantial evidence to support its existence has slowly accumulated over time. The detailed discussion of this matter in the South Africa section of this book concludes that allegations of an Israeli-South African nuclear connection deserve to be taken seriously, even if its precise contours remain unknown.

In late 1989, a major investigation by an American television news team reported that the two countries were collaborating on the development of long-range missiles.[96] The report, by NBC News, stated that Israel had provided missile technology to South Africa in return for a supply of enriched uranium for nuclear weapons and access to long-range missile test facilities on the South African mainland at Overberg and on the Prince Edward Islands in the Antarctic. The report stated that the two countries had conducted a test of a missile from the Overberg site on July 5, 1989, to a range of 900 miles—a range that suggests the system tested was one version of the Jericho Follow-on. An important benefit to Israel of gaining access to a test site in South Africa is that this would permit it to test missiles to a considerably longer range than is feasible from Israel.[97]

Both Israel and South Africa denied the charges in the NBC report, although some Israeli statements left open the possibility that collaboration in the missile field might be continuing under contracts signed by the two states some years ago.[98] Moreover, as the story unfolded, additional press reports, quoting unnamed U.S. officials, basically confirmed the existence of joint missile activities between the two states.[99] There appeared to be no additional support, however, for the view that Israel had received a nuclear *quid pro quo* for providing missile technology to Pretoria, and even NBC News began to downplay this aspect of the story.

The reported missile links to South Africa exposed Israel's missile and nuclear activities to public scrutiny in the United States for the first time in many years. Bush Administration officials, for example, publicized the fact that President Bush had raised Israel's transfers of missile technology with Israeli Prime Minister Yitzhak Shamir, during the latter's mid-November 1989 visit to Washington shortly after the story broke.[100] Moreover, during the visit, Shamir was repeatedly questioned by the press about the matter and about Israel's nuclear program, issues which had traditionally been treated as off-limits.[101] In addition, during late 1989, Israel's missile ties to South Africa augmented opposition to the sale of U.S. supercomputers to Jerusalem. Supercomputers can simulate nuclear explosions and ballistic missile reentry in order to reduce the need for costly and politically controversial full-scale tests.[102] Concerns that Israel might have transferred U.S.-origin missile technology to South Africa, and Israel's possible involvement in a scheme to smuggle advanced U.S. missile gyroscopes to that country, added to the controversy.[103]

Heavy water controversies. As discussed above, in 1959, Norway provided Israel with 22 tons (20 metric tons) of heavy water, essential for running the Dimona reactor. Israel had pledged to use the material only for peaceful purposes and had granted Norway the right to inspect the material to verify this pledge. Norway supplied an additional 1.1 tons (1 metric ton) of the material in 1970. Norway inspected its heavy water only once, however, in 1961, before the Dimona reactor began operating and took no steps to check further about how Dimona's spent fuel was being used.[104]

Although Norway acknowledged in 1979 that it had provided the heavy water to Israel, it was not until November 1986 that its right to inspect the material was publicly revealed as the result of an investigation by University of Wisconsin law professor Gary Milhollin.[105] In March 1987, the Norwegian government formally requested that Israel submit the material to IAEA inspection. Although Israel rejected the Norwegian initiative, it later acknowledged that the heavy water had in fact been used in the Dimona reactor, but insisted that it had not violated its agreement with Norway.[106]

In principle, to account for the heavy water, Israel would need to show how the material had been used since 1961. Since Israel has acknowledged that the material has been in the Dimona reactor for at least part of this time, the next steps would be to determine how much plutonium had been produced in the reactor during this period (by reconstructing operating records) and to verify that the plutonium was being used for peaceful purposes, as promised.

In mid-April 1988, negotiators from the two countries announced that they had reached a compromise agreement: Israel would permit inspection (initially to be conducted by Norway and subsequently by the IAEA) of 9 tons of heavy water—the amount that it claimed remained of the 22 tons Norway originally supplied.[107] This would have permitted Oslo to verify that in the future the material would not be misused. The arrangement, however, did not provide an account of how Norway's heavy water had been used over the years.

Because of this obvious flaw, the Norwegian parliament rejected the proposal in October 1988.[108] Discussions with Israel were renewed in April 1989. A year later the two sides announced that they had reached agreement in principle that Norway would repurchase for $2 million the quantity of Norwegian heavy water still in Israel's possession, approximately 11 tons.

This formula, however, as before, would leave Israel's prior use of the material unresolved.[109]

Political and strategic developments. With neither the conservative Likud Party nor the liberal Labor Party gaining a clear parliamentary majority in Israel's October 1988 elections, Israel continued to be governed by a fragile Labor-Likud coalition under Likud Prime Minister Yitzhak Shamir. This government fell in March 1990, but Shamir retained his office as the head of a new right-wing coalition government. This continuity of leadership and the more general continuity of Israeli nuclear policy over the past decade—which saw both Labor and Likud leaders pursue the development of the advanced nuclear weapons and new missile systems discussed above—suggest that the current Israeli government is unlikely to make dramatic shifts in Israel's military nuclear program.[110]

During 1989 and the first half of 1990, the security environment that has motivated Israel to acquire and expand its nuclear capabilities underwent important changes. On the positive side, two of Israel's most hostile Arab adversaries, Libya and Iraq, aligned themselves increasingly with the more moderate Arab states, and there were hints in early 1990 that Iraq might have been prepared to open discussions with Israel on ending the state of war between the two countries.[111] Similarly, the Palestine Liberation Organization (PLO), in a major reversal, recognized Israel's right to exist.[112] In November 1989, moreover, the Soviet Union advised Syria that it would not support the latter's bid to achieve "strategic parity" with Israel—a posture that implies the acquisition of threatening offensive capabilities—but would press Damascus to adopt the new Soviet strategic outlook based on "reasonable defensive sufficiency."[113] The new line, which Moscow began pressing on Syria in 1988, not only promises to reduce the threat to Israel from Syria itself, but also to lessen significantly the likelihood of Soviet support for possible Syrian aggression—a tendency which is further strengthened by Israel's improving bilateral ties with Moscow.

These positive trends were partially reversed in the spring of 1990, however, as Iraq became increasingly belligerent and as the PLO refused to distance itself from an attempted terrorist attack on a crowded Tel Aviv beach. Moreover, increasing evidence that Iraq had revitalized its program to develop nuclear weapons became a growing cause for concern for Israeli strategists. Though it is likely to be five to ten years before Iraq could hope to manufacture nuclear arms, this development potentially threatens Israel's regional nuclear monopoly.

Following the Israeli 1981 raid on the Osiraq reactor, then-Israeli Prime Minister Menachem Begin declared that Israel would never tolerate a second nuclear power in the region. However, it will be far more difficult for Israel to pre-empt Iraq's nuclear aspirations in the future than it was in 1981.

Iraq, for example, has reportedly taken precautions to locate key laboratories underground and disperse them to protect them from possible air strikes, and has also developed extensive air defenses—all of which would make it difficult for Israel to be confident that it could actually destroy Iraq's nuclear capability *in toto*.[114] Moreover, Iraq's extensive ballistic missile forces now give it the ability, which it lacked in 1981, to retaliate powerfully against Israeli territory in response to an attack on one of its nuclear installations. Thus Israel could not assume that if it mounted such an attack today, it would escape unscathed, as it did nine years ago. Indeed, in April, 1990, Iraqi President Saddam Hussein declared that he would destroy "half of Israel" with chemical weapons if it were to mount such an attack.[115] Nonetheless, Baghdad would have to calculate carefully before taking retaliatory action for fear that this, in turn, might provoke still more devastating counter-retaliation or trigger a wider conflict. Presumably military action would be a last resort for Israel, to be taken only after it had exhausted diplomatic efforts to slow Iraq's nuclear program—and possible covert operations, akin to the campaign of sabotage, intimidation, and assassination many believe Israel pursued in the late 1970s to stymie completion of the Osiraq reactor.[116]

Apart from Iraq's emerging nuclear potential, Jerusalem also confronts a threat from Iraq's growing offensive missile and chemical weapons capabilities and from those of Israel's other regional adversaries.[117] Syrian, Libyan, and Iranian chemical weapon capabilities, for example, grew substantially during the latter part of 1988 and 1989.[118] Several of Israel's antagonists are also pursuing the development of biological weapons.[119]

In recent years, Israel has become increasingly concerned about these growing capabilities, particularly following their extensive use in the Iran-Iraq War.[120] One scenario that is repeatedly cited by Israeli strategists, for example, is the possibility that Syria will arm its highly accurate Soviet SS-21 missiles with chemical warheads and launch a sudden attack on Israeli airfields, incapacitating them for perhaps 24 hours. With Israel denied air superiority, Syrian troops could then theoretically seize the Golan Heights, leaving Israeli forces the difficult task of retaking this ground, while Israel's leaders fended off the inevitable calls by outside parties for a ceasefire-in-place to avoid an escalation of the conflict.[121]

Although none of the missile systems currently being developed or deployed by Iran, Iraq, Libya, Saudi Arabia, or Egypt is likely to have the accuracy needed to hit military targets, they—and Syria's less accurate Scud-B missiles—might be used as weapons of terror against civilians, particularly if coupled with lethal chemical agents, which all of these states except Saudi Arabia can now produce in quantity. Without predicting which of these countries might participate in a future conflict with Israel, the dangers that confront the latter are obvious.

To meet these challenges, Israel is improving civil defense measures, working on strategies to destroy enemy missiles during an unfolding crisis before they can be used, and with U.S. assistance, developing the Arrow anti-tactical-ballistic-missile system. In the meantime, however, the country is relying increasingly, and more explicitly than ever before, on nuclear threats. Israel is believed by the United States to possess chemical arms and a number of ways to deliver them, including missiles.[122] These could well deter a wholesale chemical attack, but more obviously, Israel's adversaries would also have to fear that Israel might retaliate with nuclear arms, including those placed on the Improved Jericho. Since early 1988, Israeli leaders have been openly alluding to the country's potent retaliatory capabilities; Defense Minister Yitzhak Rabin, for example, has repeatedly invoked the threat that if any adversary dared to use chemical weapons against Israeli cities, it would be "clobbered 100 times harder, if not more," in return.[123] Prime Minister Yitzhak Shamir, Shimon Peres, and other senior Israeli leaders have made comparable statements. Although the word "nuclear" is never used, there can be little question—particularly given the wide publicity that the 1986 revelations of Israeli nuclear technician Mordechai Vanunu have received—that the invocation of Israel's ability to retaliate "one hundred times" over is intended as a reminder of the country's nuclear might.

Prospects

Vanunu's revelations—and the added details concerning them published by Frank Barnaby—together with the confirmation in authoritative U.S. government documents that Israel had deployed nuclear-capable missiles and nuclear warheads, have persuasively established that Israel possesses a potent nuclear force. Moreover, with its launch of the Offeq I and Offeq II satellites, Israel publicly demonstrated its considerable capabilities in the missile field, in effect partly removing the veil of ambiguity it had placed around its overall nuclear capability for some 20 years.

Israel's potential adversaries fully recognize Israel's nuclear might. Indeed, at the January 1989 Paris Conference on Chemical Weapons, a bloc of Arab states publicly opposed measures to slow the spread of chemical weapons because of Israel's continued possession of nuclear arms.[124]

The following month, during a visit to the Middle East, Soviet Foreign Minister Eduard Shevardnadze proposed that a ban on all nuclear and chemical weapons, to be verified by on-site inspections, be included as part of a future comprehensive peace plan for the region. Unfortunately, progress on such a peace plan between Israel and its Arab foes is at a standstill, and as of mid-1990, talks on the future status of the West Bank and Gaza Strip also appeared to be stymied. In this climate, Israel's decades-long commitment to acquiring nuclear arms, the vital role nuclear deterrence appears to play in Israeli strategic thinking, and the recent advances Israel has apparently made in its nuclear capabilities—surely at great economic cost—leave little doubt that its nuclear forces will remain a major feature of the Middle East security landscape for the foreseeable future.

As the chemical and missile capabilities of Israel's enemies grow and its cities become more vulnerable to attack, the importance of Israel's nuclear deterrent in the minds of Israeli planners will increase. If, as a result of the Iran-Iraq conflict, inhibitions against the use of chemical weapons and missiles against civilians are overcome, the risk that nuclear weapons might be used in the Middle East could be commensurately heightened. Indeed, the missile and chemical threat to Israel has now grown so serious that it is possible to imagine circumstances short of threatened national annihilation in which Israel might consider escalation to the nuclear level.

Similarly, as it progresses in the years ahead, Iraq's revitalized nuclear program will pose a security threat to Israel of increasing gravity. The prospect that Israel may choose to respond to this threat through preventive military action could greatly increase the risk of regional conflict and with it the possibility of open-ended escalation.

The combination of Israel's nuclear capability and the chemical, missile, and emerging nuclear threat to that country promises to be extremely dangerous. Dampening the risk that these capabilities might become the trigger for serious hostilities in the region must be an urgent priority for international diplomacy.

LEBANON
SYRIA
Golan Heights
Sea of Galilee
Haifa
Mediterranean Sea
Tel Aviv
Rehovot
WEST BANK
Amman
Nahal Soreq
Jerusalem
Dead Sea
Gaza Strip
ISRAEL
Beersheba
Dimona
JORDAN
Negev Desert
EGYPT
ISRAEL
phosphate/uranium mining area
0
50
MILES
Gulf of Aqaba

Israel

Research Reactors

IRR I Nahal Soreq (light-water/highly enriched uranium, 5MWt)[a]

- supplier: United States
- start up: 1960
- fuel source: United States, through 1977 (expiration of U.S. agreement).
- safeguards: yes

IRR 2 Dimona (heavy-water/natural-uranium, 40? 70? 150? MWt)[b,c]

- supplier: France
- start up: 1963
- fuel source: Israel, Argentina, South Africa(?), Belgium(?), France(?), Niger(?), Central African Republic(?), Gabon(?)[d]
- heavy water: Norway[a,m], Israel(?)[a], France [f]
- safeguards: no

Uranium Resources/Active Mining Sites/Uranium Mills

- reasonably assured reserves: 30-60,000 tons is available from processing phosphate ores.[a]
- currently active site: phosphate deposits in the Negev near Beersheba.[e]
- mills in operation: phosphoric acid plants producing yellowcake as by-product, two in Haifa, one in southern Israel. Combined output, 100 tons.[a]

Uranium Purification (UO_2)

Dimona[n]

- capacity: sufficient to supply Dimona reactor
- supplier: Israel
- start up: ?
- safeguards: no

Heavy Water Production

Rehovot

- capacity: pilot scale[a]; sufficient to replace Dimona reactor losses (presumed).
- supplier: Israel
- start up: 1954(?)[a]
- safeguards: no

Fuel Fabrication

Dimona[n]

- capacity: sufficient to supply Dimona reactor (presumed)
- supplier: France[f]
- start up: mid-1960s(?)
- safeguards: no

Reprocessing (Plutonium Extraction)

Dimona[c,f,l]

- capacity: Probably 15[f] - 40[c] kilograms of plutonium per year
- supplier: Israel and France[c,f,h]
- start up: probably 1966, full operation sometime later[c,f]
- safeguards: no

Nahal Soreq

- capacity: laboratory-scale[j]
- supplier: Great Britain; some U.S. equipment(?)[j]
- start up: 1960(?)
- safeguards: no[k]

Enrichment

Dimona

- type: centrifuge and laser[n]
- capacity: 2-3 kilograms highly enriched uranium per year(?)[g]
- supplier: Israel
- start up: 1979-1980 (centrifuge facility); 1981 (laser unit)[n]
- safeguards: no

Uranium Conversion (UF_6)

Dimona[n]

- capacity: Capacity sufficient to permit enrichment activities.
- supplier: Israel (?)
- start up: mid-1970s(?)
- safeguards: no

Tritium, Lithium Deuteride

Dimona (1983?-1985)

- capacity: 375 pounds of lithium-6 produced over two years, allowing production of sufficient tritium or lithium-deuteride for tens of advanced n-weapons.[n]
- supplier: Israel
- start up: Pilot lithium-6 unit, 1977; full-scale lithium-6 unit 1983 (?) (full-scale production 1984-1985).[n]
- safeguards: no

Sources and Notes

a. *Israeli Nuclear Armament*, Report of the Secretary General, United Nations General Assembly A/36/431, September 18, 1981, pp. 8-11, 15-18.

b. Pierre Péan, *Les Deux Bombes* (Paris: Fayard, 1982), p. 96 (75 megawatts); "The Middle East's Nuclear Arms Race," *Foreign Report*, August 13, 1980 (reactor upgraded from 26 megawatts to 70 megawatts in 1980). Traditional estimate of size is 24-26 megawatts; see reference a. Estimate of 150 megawatts given in reference c. U.S. officials believe the facility has operated at 40-70 megawatts (see discussion in text).

c. "Revealed: The Secrets of Israel's Nuclear Arsenal," *Sunday Times* (London), October 5, 1986.

d. John R. Redick, *The Military Potential of Latin American Nuclear Energy Programs* (Beverly Hills, CA: Sage Publications, 1972), p. 13 (Argentina); personal communication with informed former U.S. official (Argentina); Richard Kessler, "Argentine Officials Deny Rumors of Trade With Israel," *Nucleonics Week*, May 29, 1986, p. 6. Bertrand Goldschmidt, *Le Défi Atomique* (Paris: Fayard, 1980), pp. 205-206 (initially, uranium bought on world market from a number of sources, mainly Western and African); Fuad Jabber, *Israel and Nuclear Weapons* (London: Chatto and Windus, 1971), p. 89 (first load of Dimona reportedly comprised of 10 tons from South Africa, 10 tons from Dead Sea phosphates, 4 tons from French sources); Christopher Raj, "Israel and Nuclear Weapons," in K. Subrahmanyam, ed. *Nuclear Myths and Realities* (ABC: New Delhi, 1981), p. 105 (South Africa, France, French-controlled uranium mines in Gabon, Central African Republic, and Niger); William Drozdiak, "Uranium Loss is Admitted," *Washington Post*, May 3, 1977 (200 tons of yellowcake diverted on Mediterranean Sea from Belgium thought to wind up in Israel).

e. "Yellowcake from Phosphates," *Business Review and Economic News from Israel* (Tel Aviv), reprinted in *Joint Publication Research Service-Nuclear Development and Proliferation*, March 14, 1983, p. 28.

f. Péan, *Les Deux Bombes* (see note b).

g. Robert Gillette, "Uranium Enrichment: Rumors of Israeli Progress with Lasers," *Science,* No. 183, March 22, 1974, p. 1172; U.N., *Israeli Nuclear Armament* (see note a), p. 15. It is not certain whether Israel possesses an operating enrichment facility.

CIA officials were convinced that Israel diverted approximately 100 kilograms of highly enriched uranium from the NUMEC plant in Apollo, Pennsylvania, between 1964 and 1966. Transcript of "Near Armageddon: The Spread of Nuclear Weapons in the Middle East," ABC News Close-Up, April 27, 1981, pp. 13-14 (Statement by former CIA official Carl Duckett). No conclusive evidence that the material was diverted has come to light, however.

h. Steve Weissmann and Herbert Krosney, *The Islamic Bomb* (New York: Times Books, 1981), p. 113, citing Charles de Gaulle, *Memoirs of Hope: Renewal and Endeavor* (New York: Simon and Schuster, 1971), p. 266.

i. Fuad Jabber, *Israel and Nuclear Weapons* (London: Chatto and Windus, 1971), p. 45.

j. *Ibid.*, p. 43.

k. Not listed in 1985 IAEA Annual Report among safeguarded installations.

l. IAEA Bulletin, Vol. 19, No. 5, p. 2; Weissman and Krosney, *The Islamic Bomb* (see note h), p. 110.

m. Charles R. Babcock, "Norway Eyes Israel's Use of Nuclear Ingredient," *Washington Post,* November 10, 1986.

n. Frank Barnaby, *The Invisible Bomb* (London: I. B. Taurus, 1989), pp. 25-45.

10
Libya

The image of a nuclear-armed Libya has become a symbol of the dangers posed by unrestrained proliferation. Libya has made little progress toward this goal, in part because the country is the object of a *de facto* international embargo of nuclear transfers. Nevertheless, by his actions and statements over the years, Libyan leader Colonel Muammar Khadafi has made clear his interest in acquiring nuclear arms. In 1986 and 1987, moreover, he demonstrated his willingness to engage in "unconventional" warfare by using chemical weapons against Chadian forces. During 1989, Libya's efforts to complete a large-scale chemical weapons manufacturing complex at Rabta and to acquire long-range delivery systems triggered international concern. Although Tripoli also appears to have begun to moderate its foreign policy somewhat by easing tensions with a number of Arab countries, its continuing military build-up, history of international mischief-making, and longstanding interest in nuclear arms mean that its nuclear activities deserve to be monitored closely.

Background: Nuclear Ambitions Constrained

During the 1970s, after its attempt to purchase nuclear arms from China was rebuffed[1], Libya established a network of nuclear trade relations—with Pakistan, India, the Soviet Union, Belgium, Argentina, and Brazil—through which it hoped to gain rapid access to nuclear-weapons-related technologies and assistance in developing an indigenous nuclear infrastructure. By the mid-1980s, however, virtually all of these links had been severely curtailed or terminated.

Libya seems to have received few if any benefits from its reported 1973 accord with Pakistan, under which Tripoli is said to have agreed to finance

Pakistan's nuclear weapons effort in return for access to technology for producing nuclear weapons material or possibly nuclear weapons themselves.[2] Between 1978 and 1980, Libya is believed to have supplied Pakistan with significant quantities of uranium concentrate, or yellowcake, not subject to International Atomic Energy Agency (IAEA) monitoring. The material, originally purchased from Niger, was apparently intended for use in Pakistan's clandestine uranium-enrichment program. (See Chapter 7, Pakistan.) Although Libya ratified the Non-Proliferation Treaty (NPT) in 1975, it did not enter into a formal safeguards agreement with the IAEA until 1980 and thus was able to import and export uranium concentrate without having to list the transactions on an IAEA-monitored inventory.[3] Since 1980, however, all Libyan yellowcake purchases have been covered by the IAEA, depriving Tripoli of additional unrestricted material that it might barter for the bomb. Presumably Libya's pre-existing stock of yellowcake was disclosed to the agency when it established its initial inventory, although it is possible that Libya secretly set aside some of the material acquired before it concluded its safeguards agreement.[4]

Similarly, Libya appears to have garnered little from its open nuclear cooperation agreement with India, concluded during the tenure of Indian Prime Minister Moraji Desai. When he was in New Delhi in July 1978 to sign the accord, Libyan Prime Minister Abdul Salam Jalloud is said to have pressed Indian nuclear officials for a commitment to assist Libya in obtaining an "independent nuclear capability" —a veiled reference to plutonium extraction and nuclear explosives technology.[5] India, over the course of the following year, limited nuclear cooperation to strictly innocuous areas. In response, Libya abruptly terminated its oil shipments to New Delhi—expected to amount to one million tons per year—in an attempt to force India to share more sensitive nuclear technology. India refused and recalled its ambassador from Tripoli for consultation.[6]

In August 1984, Indian Prime Minister Indira Gandhi, who succeeded Desai, revealed in response to a parliamentary question, that she had "some reservations" about the Indo-Libyan accord, which had been negotiated by Desai without consulting senior Indian nuclear aides. Accordingly, she stated, she had confined activities to the exchange of scientists, before allowing the agreement to lapse. This foreclosed further Indo-Libyan cooperation of any kind.[7]

Tripoli's gains from its nuclear dealings with Moscow have also been modest. In 1975, the two signed a nuclear trade pact under which the Soviets agreed to provide a small research reactor and help Libya build an associated atomic research center.[8] The reactor, located at Tajoura, is believed to have begun operating in 1981.[9] In 1977 Tripoli and Moscow signed a second accord for the construction of two 440-megawatt nuclear-power reactors.[10] In early 1986, after years of inconclusive negotiations, the deal appeared to have been shelved indefinitely, only to be rekindled in May of that year. Nonetheless, it seems that the Soviet Union, like India, is wary of extensive nuclear trade with Colonel Khadafi, and no further progress on the nuclear power plant sale has been reported.[11]

The United States has not made any sales to Libya of nuclear equipment or materials. In 1983 Washington prohibited the training of Libyan nationals in nuclear sciences at U.S. universities and placed Libya on a list of states subject to especially strict nuclear export controls.[12]

During the early 1980s two Belgian firms, Belgatom and Belgonucléaire, provided Libya with extensive technical assistance on the Tajoura Research Center, the then-pending Soviet nuclear power reactor sale, and the development of a number of Libyan uranium deposits.[13] In 1982, Libya began negotiations with Belgonucléaire for the purchase of a plant for producing uranium tetrafluoride. The material itself is not usable for nuclear weapons or as nuclear fuel; however, both uranium metal, which can be irradiated to produce plutonium, and uranium hexafluoride, which can be enriched to produce highly enriched uranium, are made from uranium tetrafluoride.[14]

In 1984, Belgian nuclear officials announced that the two countries would shortly sign an agreement for nuclear cooperation. This would have provided the legal framework for the sale not only of the uranium tetrafluoride plant, but also for a $1 billion contract under which Belgium was to sell Libya non-nuclear equipment and architectural-engineering services for the twin Soviet nuclear power plants.[15] Concerned that there appeared to be no logical use for the tetrafluoride plant's output within Libya's nuclear energy and research program, Washington pressed Brussels to cancel the pending accord.[16] Belgium ultimately rejected the Libyan nuclear trade pact in early 1985.[17]

Argentina and Libya have also had a long-standing nuclear relationship, which Tripoli may have hoped to strengthen by its reported sale of arms, valued at $100 million, to Buenos Aires during the 1982 Falklands/Malvinas War.[18] In mid-1983 an Argentine nuclear trade delegation reportedly visited Tripoli, and at approximately the same time Libya tried unsuccessfully to purchase an Argentine research reactor.[19] There have been no reports of continuing Libyan-Argentine nuclear trade, however, since the civilian government of Raul Alfonsín took power in late 1983.

Nuclear relations with Brazil may run counter to this overall trend. In late 1984, Brazil began negotiations to supply Libya, a major purchaser of Brazilian armaments, with uranium prospecting and development services.[20] No further nuclear commerce between the two countries has been reported, but in early 1988 Libya is said to have agreed to purchase $2 billion worth of arms from Brazil, including intermediate-range ballistic missiles now under development.[21] No nuclear commodities are apparently involved, but Libya presumably could use such lucrative arms contracts as inducements to obtain such items in the future. Given Libya's long-standing interest in acquiring sensitive enrichment and reprocessing technology and Brazil's emerging capabilities in these fields, nuclear cooperation between the two could conceivably pose significant proliferation dangers in the years ahead.

In addition to these dealings with the various nuclear-supplier countries noted above, in 1981 Libya sought to obtain nuclear weapons on the black market from former CIA employee, Edwin Wilson. Wilson's offer was plainly a hoax; nevertheless, the seriousness with which Libya approached the negotiations is testimony to Khadafi's desire to acquire such arms.[22]

Nuclear goals described. Despite the obvious limitations on Libya's nuclear capabilities, Colonel Khadafi, in two major statements in 1987, strongly called for the Arab states to acquire nuclear arms—presumably under Libyan leadership. In a televised address to Libyan university students in June, he declared:

> . . . They make bombs daily; why should we not do likewise. The Arabs must possess the atomic bomb to defend themselves until their numbers reach one billion, until they learn to desalinate seawater, and until they liberate Palestine. We undertake not to drop the atomic bomb on any state around us, but we must possess it.
>
> If there is going to be a game using atomic bombs, then it should not be played against the Arab nation. The Arabs should have it, but we undertake not to drop it on anyone. However, if someone is going to drop one on us, or if someone is going to threaten our existence and independence even without the use of an atomic weapon, then we should drop it on them. This is an essential defensive weapon.
>
> We should be like the Chinese—poor and riding donkeys, but respected and possessing an atom bomb. Then they will not be able to fool anyone with this weapon.[23]

Khadafi reiterated his stand at a press conference in November 1987, at which he declared that he intended to boycott an Arab summit meeting to be held later that month in Amman, Jordan. Challenging the proposed agenda for that meeting, he stated:

> . . . I will work for attendance at an Arab summit, one whose agenda will include the fight against the United States; the boycott of the United States; the fight against France; the boycott of France; the liberation of Palestine; the creation of an Arab federation according to a charter I will present to them; and the manufacture of an Arab atomic bomb, since the Israelis, with the assistance of the United States, France, Western nations, and Britain possess it and target it against every Arab country. They are now working day and night to manufacture missiles with a range of 4,000 km that can strike every Arab city from the [Atlantic] [O]cean to the [Persian] Gulf. The Arabs thus have the right to manufacture nuclear weapons and to acquire the atomic bomb to defend their existence. After all, their enemy possesses this weapon, and atomic bombs are now found in the Middle East.[24]

At the time, Libya's prickly relations with other Arab countries and their lack of the necessary nuclear infrastructure for such an endeavor made it unlikely that they would soon take part in a collaborative nuclear weapons program with Libya, and no evidence of such joint efforts has since emerged.

Nonetheless, Khadafi's open calls for nuclear arms are disturbing and belie Libya's renunciation of nuclear weapons under the NPT.

Political and Technical Developments

Little nuclear progress. During 1989, Libya's nuclear advances appeared to be stalled, and there was virtually no publicly available evidence to indicate that the country was advancing a nuclear weapons program. Nonetheless, an October 1989 report in the respected trade publication *Nucleonics Week* indicated that during the mid-1980s, Libya sought to acquire equipment for producing weapons-usable nuclear materials. In a report reviewing Argentine nuclear export policy, the weekly quoted the new head of the Argentine nuclear program, Manuel Mondino, as stating that "several years ago" the Italian subsidiary of an Argentine firm agreed to sell Libya a small-scale plutonium extraction unit, or "hot cell." Mondino stated that the Argentine President Raul Alfonsín "had to personally intervene," with the parent company in Argentina to halt the sale.[25] How Libya intended to obtain plutonium-bearing spent nuclear fuel to process in the unit is not known; the country's only current source of irradiated material is the small Soviet-supplied research reactor at Tajoura, which is subject to International Atomic Energy Agency inspection.

While Tripoli has made little headway in achieving its nuclear ambitions, it has rapidly advanced its chemical weapons production capabilities and is working assiduously to develop longer range systems that could deliver chemical ordnance to states as far away as Israel or southern Europe. These military advances are adding to the unconventional threat facing Israel.

Chemical weapons. In late 1988, it was disclosed that Libya was building one of the world's largest chemical weapons production plants near the town of Rabta and was relying heavily on assistance from a number of West German firms.[26] Although the outcry in the United States and Western Europe generated by these disclosures ultimately led Bonn to prosecute a number of the West German concerns involved and to propose a significant tightening of its export controls, Libya apparently continued construction of the facility. According to one report, the plant was nearing completion in late 1989, and full-scale production was planned for mid-1990.[27] It was expected that the facility would be able to produce 40 to 45 tons per day of various chemical agents, including mustard gas and the nerve gasses Sarin and Tabun. A fire at the facility on March 14, 1990, was initially thought to have destroyed it. Subsequently, however, U.S. officials concluded that the event was a ruse by Khadafi intended to deflect international criticism and avert the possibility of a military attack on the installation.[28]

Libya's demonstrated willingness to use chemical weapons (against Chadian forces in late 1986 and again in the summer of 1987) has contributed to international concerns over the plant.[29] Reportedly, Libya obtained the agents used in that conflict from Iran, in exchange for Soviet-made mines or, possibly, for Soviet-origin Scud-B missiles.[30] By early 1988, however, Libya was thought to have acquired its own chemical weapons manufacturing capability and, presumably, had begun work on the Rabta plant. The massive capabilities it is acquiring in this field, as seen in the construction of the Rabta facility, will pose a considerable threat to neighboring states and will undoubtedly stimulate the further proliferation of such arms in the region. In addition, given Libya's past support for international terrorism, there is considerable concern that it might supply chemical arms to terrorist groups.

As in the case of Iraqi and Iranian use of chemical arms in the Gulf War, Libya's resort to these weapons violates the 1925 Geneva Protocol, which Libya ratified in 1971—after Khadafi took power. Khadafi's readiness to disregard this arms control pact to which he, himself, committed his country, raises further questions about his commitment to the NPT, which Libya also ratified under his leadership.

Ballistic missiles. For many years, the Libyan arsenal has included the 190-mile range Scud-B missile, obtained from the Soviet Union, and, according to the U.S. Arms Control and Disarmament Agency, Tripoli may also possess the far more accurate, 75-mile range Soviet SS-21.[31] In addition, since the late 1970s, Libya has pursued its own missile development program. According to congressional testimony by then-U.S. Secretary of Defense Frank Carlucci:

> Libya has been attempting to establish its own ballistic missile development capability, and has been receiving assistance from German-owned firms, including OTRAG, which has built missile facilities in Libya. . . . They have established a secret missile test range in the Libyan desert in Tauwlwa, where work has focused on development of a 500-kilometer-range [300-mile] ballistic missile.[32]

During 1989, reports in the British press indicated that Libya was attempting to develop two indigenous missiles, the Ittisallat or Ittisalt, with a range of 50 miles, and the Al-Fatah (the "Conqueror")—apparently the system referred to by Secretary Carlucci—with a range of over 200 miles.[33] These reports also cited the involvement of West German firms and specialists in the project. However, the 1988 annual report of world-wide military expenditures and arms transfers released by the U.S. Arms Control and Disarmament Agency in mid-1989, lists Libya as developing only one missile, a 300-mile range system of "West German design," which, the report states, is still in the research and development stage.[34]

Given their range, it would appear that the missiles currently under development in Libya will pose a threat to other North African states, including Egypt and Chad, and possibly to NATO military installations in Sicily—but not to Israel.[35] In early 1988, however, Libya is said to have entered into a $2 billion arms agreement with Brazil, which included the purchase of Brazilian longer-range missile systems.[36] Brazil's longest range system, the MB/EE-1000, which is still at the research and development stage, would have a range of approximately 625 miles and would expose Israel to a missile attack from the easternmost parts of Libya.

Libya is also acquiring long-range strike aircraft, including the Soviet Sukhoi-24D "Fencer" advanced bomber, and aerial refueling capabilities.[37] Such systems could also permit Libya to attack Israel, as well as other states in the region. Despite Israeli concerns, senior U.S. officials believe Israeli air defenses would have little difficulty in meeting this challenge.[38]

Changing strategic environment. In a move that contrasted sharply with the threatening build-up of Libya's military forces, Colonel Khadafi launched a series of initiatives in 1989 that were aimed at improving relations with neighboring states and easing his country's international isolation. In February, Libya joined Algeria, Mauritania, Morocco, and Tunisia in the establishment of the Arab Maghreb Union, a grouping aimed at forging the Maghreb states into a coherent economic and political bloc.[39]

Similarly, at the Arab League summit in May, Khadafi began a process of rapprochement with Egypt that shortly led to the reopening of the Egyptian-Libyan border. That border had been closed in 1977, following a brief frontier war that left relations between the two countries tense for over a decade. In mid-October 1989, Colonel Khadafi held a bilateral summit with Egyptian President Hosni Mubarak, and the following month, the two countries signed a joint protocol on industrial cooperation. On a separate front, in August 1989, Libya entered into a peace agreement with Chad, ending the 16-year war between the two countries.

Tripoli even sought to ease tensions with the United States in 1989; these tensions had intensified in past years following Washington's condemnation of Libya's military adventurism and support of international terrorism. The Reagan Administration had imposed anti-terrorism sanctions against Libya in January 1986, prohibiting U.S. companies from trading with the country and U.S. nationals from traveling there. U.S.-Libyan relations reached their lowest point in mid-April of that year, when the United States conducted air strikes against a number of targets in Libya in retaliation for Libyan involvement in a terrorist bombing of a Berlin discotheque frequented by American servicemen. Late 1988 and early 1989 saw a major diplomatic confrontation between the two states, as the Reagan Administration's efforts to halt Western assistance for the Rabta chemical weapons plant became an international *cause célèbre*. In January 1989, as this crisis unfolded in the last days of the Reagan Administration, the United States also shot down two Libyan jets over the Mediterranean after they appeared to threaten several carrier-based U.S. aircraft on routine patrol. Nonetheless, by June, prospects for improved relations seemed to brighten, as Khadafi referred to the new Bush Administration as "wise and mature" and several months later expressed the hope that

Libyan-American relations could be conducted "on the basis of mutual respect and common interest."[40] In the fall of 1989, Khadafi also announced that Libya would begin reducing support for groups associated with international terrorism, a step essential for a thaw with Washington. Some U.S. officials greeted the announcement with skepticism, however.[41] (Though Washington remained cool to Khadafi's new moderate posture, Italy, Belgium, and France had eased sanctions against Tripoli by the end of the year.[42])

Prospects

By mid-1990, it was difficult to harmonize these diplomatic initiatives with Colonel Khadafi's construction of the massive Rabta chemical weapons plant and his efforts to acquire long-range delivery systems. Although improving relations with various regional states will reduce the likelihood that Libya might actually use its new armaments against them, Libya's neighbors undoubtedly perceive the country's growing military might as a serious threat—particularly given the past unpredictability of Libyan foreign policy. As a result, Libya's expanding chemical warfare and missile capabilities will add to pressures for the further proliferation of these military systems in the region. Israel, moreover, must be concerned that as Khadafi consolidates his diplomatic base in North Africa he may attempt to make mischief further to the east.

For the foreseeable future, nuclear weapons are likely to remain beyond Libya's grasp. Its indigenous capabilities are too rudimentary to permit the production of weapons-grade nuclear materials, and the embargo that the major nuclear suppliers have placed on nuclear transfers to Libya is likely to prevent it from making further advances for some time. Moreover, the seeming improvement in Libya's external relations, together with the country's newly acquired chemical warfare and delivery capabilities, would appear to reduce the need for Libya to acquire nuclear arms for reasons of national security.

Nonetheless, Colonel Khadafi's nuclear ambitions have been well documented, and the construction of the Rabta plant demonstrated his ability to obtain sensitive military technology clandestinely from Western Europe—a skill that might also be applied in the nuclear area. Moreover, the growing ability of emerging nuclear supplier states, such as Argentina, Brazil, China, and North Korea, to provide weapons-relevant nuclear technology could offer new avenues for Khadafi to pursue his nuclear objectives. As noted above, Libya already has significant military trade links to one such supplier, Brazil. These factors suggest that while nuclear-weapons related activities in Libya appear to be at a standstill, the country's nuclear program deserves to be monitored closely.

As if to confirm these fears, in a mid-April 1990 television address, Khadafi called for the development of a multifaceted Libyan deterrent force that would include a nuclear component. Referring to the April 1986 U.S. air strike against targets in Libya, Khadafi declared:

> If we had possessed a deterrent—missiles that could reach New York—we would have hit it at the same moment. Consequently, we should build this force so that they and others will no longer think about an attack. Whether regarding Libya or the Arab homeland, in the coming twenty years this revolution should achieve a unified Arab nation. . . . This should be one homeland, the whole of it, possessing missiles and even nuclear bombs. Regarding reciprocal treatment, the world has a nuclear bomb, we should have a nuclear bomb.[43]

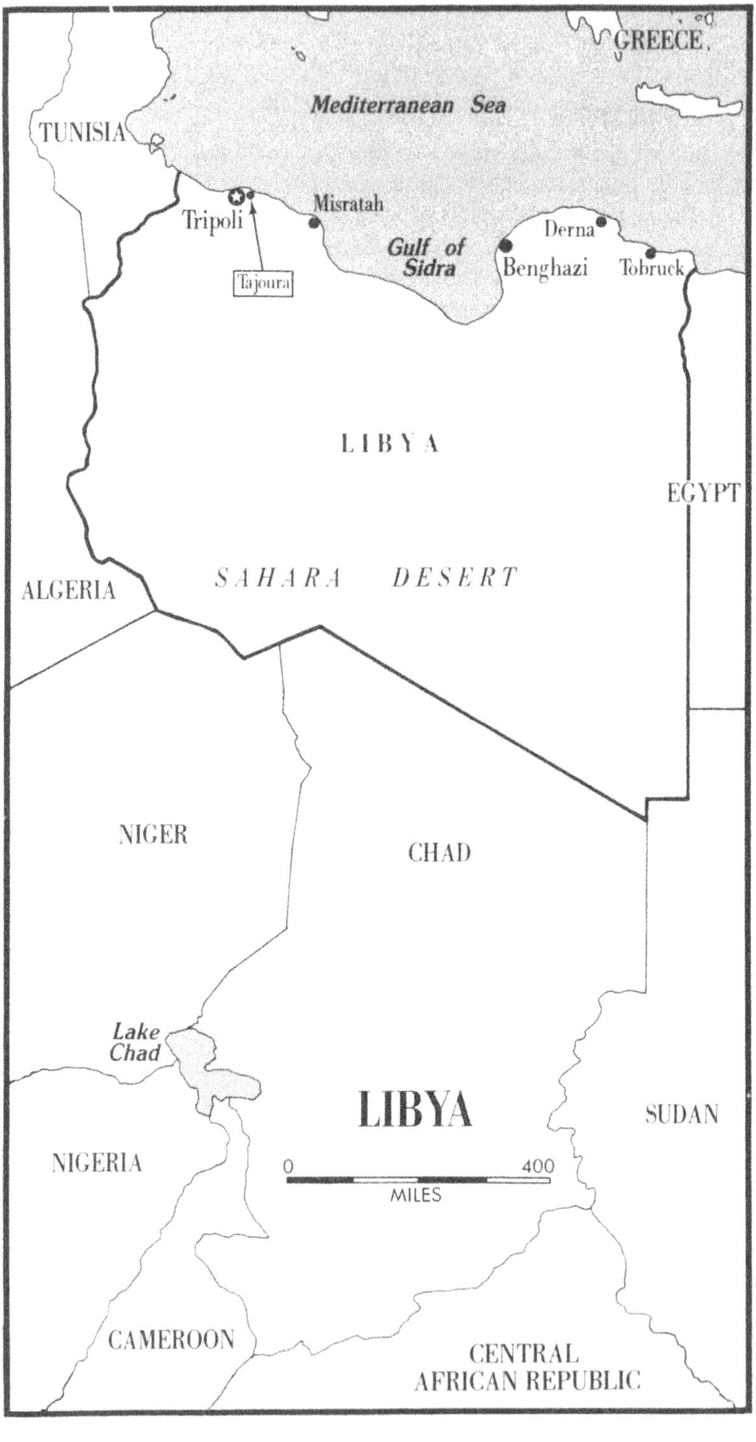
GREECE
Mediterranean Sea
TUNISIA
Tripoli
Misratah
Derna
Gulf of Sidra
Benghazi
Tobruck
Tajoura
LIBYA
EGYPT
ALGERIA
SAHARA DESERT
NIGER
CHAD
Lake Chad
LIBYA
SUDAN
NIGERIA
0
400
MILES
CAMEROON
CENTRAL AFRICAN REPUBLIC

Libya

Research Reactor

Tajoura (light-water/highly-enriched uranium, 10 MWt)[a]

- supplier: Soviet Union[a]
- fuel source: Soviet Union[a]
- start up 1981[b]
- safeguards: yes[c]

Power Reactor Planned

Gulf of Sidra (light-water/low-enriched uranium, 440 MWe)[d]

- supplier: Atomenergoexport (USSR)[e]
- fuel source: Soviet Union?
- start up: (?)
- safeguards: yes

Uranium Resources/Active Mining Sites/Uranium Mills

- reasonably assured reserves: none assessed, some known past exploration.[f]
- currently active sites: none
- mills in operation: none

Sources and Notes

a. Senate Foreign Relations committee, *Hearings on the Israeli Air Strike,* June 18, 19, and 25, 1981, 97th Congress, 1st Session (Washington, D.C.: Government Printing Office, 1981) p. 40.

b. Claudia Wright, "Libya's Nuclear Program," *The Middle East,* February 1982, p. 47; *Hearings on the Israeli Air Strike,* p. 50.

c. International Atomic Energy Agency, *The Annual Report, 1985,* (Vienna, Austria: International Atomic Energy Agency, 1986), p. 81.

d. Claudia Wright, "Libya's Nuclear Program," p. 50; "Envoy Says Soviets May Build Nuclear Plant in Libya," Associated Press, May 10, 1986, AM cycle.

e. Robin Miller, "Nuclear Power Plans Outlined," *Jamahiriyah Review, Joint Publication Research Service/Nuclear Development and Proliferation,* March 22, 1982, p. 17.

f. OECD Nuclear Energy Agency and the International Atomic Energy Agency, *Uranium Resources, Production, and Demand* (Paris: OECD, 1983), p. 208. Libya reportedly purchased several hundred tons of uranium concentrate from Niger between 1978 and 1980. Steve Weissman and Herbert Krosney, *The Islamic Bomb,* (New York: Times Books, 1981), p. 210.

11
Iraq

Although Iraq's victory over Iran in the eight-year (September 1980-August 1988) Persian Gulf War brought it few concrete benefits, the conflict left Iraq as the strongest military power in the region. During the war, Iraq initiated the use of chemical weapons and surface-to-surface missiles and built up a substantial production capability for both types of weapons. In contrast, Iraq's nuclear program remained dormant through mid-1988, stymied by Israel's 1981 bombing raid which destroyed the program's centerpiece, the large French-supplied Osiraq research reactor outside Baghdad.

Sometime after the cessation of hostilities with Iran in August 1988, Iraq is believed to have renewed its quest for nuclear weapons, apparently initiating efforts to acquire a uranium enrichment capability. Given this revived interest in nuclear arms, Iraq's continued possession of 27.5 pounds (12.5 kilograms) of International Atomic Energy Agency-inspected highly enriched uranium fuel—possibly enough for one carefully designed nuclear weapon—presents an additional proliferation threat. Despite Iraq's status as a party to the Non-Proliferation Treaty (NPT), its nuclear activities thus deserve to be followed closely.

Background

Iraq's nuclear program. There is a growing consensus, based on strong circumstantial evidence, that the hidden objective behind Iraq's expansion of its nuclear sector in the mid-1970s was the gradual acquisition of a nuclear weapons capability. This capability was presumably intended to advance three primary objectives: to counter Israel's existing, but unacknowledged nuclear arsenal; to advance Iraqi president Saddam Hussein's claim to Arab leadership; and to hedge against Iran's nuclear potential.[1] (New evidence also indicates that Iraq first sought to acquire a chemical weapons capability during the mid-1970s.[2])

The first indicator of Iraq's intentions was that the Osiraq research reactor, purchased from France in 1976, was unusually large and was therefore capable of irradiating uranium specimens to produce significant quantities of plutonium. Iraq had simultaneously acquired from Italy three radiologically shielded "hot cells," in which at least small quantities of plutonium could be extracted from irradiated uranium specimens—perhaps adequate for a single weapon over a period of years—and a model of a larger plutonium extraction, or "reprocessing" unit that might have permitted a subsequent expansion. In 1981 Baghdad was also negotiating with Italian vendors to buy a larger heavy-water power reactor (an efficient plutonium producer), and a sizeable reprocessing facility; these purchases would have been highly premature additions to Baghdad's nascent nuclear energy program, but an essential part of a nuclear weapons effort.[3] All of these negotiations were conducted in secret, or at least outside of public view.[4]

In 1980 and 1981, Iraq also bought inexplicably large amounts of natural uranium from Brazil, Portugal, Niger, and Italy, suggesting a secret intention to irradiate the material in Osiraq and/or in a subsequently purchased heavy water reactor to produce plutonium. (The natural uranium could not be used to fuel Osiraq because the reactor required enriched uranium fuel.) A number of statements by Iraqi officials, moreover, strongly suggested that Baghdad had a deep interest in acquiring nuclear arms.[5]

Perhaps most revealing of Iraq's intentions to use Osiraq to produce plutonium was that in early 1980 Baghdad placed an order with a West German firm, NUKEM, for 25,000 pounds (11,364 kilograms) of depleted-uranium-metal fuel pins.[6] The pins were sized to fit into Osiraq, where they could be irradiated to yield plutonium, and were unsuited to any other purpose—including use in a lab-scale "subcritical assembly," which was the stated end-use for the NUKEM pins. The deal, which could have netted Iraq one or two bombs' worth of plutonium, was ultimately aborted when NUKEM's subcontractors in the United States and Canada were told by regulatory officials that export licenses would not be issued for the material.

Whatever its covert intentions, Iraq, as a party to the NPT, had to place all of its nuclear installations under International Atomic Energy Agency (IAEA) inspection, and as Osiraq neared completion, the agency was preparing plans to apply stringent controls to the facility.[7] In 1978, Iraq had also agreed to permit French technicians to remain at the Osiraq plant until 1989; so it seems highly unlikely that by the time of the Israeli raid in 1981 Osiraq could have secretly produced plutonium in the quantities needed for a clandestine nuclear weapons program. It is more likely that Iraq intended to pursue a plan, similar to Sweden's in the 1960s, of mastering plutonium-production technology in the open and building a nuclear infrastructure, perhaps with a stockpile of plutonium, which could later be turned into nuclear armaments, if such a decision were taken.[8] (The NPT allows parties to withdraw from the pact on ninety days' notice if their "supreme interests" are threatened.)

When it appeared that the Osiraq reactor shortly would go into operation, despite a series of setbacks and intensive Israeli lobbying for the imposition of additional controls by Iraq's European suppliers, Israel mounted its highly controversial June 7, 1981, air attack.[9] Planning for the raid had apparently begun some eighteen months earlier, and its success is said to have depended on the availability of U.S.-supplied F-16 aircraft and access to detailed U.S. reconnaissance satellite photos.[10]

Caught up in its deadly war with Iran and lacking warplanes or missiles with sufficient range to reach Israel, Iraq was unable to respond militarily to the attack. Baghdad, however, subsequently launched an aggressive diplomatic campaign against Israel in the United Nations and the IAEA, seeking to impose sanctions against Jerusalem for its refusal to rule out future military action against nuclear installations in the Middle East. When the 1982 IAEA General Conference, at Iraqi instigation, refused to accept Israel's credentials, the United States temporarily withdrew from the agency. This precipitated a crisis that was not finally resolved until the 1985 General Conference, when Israel's assurance that it would not "attack or threaten to attack any nuclear facilities devoted to peaceful purposes either in the Middle East or anywhere else" was accepted as a basis for its continued participation.[11]

After Osiraq was destroyed, it appears that the Hussein government pursued a separate track for obtaining a nuclear weapons capability. According to evidence obtained in a 1984 Italian prosecution, senior Iraqi military figures expressed interest in obtaining 74.6 pounds (33.9 kilograms) of plutonium—enough for several weapons—from an Italian arms smuggling ring purporting to have such material for sale.[12] One member of the ring, interviewed in 1985, claimed that he met with members of the Iraqi military, first in Baghdad and then in Rome, to discuss the sale; 1982 telexes between members of the smuggling ring contained in the prosecution file are consistent with this claim.[13] The deal fell through when, after a third meeting in Baghdad, the smugglers were unable to produce samples of the nuclear material. Although it is virtually certain that the plutonium offer was a hoax, possibly intended as a come-on to pave the way for sales of conventional arms, the episode suggests that through mid-1982, when negotiations on the matter ended, at least some in the Iraqi government remained interested in nuclear arming.

Strategic environment. During the late 1970s, as it began to build its nuclear infrastructure—and apparently made its first attempts to acquire a chemical weapon capability—Iraq perceived nuclear-armed Israel to be its principal antagonist. Following the rise to power of Ayatollah Ruhollah Khomeini in 1979, however, Iran soon displaced Israel in this role. In September 1980, hoping to capitalize on the turmoil caused by the Iranian Revolution, Iraq invaded its eastern neighbor and seized the Shatt al Arab waterway separating the two states, precipitating a bitter eight-year conflict.

Iraq lost much of the territory it seized after Iranian counter-offensives in 1981 and, by 1982, was thrown on the defensive, as the Khomeini government raised the stakes by demanding the end of the Saddam Hussein regime as a condition for peace. In February 1984, an Iranian invasion threatened Iraq's second largest city, Basra, but Iraq was able to hold back Iran's forces, in part by using chemical weapons.[14]

Iraq lost additional territory to major Iranian offensives in 1985, 1986, and 1987, but held fast at Basra, inflicting heavy casualties. During this period, Iraq was able to increase its military capabilities substantially, bolstered by massive arms transfers from the Soviet Union and economic aid from Saudi Arabia, Kuwait, and other Arab states in the Persian Gulf. Indeed, throughout the conflict, Iraq maintained unquestioned air superiority, its air force outnumbering Iran's by a margin of six to one.[15]

As the war dragged on, Iraq used chemical weapons with increasing ferocity. According to Iranian reports, after first using chemical weapons in 1983, Iraq subsequently employed them on 195 additional occasions before hostilities ceased, with the number of chemical attacks increasing during 1987 and 1988.[16] Moreover, although many such attacks were directed predominantly against Iranian forces, by early 1988 Iraq was also using chemical agents against civilian targets.

In a particularly infamous March 1988 episode in the town of Halabja, Iraqi forces killed hundreds, and possibly thousands, of Iraqi Kurdish civilians with chemical weapons, apparently in retaliation for their having sided with Iran when the latter gained control of the area during an earlier phase of the conflict.[17] In September 1988, when Iraq's Kurds openly revolted against the Hussein regime in a bid for independence, Iraqi forces again used chemical weapons extensively in the ensuing campaign against the rebels, causing widespread civilian casualties according to many reports.[18] Iran responded in the mid-1980s by producing its own chemical weapons and later using them in retaliation against Iraqi troops.[19] The use of such weapons by both states violated their respective commitments under the 1925 Geneva Protocol.[20]

A number of United Nations investigations confirmed that Iraq had used these agents and, in 1988, the United States repeatedly condemned this conduct in a series of strong public statements.[21] Moreover, although Iraq issued numerous official denials to the charges that it had used chemical arms, in mid-1988, Iraqi Foreign Minister Tariq Aziz ultimately acknowledged in interviews in West Germany and Brazil that the country had, indeed, employed these weapons.[22] Nonetheless, neither the United States, the Soviet Union (Iraq's long-time ally), nor other members of the international community sought to impose sanctions against Iraq for its actions, largely for fear that this might work to the advantage of Iran, whose ascendency in the region was opposed by Washington and the Arab states of the Persian Gulf.[23]

Iraq also used surface-to-surface missile attacks against Iranian cities with increasing effectiveness. Both Iran and Iraq had engaged in such attacks in earlier stages of the war, Iraq using Soviet-supplied 40-mile FROG and 190-mile Scud-B rockets. In early 1988, however, Iraq introduced the 375-mile Al-Husayn—a version of the Scud-B that Iraq had modified to increase its range. The new missile enabled Iraq to strike Tehran for the first time, with devastating effect on Iranian morale.[24] Several months later, Iraq mounted a series of offensives that routed Iranian troops from Iraqi soil and made limited inroads into Iranian territory. On July 18, 1988, Iran accepted a U.N.-sponsored ceasefire and hostilities ended the following month.

Attacks on Iranian reactor. During the war, with its own nuclear prospects temporarily stymied by Israel's 1981 attack on the Osiraq reactor, Iraq apparently became concerned about the possibility of Iranian advances in this field. According to a prominent civilian defense advisor to the Iraqi government, these misgivings led to Iraq's repeated attacks on Iran's Bushehr nuclear power reactor construction site, although Baghdad, no doubt, also considered this to be a lucrative economic target.[25] Iraq bombed the Bushehr site seven times between March 1984 and July 1988, causing considerable damage to the two partially completed nuclear installations there and killing a total of thirteen individuals, according to Iranian reports.[26] The last attack occurred the day after Iran accepted the U.N.-sponsored ceasefire in the Gulf War, but before hostilities had ended. According to the Iraqi defense strategist cited above, Iraq sought to take advantage of the "cover of war" to thwart Iran's nuclear capabilities, recognizing that once the conflict ended, preventive attacks on the nuclear site would carry far greater political costs and could potentially reignite the war between the two states. (The attacks and their aftermath are discussed in greater detail in Chapter 12 on Iran.)

Political and Technical Developments

Changing security environment. By the war's end, Iraq had become a potent military power. It possessed a highly trained, battle-hardened army, with more main battle tanks than Britain and France combined. Iraq's air force, similarly, included the advanced Soviet MiG-29, and had gained extensive combat experience, which included the prosecution of numerous long-range operations. Nevertheless, with its vast superiority over its Iranian counterpart, Iraq's air arm never confronted a serious challenge of the type it might expect from Israel, or even Syria.

Iraq had also strengthened its diplomatic position during the conflict, solidifying its ties with the Soviet Union and the Arab Gulf states and bolstering its links to Western Europe. Moreover, relations with the United States had improved substantially. Washington had reopened full diplomatic relations with Baghdad in 1984 and, though officially neutral, had helped the Iraqi cause during 1987 and 1988 by protecting Saudi and Kuwaiti shipping in the Persian Gulf from Iranian attacks.

Despite its current advantage over Iran, Baghdad must be concerned that, over the longer term, its larger and more populous eastern neighbor will reverse the current bilateral military imbalance. Baghdad must also take account of other regional antagonists. It remains technically in a state of war with Israel, a state which is known to possess nuclear weapons. Indeed, at the January 1989 Paris Conference on Chemical Weapons, Iraq rejected calls for a ban on chemical arms, arguing that it would only accept such restrictions if accompanied by comparable restrictions on Israel's nuclear capability.[27] Israeli strategists have expressed considerable concern over Iraq's growing military capabilities, raising the possibility that Israel might contemplate another preventive attack against sensitive Iraqi military installations, such as chemical weapons or missile production facilities. Longstanding Iraqi frictions with Syria, which sided with Iran during the Iran-Iraq War, are another potential source of conflict, and, in 1989, led Iraq to increase support for anti-Syrian elements in Lebanon.

Apparently with these challenges in mind, Iraq has continued to build up its military strength. Among other measures, Baghdad has acquired an airborne early warning capability; continued work on a variety of missiles, some with a range of 1,250 miles; sought to develop an anti-missile missile; expanded its chemical weapons production capabilities; and pursued the development of biological weapons.[28]

Renewed nuclear weapons effort. By early 1989, it appeared that as part of this overall trend, Iraq had revitalized its efforts to acquire nuclear arms. In testimony before a congressional committee in February 1989, Director of Naval Intelligence Rear Admiral Thomas A. Brooks declared that Baghdad was "actively pursuing" a nuclear weapons program, though he provided no details.[29] Press reports quoting Israeli and U.S. sources, however, stated that Iraq had been working to develop a warhead for an intermediate range missile, possibly the 600-mile Condor II or the 1,250-mile Tammuz-1, discussed below.[30] One account suggested that Baghdad might be able to produce nuclear arms within two to five years,[31] but U.S. government sources have estimated that an Iraqi nuclear device is at least five to ten years distant, and Israeli Defense Minister Yitzhak Rabin offered a similar estimate in May 1989.[32] Iraq has denied the allegations that it has launched a nuclear weapons program.[33]

Nonetheless, it appears that Iraq is indeed attempting to acquire hardware for a facility to enrich uranium to weapons grade.

- In February, 1989, the U.S. Commerce Department blocked an attempt by Iraq to obtain specialized vacuum pumps. The units are used for a number of non-nuclear purposes, but are also potentially useful in uranium enrichment plants, according to U.S. officials.[34]

- Iraq has also attempted to obtain "maraging steel-350", a particularly strong, high-nickel-content steel, whose principal industrial use is for the manufacture of uranium enrichment centrifuges.[35]
- In December 1989, the West German weekly *Der Speigel* reported that West German authorities were investigating allegations that between 1987 and 1989, the firm of H. and H. Metalform, Ltd., of Drensteinfurt, had supplied Iraq with special rolling machines or lathes for manufacturing enrichment centrifuges. The report stated that a centrifuge manufacturing facility had been built at Tuwaitha, the site of the Osiraq reactor, and that Walter Busse, a former employee of MAN Technologies, Ltd., had arranged the sale of the H. and H. equipment and provided additional technical assistance. In 1979, MAN had built a uranium-enrichment-centrifuge production plant at Gronau, FRG, for URENCO, the West German-British-Dutch uranium enrichment consortium.[36] Presumably, the Iraqi centrifuge effort is based on that facility.
- Finally, also in late 1989, a newsletter published by the *Financial Times* reported that Iraq had been attempting to purchase in Western Europe special magnets made of samarium cobalt that are used to hold enrichment centrifuges in place as they rotate at high speed.[37] The report stated that Iraq had not been able to obtain the items but that Baghdad was now attempting to manufacture the magnets itself and was receiving help from the People's Republic of China. The report also alleged that Iraq has been attempting to obtain scoops used at the top and bottom of the units to collect enriched and depleted product. The report stated that Iraq was conducting work on centrifuge enrichment not only at Tuwaitha, but also at the Saad-16 facility in Mosul, where the country also conducts missile research.[38]

As of late 1989, Iraq was not known to have begun construction of a uranium enrichment plant, nor was it clear how long it might take Baghdad to complete such a facility once work commenced. Pakistan required more than ten years to design and build the Kahuta enrichment plant, which is the backbone of its nuclear weapons capability.

Nuclear weapon triggers. On March 28, 1990, five persons were arrested in London for attempting to export to Iraq military-grade capacitors used to trigger nuclear weapons.[39] The arrests were the result of an 18-month "sting" operation conducted by U.S. Customs Service officials. The operation was initiated when the U.S. manufacturer of the capacitors notified Customs that Iraq was attempting the purchase through a London-based front company, Euromac. On March 29, 1990, Euromac, a related company, two Euromac officials, and three employees of Iraq's Al-Qaaqaa military research and development institute were indicted in San Diego for attempting to export the capacitors without obtaining the necessary export license from the State Department's Office of Munitions Control.[40]

Iraqi officials have not denied that Euromac was attempting to acquire the capacitors for use in Iraq, but they have claimed the units were intended for use in laboratory-based laser research.[41] The "ruggedized" military specifications of the capacitors belie this claim, however. Moreover, the U.S. Customs officials, apparently relying on statements to U.S. undercover agents by various defendants in the case and related evidence, have insisted that the units were intended for use in nuclear weapons.[42]

It is not clear why Iraq might be seeking components for nuclear weapons when it remains so far from completing the facilities needed to produce the nuclear materials that would be essential for the core of a nuclear device. Possibly, like other would-be nuclear states, Iraq has divided responsibility for various parts of its weapons program among several teams, one of which is responsible for designing and fabricating the weapon, while another is in charge of producing nuclear weapons material. Conceivably the design team has progressed to the point that it wants to build and test the non-nuclear components that would be used to detonate a device, and for this purpose sought to obtain the actual electronic hardware that would be used in a real nuclear weapon.

Example of chemical weapons, missile production. Like Pakistan, Iraq appears to be using clandestine purchasing networks in the West to obtain nèeded technology and equipment. Iraq, however, has had considerable experience in using such covert mechanisms to support its chemical weapon and ballistic missile programs, and these techniques and networks could well permit Baghdad to accelerate its nuclear plans.

Indeed, by the end of the Iran-Iraq War, Iraq had become nearly self-sufficient in the manufacture of certain chemical agents and was reported to have built five major chemical weapons plants.[43] Its missile production program included the Saad-16 facility at Mosul for missile research, development, and static testing; a plant at Fallujah for modifying Soviet-supplied Scud missiles to make the Al-Husayn and the Al-Abbas missiles (the latter having the same range as the Al-Husayn, but with a heavier payload) used against Tehran; and three additional plants for the manufacture and static testing of a new missile, possibly the 600-mile Condor II, which, until mid-1989, was being pursued as a joint project with Argentina and Egypt, or the Tammuz-1.[44] In early December 1989, Iraq conducted a partial flight test of a satellite launching rocket and simultaneously announced that it had developed two new surface-to-surface missiles with a range of 1,250 miles (2,000 kilometers).[45] The developments took Western observers by surprise and demonstrated Baghdad's ability to progress rapidly in high-technology military fields—an ability that may carry over into the nuclear realm.

To advance its capabilities in both the chemical weapons and missile fields, Iraq relied on a network of purchasing agents and compliant business concerns based in Western Europe and the United States. From these Iraq obtained not only the necessary hardware and materials (such as precursors for chemical weapons), but also the engineering and project management services needed to construct the various facilities noted above and to master the intricacies of chemical weapons and missile production.[46] It now appears that Baghdad is employing these very techniques in the nuclear sphere.[47]

NPT restrictions. As a party to the Non-Proliferation Treaty, Iraq would be obligated to place any uranium enrichment facility under IAEA inspection. It would not have to take this step, however, until it were ready to introduce nuclear material into the facility. Thus the clandestine construction of such a plant would not constitute a violation of the country's NPT commitments.

If Baghdad decided to place the facility under inspection at the appropriate time, moreover, it would still be entitled to produce weapons-grade uranium, as long as this material, too, was under IAEA supervision to ensure that it was not being diverted to military uses. This option, as discussed earlier in relation to the possible production of plutonium from the Osiraq reactor, would allow Iraq to stockpile nuclear weapons material. If it also secretly readied the non-nuclear components for nuclear arms, it could manufacture such weapons in a matter of weeks after abrogating the treaty or withdrawing from it. The pact allows parties to withdraw on ninety days' notice if their "supreme interests" are threatened. Iraq might also simply choose to disregard the treaty from the outset, hoping to keep its enrichment activities undiscovered and, if caught, being prepared to face international disapproval and possible sanctions in order to achieve its strategic goals.

Baghdad's repeated use of chemical weapons notwithstanding its commitments under the Geneva Protocol, and the failure of the international community to respond forcefully to its repeated violations of the pact are also relevant. Given this history, there is good reason to fear that, in the end, the Hussein government will not feel constrained by the Non-Proliferation Treaty—although it may continue to abide by the accord until it has built up its nuclear infrastructure to the point where the acquisition of nuclear weapons is within its grasp.

Osiraq reactor and highly enriched uranium fuel. Despite occasional reports of renewed activity at the Osiraq site and press stories stating that Saudi Arabia has agreed to finance reconstruction of the unit, Iraq has not succeeded in rebuilding the reactor, and France remains unwilling to assist in such an effort.[48] Since plans to purchase a Soviet nuclear power reactor similarly have not proceeded, it appears that for the foreseeable future, Iraq will lack the ability to produce militarily significant quantities of plutonium. However, the Iraqi Atomic Energy Commission has stated that it has conducted research on the production of plutonium in spent reactor fuel, using its small, Soviet-supplied IRT-2000 research reactor.[49] Iraq possesses radiologically shielded hot cells, supplied by Italy as noted above, in which it could undertake experiments to separate sample quantities of this material. (Since the fuel for the IRT-2000 reactor is under IAEA inspection, any such experiments would also have to be under IAEA safeguards.)

Given the apparent interest of the Hussein regime in acquiring nuclear weapons, Iraq's continued possession of approximately 27 pounds (12.5 kilograms) of highly enriched uranium fuel, supplied by France in 1980, could pose a proliferation threat. The material represents only half of the quantity deemed necessary for a crude nuclear device under International Atomic Energy Agency standards. Nonetheless, it could be fabricated into a nuclear weapon if a somewhat more sophisticated design were used employing techniques that have been discussed widely in unclassified sources. Reports in early 1989 indicated that Israeli observers were concerned that Baghdad might misuse the material in this way.[50] The material is under periodic IAEA inspection, however, and through early 1990 the IAEA had confirmed that Iraq had not tampered with it.[51] In addition, to use the highly enriched uranium for nuclear weapons, sophisticated machining equipment would be needed to shape it. As of the fall of 1989, Iraq was not known to possess such equipment, according to U.S. government sources.

Potential foreign sources of nuclear assistance. If Iraq continues to pursue the development of nuclear arms, as appears likely, it will undoubtedly continue to exploit clandestine sources of technology and equipment in the West. In addition, however, it could turn to certain developing countries for assistance. China already appears to be aiding Iraq in the manufacture of special magnets needed for uranium enrichment centrifuges. China is believed to have sold arms to Iraq during the Gulf War and has a history of lax nuclear export controls.[52]

Brazil is a second potential source of nuclear assistance. It has an indigenously built pilot-scale centrifuge enrichment plant, which it is currently expanding and which is not subject to any form of international non-proliferation controls. Brazil thus possesses know-how of the type Iraq apparently is seeking and is legally free to share it. Through clandestine dealings, both Iraq and Brazil appear to have obtained special rolling machines or lathes for manufacturing enrichment centrifuges from the same West German firm, H. and H. Metalform, Ltd.[53] It is not clear whether this is mere coincidence or whether it reflects some degree of Iraqi-Brazilian collaboration in this field.

In the late 1970s, Brazil sold Iraq large quantities of natural uranium, and the two countries signed a nuclear cooperation agreement, under which Brazil agreed to assist Iraq in uranium exploration and nuclear facility construction and to supply it with low-enriched uranium (for fueling nuclear power plants).[54] This agreement was due to expire in October 1989. It had not been possible to ascertain whether the accord had been extended when this book went to press in late June 1990. In March 1989, the two countries signed an agreement for cooperation in space technology, including the development of rockets and reconnaissance satellites, and Iraq has signed separate undertakings with Brazilian aerospace firms.[55]

Importantly, Iraq has purchased sizeable quantities of conventional arms from Brazil. Following the ceasefire with Iran, Iraq was said to have entered into a contract to buy $1 billion worth of Brazilian arms.[56] It is possible that Iraq might attempt to use the economic leverage from these purchases to gain Brazil's agreement to provide sensitive nuclear assistance, as well.

Argentina might be a third source of unrestricted nuclear know-how for Baghdad, and in 1986, there were rumors of possible Argentine uranium sales to Iraq.[57] Like Brazil, Argentina possesses indigenously developed uranium enrichment technology, although its non-IAEA-inspected enrichment plant uses the older gaseous diffusion method, rather than the more advanced gas centrifuge. Argentina also has an aggressive nuclear export program. So far, it has been Argentina's policy to place all exports under IAEA safeguards, and there have been no reported Argentinian exports of sensitive uranium enrichment or plutonium extraction equipment. However, it appears that owing to weak export control laws and procedures during the mid-1980s an Argentine firm nearly succeeded in exporting a radiologically shielded plutonium extraction "hot cell" to Libya. The export was apparently prevented only by the personal intervention of Argentine President Raul Alfonsín.[58] Argentina's new president Carlos Menem, whose Peronist party has favored an aggressive nuclear development and export policy, could ease the country's nuclear transfer policies, but there has not yet been evidence of movement in this direction by mid-1990.

Iraq and Argentina have had close ties in at least one sensitive area of military technology. Together with Egypt, the two countries collaborated throughout the 1980s on the 600-mile Condor II missile. Although joint work on this project appears to have ended, important links between the two countries undoubtedly have been established, which could pave the way for sensitive nuclear commerce in the future.

Pakistan is another potential source of nuclear assistance. It has close ties to Iraq's allies, Saudi Arabia and Kuwait, and has built a non-IAEA-inspected centrifuge uranium enrichment plant, which is the backbone of its own nuclear weapons program. In 1985, Iraq established a joint nuclear cooperation program with Pakistan and Egypt. The focus of the program was to be the construction of an experimental nuclear reactor in Al-Wadi Al-Jadid, Egypt, and the establishment of a Cairo-based nuclear safety authority with offices in all three countries. Though apparently oriented toward civilian nuclear research, the enterprise will foster links between the Pakistani and Iraqi nuclear establishments that could facilitate militarily useful nuclear transfers to the latter.

The tripartite nuclear commission could also provide a vehicle for valuable nuclear assistance to Iraq from Egypt, Iraq's erstwhile Middle East partner on the Condor II missile and with whom Baghdad has had a longstanding military production agreement.[59] Although Egypt's nuclear infrastructure is less developed than Iraq's, Iraq has obtained the help of Egyptian nuclear specialists in the past and might hope to do so again.[60]

On the other hand, Iraq's nuclear aspirations appear unlikely to benefit substantially from the country's participation in the Arab Atomic Energy Commission, formed in December 1988.[61] Of those involved (Iraq, Kuwait, Saudi Arabia, Libya, Jordan, Lebanon, Tunisia, Syria, and Palestine), Iraq is probably the most advanced in terms of nuclear development.[62]

The Soviet Union's promise to provide Iraq with a nuclear power plant has not gone forward, and Moscow has tended to be very restrictive in making nuclear transfers that might contribute to the development of nuclear weapons.[63]

New missile capabilities. On December 5, 1989, Iraq tested a large rocket potentially capable of orbitting a space satellite. In announcing the flight two days later, an Iraqi spokesman stated that the rocket was a three-stage system that weighed 48 tons and was 80 feet in length.[64] The announcement referred to the rocket, which was launched from Iraq's Al-Anbar space center, as the "Al-Abid," Arabic for "the worshipper." Simultaneously, Iraq announced that it had developed two different surface-to-surface missile systems, each of which had a range of 1,250 miles. However, only a single name was given to the systems—"Tammuz-1," a reference to July, "the month of giving, well-being, and revolution"[65] —creating some confusion as to whether two separate missiles, or only one, had been developed. Assuming that the 1,250-mile Tammuz-1 is based on the liquid-fuel Al-Abid, it apparently is very different from the 600-mile, solid-fuel Condor II and reveals the considerable breadth of Iraqi missile activities.

U.S. officials subsequently confirmed that Iraq had indeed flight tested a rocket with a satellite launching potential and indicated that the system could ultimately be configured as an intermediate-range missile, and possibly as one with an intercontinental reach.[66] As more information about the launch came to light, however, it appeared that Iraq had tested only the first stage of the system, using dummy second and third stages for the test flight. In addition, the first stage of the rocket—which was said to use liquid fuel—was thought to be made of five Soviet-designed Scud-Bs that had been "strapped" together, rather than a totally new system.[67] Nonetheless, the new development was considered a significant achievement by most observers, particularly since, in 1987, seven Western countries had imposed restrictions—collectively known as the Missile Technology Control Regime—on missile-related transfers to developing countries, including Iraq. On December 14, Iraq announced that it tested a version of the 1,250-mile range missile, but no further details on the event have come to light.[68]

With its 375-mile range Al-Husayn and Al-Abbas missiles, Iraq already possessed the ability to strike many strategic locations in Iran, including Tehran and, if it launched the missiles from southwestern Iraq, virtually all of Israel. A missile able to travel 1,250 miles, however, would permit Iraq to strike anywhere in Iran or Israel from any location in Iraq. This would bring key Iranian oil facilities in the south of that country, such as Siri Island and Kharg Island, within range. Moreover, it would make pre-emptive attacks against Iraqi missile launchers far more difficult, since the missiles could be placed anywhere in the country. Indeed, Israel reacted with considerable concern to the Iraqi missile advances.[69]

Prospects

Iraq's continuing tensions with Iran, its longstanding antagonisms with nuclear-capable Israel and with Syria, and Saddam Hussein's aspirations for leadership in the Arab world suggest that the country's incentives for acquiring nuclear weapons will remain strong for the foreseeable future. Baghdad's increasing alignment with more moderate Arab states could eventually open the door to accommodation with Israel. In the Spring of 1990, however, Iraq resumed its belligerent anti-Israel stance of years past. Nor does a change of national leadership that might alter Iraq's nuclear thinking appear likely, given Saddam Hussein's grip on the country's political system.

In view of the long lead times required to build nuclear arms, it is highly unlikely that Iraq will be able to acquire them before the mid- to late- 1990s. Nonetheless, the evidence cited above strongly indicates that Saddam Hussein—who two weeks after Israel's 1981 raid called on the international community to "assist the Arabs in one way or another to obtain the nuclear bomb in order to confront Israel's existing bombs"[70] —has indeed launched a program to achieve this goal. Moreover, it will be far more difficult for Israel, Iran, or Syria to thwart this program through military action than was the case in 1981.

Iraq, for example, has reportedly taken precautions to locate key laboratories underground to protect them from possible air strikes and has also developed a potent air defense capability.[71] Moreover, Iraq's extensive ballistic missile force (and its current air superiority vis-à-vis Iran) now give it the ability, which it lacked in 1981, to retaliate powerfully against any future attack on one of its nuclear installations. Thus neither Iran, Syria, nor Israel could assume that if it mounted such an attack today, it would escape unscathed. Indeed, in April, 1990, Iraqi President Saddam Hussein declared that he would destroy "half of Israel" with chemical weapons if it were to mount such an attack.[72] The Iraqi president would no doubt have to calculate carefully before taking retaliatory action—for example, by launching a missile attack on an Israeli or Iranian city—for fear that this might, in turn, provoke a still more devastating counter-retaliation or trigger a wider conflict.

These uncertainties mean that if Israel or Iran ultimately feel compelled to take military action against Iraq's emerging nuclear capability, the risk of regional hostilities will be greatly increased, and with it, the possibility of open-ended escalation. On the other hand, the advent of a potent Iraqi missile capability has clearly increased the risks to its adversaries of a pre-emptive strike and may ultimately be the key factor that leads to the emergence of a second nuclear power in the Middle East.

Finally, along with Iraq's quest for nuclear arms, its pursuit of chemical and biological weapons and of increasingly powerful ballistic missiles is likely to have a continuing impact on Middle East nuclear affairs. These systems—and similar Libyan and Syrian capabilities that have been stimulated in part by Iraq's example—pose a growing threat to Israel, which has led it to hint that it might retaliate to a major chemical, biological, or missile attack on its cities with nuclear arms. In effect, though Iraq may be five to ten years away from acquiring nuclear weapons of its own, open-ended expansion of its non-nuclear capabilities has seriously heightened the risk that nuclear arms may someday be used in the region.

SOVIET UNION
TURKEY
SYRIA
Mosul
Arbela
Kirkuk
Sulaymaniyah
Halabja
IRAN
Tigris River
Euphrates River
Baghdad (Tuwaitha)
IRAQ
Karbala
Hillah
Najaf
JORDAN
Basra
KUWAIT
Persian Gulf
SAUDI ARABIA
IRAQ
0
300
MILES

Iraq

Research Reactors

Osiraq, Tammuz I (light-water/highly enriched uranium, 40 MWt)

• supplier:	France
• start up:	Destroyed by Israel June 1981, prior to start up; future status uncertain.
• fuel source:	France
• safeguards:	yes

Isis, Tammuz II (light-water/highly enriched uranium, 800 KWt)[a]

• supplier:	France
• start up:	1980 (?)
• fuel source:	France
• safeguards:	yes

IRT-2000 (light-water/highly enriched uranium, 5 MWt)[b]

• supplier:	Soviet Union
• start up:	1968
• fuel source:	Soviet Union[b]
• safeguards:	yes

Uranium Resources/Active Mining Sites/Uranium Mills

• reasonably assured reserves:	none
• currently active site:	none
• mills in operation:	none[c]

Uranium Purification (UO_2)

Tuwaitha (laboratory scale)

• capacity:	?
• supplier:	Italy
• start up:	not operating (?)
• safeguards:	yes

Fuel Fabrication

Tuwaitha

• capacity:	?
• supplier:	Italy
• start up:	not operating (?)
• safeguards:	yes

Reprocessing (Plutonium Extraction)

• capacity:	8 kg plutonium per year(?)[d]
• supplier:	Italy
• start up:	not operating
• safeguards:	yes

Enrichment

Tuwaitha? Mosul?

• capacity:	Possible construction of one or several prototype uranium enrichment centrifuges[e]
• supplier:	Equipment allegedly obtained from PRC & W. German firms; design allegedly based on W. German plant at Gronau.
• start up:	1989? (prototype centrifuge)
• safeguards:	no

Sources and Notes

a. Steve Weissman and Herbert Krosney, *The Islamic Bomb* (New York: Times Books, 1981), pp. 94, 100, 266. See also, Congressional Research Service, *Analysis of Six Issues About Nuclear Capabilities of India, Iraq, Libya, and Pakistan,* prepared for the Subcommittee of Arms Control, Oceans, and International Operations and Environment of the Senate Foreign Relations Committee (1982).

b. Jed C. Snyder, "The Road to Osiraq: Baghdad's Quest for the Bomb," *The Middle East Journal* (Autumn 1983): 1.

c. Yellowcake supplied from Portugal, Brazil, Niger, and Italy. *Congressional Research Service, Six Issues* (see note a), p. 46; see notes 6 and 8 of this section.

d. Richard Burt, "U.S. Says Italy Sells Iraq Atomic Bomb Technology," *New York Times,* March 18, 1980, p. 1. Five to ten kg of plutonium per year. Weissman and Krosney, *The Islamic Bomb* (see note a), p. 268; Italian and U.S. officials state only hot cells, not reprocessing equipment, supplied.

e. "Nuclear Equipment Sold to Iraq by FRG Company," *Der Spiegel,* December 18, 1989; "Iraq and the Bomb," *MidEast Markets,* December 11, 1989, p. 13.

12
Iran

In the wake of its defeat in its eight-year war with Iraq (September 1980 - August 1988), Iran, whose military capabilities already include ballistic missiles and chemical weapons, has launched a major rearmament program. Given the country's efforts to sustain its nuclear research program throughout the Iran-Iraq War and recent reports that Iraq has revitalized its quest for nuclear arms, there is reason for concern that Iran, too, will launch a secret nuclear weapons effort to avoid being caught at a strategic disadvantage. Indeed, such a program, which would likely require ten years to achieve success, might already be under way. Thus, despite Iran's adherence to the Non-Proliferation Treaty (NPT), its nuclear activities deserve continuing scrutiny.

Background

Acceding to power in late 1979 following the fall of Shah Mohammed Reza Pahlavi, the revolutionary Muslim fundamentalist government of Ayatollah Ruhollah Khomeini inherited a sizeable nuclear infrastructure, albeit one that was well short of being able to manufacture nuclear weapons. Iran's revolutionary government has attempted to pursue various elements of the Shah's program, but many of the earlier accomplishments were dissipated during the turmoil of the Iranian Revolution and the country's subsequent war with Iraq.

Under the Shah, Iran's nuclear activities had included two major components: a widely publicized commercial nuclear power program and a partially clandestine nuclear research effort. Some activities under the latter were oriented towards the development of nuclear weapons.

Commercial nuclear power program. Until it was abruptly truncated in 1979, Iran's nuclear power program was the most ambitious in the Middle East.

Drawing on a scientific and technical base that had been growing slowly since the 1950s, by 1974, the Shah had established the Atomic Energy Organization of Iran (AEOI) to implement a grandiose nuclear energy plan that called for the construction of 23 nuclear power stations by the mid-1990s.[1] At the time, Iran had only one operating nuclear installation, a U.S.-supplied research reactor at the Tehran Nuclear Research Center. Within months of its creation, however, the AEOI, backed by Iran's enormous oil wealth, was negotiating with the United States, France, and West Germany for the purchase of over a dozen large nuclear power plants. By the time of the Shah's fall in January 1979, two West German units were more than half completed at Bushehr, site preparation work had begun for two additional French-supplied power reactors near Darkhouin, and thousands of Iranian nuclear specialists were being trained in these countries and in the United States, the United Kingdom, and India.[2]

In addition, Iran had negotiated an elaborate set of arrangements to ensure long-term supplies of non-weapons-grade low-enriched uranium fuel for the nuclear power plants it was building. These arrangements included extendable ten-year fuel contracts with the United States, Germany, and France, concluded in 1974, 1976, and 1977, respectively, and the purchase in 1975 of a ten-percent share in a major enrichment plant being built in France by the EURODIF consortium, which included France, Belgium, Spain, and Italy.[3]

All of these reactor and fuel imports, like Iran's U.S.-supplied research reactor, would have been subject to International Atomic Energy Agency (IAEA) inspection. This was required both by the supplying countries and by the fact that Iran was a party to the Nuclear Non-Proliferation Treaty, which stipulates that all nuclear activities within a non-nuclear-weapon-state party must be placed under the IAEA system.

Post-revolution construction freeze. After the Khomeini government took power in late 1979 and launched its anti-modernization, Muslim fundamentalist revolution, work on both the partially completed Bushehr nuclear power reactors and the Darkhouin reactor site was halted.[4] In addition, many of the Iranian technicians involved in the projects fled the country.

Tehran sought to restart work on the Bushehr plants in 1984, but West Germany stated it would not renew construction activities until the Iran-Iraq War, then in its fourth year, had ended.[5] In March 1987, Iranian nuclear officials pursued contract negotiations with a consortium of West German, Spanish, and Argentinian firms on restarting the Bushehr project—presumably, after peace returned to the region—and through the fall of that year, Iran took preliminary steps at the construction site to prepare for renewed work on the plants.[6] Nonetheless, through late 1989, more than a year after the cease-fire in the Iran-Iraq conflict, the project remained dormant. Moreover, additional work would be required on the partially completed facilities since they were damaged by repeated Iraqi bombing attacks. Iran's fuel supply arrangements with the United States, France, and West Germany, and its participation in the EURODIF enrichment consortium have also been in limbo since the Iranian Revolution.[7]

Undisclosed uranium purchases and domestic resources. Although most aspects of the Shah's nuclear power program were well publicized, one element was not disclosed: a secret contract for the purchase of natural uranium concentrate, known as "yellowcake," from South Africa. The agreement is documented in a confidential 1976 U.S. State Department cable, which notes:

> The Western press has reported that Iran has signed an agreement to purchase $700 million worth of uranium from the Republic of South Africa and that it will invest in a South African uranium enrichment plant. This story has been denied publicly, but in confidence an AEOI [Atomic Energy Organization of Iran] official has confirmed that there is a secret agreement to purchase uranium from South Africa. It is evidently being kept under wraps at the insistence of South Africa. The exact source of the ore may be in Namibia, rather than the Territory of the Republic itself. . . .[8]

Presumably, the South African material was to have been shipped to the United States, West Germany, or France for enrichment to make it suitable for use as fuel in Iran's nuclear power plants. This would have been a normal commercial arrangement.

It is not known whether Iran obtained yellowcake under this agreement during the Shah's reign. If so, this material could now be in the hands of the Iranian Revolutionary Government. In addition, it is possible that the Khomeini government, itself, has made secret uranium purchases from South Africa since coming to power, as indicated in an unconfirmed 1987 press report. Iran, the report states, would have had the right to purchase up to 1,000 metric tons of yellowcake annually during this period.[9] As a party to the NPT since 1970, Iran has been required to advise the IAEA of any imports of such uranium concentrate, and in 1984 South Africa declared that it would require all of its nuclear exports to be subject to IAEA requirements. Yellowcake is not subject to formal IAEA inspection, however, and even if Iran had notified the agency of any imports from South Africa, it could probably have diverted the material to a covert nuclear weapons program without detection. Moreover, the possibility of unreported transfers of South African uranium to Iran cannot be ruled out. South African yellowcake could contribute to a future Iranian nuclear weapons effort if Iran were to acquire clandestine uranium enrichment or plutonium production facilities.

Iran also may have obtained uranium dioxide secretly from Algeria. Uranium dioxide—uranium that has been refined one step beyond the yellowcake stage—could serve as the raw material for the production of plutonium or highly enriched uranium if Iran ultimately acquired the facilities to produce these substances. Algeria purchased a minimum of 6,600 pounds (three metric tons) of the material from Argentina in January 1986, and U.S. officials are said to believe that Algeria retransferred a portion of the uranium to Iran.[10]

Because it is more useful for the production of nuclear weapons material than yellowcake, under the IAEA safeguards system, uranium dioxide stocks are normally subject to IAEA inspection, while yellowcake stocks need only be reported to the agency. Iran would therefore have been obligated under the NPT to place the material under IAEA safeguards. U.S. officials were reportedly concerned, however, that the transfer was, in fact, not subject to the IAEA system and that Iran, therefore, might be considering the use of the material in a possible clandestine nuclear program.[11]

Iran will also soon be able to produce yellowcake domestically. In 1987, the Khomeini government announced plans to set up a plant for producing uranium concentrate from uranium ore in Yazd province, the location of substantial ore deposits.[12] As of late 1989, construction of the facility was continuing.[13]

Inherited nuclear weapons research. While Iran's nuclear power program has been beset by reverses since 1979, the nuclear research activities inherited from the Shah have suffered less upheaval. Work at the Tehran Research Center apparently continued without major interruption after Khomeini took power.[14] This means that Iran has been able to train specialists and perform experiments using the small U.S.-supplied research reactor, which remains under International Atomic Energy Agency safeguards.

Moreover, specialists at the center have presumably had access to the research results obtained during the years of work under the Shah, including the fruits of what appears to have been a three-pronged undeclared nuclear-weapons research effort. According to several former U.S. officials, one component of this effort was the investigation of nuclear weapon designs undertaken by a secret research group. One knowledgeable U.S. specialist stated that the unit was a "nuclear weapons design team" whose existence so troubled U.S.non-proliferation aides that a then- pending agreement for cooperation with Iran in the areas of nuclear research and power was put on hold. Another former U.S. official recalled that "at one of their research centers we were concerned that paper studies and computer analyses of nuclear weapons were under way;" it appeared, he continued, that "they had gotten a charter from the Shah." A third U.S. insider described the unit as a group doing "advanced research that didn't look too good," but he did not recall conclusive evidence of work being performed on nuclear arms. It is not clear when these research activities were initiated.[15]

According to Akbar Etemad, the former head of the Iranian Atomic Energy Organization, in the last years of the Shah's reign, Iranian investigators also undertook studies on the extraction of plutonium from spent reactor fuel, a second component of research relevant to the development of nuclear arms.[16] In 1988, an Iranian nuclear engineer who participated in this program at the Tehran Research Center was quoted as stating that a plutonium extraction laboratory had been built at the facility, although it had not operated with plutonium-bearing materials.[17]

The third strand of weapons-relevant research under the Shah was an effort by the Tehran Center to underwrite the development by a U.S. firm of lasers supposedly capable of enriching uranium to weapons grade. The firm was headed by an American scientist, Dr. Jeffery Eerkens, who had previously worked on classified enrichment research for a U.S. defense contractor. Eerkens ultimately shipped four lasers to Iran, claiming on his export license application that they were intended only for laboratory plasma research.[18] In a 1986 interview, Eerkens stated that subsequent research had convinced him that his 1979 design was unworkable as a means for enriching uranium—as a number of U.S. government specialists had concluded when they had reviewed the 1979 export license.[19]

Continuing research. The status of the Shah's undeclared nuclear weapons research program remains unknown. Many nuclear experts associated with the Iranian nuclear power program fled the country after Khomeini's ascendancy. According to the former head of the AEOI, however, most specialists working at the Tehran nuclear Research Center remained.[20] It is possible that these included the experts engaged in weapons-related activities. Some work on laser enrichment may also have continued, although not, apparently, based on the Eerkens units.[21]

The Khomeini regime apparently pursued the research side of its nuclear program with some vigor. In 1984, it opened a new research center at Isfahan, for which ground had been broken under the Shah.[22] In 1987, moreover, it signed a $5.5 million contract with Argentina under which the latter agreed to supply new, non-weapon-grade, 20-percent enriched uranium fuel for the Tehran research reactor.[23] (The reactor has been using five kilograms (11 pounds) of weapons-grade, 93-percent-enriched supplied by the United States in the mid-1970s—an amount insufficient for a nuclear device—but this is nearly exhausted and the unit has been able to operate only intermittently.) The agreement with Argentina also provides for the training of Iranian nuclear technicians at Argentina's José Balaseiro Nuclear Institute. More disturbing, however, is the possibility that the agreement may provide for the transfer of Argentine uranium enrichment technology as well.[24] In March 1988, finally, a report in the British press stated that Iran was attempting to rebuild the plutonium extraction laboratory mentioned above.[25]

Interest in nuclear arms within Revolutionary Government. There is also convincing evidence that in 1979 some elements of Iran's revolutionary government were interested in acquiring nuclear arms. In May of that year, during the turmoil that would soon lead to Khomeini's seizure of power, Dr. Fereidun Fesharaki, a former energy advisor to the Shah, was called before Ayatollah Mohammed Beheshti, one of Ayatollah Khomeini's closest advisers. Fesharaki had headed a group of government energy specialists reporting to Jamshid Amouzegar, one of the Shah's last prime ministers, which had prepared a proposal for developing nuclear arms using clandestine, dedicated facilities.[26] Beheshti, who had acquired a copy of Fesharaki's plan, told him, "It is your duty to build the atomic bomb for the Islamic Republican Party. . . . Our civilization is in danger and we have to have it."

When Fesharaki explained the project would be extremely costly, Beheshti replied, "Fine, but the costs are tolerable and we must start. It is our duty to start."[27]

Beheshti was assassinated in June 1980. It is possible, however, that initial planning in pursuit of a nuclear weapons effort was undertaken before his death. A March 1980 U.S. Defense Intelligence Agency internal report contains the following passage, which begins by quoting an intelligence source in Iran:

> "The current reactor [at Bushehr] is completed nearly 80 percent. We are planning to go into new smaller size reactors, but this is also indefinite at this time because of the political situation."
>
> Source speculated that Iran's interest in new smaller reactors triggers him to believe that Iran is interested in developing their own weapons.[28]

Since Iran is not known to have pursued the smaller reactor option, it appears that plans to use this route to develop nuclear weapons were abandoned. Nonetheless, the document further supports the view that elements of the Khomeini government were interested in exploring the acquisition of nuclear arms.

In February 1987, another prominent Iranian official hinted that the country was seeking to develop nuclear weapons. In a speech at the AEOI, Ali Khamenei, then Iran's president, reportedly declared:

> Regarding atomic energy, we need it now. . . . Therefore, we need a tireless effort from all of you brothers in this field. Our nation has always been threatened from outside. The least we can do to face this danger is to let our enemies know that we can defend ourselves. Therefore, every step you take here is in defense of your country and your revolution. With this in mind, you should work hard and at great speed.[29]

Despite this call for action and the earlier expressions of interest in nuclear arms by some Iranian officials, it does not appear that Iran pursued this goal aggressively during the Gulf War, and its nuclear infrastructure remained rudimentary.

Iraqi attacks on Bushehr construction site. Between March 1984 and July 1988, Iraqi warplanes struck Iran's nuclear reactor construction site at Bushehr seven times, causing heavy damage and, according to Tehran radio, killing a total of thirteen individuals.[30]

Following the first three attacks in 1984 and 1985, Iran had sought to have Iraq condemned at the IAEA, but because the reactors were unfinished and contained no nuclear fuel, the agency did not apply its safeguards at the site and claimed it had no jurisdiction over the matter. Apparently to ensure IAEA involvement in the event of any future attack, in February 1987 Iran moved a small quantity of safeguarded nuclear material to the Bushehr facility. Although the amount involved was apparently too small to trigger formal IAEA safeguards there, following three air raids in November 1987, Iran demanded that the agency conduct an inspection at the site for radiological consequences, publicly comparing the risks to those from the Chernobyl disaster. The agency initially denied the request, however, after concluding "that the small amount of nuclear materials stored at the site . . . could not create a significant hazard to the public."[31] Nonetheless, it later agreed to visit the site to help assess whether the reactors were potentially salvageable.[32] After the July 1988 Iraqi attack, Iran again protested to the IAEA, but without concrete result.[33]

Iraq has never disclosed its motives for the air raids. According to a prominent civilian defense adviser to the Iraqi government, Iraq was concerned about the long-term strategic threat inherent in Iran's expansion of its nuclear infrastructure and attacked Bushehr in order to prevent Tehran from advancing this capability.[34] (Undoubtedly, Baghdad also perceived the reactor construction site, where Iran has probably invested more than $1 billion, to be an attractive economic target.)

Iraq's Osiraq research reactor outside Baghdad was also the target of bombing attacks, one by Iran on September 30, 1980, and the second by Israel on June 7, 1981, which destroyed the facility soon before it was scheduled to become operational. Such strikes set precedents that have surely increased the risk of future attacks on potentially threatening nuclear facilities in the region, and possibly elsewhere, at least preceding facility start-up and the creation of large inventories of radioactive materials.

Strategic environment. From the time it took power, the Khomeini regime was beset by civil turmoil, international isolation, and, after Iraq's invasion of the country on September 22, 1980, war with an embittered, heavily armed foe. During 1979 and much of 1980, moreover, the threat of invasion by the Soviet Union also loomed as a major concern. An arms embargo by Iran's principal arms supplier, the United States—triggered when Iranian revolutionaries seized more than 50 American diplomats in November 1979 and held them hostage for over a year—added to Iran's vulnerability, as did the Khomeini regime's purge of the country's traditional armed forces and the advent of the untrained, internally fragmented Revolutionary Guards militia as a rival center of military power.

As Khomeini gradually solidified his domestic political position, Iran gained the upper hand in the conflict, leading Iraq to employ chemical weapons against Iranian forces beginning in 1983. (See Chapter 11.) Iran ultimately responded in kind, and apparently used chemical weapons during the battle for Halabja, killing many civilians.[35] Iranian use was more limited, however, directed mainly against enemy forces, and, in the view of most observers, was a response to Iraq's intial use of these weapons. The use of such weapons by both states violated their commitments under the 1925 Geneva Protocol.[36]

Iran also came under pressure from escalating Iraqi surface-to-surface missile attacks against its cities. Both Iran and Iraq had engaged in such attacks in earlier stages of the war, Iran using Soviet-designed 40-mile FROG and 190-mile Scud-B rockets (the latter thought to have been obtained from Libya and North Korea) and, later, two new indigenously developed systems, the 25-mile Oghab and 80-mile Iran-130. In early 1988, Iraq introduced the 375-mile Al-Husayn, which enabled it to strike Tehran for the first time with devastating effect on Iranian morale.[37] When, in April and June 1988, Iraq mounted a series of offensives that routed Iranian troops from Iraqi soil and again threatened Iranian territory, an exhausted Iran accepted a U.N.-sponsored ceasefire on July 18, 1988, and brought the war to a close the following month.

Given the intensity of the conflict and the readiness of both sides to develop new classes of weapons systems, including chemical arms and ballistic missiles, there is every reason to believe that both countries would have sought to acquire nuclear arms had this been feasible. Iran's international isolation during the conflict would have given it an added incentive to move in this direction, as would its grandiose war goal of toppling Iraqi leader Saddam Hussein and installing a Muslim fundamentalist regime in Baghdad. Tehran's efforts to sustain the country's nuclear research effort may well have been a start down this road.

Political and Technical Developments

The conflict with Iraq left a defeated Iran at a decided military disadvantage vis-à-vis its neighbor. Both states have since launched rearmament programs, but for Iran this effort probably has had a greater urgency, as it attempts to rectify this military imbalance. According to one estimate, in mid-1989, Iran was seeking to acquire some $20 billion worth of new military equipment.[38]

Both countries are known to be expanding their chemical weapons capabilities and both also are pursuing the development of improved ballistic missiles.[39] Indeed, in April 1989, Iran announced that it had developed a new missile with a range of 125 miles, which would bring Baghdad easily within range.[40] In addition, it has been reported that Iraq is working on biological weapons, and there is a growing consensus that it has launched a nuclear weapons program.[41] In this context, it is hard to imagine that Iran is not attempting to follow on both fronts.[42]

Call for nuclear arms. Indeed, in October 1988, Hojat ol-Eslam Akbar Hashemi-Rafsanjani, then the speaker of the Iranian parliament and commander-in-chief of Iran's armed forces, openly called for the development of nuclear and other unconventional weapons. In an address to a group of Iranian soldiers, he declared:

> With regard to chemical, bacteriological, and radiological weapons training, it was made very clear during the war that these weapons are very decisive. It was also made

clear that the moral teachings of the world are not very effective when war reaches a serious stage and the world does not respect its own resolutions and closes its eyes to the violations and all the aggressions which are committed in the battlefield.

We should fully equip ourselves both in the offensive and defensive use of chemical, bacteriological, and radiological weapons. From now on you should make use of the opportunity and perform this task.[43]

On July 28, 1989, following the death of Ayatollah Khomeini the previous month, Rafsanjani was elected Iran's president under a new constitution that gave the office greatly expanded powers. The direct call for nuclear arming by such an authoritative figure would seem to be an important indicator of Iran's long-term nuclear intentions. Rafsanjani's disregard for international arms control limitations, apparently including the Non-Proliferation Treaty, is also evident from his remarks.

In early 1989, moreover, Admiral Thomas E. Brooks, Director of Naval Intelligence, told Congress that Iran was "actively pursuing" a nuclear weapons capability, but provided no details in his public testimony.[44] Through mid-1990, no specific information had come to light offering further particulars of Iran's efforts in pursuit of nuclear (or biological) weapons.[45] Undoubtedly, Iran has given a higher near-term priority to the acquisition of conventional arms, such as advanced warplanes. Given Rafsanjani's statement and the earlier interest of certain elements of the Iranian Revolutionary Government in nuclear arms, however, it would be surprising if Iran were not devoting its nuclear research program increasingly to the military applications of nuclear energy and attempting to advance key parts of its nuclear infrastructure, such as uranium production.

Potential foreign nuclear assistance. Tehran's nuclear infrastructure remains rudimentary, but it may be able to accelerate its nuclear program with assistance from a wide range of foreign sources. Despite an effort by the United States to establish an international embargo on nuclear transfers to Iran, the country has already been able to obtain certain nuclear assistance from abroad and acquisitions from other suppliers may also be possible.

As noted earlier, in 1987, Argentina agreed to supply fuel for the research reactor at the Tehran Research Center. It was also rumored at the time that Argentina was considering the transfer to Iran of uranium enrichment technology, know-how that could make a critically important contribution to an Iranian nuclear weapons program. The 1987 deal, however, was made during the term of Argentine president Raul Alfonsín, who was committed to nuclear non-proliferation, and no evidence has emerged to indicate that Argentina has shared enrichment technology with Iran. However, Argentina's new president, Carlos Menem, whose Peronist Party has opposed restrictive nuclear policies, might be more inclined to permit such sensitive exports.[46]

A consortium of West Germany, Argentina, and Spain has also bid to complete the Bushehr nuclear power plants. With the cessation of hostilities between Iran and Iraq, the Bushehr project may once again be revitalized. Indeed, in June 1989, IAEA Director General Hans Blix visited the Bushehr construction site and indicated that the IAEA was prepared to help Iran assess the feasibility of completing the units in light of the damage they had sustained from Iraq's attacks.[47] (Both the Tehran research reactor and the Bushehr plants are subject to IAEA inspections and could not be readily used for nuclear weapons, but both projects could assist Iran in gaining valuable nuclear expertise.)

Iran is also believed to have signed a nuclear cooperation agreement with the Soviet Union in June 1989.[48] Since the latter is a party to the Non-Proliferation Treaty and has been very cautious in making nuclear exports to the Middle East, it is unlikely that Soviet nuclear transfers would contribute to an Iranian nuclear weapons effort.

Iran may also be receiving nuclear aid from Pakistan, which, by most estimates, achieved the capability to manufacture nuclear weapons in 1986, using a uranium enrichment facility that is not subject to IAEA safeguards. Pakistan also possesses an unsafeguarded plutonium extraction plant, although it is not clear whether the facility has ever operated. According to a report in the British press, Pakistan and Iran signed a nuclear cooperation agreement in 1987, following a trip to Iran in January of that year by Dr. Abdul Qadeer Khan, the head of Pakistan's clandestine uranium enrichment program.[49] A subsequent report in an Arab paper stated that 31 nuclear specialists from the Atomic Energy Organization of Iran were receiving training in Pakistan, some at an advanced level, and that the subjects of the training included plutonium extraction, among others.[50] In 1988, U.S. government sources privately confirmed that nuclear cooperation between the two neighboring countries was being conducted, but offered no details.[51] Finally, a 1990 report stated that Pakistani engineers were helping Iran to build a reactor to produce plutonium, but provided no specifics.[52] Apart from their common Islamic heritage, Iran and Pakistan have each been threatened by Soviet encroachment in the region, have supported anti-Soviet guerrilla forces in Afghanistan, and have cooperated in other defense areas.[53]

The United States has long feared that Pakistan, having acquired the ability to manufacture nuclear weapons, might share this technology with other states, such as Iran. The Pakistani government has declared that it will not engage in such transfers, but given that Pakistan continues to conduct a range of covert nuclear activities, including smuggling operations in the West, the possibility of clandestine nuclear commerce with Iran cannot be ruled out.

Iran might also seek nuclear assistance from the People's Republic of China and North Korea, both of which became major arms suppliers to Iran during the Iran-Iraq War. During the early 1980s, China made a series of nuclear exports under lax controls, and there is growing evidence that some Chinese nuclear materials potentially useful in nuclear weapons programs wound up in such unlikely countries as South Africa and India.[54] China is also believed to have supplied the design of a nuclear device to Pakistan, its long-standing ally, and U.S. officials believe that some weapons-related nuclear cooperation between the two countries is continuing.

Although Chinese ties to Iran are not as strong as they are to Pakistan, Iran's massive defense purchases from China included such sensitive items as Silkworm anti-ship missiles. In addition, Western diplomats have been quoted as stating that China is assisting Iran to build a plant for manufacturing ballistic missiles, possibly with a 500-mile range.[55] These links could create the basis for Chinese nuclear transfers as well, particularly in view of Beijing's seemingly cavalier attitude towards nuclear exports. Although China is alleged to have sold Iraq components for its sensitive uranium enrichment program in 1989, such assistance undoubtedly would have been provided for economic, rather than strategic reasons, and would not rule out comparable nuclear assistance for Iran; indeed, China is thought to have supplied arms to both countries during the Gulf War. Through mid 1990, however, no reports have surfaced indicating that such transfers have taken place.

North Korea appears to have launched a nuclear weapons development program of its own and could soon be in a position to share relevant know-how with Iran or, indeed, to collaborate with that country on such a project. As part of its extensive arms trade with Iran, North Korea is reported to have become Iran's principal supplier of Scud-B missiles, defying U.S.-led efforts to halt international sales of these nuclear-capable systems. Under an Iranian-North Korean agreement signed in the mid-1980s, Iran is said to have financed North Korea's production of the missiles and then, in 1987, to have purchased 90 to 100 of them, as part of a broader, $500 million arms sale agreement.[56] In December 1989, the two countries signed an agreement for defense cooperation that will cover exchanges of military technology and operational experiences.[57] North Korea and Iran also share a similar international status: both are insular societies led by theocratic leaders; both have been largely isolated diplomatically (although this is changing somewhat in the case of Iran); both have supported international terrorism to advance their foreign policy objectives; and both perceive themselves to be confronted by intensely hostile neighbors. To date, however, there are no public reports of significant nuclear cooperation between the two states, although evidence suggests that some cooperation may exist in the area of uranium exploration and processing. (See chapter 8 on North Korea for a full discussion of possible ties between these two countries in this area.)

India, which tested a plutonium-using nuclear explosive in 1974 and has built a sizeable infrastructure of unsafeguarded nuclear facilities, is another potential source of nuclear assistance for Iran. As noted above, during the Shah's reign, numerous Iranian nuclear specialists were receiving training in

India. Although India historically has been quite cautious in making nuclear transfers, it is possible that this attitude could change. India has openly defended the sale to Iran of materials useful for the manufacture of chemical weapons—asserting that the materials have other non-weapons-related industrial uses—and also plans to become a major exporter of conventional arms.[58] While India may continue to treat nuclear transfers more restrictively, the trends in these related areas suggest that its commitment to restraining the spread of advanced weapons may be weakening.

South Africa's potential nuclear assistance to Iran has already been described. Given Pretoria's nuclear links to Israel, however, it is likely to be cautious in dealing with Tehran in this sphere and transfers are unlikely to go beyond uranium sales in any event. (Although Israel transferred conventional arms to Iran during the Iran-Iraq conflict, it is most unlikely to contribute to the advent of a second nuclear power in the Middle East.)

Finally, Iran could attempt to bolster its nuclear capabilities by obtaining nuclear goods from the West illicitly, much as Pakistan has done. Iran has already employed clandestine purchasing networks in Western Europe and the United States to obtain embargoed conventional arms and to advance its chemical weapons program.[59] These networks could be employed in pursuit of nuclear goods and technology, and indeed, according to U.S. Director of Central Intelligence, William Webster, some states are already employing such pre-existing networks for this purpose.[60]

In sum, given its strong motivation to acquire nuclear weapons, its currently limited indigenous capabilities, and these diverse sources of potential assistance, it is reasonable to assume that Iran will seek to exploit its opportunities to obtain aid from abroad and, indeed, may already be doing so.

Changing security and political climates. The end of the Iran-Iraq War raised hopes in the outside world for a new, more moderate period in Iranian diplomacy, including easier relations with Iran's Arab neighbors in the Persian Gulf and improved ties with the West. Despite some positive developments in this regard,[61] a series of incidents stemming from the continued sway of Muslim fundamentalism on Iran's foreign policy prevented an easing of long-standing tensions with two major powers, the United States and Great Britain. These incidents included the continued holding of American hostages by the Iran-backed Party of God in Lebanon; the placing of a bomb aboard Pan Am flight 103 by an Iranian-backed terrorist group, which destroyed the aircraft over Scotland on December 21, 1988, with the loss of hundreds of lives; and the demand by Ayatollah Khomeini in February 1989 for the death of Salman Rushdie, the British-based Indian author of *Satanic Verses*, a novel the Ayatollah declared to be blasphemous.

Surprisingly, however, Iran was able to bring about a dramatic rapprochement with the Soviet Union, a development no doubt facilitated in part by the withdrawal of Soviet forces from Afghanistan in February 1989. Indeed, by

May 1989, relations had improved to the point that Iran announced that the Soviets had agreed to sell it advanced armaments, despite Moscow's role as Iraq's long-term ally and its principal source of arms during the Iran-Iraq War.[62]

Although these improved ties have reduced the threat to Iran from the north and may well help the Iranian rearmament effort, they do little to reduce the Iraqi threat, by far Tehran's principal security concern. For this reason, Iran's interest in nuclear weaponry is likely to continue largely unabated.

The death of Ayatollah Khomeini on June 3, 1989, and the election of Hashemi-Rafsanjani as the country's new president the following month may lead to a moderating of Iran's foreign policy that could eventually open the way to reduced tensions with Iraq. Indeed, in June 1990, both sides hinted that they were ready to begin peace talks in earnest. Even should such talks begin, however, the military build-ups in both states are likely to continue for some time to come.

Prospects

Although no details of Iranian efforts to develop nuclear arms had emerged by the middle of 1990, the evidence cited above provides good reason for concern that Tehran has secretly taken the first steps down this road by dedicating an increasing portion of its nuclear research to military objectives, working to develop domestic uranium resources, and examining the prospects for foreign nuclear assistance. However, since Iran apparently has not even begun the construction of the key installations that would be needed for the production of nuclear arms, at least ten years would probably be required to bring a weapons program to fruition.

The actual manufacture of nuclear weapons would violate Iran's NPT commitments, but the country could take many steps towards this goal without violating the pact. As long as its activities and facilities using nuclear materials were subject to IAEA inspection, Tehran could build and operate all the installations necessary for the production of plutonium or highly enriched uranium and could stockpile such material. If it also secretly manufactured the non-nuclear components for nuclear arms, it could rapidly assemble such weapons in a crisis, possibly after withdrawing from the NPT, as the treaty itself permits on 90 days' notice if a country declares that its "supreme interests" are threatened.

Iran could also, of course, disregard the NPT—which requires Iran to place all of its nuclear facilities under IAEA inspection—and attempt to build the necessary installations clandestinely. Such a violation of the IAEA system would expose Iran to the threat of international opprobrium and sanctions if discovered. Iran might consider this risk worth taking, however, since acknowledging the existence of facilities with obvious weapons potential could invite Iraqi (or Israeli) efforts to destroy them. The memory of Iraq's repeated attacks on the Bushehr facility during the Iran-Iraq War would undoubtedly be a factor in Iranian thinking. Iran's Muslim fundamentalist government has been untroubled by past international criticism and has shown little regard for another arms control treaty, the Geneva Protocol, at least after that accord was violated by Iraq. If Tehran believed Baghdad were violating the NPT by pursuing a secret nuclear weapons program, it might have few qualms about following suit.

Indeed, Iran's nuclear strategy must also consider the Iraqi threat from a different perspective. Although Baghdad may be five to ten years away from achieving a nuclear capability, even a nascent program aimed at this objective must be of grave concern to Iranian strategists, who might well consider military action to delay such an effort. However, given Iraq's enormous advantage in the air and the steps it is reported to have taken to protect its nuclear activities from attack, Iran might have considerable difficulty destroying Iraqi nuclear targets. Moreover, Iran would have to be concerned about Iraq's considerable retaliatory capabilities and the possibility that Baghdad would view an attack on a nuclear facility to be a basis for reopening hostilities more broadly. These factors, while not ruling out military action, could also prompt Iran to pursue a nuclear deterrent of its own all the more vigorously, much as it is expanding its chemical warfare and missile capabilities to respond to those of its neighbor.

Apart from their deterrent value, nuclear arms would also enhance Iran's authority in the Persian Gulf, increasing its ability to intimidate neighboring states and to advance its brand of messianic Muslim fundamentalism. In addition, such weapons would enable Tehran to challenge the superpowers' political and military involvement in the region, by increasing the potential costs of confrontation. Nuclear weapons would also support Iran's claims to leadership in the Muslim world, offsetting the prestige that Pakistan—and, eventually, Iraq—may garner for attaining nuclear status.[63]

In sum, while it is not yet a foregone conclusion that Iran will build nuclear arms, political, technical, and regional security considerations point to the importance of keeping the Iranian nuclear program under close scrutiny.

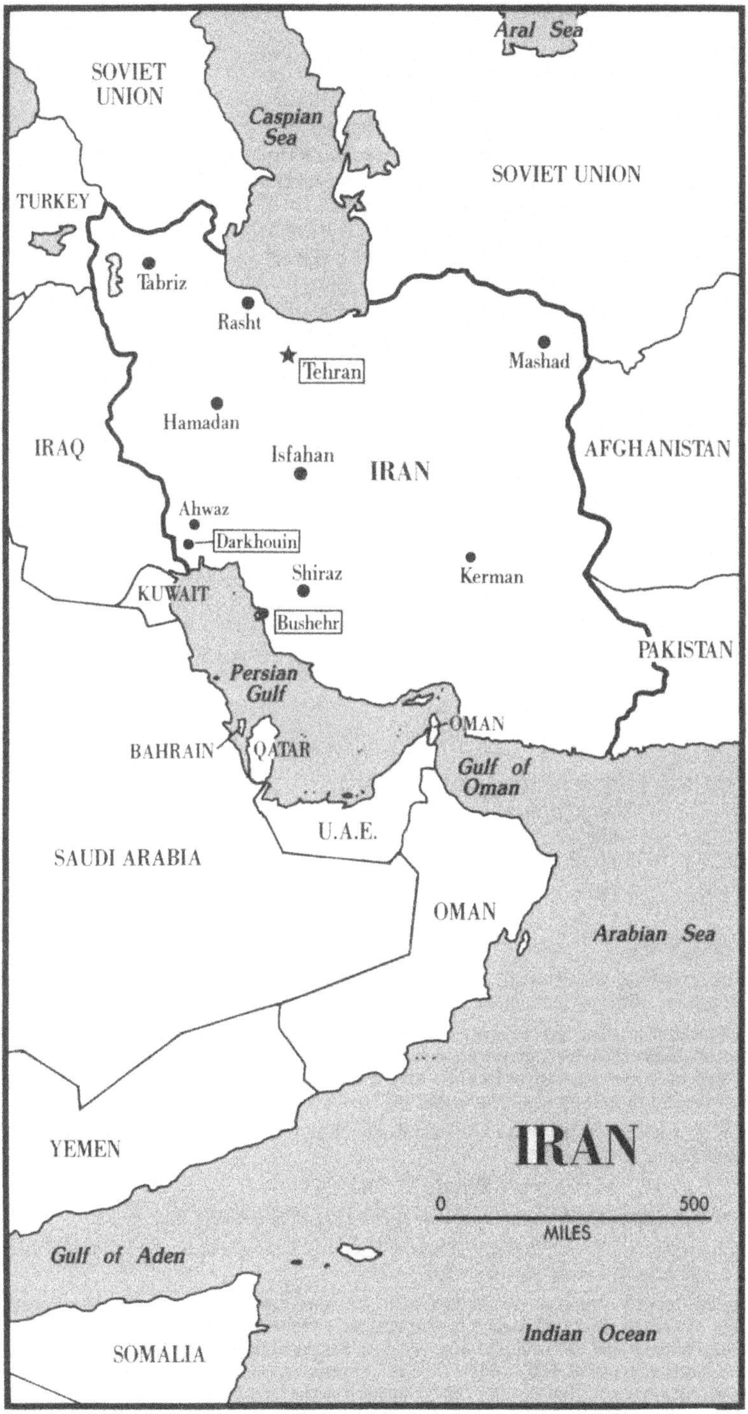
Aral Sea
SOVIET UNION
Caspian Sea
SOVIET UNION
TURKEY
Tabriz
Rasht
Tehran
Mashad
Hamadan
IRAQ
Isfahan
IRAN
AFGHANISTAN
Ahwaz
Darkhouin
Shiraz
Kerman
KUWAIT
Bushehr
PAKISTAN
Persian Gulf
OMAN
BAHRAIN
QATAR
Gulf of Oman
U.A.E.
SAUDI ARABIA
OMAN
Arabian Sea
YEMEN
IRAN
0
500
MILES
Gulf of Aden
Indian Ocean
SOMALIA

Iran

Power Reactors/Operating or Under Construction

Bushehr I (light-water/low-enriched uranium, 1300 MWe)[a]
- supplier: Kraftwerk Union (West Germany)
- start up: Partially completed; construction suspended in 1979;[b] damaged by Iraqi air strikes in November 1987 and July 1988.
- fuel source: West Germany (no fuel supplied)
- safeguards: yes

Bushehr II (light-water/low-enriched uranium, 1300 MWe)[a]
- supplier: Kraftwerk Union (West Germany)
- start up: Partially completed; construction suspended in 1979;[b] damaged by Iraqi air strikes in November 1987 and July 1988.
- fuel source: West Germany (no fuel supplied)
- safeguards: yes

Darkhouin (light-water/low-enriched uranium, 935 MWe)[c]
- supplier: Framatome (France)
- start up: Only minimal site work completed before termination in 1979[b]
- fuel source: France (no fuel supplied)
- safeguards: yes

Uranium Resources/Active Mining Sites/Uranium Mills

- reasonably assured reserves: 5,000 tons discovered in Yazd Province.[d]
- currently active sites: none
- mills in operation: none

Research Reactor

Tehran (light-water/highly-enriched uranium, 5 MWt)[e,g]
- supplier: United States
- start up: 1967[f]
- fuel source: United States (Argentina)[g]
- safeguards: yes

Sources and Notes

a. Office of Technology Assessment, *Technology Transfer to the Middle East* (Washington, D.C.: U.S. Congress, Office of Technology Assessment, 1984), p. 353.

b. Daniel Poneman, *Nuclear Power in the Developing World* (London: George Allen & Unwin Ltd., 1982), p. 96. Since 1984, Iran has unsuccessfully sought outside assistance to restart construction. Iran's efforts to complete these facilities were seriously set back when Iraqi warplanes bombed the Bushehr site several times in November 1987 and July 1988.

c. James Katz and Onkar Marwah, *Nuclear Power in Developing Countries* (Lexington, MA: Lexington Books, 1982), p. 206.

d. "Briefly . . . Iran," *Nuclear Fuel*, January 28, 1985, p.11.

e. Office of Technology Assessment, *Technology Transfer to the Middle East*, p. 355.

f. International Atomic Energy Agency, *Nuclear Research Reactors in the Third World* (Vienna: International Atomic Energy Agency, 1986), p.11.

g. In May 1987 Iran entered into a $5.5 million contract with Argentina under which the latter agreed to modify the Tehran research reactor to permit it to be fueled with 20 percent-enriched uranium, which Argentina would then supply. See, "Iran to Receive Nuclear Technology, Know-How," Noticias Argentinas, 1800 GMT, May 18, 1987, translated in *Foreign Broadcast Information Service-Latin America*, May 19, 1987; "Argentina Confirms Deal for Work on Bushehr," *Nuclear News*, (July 1987): 54; Richard Kessler, "Argentina to Enforce Curbs on Nuclear Trade with Iran," *Nucleonics Week*, March 19, 1987, p. 12.

Part IV
Latin America

The advent of civilian governments in Argentina and Brazil in late 1983 and early 1985, respectively, and a series of subsequent bilateral initiatives aimed at easing the two countries' long-simmering nuclear rivalry have significantly reduced the threat that either will build nuclear weapons. These two are the only nations in the region with programs that might lead to the development of nuclear arms before the end of the century.[1]

Both states, however, have refused to accept binding non-proliferation commitments under the Non-Proliferation or Tlatelolco treaties and are operating uranium enrichment facilities not subject to International Atomic Energy Agency (IAEA) safeguards. In the recent past, moreover, both countries have begun to produce uranium enriched to 20 percent at these facilities, thereby demonstrating the technical capability to produce unsafeguarded material enriched to the 90-percent or greater level normally used in nuclear arms.

Neither Argentina nor Brazil has faced an external security threat that might arguably call for the development of a nuclear deterrent. Rather, the risk of a nuclear arms race between them appears to be the result of the ardent nationalism that characterized the military regimes in both states during the 1970s and early 1980s. This orientation, together with the suppression of possible dissenting voices, led both countries to pursue costly programs to "master" all phases of nuclear energy production, including those with potential military applications. At the same time, nuclear achievement became a factor in the heightened competition between Buenos Aires and Brasilia for regional preeminence. As each country advanced its nuclear program, this rivalry led to a process of action and reaction, gradually propelling the two countries ever closer to nuclear weapon capabilities. In the early 1980s, mutual suspicions intensified as advocates of nuclear arming in each country appeared to become more outspoken. Today, with both nations under civilian rule, these tendencies appear to be in decline, although influential elements in both societies remain interested in pursuing an open-ended nuclear agenda.

In mid-1987, the two countries began an exchange of visits to each other's nuclear installations, including those not subject to IAEA safeguards. This has provided mutual reassurance that neither state is building nuclear weapons. The visits are not formal inspections, however, and with both countries now able to produce 20-percent enriched uranium, ambiguities could arise concerning whether one or the other were using its unsafeguarded enrichment plant to produce weapons-grade material.

Carlos Menem, who took office as Argentina's new president in June 1989, and Fernando Collor de Mello, who was inaugurated as Brazil's new president in March 1990, appear to be continuing the exchange of visits and other measures to promote cordial bilateral relations. In Brazil, moreover, where the most sensitive nuclear activities have been under military control, Collor is taking steps to improve civilian oversight.

For these reasons, the prospects for limiting the spread of nuclear weapons in Latin America remain promising.

13
Argentina

The risk that Argentina might develop nuclear arms diminished significantly when Raul Alfonsín was elected president in November 1983, following seven years of military rule.[1] Alfonsín, who served until June 1989, was strongly opposed to the pursuit of nuclear weapons and brought the country's nuclear program under civilian control. While maintaining Argentina's longstanding refusal to place all of its nuclear facilities under International Atomic Energy Agency (IAEA) safeguards, he initiated an unprecedented program with Brazil for reciprocal visits to nuclear installations in an effort to establish mutual confidence that neither country was building nuclear arms.

Even so, during Alfonsín's tenure, Argentina effectively acquired the ability to produce nuclear weapons material free from international non-proliferation controls at its Pilcaniyeu uranium enrichment plant. Moreover, the Argentine Air Force pursued the development of the nuclear-capable Condor II medium-range ballistic missile, and Argentina's aggressive nuclear export policy sometimes resulted in questionable nuclear transfers.

In June 1989, Carlos Menem, candidate of the opposition Peronist Party, succeeded Alfonsín as president. Despite his different political orientation, Menem continued Alfonsín's generally cautious nuclear policies, including the innovative reciprocal facility visits with Brazil. Menem also halted work on the Condor II. Like Alfonsín, Menem has faced a continuing economic crisis and a restive military, which have eroded his ability to govern and have introduced an element of unpredictability into the country's future, including the future of its nuclear affairs.

Background

Argentina's nuclear program, which dates back to the late 1940s, has long been a symbol of the nation's technological prowess and its aspirations for continental leadership. When Raul Alfonsín took office in December 1983, that program was the most successful in Latin America and included two operating nuclear power plants,[2] along with uranium-production and nuclear-fuel manufacturing facilities, which gave Argentina increasing independence from outside nuclear suppliers.[3]

Two nuclear projects initiated by Argentina's former military government, the Ezeiza plutonium extraction (reprocessing) plant and the Pilcaniyeu (pronounced Pil-can-ee-jay-oo) enrichment plant, triggered fears that Argentina might have launched a program to develop a nuclear weapon capability. The decision to build the Ezeiza plutonium plant was announced by Argentina's military leaders in 1978, and it appeared that the facility would provide Buenos Aires with direct access to weapons-usable plutonium by the early 1980s.

The Pilcaniyeu plant, which uses the gaseous diffusion method of enrichment, was also begun in 1978, but the project was kept secret for five years and revealed by the military government only weeks before Alfonsín's inauguration. Although Argentine nuclear aides asserted at the time that it had been built to produce only reactor fuel that would not be usable in nuclear weapons, there are strong grounds for believing that the secret project, which is not subject to IAEA safeguards and was begun at a time of growing Argentine militarism, was initiated to provide Argentina with a nuclear weapons option.[4] Indeed, the revelation of the plant's existence led several top officials of Brazil's military government to declare in December 1983 that it would acquire a nuclear weapons capability of its own by 1990.[5]

In a June 1990 interview with the principal author of this book, former President Alfonsín declared that he was taken by surprise when the existence of the facility was announced prior to his inauguration. He stated, however, that he had been assured subsequently by political allies within the CNEA that the facility was not intended for military purposes.

Nonetheless, in a seeming attempt to end the military overtones of the nation's nuclear program, shortly after taking office, Alfonsín ousted the former head of the Argentine Nuclear Energy Commission (CNEA), Admiral Carlos Castro Madero, and replaced him with a civilian, Alberto Costantini. From this point on, however, Alfonsín's efforts to change the direction of the Argentine nuclear program seemed to founder. During 1984 and the first half of 1985, Costantini, backed by nationalistic parliamentarians from the opposition Peronist Party, successfully challenged the president's efforts to cut the nation's nuclear budget and, specifically, to kill funding for the Ezeiza and Pilcaniyeu plants.[6] Similarly, apparently fearing a divisive fight with these same elements, Alfonsín refrained from pursuing legislation to reorganize the CNEA and make it strictly accountable for materials under its control.[7]

Early confidence-building efforts. Despite this seeming paralysis over nuclear policy, in mid-March 1985, Argentine sources revealed that Alfonsín and Brazilian President-elect Tancredo Neves had agreed in principle to open their countries' nuclear installations to mutual inspections.[8] (Like Alfonsín, Neves, who was selected through an indirect election, was scheduled to become his country's first civilian leader following many years of military rule.) The two leaders were said to have agreed to this at a meeting on February 7, 1985, in Buenos Aires. Neves' tragic death prior to taking office cast the proposal into limbo, however. His successor, José Sarney, did not hold a clear-cut political mandate, and it was uncertain whether he would have the power or the inclination to make the proposal a reality.

Against this background, Alfonsín and Sarney met in late November at Foz de Iguacu, on the Brazil-Argentina border. The two leaders signed a number of bilateral initiatives in the realm of economic and foreign affairs, including a joint declaration on nuclear policy, which reiterated the commitment of both nations to develop nuclear energy solely for peaceful purposes. The statement also established a bilateral commission to further nuclear cooperation between them and to "create mechanisms to insure the supreme interests of peace, security and development in the region, without detriment to the technical aspects of nuclear cooperation, which will continue to be governed by the regulations in force *viz.*, a nuclear cooperation agreement the two countries signed in 1980."[9] Although this part of the declaration did not refer explicitly to mutual nuclear inspections, apparently because of Sarney's reluctance to embrace this measure publicly, it was understood that inspections would be on the agenda of the commission.[10] By addressing the nuclear issue openly the declaration constituted an important step toward reducing nuclear tensions between the countries and seemed to signal that the two leaders were intent on ending the suspicions built up during the years of military rule.[11]

The joint nuclear commission met in late March 1986 and again in mid-July. At their first meeting, the participants appear to have endorsed the mutual inspection concept, but decided that any such system would operate independently of the safeguards regime of the IAEA and that data obtained in the course of mutual inspections would not be shared with the agency.[12] A separate internal Brazilian review of nuclear policy, completed in April, also supported the reciprocal inspection concept.[13]

The mid-July joint commission meeting, coming in the wake of the April 28 accident at the Chernobyl reactor in the Soviet Union, focused on nuclear safety issues. Later in the month, Presidents Sarney and Alfonsín agreed to create an early warning system to notify each other promptly in the event of a nuclear accident.[14]

In a December 1986 meeting in Brasilia, Alfonsín and Sarney entered into several understandings for increased collaboration in the field of civilian nuclear energy.[15] A report in the Brazilian press several months earlier that a Brazilian nuclear test site was being built in Cachimbo apparently did not mar the growing trust between the two leaders, and Alfonsín appeared to accept Sarney's assurances that the site was intended only for the testing of conventional munitions.[16]

Nuclear visits. Also in December, it was disclosed that Brazil had permitted Argentine nuclear aides to visit key parts of Brazil's classified nuclear research center at the Institute for Nuclear and Energy Research (IPEN), located at the University of São Paulo. The facilities visited included an experimental centrifuge uranium enrichment unit and an as yet unused laboratory-scale plutonium extraction facility, neither of which was subject to IAEA safeguards.[17] Given the sensitivity of these activities, which could eventually permit Brazil to produce nuclear weapon material and which were controlled by the Brazilian Navy, the opening of the installations was a powerful demonstration that Brazil did not intend to use the facilities to develop nuclear arms and represented a significant step in building mutual trust on nuclear matters between the two countries.

Alfonsín and Sarney next met in July 1987, in Buenos Aires. During the summit, Alfonsín took Sarney on a forty-minute tour of Argentina's previously classified Pilcaniyeu enrichment plant in an unprecedented gesture of openness. The visit, said to have covered all sensitive parts of the plant, was intended by both leaders as a major step toward establishing mutual trust and a noncompetitive nuclear relationship between the two countries.[18]

Although the Argentines did not disclose technical details concerning the facility's capabilities and operations, and although no arrangements were established for monitoring the output of the unit, the opening of the plant to Brazil's president was an act of great symbolic importance, demonstrating that Argentina had no intent to use Pilcaniyeu for nuclear arms.[19] Indeed, one has only to compare the secrecy surrounding comparably sensitive unsafeguarded installations in India, Israel, and Pakistan to appreciate the historic importance of the gesture.[20] These first reciprocal visits—to Pilcaniyeu and to IPEN's research areas in December 1986—demonstrated the progress made by the two civilian leaders in overcoming the intense nationalism that had characterized the nuclear policies of their military predecessors. Alfonsín, it may be noted, faced criticism for his actions from nationalist elements associated with the opposition Peronist Party and from some CNEA technocrats, but brushed it aside.[21]

By March 1988, Argentina had obtained access to all operating nuclear facilities in Brazil not covered by IAEA safeguards, including a laser enrichment research project run by the Brazilian Air Force. Moreover, Argentinian nuclear officials indicated that IPEN had been visited several times.[22] The capstone to these growing ties came in April, when Alfonsín participated in the dedication of the Aramar centrifuge enrichment plant in Ipero—a visible expression of trust in Brazil's assurances that the plant would not contribute to the manufacture of nuclear arms.[23]

Broadened presidential control; waning military interest. These diplomatic achievements were accompanied by Alfonsín's growing control over nuclear affairs at home. On April 9, 1987, the head of the CNEA, Alberto Costantini, resigned in protest after the Economic Ministry severely cut funding for a number of major nuclear projects.[24] The resignation appeared to symbolize the decline of the once-powerful CNEA as a political force influencing the overall direction of Argentine nuclear affairs. Under the nation's previous military government, the nuclear bureaucracy had been partly protected from austerity measures and, even after Alfonsín had taken office, had been treated deferentially because the nuclear program was an important symbol of national prestige. Indeed, in May 1986, under pressure from Costantini and his allies in the Peronist Party, Alfonsín had embraced the nuclear plan of his military predecessors, rejecting proposals for a broad retrenchment of the nation's nuclear construction projects.[25]

As Argentina's economic crisis worsened, however, these projects became less supportable. The combination of inflation and construction delays, together with a preference for non-nuclear energy sources among some senior Argentine planning officials, prompted cuts that, according to Costantini, paralyzed the CNEA and triggered his resignation.[26] In particular, the budgetary cutbacks delayed completion of Argentina's third nuclear power reactor (Atucha II) and the Swiss-supplied heavy-water production plant being built at Arroyito, and caused work on a smaller, indigenously designed heavy water plant at Atucha to be suspended.[27] The fact that Alfonsín continued to give funding priority to the unsafeguarded Pilcaniyeu enrichment plant, however, was an apparent reflection of the widespread political support this sensitive project enjoyed.[28]

In contrast to the pervasive behind-the-scenes involvement of Brazil's military in national political affairs and in nuclear matters, in particular, the concerns of the Argentine military at this juncture appeared to be confined to two discrete sets of issues. The first involved Alfonsín's prosecution of those responsible for incompetence during the 1982 Falklands/Malvinas War and for the brutal suppression of dissidents during the years of military rule, when indiscriminate imprisonment, torture, and "disappearances" had become rampant. Opposition to these prosecutions within the military led to mutinies and political instability in the country and obliged Alfonsín to introduce legislation to end legal action against all but the highest ranking officers involved.[29] The second set of issues were more narrowly focused, involving military budgets, equipment modernization, and pay.[30] Surprisingly, the navy remained virtually silent on Alfonsín's slowdown of a number of the nuclear projects it had initiated as well as on his confidence-building initiatives with Brazil, despite its role in supervising Argentina's nuclear program in the 1970s and early 1980s.

Political and Technical Developments

On May 14, 1989, Carlos Saul Menem of the opposition Peronist Party was elected president of Argentina, defeating Eduardo Angeloz of Alfonsín's Radical Civic Union.[31] Although Alfonsín's term was not due to expire until December, a deepening economic crisis, which led to food riots and anti-government bombings, prompted him to resign six months ahead of schedule to permit Menem to take office on June 30.[32]

Given the ardent nationalism long associated with the Peronist cause, many observers expected Menem to accelerate Argentina's nuclear activities once again and restore the country's nuclear program to the exalted status it had enjoyed during the military dictatorship of 1976 to 1983. The new president, however, chose to continue the more cautious nuclear policies of his predecessor.

Alfonsín nuclear legacy; weapons potential demonstrated. The prestige of Argentina's nuclear establishment continued to wane as a series of technical and financial problems plagued the country's key nuclear installations during the latter part of Alfonsín's presidency. During late 1987 and early 1988, both of the country's operating nuclear power plants, Atucha I and Embalse, were out of service, leading to rotating black-outs in Buenos Aires. In August 1988, soon after both plants had been brought back on line, Atucha I suffered yet another breakdown, which kept the reactor from operating for seventeen months, until January 1990. In October 1988, a shortage of funds interrupted work on Argentina's two largest nuclear construction projects, the West-German-supplied Atucha II nuclear power reactor and the Swiss-supplied Arroyito heavy water plant.[33] Only after West Germany agreed to refinance the Atucha II reactor and other funds were obtained did work recommence later in the year.

Despite these setbacks, work continued apparently without serious interruption to complete Argentina's most sensitive nuclear installation, the unsafeguarded Pilcaniyeu enrichment plant. In late 1988, the facility passed an important milestone when it began to produce uranium enriched to 20 percent.[34] Twenty-percent enriched uranium is used to fuel research reactors and cannot be used for nuclear weapons.[35] However, since most of the enrichment work required to improve uranium to weapons grade, *i.e.*, to a level of 90-percent enrichment or more, is expended in reaching the 20-percent level, attaining this capability meant that Argentina could potentially produce weapons-grade uranium, as well.

Theoretically, a plant with the overall capacity of Pilcaniyeu could produce 220 pounds (100 kilograms) of weapons-grade uranium annually, enough for four to six nuclear weapons. In practice, the actual output of the facility is probably considerably less because, presumably, it has been set up to maximize the production of 20-percent enriched uranium rather than weapons-grade material. Changing this configuration would be difficult, because the facility uses the relatively inflexible gaseous diffusion method of enrichment. Nevertheless, it would still be possible to produce weapons-grade uranium at Pilcaniyeu if the plant's managers employed a series of special procedures, such as withdrawing enriched material from the plant at a lower than normal rate and recycling partially enriched material through the facility for a second time.[36]

Whatever the plant's capabilities, Argentine nuclear officials have declared that the facility is producing material enriched to no more than 20 percent, and there has been no evidence to suggest the plant has gone beyond this level.[37] The fact that Argentina has opened the facility to Brazilian specialists as part of the two countries' on-going exchange of nuclear-site visits has made it more unlikely that the country might be producing weapons-grade material there.

Also reducing the plant's strategic potential is the fact that its output appeared to be far below published projections.[38] Indeed, in the latter part of 1988, Argentina found it necessary to purchase 330 pounds (150 kilograms) of 20-percent enriched uranium from the Soviet Union in order to satisfy the requirements of Argentina's domestic research reactors and to meet pre-existing contractual obligations to provide fuel to several foreign customers.[39]

Bilateral ties strengthened. The strategic importance of Argentina's accomplishments at Pilcaniyeu had little impact on relations with Brazil, which continued on an even keel. In late November 1988, Sarney visited Buenos Aires for a fifth summit with Alfonsín, and the two leaders signed a series of accords aimed at removing tariff barriers between the two countries and enhancing their efforts to integrate the two economies. During this meeting, Sarney toured the still incomplete Ezeiza plutonium extraction plant, apparently the only nuclear facility in Argentina or Brazil that had not been visited by scientists or political leaders of the other country. There the two leaders signed a joint declaration on nuclear policy, reiterating that their respective nuclear programs were dedicated exclusively to peaceful ends and emphasizing their commitment to continued exchanges of nuclear information and to the pursuit of joint nuclear projects, including the development of plutonium-using breeder reactors. Indeed, mutual confidence was apparently so great that the declaration specifically endorsed Argentina's decision to use the Ezeiza facility to obtain plutonium for use in existing and future nuclear power reactors.[40]

Condor missile. Alfonsín was opposed to Argentina's development of nuclear weapons and was a prominent figure in international nuclear disarmament efforts. Somewhat surprisingly, then, recently published reports have established that Argentina actively pursued the development of a medium-range nuclear-capable ballistic missile, known as the Condor II, during his administration. The project had its roots in the late 1970s, when Argentina's military government had initiated work on a precursor system, the Condor I, a single-stage weather rocket with the potential to serve as a short-range tactical missile.[41] In 1984, Argentina concluded an agreement with Egypt and Iraq to develop the 600-mile, two-stage, Condor II system. According to documents and testimony obtained by U.S. prosecutors in a case involving an Egyptian smuggling operation related to the project, the missile:

> was intended to be equivalent to the American Pershing missile and to have a range of approximately 1,000 kilometers [625 miles] and a payload of approximately 1,100 pounds. . . . [T]he initial plans for the Condor II called for the production of ten missiles. Five of these were to be retained by the MOD [Egyptian Ministry of Defense] and five delivered to Iraq. Iraq agreed to fund the project and Egypt agreed to provide the necessary technical expertise. The initial construction and testing of the ten missiles was to be done in Argentina. . . .[42]

Presumably Argentina would have obtained its own missiles in subsequent production runs.

Extensive technical support and equipment for the effort was provided through an elaborate network of European corporate entities under two primary companies, IFAT Corp., Ltd., based in Zug, Switzerland, and Consen S.A.M., located in Monaco. These companies, in turn, obtained the assistance of a number of top European aerospace firms, often in violation of national export laws.[43] In view of the project's start-up date, it appears that the Alfonsín Administration was responsible for initiating the Argentinian component of the Condor II program, and there is little question that it was pursued aggressively throughout Alfonsín's years in office.[44]

The project was under the control of the Argentine Air Force, the only branch of the armed services whose reputation was relatively unsullied by Argentina's defeat in the 1982 Falklands/Malvinas War. Although Alfonsín removed the country's nuclear program from the control of the Navy after taking office in late 1983, his decision to allow the Air Force to retain control and expand Argentina's missile program may have been intended as a reward for its relative moderation and as a means of ensuring its continued loyalty.

In April 1987, the United States, Great Britain, France, Japan, Canada, Italy, and West Germany created the Missile Technology Control Regime to slow the spread of missiles with capabilities like that of the Condor II. In September 1988, when the group met in Rome, they singled out the Condor II as a program of particular concern.[45] Washington feared that the missile would be exported to the Middle East, where it could be matched with chemical, and eventually nuclear, warheads. Britain, in turn, was apprehensive that Argentina might use the system against British forces in the Falklands; Argentina's nuclear potential contributed an additional reason for anxiety.

Responding to these pressures in late 1988 and early 1989, Argentine Air Force officials acknowledged the existence of the project (though not its ties to Iraq and Egypt), but defended the Condor II as a rocket exclusively intended to launch peaceful space satellites.[46] The secrecy and illicit acquisition of technology that were the hallmarks of the project were hardly consistent with a peaceful space program, however, and there is no doubt that Argentina's partners, Egypt and Iraq, viewed the Condor II as a military system. When Menem took office in June, the project was still active.

Queried about the Condor project in a June 1990 interview with the principal author of this book, Alfonsín stressed that the project had no connection with the country's nuclear program and was oriented toward space research. He characterized the effort as a special project of the Air Force, which he tolerated but did not actively support. When asked why Argentina had pursued a joint rocket program with Iraq and Egypt, which they saw as an intermediate-range missile project, and why Argentina had used clandestine channels to acquire needed technology, he deflected further inquiry and said that concerns about the Condor project were overblown.

Continuity under Menem. As noted above, during Alfonsín's tenure, Peronist parliamentarians had expressed reservations about the nuclear rapprochement with Brazil and had criticized the budget cuts forced on Argentina's nuclear program.[47] During his campaign for the presidency, moreover, Menem himself appeared to embrace the idea of a resurgent national nuclear program. Declaring that he did not see a need for nuclear weapons "at the moment," he did not rule them out for the future.[48] Given this background and the nationalistic fervor long associated with Peronism, many observers anticipated that Menem would appoint a staunch advocate of open-ended nuclear development to head the country's nuclear program.

Rather unexpectedly, however, Menem rejected a highly visible champion of such policies, Admiral Carlos Castro Madero, who had served as Chairman of the Argentine Atomic Energy Commission during the years of military rule. Instead, Menem appointed a little-known technocrat, Dr. Manuel Mondino, to the post in July. Mondino has concentrated on keeping the country's existing nuclear facilities functioning and on completing the two major construction projects previously noted, the Atucha II reactor and the Arroyito heavy water plant, both of which have faced severe financing constraints.

Importantly, Menem sustained Alfonsín's close ties with Brazil and continued the exchange of nuclear plant visits—despite revelations in mid-1989 that parts of the Brazilian military were apparently pursuing elements of a nuclear weapons program. (See Chapter 14.) The new Argentine leader visited Brazil in August 1989, less than two months after he took office, signing a number of agreements, including the extension of a prior accord for joint civilian nuclear research activities and a new agreement for cooperation on space-related projects.[49] By October 1989, moreover, Argentine officials were quoted as stating that the on-going reciprocal nuclear facility visits had made considerable progress and had become relatively frequent, with one aide declaring that "neither side wants to call the visits inspections, but in effect that is exactly what they are."[50]

By this time, both countries had publicly announced that their unsafeguarded enrichment plants were producing 20-percent enriched uranium—announcements which implied that both states could also produce weapons-grade uranium free from IAEA oversight at these installations, if they chose to do so. In this context, the reciprocal nuclear facility visits have become all the more important in offering a measure of reassurance that neither side is taking such a step. It is not clear, however, whether the visits have been sufficiently rigorous to provide each state complete confidence as to the other's activities, leaving the possibility that troubling ambiguities could arise.

Opposition to full-scope safeguards. As noted earlier, Argentina has long rejected the Non-Proliferation Treaty (NPT) and has refused to ratify the Treaty of Tlatelolco, both of which are perceived as unduly impinging on Argentinian sovereignty and as imposing restrictions on Argentina's nuclear program.[51] Both treaties would require the country to place all of its nuclear installations under IAEA safeguards and to renounce nuclear arms. Given Menem's nationalistic orientation, it is not surprising that, like his predecessors, he has publicly declared that Argentina will not accept either accord.[52]

Condor II and Ezeiza projects halted. Nevertheless, Menem has taken a number of steps that have reduced concerns over the country's nuclear program. In September 1989, U.S. officials advised the press that Egypt had cancelled its participation in the Condor II project, apparently under U.S. pressure.[53] The development, at first, appeared to have little impact in Argentina, where the commitment of the Air Force to the missile remained strong.[54] Nonetheless, by March 1990, Argentina had apparently frozen the project because of lack of funds and concerns over further confrontations with the United States; in April, it publicly announced that work on the missile had been halted.[55]

Also in March, Atomic Energy Commission Chairman Mondino announced that the controversial Ezeiza plutonium extraction plant, then still a number of years from completion, would be postponed indefinitely. During the late 1970s, the plant had been a major source of friction between Argentina's military regime and the Carter Administration. At the time, Washington feared that Argentina would use the facility to extract weapons-usable plutonium from a planned unsafeguarded research reactor, which has since been cancelled. Argentine nuclear planners have argued that the facility was needed to recycle plutonium as power reactor fuel, but critics have questioned the economics of this option, noting that Argentina's reactors can operate on inexpensive natural uranium fuel and that, with only a handful of nuclear power plants to use the recycled plutonium, the process would never be able to achieve economies of scale. Mondino cited the country's severe economic crisis as the reason for indefinitely freezing work on the facility and stated that the project was no longer a priority—a dramatic departure from nearly fifteen years of previous Argentine nuclear policy.[56]

Nuclear exports. As discussed in Chapter 4, Argentina has become a significant exporter of nuclear equipment, materials, and services, and has concluded nuclear cooperation agreements or established other nuclear ties with many states in both the industrialized and the developing world.[57] Its sales to Peru and Algeria of indigenously designed and manufactured research reactors and the enriched uranium needed to fuel them were the first such transactions by a non-industrialized country and demonstrated Argentina's significant capabilities in the nuclear field.[58]

In 1985, the Alfonsín Administration declared that Argentina would require recipient countries to accept IAEA safeguards on all nuclear material and equipment that they imported from Argentina, the key non-proliferation restriction required by the Nuclear Suppliers Group of industrialized countries, which has gradually become a norm for international nuclear transfers. (See Appendix F.) Alfonsín also showed his desire to avoid contributing to nuclear proliferation by personally intervening in the mid-1980s to dissuade an Argentine corporation from allowing its Italian subsidiary to conclude the sale of a "hot cell" to Libya; the unit could have been used to extract plutonium, which Libya is known to want for nuclear arms.[59]

Nevertheless, recently published reports have revealed that during Alfonsín's tenure, Argentina entered into a number of troubling nuclear arrangements. Growing ties between Algeria and Libya in 1988, for example, triggered fears in Washington that Libya might gain access to Argentine nuclear technology through this channel.[60] U.S. officials were also said to believe that in 1986 Algeria, which is not a party to the Non-Proliferation Treaty, retransferred over three tons of Argentine-origin uranium dioxide to Iran without IAEA safeguards; the material is a purified form of uranium that can be used directly as fuel in some reactors or can be transformed into uranium hexafluoride for enrichment.[61] Argentina's agreement to modify a U.S.-supplied research reactor in Tehran to enable it to use 20-percent enriched uranium and to provide fuel for the unit also troubled the United States, since the re-start of the facility, which had been dormant for lack of fuel, would enable Iran to improve its nuclear expertise.[62] Planned Argentinian cooperation with Albania (possible research reactor sale), Romania (possible sale of fuel fabrication machinery), and Syria (modification of research reactor to accept 20-percent enriched uranium and fuel sale) could similarly enhance nuclear skills of governments unfriendly to U.S. interests.[63]

In mid-1988, moreover, Alfonsín apparently lifted a *de facto* prohibition on nuclear transfers to India and Pakistan. This permitted Argentina to sell an automatic welding machine used in fabricating fuel for CANDU-style reactors to India. India possesses four unsafeguarded CANDU reactors, which are considered to be part of the country's potential nuclear-weapons production infrastructure.[64] The possibility of cooperation with Pakistan was opened in mid-1988, when officials from the Pakistan Atomic Energy Commission visited Argentina to discuss possible Argentine technical assistance for Pakistan's

CANDU-style Karachi Nuclear Power Plant (KANUPP). The KANUPP facility is under IAEA safeguards, but Western suppliers have refused to supply material or equipment for it because of Pakistan's pursuit of nuclear arms.[65]

President Menem appears likely to continue this pattern of nuclear transfers. In late 1989, he declared in an interview that, "an aggressive policy of nuclear exports will be promoted," and new exports to Turkey, Egypt, and Algeria subsequently received active consideration.[66] In addition, in October 1989, Argentina enacted a law ratifying the country's 1985 nuclear cooperation agreement with China.[67] Atomic Energy Commission head Mondino has declared that Argentina will continue to require IAEA safeguards on all of its nuclear exports, however.[68]

The Menem government has also announced that it will pursue an "aggressive and non-restrictive policy" in marketing arms and other defense items.[69] Such transfers could pave the way for Argentine nuclear sales with military applications. Arms transfers to Pakistan and Iran have already been announced, and military ties to Iraq through the Condor II project could eventuate in nuclear cooperation with Baghdad, as well.[70]

Continuing economic crisis; resurgent military. Menem took office in the midst of a grave economic crisis, which eased briefly during the fall of 1989, only to return with additional surges of hyperinflation at the end of the year and through the first half of 1990. The crisis slowed progress on the country's domestic nuclear program, while simultaneously putting pressure on Argentina to increase its exports in all areas, including the nuclear sphere, in order to improve the country's hard currency balances. In addition, the economic turmoil bred domestic political unrest and raised the possibility of intervention by the country's military in the domestic arena.

With his nationalistic sympathies, Menem went beyond Alfonsín in placating restive elements of the country's military, issuing pardons, for example, to the vast majority of officers convicted or under prosecution for engaging in human rights abuses during the 1970s and for the Falklands/Malvinas debacle. Nonetheless, the country's deepening economic crisis triggered a military challenge from a different quarter, as the threat of anarchy led to hints of a possible coup d'état by mainstream elements of the country's armed forces in January 1990. In March, these pressures led Menem to authorize the military to act to restore order in the event of "social upheaval," a potentially serious setback for Argentina's fledgling democracy.[71] These developments opened the prospect that the military soon might regain a prominent role in running the affairs of the country.

Although the military had been responsible for the secret construction of the Pilcaniyeu enrichment plant between 1978 and 1983, neither the rebellious officers seeking to end the prosecution of their colleagues nor those military figures expressing concern about maintaining social order had been directly involved in the country's nuclear program, and there were no indications through mid-1990 of renewed interest by the military in the country's nuclear affairs.

Prospects

Argentina's domestic preoccupations have tended to displace concerns of years past about nuclear developments in Brazil. Moreover, the continuing nuclear site visits between the two countries and the overall cordiality of relations between them have greatly eased the rivalry that long characterized their nuclear interactions. In addition, the advent in late 1989 of a civilian government in Brazil following that country's first direct presidential elections in more than twenty years, will likely reduce the influence of the Brazilian military in political affairs and can be expected to lead gradually to greater civilian control over the military's apparent nuclear ambitions. This factor, too, will reduce the risk of an open-ended nuclear competition in Latin America.

Nonetheless, both Argentina and Brazil have reached a critical stage in their respective nuclear development by achieving the implicit capability to produce nuclear weapons material free from any external non-proliferation controls at the Pilcaniyeu plant and the Ipero enrichment facilities, respectively. (See Chapter 14.) That each is apparently refraining from producing weapons-grade material and has confidence that the other is also exercising restraint is a critically important firebreak: if either country were to attempt to amass a stockpile of nuclear-weapons material—or if it came to believe that the other were attempting to do so—an undeclared nuclear arms race could follow.

For this reason, the scope and rigorousness of the mutual nuclear site visits hold particular significance. To maintain the *status quo* in an increasingly ambiguous situation, political leaders in both countries must have confidence in the procedures that have been adopted. Without such confidence, they could be tempted, should bilateral relations sour, to advance their own nuclear programs in order to avoid being caught at a disadvantage. Moreover, at some point, hawkish political or military factions in Argentina or Brazil might claim that the other country was rushing to acquire nuclear weapons and that their own nuclear program must be launched on the same path. To meet such challenges, civilian leaders must be able to show convincingly, by means of preexisting verification measures, that the neighboring state is not moving in this direction.

The most widely accepted mechanism for verifying the status of nuclear installations are IAEA safeguards, which would need to be applied to all nuclear activities in Argentina and Brazil to be useful under current circumstances. Both Latin American states have rejected such comprehensive coverage in the past. The bilateral visits the two states have adopted are not comparable to IAEA inspections, given the seemingly *ad hoc* nature of the exchanges. Existing arrangements do not, for example, appear to permit either country to monitor the quantities and enrichment level of the enriched material produced by the other or to account for its disposition.

The Alfonsín Administration proposed more stringent bilateral verification measures, but Brazil, during the years of the military-dominated Sarney government, was unwilling to embrace such enhanced controls. The new Brazilian government of Fernando Collor de Mello could be more inclined to accept such proposals, if Menem chose to pursue them.[72]

It is also possible that Argentina, itself, will reconsider the question of full-scope safeguards in the next year or two. South Africa is expected to accede to the NPT during the summer of 1990, and Brazilian President Collor de Mello may be more disposed to accept comprehensive safeguards under the NPT or the Tlatelolco Treaty than his predecessors. Such developments could leave Argentina increasingly isolated on the issue and create additional pressures for adherence to one or the other of these accords. Popular opinion, which appears to be increasingly skeptical about the objectives and achievements of the country's nuclear program, could also become a factor favoring adherence. In addition, the United States could also become more insistent on the point, following its success in persuading Buenos Aires to halt work on the Condor II. Washington will continue to have enormous potential leverage over Argentine affairs because of Argentina's need for U.S. financial support.

Another scenario must also be considered, however: the return of a military government in Buenos Aires. If this should occur, or if the military begins to play a dominant role in Argentine politics behind the scenes, it could seek to restore its authority in the nuclear sector. These developments could lead to a breakdown of relations with Brazil and a resurgence of Argentine interest in nuclear arms.

BOLIVIA
PARAGUAY
Salta
Tucuman
Resistencia
Corrientes
Parana
BRAZIL
Los Gigantes
Córdoba
Santa Fe
Río Tercero (Embalse)
Rosario
URUGUAY
Mendoza
Pacific Ocean
La Estela
Atucha
San Rafael
Sierra Pintada
Malargüe
Buenos Aires (Constituyentes) (Ezeiza)
Río de la Plata
ARGENTINA
Mar del Plata
CHILE
Bahia Blanca
Neuquén
Atlantic Ocean
Pilcaniyeu
San Carlos de Bariloche
Rawson
Comodoro Rivadavia
ARGENTINA
uranium mining area
0
400
MILES
Río Gallegos
Falkland/Malvinas Islands

Argentina

Power Reactors/Operating or Under Construction

Atucha I (heavy-water/natural-uranium, 320 MWe)

- supplier: Siemens AG (West Germany)
- start up: 1974
- fuel source: Argentina and West Germany (for some process steps); Argentina, principally, after 1984.[a,z]
- heavy water: initial charge, U.S., West Germany.[b]
- safeguards: yes

Embalse (heavy-water/natural-uranium, 600 MWe)

- supplier: Atomic Energy of Canada Ltd. (Canada) and Italimpianti (Italy)
- start up: 1983
- fuel source: Canada and Argentina through 1985; Argentina, principally, after 1985.[a,z]
- heavy water: initial charge, Canada[e]
- safeguards: yes

Atucha II (heavy-water/natural-uranium, 745 MWe)

- supplier: Kraftwerk Union (West Germany)
- start up: 1993;[v] project may be suspended [cc]
- fuel source: Argentina
- heavy water: Argentina and West Germany[r]
- safeguards: yes

Uranium Resources/Active Mining Sites/Uranium Mills

- reasonably assured reserves: 11,900 metric tons[f]
- currently active sites: Los Gigantes, La Estella, Sierra Pintada[f]
- mills in operation: San Rafael, La Estella, Marlagüe[f,ff]

Uranium Purification (UO_2)

Córdoba

- capacity: 150 metric tons per year[c,h]
- supplier: West Germany
- start up: 1982, full capacity 1986[c,h]
- safeguards: yes

Córdoba

- capacity: 150 metric tons per year[c,h,v]
- supplier: Argentina
- start up: status uncertain[c,h,v]
- safeguards: no

Córdoba

- capacity: 15 metric tons per year[h,s]
- supplier: Argentina
- start up: 1980;[h] may have closed after start of unsafeguarded UO_2 unit above.
- safeguards: no

Uranium Conversion (UF_6)

Pilcaniyeu

- Capacity presumed (to serve pilot enrichment plant)(unsafeguarded)[ee]

Heavy Water

Arroyito (industrial scale)

- capacity: 250 metric tons per year
- supplier: Sulzer Brothers (Switzerland)
- start up: 1991[i]
- safeguards: yes

Atucha (pilot scale)

- capacity: 2-4 metric tons per year[t]
- supplier: Argentina (with possible foreign aid)[d]
- start up: status of project uncertain; may have been suspended
- safeguards: no

Enrichment

Pilcaniyeu

- type: gaseous diffusion
- capacity: 500 kg. per year of 20% enriched uranium planned; current capacity lower[j,bb]
- supplier: Argentina
- start up: 1983 (experimental unit); 1988 (20% enrichment produced)[bb]
- safeguards: no

Fuel Fabrication

Ezeiza

- capacity: 300 metric tons of uranium/yr; substantial part of requirements for Atucha I and Embalse[z]
- supplier: Atucha I line, West Germany; Embalse line, Argentina[j]
- start up: 1982 (Atucha I line)[a]; 1985 (Embalse line)[z]
- safeguards: Atucha line safeguarded, Embalse line safeguarded only when fabricating safeguarded UO_2.[k,z]

Constituyentes

- capacity: Pilot plant for enriched uranium research reactor fuel
- supplier: Argentina
- start up: 1976[x]
- safeguards: only when safeguarded uranium is processed

Reprocessing (Plutonium Extraction)

Ezeiza

- capacity: 15 kg plutonium, 5 metric tons spent fuel per year[e]
- supplier: Argentina (possible Italian aid)[m,w,y]
- start up: project indefinitely suspended in 1990[w]
- safeguards: only when safeguarded spent fuel is processed[n]

Ezeiza

- capacity: laboratory scale
- supplier: Argentina
- start up: 1969; operated intermittently until dismantled in 1973[e]
- safeguards: no

Research Reactors [l,o]

RA-o, Córdoba (tank/medium-enriched uranium, 1 Wt)

- supplier: Argentina
- start up: 1965
- safeguards: no

RA-1, Constituyentes (tank/medium-enriched uranium, less than 1 MWt)
- supplier: Argentina
- start up: 1958
- safeguards: yes

RA-2, Constituyentes (tank/highly enriched uranium, less than 1 MWt)
- supplier: Argentina
- start up: 1966
- safeguards: yes

RA-3, Ezeiza (pool 20% enriched uranium, 3.5 MWt)[u]
- supplier: Argentina
- start up: 1967; shut down March 1988; reopened July 1990[u]
- safeguards: yes

RA-4, Rosario (homogeneous, medium-enriched uranium, less that 1 MWt)
- supplier: West Germany
- start up: 1972
- safeguards: yes

RA-6, San Rafael de Bariloche (MTR, highly enriched uranium, 500 KWt)[p, u]
- supplier: Argentina
- start up: 1982
- safeguards: only when safeguarded fuel is present.

RA-7, cancelled (heavy-water/natural-uranium 40-100 MWt)[q]
- supplier: Argentina
- start up: development suspended, 1985 [aa]
- safeguards: no

Sources and Notes

a. "Fuel Fabrication Factory Opened," *Nuclear Engineering International,* May 1982, p. 8; "Nuclear Firms Advance Level of Technology to Meet Objectives," *Energia Nuclear 1983,* translated in *Joint Publication Research Service (JPRS)-Nuclear Development and Proliferation (TND),* February 28, 1984, pp. 15-17; "Nuclear Institute's Financial Problems Reported," *Noticias Argentinas,* 1935 GMT February 5, 1985, *JPRS-TND,* February 26, 1985, p. 36. Argentina also possesses a significant zirconium alloy and tube manufacturing capability for sheathing reactor fuels. See ref. x and "Argentina Closer to Fuel Independence," *Nuclear Fuel,* May 20, 1985, p. 5.

b. John Redick, "The Military Potential of Latin American Nuclear Programs," International Studies Series, Sage Publications, 1972, p. 13; personal communication.

Between 1980 and 1982, Argentina imported 6 metric tons of heavy water each from the USSR and the People's Republic of China to make up for losses at Atucha I. Richard Kessler, "U.S. Approval for West Germany to Sell 143 Tonnes of Heavy Water to Argentina," *Nucleonics Week,* August 25, 1983, p. 7; Judith Miller, "U.S. Is Holding Up Peking Atom Talks," *New York Times,* September 19, 1982; Richard Kessler, "Argentina to Tender for Heavy Water After June Purchase from China," *Nucleonics Week,* September 25, 1986, p. 5. Additional heavy water was obtained from West Germany (with U.S. permission) in 1983. Kessler, "U.S. Approval. . . ."

c. Carlos Castro Madero, "Argentina's Nuclear Energy Program," *Nuclear Europe,* July-August 1983, pp. 31-33.

d. Burt Solomon, "Argentina Bent on Home-Grown Nuclear Program," *Energy Daily,* November 9, 1982, p. 3.

e. Powell A. Moore, Assistant Secretary of State for Congressional Relations, to Senator Gary Hart, August 19, 1982.

f. Reasonably Assured Reserves at less than $130 per kilogram. OECD Nuclear Energy Agency and the International Atomic Energy Agency, *Uranium, Resources Production and Demand* (Paris: OECD, 1988), pp. 56-7; *World Nuclear Industry Handbook, 1990* (Sutton, Surrey, U.K.: Nuclear Engineering International, 1990).

g. "Uranium Concentrate Plant," *TELAM, Noticias Argentinas* (Buenos Aires), November 16, 1982, translated in Foreign Broadcast Information Service (FBIS)-Latin America (LAT), December 9, 1982, p. B4; "CNEA Resumes Uranium Excavation," DYN, 2311 GMT, December 13, 1984, FBIS-LAT, December 17, 1984, p. B-3.

h. "Argentina to Start Producing UO_2 This Year," *Nuclear Fuel*, September 13, 1982, p. 15; "CNEA Steps Up Production of UO_2 Plant Not Under IAEA Safeguards," *Nuclear Fuel*, December 29, 1986, p. 9.

i. "Argentine Officials Optimistic on Futures of Atucha-2, Exports," *Nucleonics Week*, October 12, 1989, p. 11.

j. "CNEA Chairman Makes Announcement," *TELAM* (Buenos Aires) 2102 GMT, November 18, 1983, translated in *JPRS-TND*, December 12, 1983, p. 7; Milton R. Benjamin, "Argentina Seen Capable of 4 A-Bombs a Year," *Washington Post*, December 9, 1983; Richard Kessler, "Argentina to Delay Production," *Nucleonics Week*, February 14, 1985. See also note v.

k. Personal communication with informed U.S. official.

l. International Atomic Energy Agency, *Annual Report for 1982*, p. 84.

m. Robert Laufer, "Argentina Looks to Reprocessing to Fill Its Own Needs Plus Plutonium Sales," *Nuclear Fuel*, November 8, 1982, p. 3; Leonard S. Spector, *The New Nuclear Nations* (New York: Vintage Books, 1985), pp. 61-63; Richard Kessler, "Argentina's Alfonsín Pledges Funding for Fuel Projects," *Nucleonics Week*, June 5, 1986, p. 3.

n. Currently only safeguarded spent fuel is available.

o. Pilat and Donnelly, "An Analysis of Argentina's Nuclear Power Program and Its Closeness to Nuclear Weapons," December 2, 1982. Reactors may be operating at reduced levels. See note x.

p. "Research Reactor Inaugurated," *Nuclear Engineering International*, June 1983, p. 10.

q. Sometimes referred to as *RA-5*.

r. Some of the 143 metric tons of heavy water sold to Argentina by West Germany in 1983 (with U.S. permission) is to be available for Atucha II. Personal communication with U.S. and West German officials; some heavy water obtained from China in the early 1980s may also be available for project. (See note b.)

s. Reportedly this is a second facility in Cordoba; personal communication with knowledgeable U.S. official.

t. May not be capable of producing reactor-grade heavy water when completed.

u. Converted to 20-percent enriched uranium fuel. See Richard Kessler, "Argentina Reactor Restarted with Low-Enriched Uranium Fuel," *Nucleonics Week*, July 5, 1990, p. 5.

v. Speech by President Raul Alfonsin, Central Atomic Energy Day Ceremony, Embalse Rio Tecero, May 30, 1986; "CNEA Restructuring May Ease Problems," *Nuclear Engineering International* (January 1988): 8. One source states a plant with comparable specifications is operating at Ezeiza. See *World Nuclear Industry Handbook* (see note f).

w. "Argentina Announces Delay of Controversial Plant," *Nuclear News* (March 1990): 70.

x. Jozef Goldblat, ed., *Non-proliferation, The Why and Wherefore* (Philadelphia: Taylor and Francis, 1985), p. 295.

y. Richard Kessler, "Argentina Denies Receiving West German, Italian Reprocessing Aid," *Nucleonics Week*, May 30, 1985, p. 4.

z. Richard Kessler, "Argentina Moving Ahead with Plans for Fuel Fabrication Independence," *Nuclear Fuel*, July 15, 1985, p. 6; Richard Kessler, "CNEA Ups Production of Fuel Assemblies," *Nuclear Fuel*, March 23, 1987, p.4; fabrication of Atucha II fuel, planned; *World Nuclear Industry Handbook* (see note f).

aa. Richard Kessler, "Argentina Denies It Plans Large Unsafeguarded Research Reactor," *Nucleonics Week*, August 8, 1985, p. 5.

bb. Mark Hibbs, "Argentine Officials Optimistic on Futures of Atucha-2, Exports" (see note i); David Albright, "Bomb Potential for South America," *Bulletin of the Atomic Scientists* (May 1989): 16.

cc. Richard Kessler, "Argentina May Shutter Atucha II to Slash Mounting Cost Overruns," *Nucleonics Week*, July 5, 1990, p. 6.

dd. *World Nuclear Industry Handbook* (see note f).

ee. Mark Hibbs, "British Report Details Heavy Water, Uranium Trades by Hempel Firms," *Nuclear Fuel*, July 25, 1988, p. 7; Richard Kessler, "CNEA Officials Provide More Details About Pilcaniyeu Enrichment Plant," *Nuclear Fuel*, May 30, 1988, p. 4.

ff. "Argentine Atomic Energy Commission Shuts Down Uranium Mill Operation," *Nucleonics Week*, December 25, 1989, p. 14.

14
Brazil

Motivated by an intense nationalism, the desire for regional pre-eminence, and concerns over the nuclear advances of Argentina, Brazil has for many years pursued a sizeable nuclear program with ambiguous objectives. Brazil was under military rule between 1964 and 1985, and between 1985 and 1990, during the presidency of José Sarney, the military remained a powerful behind-the-scenes presence in Brazilian political affairs. With few restrictions on its activities in the nuclear sector, and with little political opposition to its chauvinistic agenda for national grandeur, Brazil's military leaders and their civilian allies pursued the development of technologies more suitable for nuclear weapons than for peaceful purposes.

In 1989, as Sarney's term drew to a close, Brazil, like Argentina, began to produce uranium enriched to 20 percent that was free from international nonproliferation controls, effectively demonstrating the potential to manufacture weapons-grade nuclear material. Important new evidence also came to light during this period to support earlier suspicions that the Brazilian military had a covert program to develop a nuclear weapons capability and that the military was advancing this program, in part by clandestine trade in nuclear components. In a related area, the Sarney Administration also pursued the development of space-launch vehicles with the potential to serve as nuclear-capable medium- and long-range ballistic missiles.

However, Sarney maintained cordial relations with Argentina and expanded the two countries' exchanges of visits to nuclear sites. He also supported a new Brazilian constitution containing a provision restricting the country's nuclear program to peaceful purposes and requiring congressional authorization for all nuclear activities.

Sarney was succeeded by Fernando Collor de Mello on March 15, 1990.

During his campaign, Collor was critical of certain aspects of Brazil's nuclear program and indicated his intent to restrict the wide-ranging prerogatives of the military. His nuclear policies had yet to be fully articulated by mid-1990, but appointments he has made in this sector suggest that he intends to bring the country's nuclear activities under increasing civilian supervision.

Background

Since the mid-1970s, Brazil's nuclear affairs have followed two parallel tracks. The first was managed by the state-owned Brazilian Nuclear Corporation (Nuclebras) until an August 1988 reorganization, discussed below. It has been subject to International Atomic Energy Agency (IAEA) safeguards and other non-proliferation controls and has been aimed at implementing a grandiose nuclear energy program based on imported equipment and technology.[1] Its centerpieces are Brazil's first nuclear power plant (Angra I)—a 626-megawatt reactor supplied by the United States, which entered commercial service in January 1985[2]—and a mammoth nuclear deal with West Germany, signed in 1975. Under the latter, Brazil was to purchase up to eight 1,300-megawatt nuclear power reactors, a pilot-scale plutonium extraction (reprocessing) plant, and a commercial-scale uranium enrichment facility, that would use the jet-nozzle technique.

West German agreement. Germany's agreement to sell the last two plants was highly controversial because, once completed, they would provide Brazil with the capability to produce weapons-usable fissile materials: the reprocessing facility would enable Brazil to produce plutonium, and the enrichment plant could yield highly enriched uranium. After years of major construction delays, massive cost overruns, and severe budgetary cutbacks, however, in August 1985 then-Brazilian President Sarney announced that implementation of the West German deal would be drastically curtailed. Brazil, he stated, would complete only two of the eight West German nuclear power reactors originally envisioned (Angra II and Angra III); the reprocessing plant would be indefinitely postponed[3]; and only the first stage of the jet-nozzle enrichment plant in Resende, which would produce only slightly enriched uranium, would be completed.[4]

Parallel program. The second track of Brazil's nuclear activities is officially referred to as the "parallel" or "autonomous" program. Unlike the "public" nuclear sector, which relies on foreign technology imported with safeguard requirements attached, the autonomous program relies on unsafeguarded, indigenously developed technology and is conducted largely in secret, factors which make it of much greater proliferation concern. Brazil is not legally obligated to place the parallel program under the IAEA system because the country has not adhered to the Non-Proliferation Treaty, nor has it brought the Treaty of Tlatelolco into force.[5] Both treaties would require the renunciation of nuclear weapons, as well as the submission of all of Brazil's nuclear activities to IAEA scrutiny.

The parallel program has been directed by various branches of the Brazilian military and the National Nuclear Energy Commission (CNEN), Brazil's agency for nuclear research and development. Authorized in 1979 at the end of the military government of Ernesto Geisel, the program subsequently received the active support of Brazil's next military leader, João Figueiredo.[6] Figueiredo ceded power to an indirectly elected civilian government led by José Sarney in March 1985, but Sarney remained heavily dependent on the military's political support and imposed few restrictions on those nuclear activities under military control.[7] Indeed, the parallel program made significant advances during Sarney's tenure.

The most troubling aspect of the program from the standpoint of proliferation is a joint CNEN/Navy effort, launched in 1979, to develop high-speed centrifuges for enriching uranium at the Institute for Energy and Nuclear Research (IPEN), located at the University of São Paulo.[8] Since Brazil possesses the ability to produce its own uranium and has unsafeguarded facilities to refine and convert it into uranium hexafluoride—the form of the material used in the enrichment process[9]—mastery of the centrifuge enrichment technique offered a route for Brazil to develop the ability to produce weapons-grade enriched uranium autonomously, free from any non-proliferation controls.[10] By the mid-1980s, the Navy had also built a laboratory-scale plutonium extraction unit at IPEN. As discussed below, it remains unclear whether the facility has ever operated, but it has apparently provided valuable training in this technology.

Military interest in nuclear arms. One factor causing concern about these activities was a series of statements by senior military figures during the mid-1980s urging that Brazil keep its nuclear weapons option alive by continuing to develop the necessary enrichment and plutonium extraction capabilities. In June 1985, soon after José Sarney took office, for example, Admiral José Maria do Amaral Oliveira, Chief of the Armed Forces General Staff, publicly opposed a draft amendment to the Brazilian constitution prohibiting the development of nuclear arms, declaring that, "We must not restrict ourselves in the future."[11] Two months later, Brigadier General Hugo de Oliveira Piva, director of the Air Force's Aerospace Technology Center, declared that "Brazil will be able to make the atomic bomb within five years" and that a decision on taking this step "is being discussed by the National Security Council."[12]

Some senior military aides apparently went further to speak in favor of nuclear arming itself. In September 1985, Army Minister Leonidas Pires Goncalves, according to a detailed press report, openly urged the development of a Brazilian nuclear bomb during a meeting with senior members of the Brazilian congress and solicited their support for further nuclear technology development. Asserting that Brazilian intelligence services had learned at the time of the 1982 Falklands/Malvinas War that Argentina's military regime had intended to invade southern Brazil to recover disputed territory in Rio Grande do Sul province, Pires Goncalves told the parliamentarians, in the words of the news account, "that Brazil should develop its nuclear industry only for deterrence and to prevent the Brazilian Armed Forces from remaining at a disadvantage in case of a military conflict."[13] The report, acutely embarrassing to the Sarney government, was quickly denied, but Pires Goncalves's views were said to "receive positive reaction" in other military quarters.[14] Two weeks later, Navy Minister Enrique Saboya appeared to take a stand nearly as forceful as the one ascribed to Pires Goncalves by urging support, in one Brazilian newspaper's paraphrase, for "mastering the cycle of studies for manufacturing the atomic bomb."[15]

The military's keen appreciation of, and seeming interest in, the strategic potential of the parallel program were again revealed in a series of interviews in March 1987. General Haroldo Erichsen Fonseca, Army Secretary for Science and Technology, for example, declared:

> To manufacture atomic bombs is not our objective, but if necessary, we will. With the knowledge we are gaining, obviously, we will be in a position to manufacture them. If we have the resources we can do it in 2 years time. However, Brazil does not need to build the bomb because currently it does not have any enemies.[16]

Fonseca's words were echoed by General Hermano Lomba Santoro, chief of the Army Technological Center, who stated:

> Brazil does not yet have the know-how to build atomic devices . . . but with the acquisition of knowledge, the formation of research groups, and the installation of new laboratories to gain mastery over the atom, building the bomb will be possible, but not likely, because that would depend on the government's political decision, which has not yet been made.[17]

Alleged test site. Also contributing to concerns over the military's nuclear ambitions during this period was the revelation in an August 1986 Brazilian press report that the Armed Forces General Staff and the Air Force's Institute for Advanced Studies were building a nuclear weapons test site. The facility was located at a remote military reservation in Cachimbo, deep in the Amazon jungle in the southern part of Para state.[18] The article claimed that shafts, slightly over three feet in diameter and nearly 1,000 feet deep, had been completed in July and implied that they had been built for underground nuclear tests.

The Sarney government quickly issued a denial, stating that "Brazil has no projects for the fabrication of nuclear weapons, neither does it have sufficient technological development for this, nor a testing program to this end."[19] The shafts were described as part of unsuccessful mineral exploration projects during the 1960s and 1970s, and the site was reported to have been turned over to the military subsequently for rocket testing, an application that could take advantage of the site's partial development.[20]

At the time, some U.S. non-proliferation officials discounted the possibility that the 1,000-foot shafts were intended for nuclear tests, arguing that they appeared too small in diameter for the necessary equipment and that, in any event, Brazil was still a number of years away from possessing the necessary nuclear weapons material for a test device. Other U.S. government experts and former officials, however, believed that the shafts, which were comparable to some used in American peaceful nuclear explosive tests in the 1960s, would indeed be suitable for testing a relatively small-yield device.[21] Nuclear testing, moreover, seems to be the only obvious use for shafts of such depth and diameter.

At the time, Brazil had no means for producing weapons-grade nuclear materials, and U.S. officials were in general agreement that Brazil had not launched a determined nuclear weapons program. Indeed, even those who accepted the possibility that Cachimbo was being prepared as a nuclear test site saw this as an isolated action by one faction of the Brazilian military. They speculated that this faction, in an effort to respond to Argentina's development of an unsafeguarded uranium enrichment plant in 1983, may have been hoping to deceive their Argentine counterparts by suggesting that the Brazilian nuclear program was more advanced than was actually the case.[22]

Increased military control; secret accounts. In October 1986, shortly after the revelation of the alleged test site at Cachimbo, the military acted to gain greater control over the parallel program. By presidential decree, the CNEN—a key participant in the parallel program—was brought under the direct control of the National Security Council, composed of the president and the chiefs of the armed services. The nuclear organization had formerly been under the Ministry of Mines and Energy, where its activities were receiving increasingly unsympathetic scrutiny. The shift in oversight responsibility was seen by Brazilian observers as a means for ensuring continued military control over, and support for, the parallel program.[23]

In January 1987, suspicions about the objectives of the parallel program were further aroused when the Brazilian press revealed that the program had been funded through a series of secret CNEN bank accounts since 1981.[24] The secret accounts, which were acknowledged by CNEN head Rex Nazareth, appear to have been a mechanism used by the Figueiredo government to maintain a steady source of funds for the autonomous program; at the time, the budget for Brazil's commercial nuclear energy program was subject to drastic cuts under a broad austerity plan imposed by Brazil's overseas creditors. Reliance on the secret accounts continued for the first year of the Sarney administration, apparently without the knowledge of treasury officials.

Whether the CNEN took advantage of the clandestine funding arrangements to promote activities directly related to nuclear weapons development is not known. The use of such covert schemes, however, strongly suggests that more was involved than the pursuit of purely peaceful nuclear research.

Countervailing trends: bilateral confidence-building. As these diffuse pressures to develop a nuclear weapons option continued during 1985 and 1986, a second important trend emerged that would help to attenuate tensions between Brazil and Argentina. During this period, as has been described in Chapter 13, the two countries took important steps toward integrating their economies and enhancing the coordination of their foreign policies. In late November 1985, at Foz de Iguacu, Sarney and Argentine President Raul Alfonsín signed a joint declaration pledging that both countries would develop nuclear energy solely for peaceful purposes and agreed to increase nuclear cooperation. The pledge appeared to reflect Sarney's fundamental commitment to non-proliferation. Yet he chose at the time not to embrace Alfonsín's call for the development of a system of mutual inspections of the two countries' nuclear installations, and agreed only to establish a bilateral commission to "create mechanisms to insure the supreme interests of peace, security and development in the region, without detriment to the technical aspects of nuclear cooperation."[25]

At the first meeting of the joint Argentine-Brazilian nuclear commission in March 1986, the Brazilian delegation backed the nuclear inspection concept in principle.[26] Nonetheless, Sarney remained unprepared to accept an Argentine proposal to establish such an inspection system, no doubt due to opposition among his military supporters.[27]

Sarney's preference appears to have been to increase the two countries' bilateral collaboration in the nuclear sector with a view to easing suspicions related to their respective nuclear capabilities. In December 1986, Sarney and Alfonsín signed a series of protocols in Brasilia, which established joint nuclear research projects and a nuclear accident early warning system.[28]

Also in December, Brazil permitted Argentine nuclear aides to visit key parts of IPEN, including the experimental centrifuge-enrichment unit and the laboratory-scale reprocessing facility. Given the sensitivity of these military-controlled activities, opening the installation for the visit was an impressive show of good faith, particularly since even the external dimensions of an enrichment centrifuge are often treated as a closely held industrial secret.[29]

These cooperative arrangements were extended in July, when Alfonsín, in an unprecedented gesture of openness, took Sarney on a 40-minute tour of Argentina's classified Pilcaniyeu enrichment plant while the latter was visiting Buenos Aires.[30]

Uranium enrichment capability. On September 4, 1987, President Sarney dramatically announced that Brazil had succeeded in enriching uranium by means of indigenously developed gas centrifuges. The enrichment breakthrough had been made at the laboratory-scale centrifuge unit run by the Navy at IPEN. The unit was said to house six centrifuges and to have produced a number of pounds of uranium enriched to 1.2 percent uranium-235.[31]

Sarney's announcement also stated that construction would begin immediately on a pilot-scale uranium enrichment plant at the Navy's Aramar Experimental Center in Ipero, 80 miles west of São Paulo. An indigenous enrichment capability would permit Brazil to produce enriched uranium totally free from any non-proliferation controls and, if enriched to the necessary level, potentially available for nuclear weapons.[32] Moreover, the decision to fund construction of the Aramar enrichment plant in an increasingly tight budgetary environment, the first stage of which would cost between $100 and $150 million, indicated Brazil's strong desire to master the IPEN technology on an industrial scale, a step that would bring it closer still to a nuclear weapons capability.[33]

A key question was whether the installation would enrich uranium beyond the 20-percent level. One of the principal rationales for the Aramar plant, namely to produce fuel for Brazilian nuclear-propelled submarines, suggests that a higher enrichment level may be required. The Brazilian Navy has designed a prototype, 50-megawatt submarine propulsion reactor to be installed at the Aramar center in 1991 (see Chapter 5) and, according to some accounts, the reactor is to be fueled with 70-percent enriched uranium. To produce the material, the Aramar enrichment plant would be gradually expanded by 1992 to contain 3,000 centrifuges, according to these reports.[34] If these plans are pursued, Brazil would acquire a *de facto* nuclear weapons capability, since the same facility could readily be used to upgrade the 70-percent enriched uranium to the 90-percent or greater level used in nuclear arms.[35] That the facility is not subject to IAEA safeguards or other non-proliferation controls further complicates the situation; even if Brazil did not manufacture weapons from enriched uranium produced there, outside observers could not be confident that Brazil had not taken this step without an opportunity to inspect the facilities. The resulting ambiguities could intensify nuclear competition with Argentina.

At the time of Sarney's announcement, however, CNEN chief Rex Nazareth stated that the Aramar enrichment plant would not produce more than 20-percent enriched material, in order to guarantee that the facility would not contribute to a nuclear weapons capability. Brazilian submarine reactors, he continued, would be designed to run on material enriched to the lower, more acceptable level.[36]

Shift to the right. Sarney announced the IPEN advance himself, on the eve of Brazil's independence day, and portrayed the development as a significant national achievement. This move appeared to be aimed at enhancing the prestige of the military by validating the importance of the parallel program, a practical stratagem given the plunge in Sarney's popularity in the face of domestic economic difficulties. By mid-1987, he had become increasingly dependent on the support of the military in dealing with Brazil's Congress, which was acting as legislature and constituent assembly, drafting a controversial new constitution.[37] For the remainder of 1987, Sarney confronted a worsening political crisis that greatly weakened his ability to govern and cast the country's military in an increasingly prominent political role.

Political and Technical Developments

Brazilian nuclear policy remained relatively unchanged through the remainder of Sarney's term, which ended March 15, 1990, when he ceded power to Fernando Collor de Mello. Collor, a right-of-center former state governor, won the presidency in December 1989, in the country's first direct elections for a national leader since the military had seized power in 1964.[38]

CNEN strengthened under Sarney. Impatient with the chronic delays that had beset the safeguarded nuclear program, on August 31, 1988, Sarney reorganized Brazil's nuclear energy industry. The restructuring abolished Nuclebras, the government corporation that had been responsible for implementing the various facets of the 1975 agreement with West Germany, and reassigned its responsibilities to other government agencies.

Under the new organizational set-up, the National Nuclear Energy Commission (CNEN), which had long been allied with the military and coordinated the unsafeguarded activities of the military at Ipero, IPEN, and other research centers, acquired authority over the most sensitive elements of the safeguarded program, namely, the enrichment facility and related units that had been built at Resende under the 1975 agreement with West Germany.[39] This organizational merger triggered fears on the part of some observers that Brazil might illicitly transfer technology from the safeguarded to the unsafeguarded program; this issue had become a matter of concern to the IAEA and West Germany even before the August 31 realignment.

The CNEN was to report to a new Supreme Council on Nuclear Policy composed of representatives from nineteen ministries, including each of the armed services, and non-governmental appointees. In actual practice, however, the Council was dominated by the country's military and its role appeared to be largely advisory.[40]

The reorganization also assigned responsibility for running the country's single operating nuclear power reactor, Angra I, to Brazil's electric power corporation, Electrobras. Electrobras was also placed in charge of completing the two additional nuclear power reactors Brazil had under construction, Angra II and III. Electrobras had had no previous involvement in the country's nuclear program.[41]

Constitutional restrictions. On September 3, 1988, Brazil adopted a new constitution, establishing permanent democratic institutions and defining the role of government in many other aspects of Brazilian life. In the nuclear realm, the charter provided that "any nuclear activities within the national territory will be permitted only if for peaceful purposes and if approved by the National Congress."[42] The provision appears to rule out the development of nuclear weapons and to bring all nuclear activities under civilian control. Moreover the very inclusion of a provision on nuclear matters in a national constitution is unusual and appears to signal a serious commitment to non-proliferation.

In practice, however, the impact of the provision may be more limited. First, it does not rule out the accumulation of weapons-grade nuclear materials free from external non-proliferation controls and thus would permit Brazil to develop a ready nuclear weapons option. Nor does the provision prohibit the development of so-called peaceful nuclear explosives or even the conducting of nuclear tests under this rubric.

Moreover, the provision does not establish mechanisms for day-to-day civilian control of nuclear affairs by, for example, requiring that all nuclear activities be conducted by civilian-led agencies. This would seem to leave the military free to continue its various programs at IPEN, the Aramar Experimental Center, the São José dos Campos Aerospace Technology Center, and the base at Cachimbo. Although the Brazilian Congress will in principle exercise budgetary control over these activities, the military is likely to be a powerful lobby within Brazil's democratic structures, and there is reason to question whether this oversight mechanism will be effective. It should also be recognized that the military lobbied *for* this provision—which, in effect, codified the country's declared nuclear policy—in preference to another proposal. The alternative would have explicitly prohibited the manufacture and storage of nuclear weapons and would have required a national plebescite on the construction of additional nuclear power plants.[43]

Sarney nuclear legacy; weapons potential demonstrated. After the unsafeguarded Ipero uranium enrichment plant was inaugurated in April 1988, work at the facility continued, and by April of the following year, some uranium enriched to 20 percent had been produced on a laboratory scale.[44] At that time, Rear Admiral Othon Pinheiro da Silva, the Navy officer responsible for the project, stated that a new wing of the facility, that would house Brazil's most advanced gas centrifuges, was being built. It would take the 5-percent enriched output from the plant's existing units and upgrade it to 20 percent on a commercial scale beginning in 1990.[45] Although this construction was apparently completed, budget cuts delayed the full-scale operation of the new units at least through early 1990 and held the amount of 5-percent enriched material processed at the Center in 1989 to 880 pounds (400 kilograms), instead of the 8,880 pounds (4,000 kilograms) that had been planned; some additional small quantities of 20-percent enriched material were apparently produced however.[46] Brazilian nuclear aides have stated that 20-percent material is needed to fuel Brazil's existing research reactors and for future nuclear-powered submarines (see Chapter 5) and have declared that the Ipero plant will not produce material enriched to higher levels.

Twenty-percent enriched uranium cannot be used for nuclear weapons.[47] However, most of the enrichment work required to improve uranium to weapons grade, *i.e.*, to a level of 90-percent enrichment or more, is expended in reaching the 20-percent level. Thus, for all practical purposes, Brazil had demonstrated the ability to produce weapons-grade uranium, as well.

Enrichment plants like Ipero that use the gas centrifuge method can be readily reconfigured to produce higher enrichment levels than that for which they were originally designed. It is also relatively simple to recycle partially enriched material through such a facility to increase its enrichment level. Thus there is little question that the Ipero plant could be used to produce weapons-grade uranium, if Brazil chose to do so. The exact size of the facility has not been disclosed. If, however, it can produce 8,800 pounds (4,000 kilograms) of 5-percent enriched uranium annually, and has an additional unit to upgrade some of this material to the 20-percent level, it could certainly produce enough weapons-grade product for several weapons in the course of a year.[48]

Despite Brazil's progress in this area, it appears that Argentina maintains a slight lead. Argentina was apparently able to produce more than laboratory-scale quantities of 20-percent enriched uranium in the latter half of 1988, when Brazil was still limited to laboratory-based production. Argentina also claimed to be able to supply kilogram quantities of 20-percent enriched uranium fuel for research reactors in Peru and Algeria in late 1988 and early 1989, respectively, while in July 1989, Brazil declared that its goal was to produce its first 26 pounds (12 kilograms) of 20-percent material by January 1990.[49] Brazil's enrichment technology is more advanced than Argentina's, however, which would give it an advantage over the longer term.

Mounting evidence of nuclear weapons program. Brazil's efforts to expand its unsafeguarded uranium enrichment capability are particularly disturbing in light of new evidence suggesting that the country's military secretly pursued a nuclear weapons development program throughout the 1980s. Among the recent disclosures that have heightened concerns in this regard have been statements by top Brazilian nuclear aides declaring that they have engaged in clandestine and apparently illicit nuclear trade to build the Ipero facility. In June 1988, for example, Admiral Pinheiro da Silva revealed in an interview that he had personally obtained under false pretenses a special West German lathe for machining cylinders for the Ipero centrifuges. He misrepresented the intended use of the lathe to its German vendors, claiming that it was for the manufacture of rocket shells.[50]

Similarly, then-CNEN Chairman Rex Nazareth, responding to allegations that Brazil had obtained a wide variety of nuclear equipment by smuggling it out of foreign countries as "tractor parts," agreed that his country had illicitly obtained essential components abroad. This, he stated, was probably known to the United States, but, he continued:

> The catch is that it [the United States] has no way to prove it. I will not say a word about it, even under torture! Perhaps some day, when I write my memoirs, I will reveal just how it was done.[51]

In August 1989, new details also came to light concerning the secret bank accounts that were apparently used to finance much of this activity. According to a Brazilian press report, since the early 1980s, the CNEN had deposited $700 million in an account in the Luxembourg branch of the Banco di Roma for unknown purposes. One set of transactions involving the account were uncovered, however: CNEN was said to have written yearly drafts for $1.2 million from that account to pay interest on a debt to a Luxembourg firm, Gradel, supposedly for the purchase of nuclear material in 1981.[52] These machinations are hardly consistent with a program whose sole purpose is the development of nuclear energy resources for peaceful research or even for an acknowledged nuclear-powered submarine program.

Indeed, documents obtained by a German parliamentary committee investigating illicit nuclear exports indicate that by the late 1980s, both the British government and the West German intelligence service (the *Bundesnachrichtendienst*, or BND) had deep suspicions about the objectives of the parallel program. In mid-1986, the British Embassy in Bonn sent a diplomatic note seeking to dissuade West German export officials from authorizing the sale of a mass spectrometer—used to measure the enrichment level of uranium—to the Brazilian Navy. The note stated, "The Brazilian Navy, *which is known to be involved in the weapons program*, has given an end utilization that is clearly suspect."[53] Similarly, a 1987 report by the West German BND concluded that, "The parallel program is clearly directed towards military goals."[54]

The BND report also stated that IPEN and a pair of other military-run research centers were working to develop nuclear reactors designed specifically to produce plutonium. Since plutonium would have no utility in a nuclear submarine program and since, in recent years, Brazil has not claimed to have any need for plutonium in either its safeguarded or unsafeguarded nuclear program, it would appear that the projects were being pursued as part of a clandestine nuclear weapons effort.

The 22-page intelligence agency report stated that IPEN was designing a 20-megawatt natural uranium/heavy water-moderated reactor, while the Technical Development Center (CETEX) of the Brazilian Army and the Institute of Nuclear Energy (IEN) in Rio de Janeiro were developing a gas-cooled, graphite moderated unit.[55] In October 1989, a report in the Brazilian press confirmed the existence of the second project, which it said was launched in 1981. It noted that high-purity graphite bars had been manufactured for the reactor and implied that construction had begun, with the project scheduled for completion by 1994.[56] To obtain separated plutonium, spent fuel from these reactors would have to be processed at a plutonium extraction plant. As noted earlier, Brazil possesses a small plutonium extraction laboratory at IPEN not subject to IAEA safeguards. Although it is not clear whether this facility has ever operated, the 1987 West German intelligence report concluded that it was "most probable" that unit was in use in 1987, producing one gram of plutonium per day and providing valuable training in this difficult technology.[57] In late 1989, Brazilian officials denied that the unit had ever operated with radioactive materials.[58]

Under the 1975 agreement with West Germany, however, Brazil acquired engineering plans for a plutonium facility capable of producing 55 pounds (25 kilograms) of plutonium per year, enough for three to five nuclear weapons.[59] The agreement stipulates that any facility built with technology transferred to Brazil is to be subject to IAEA inspection. Since the IPEN plutonium extraction unit appears to fit this criterion, questions have been raised as to the adequacy of IAEA monitoring of this provision.[60] To date it does not appear that Brazil has begun construction of such a full-scale plutonium extraction plant, but there are indications that one may be planned: according to the West German intelligence agency report, by 1987, Brazil had begun to manufacture tributyl phosphate, the key chemical used in plutonium extraction plants.[61] These inconsistencies add to the circumstantial evidence of a clandestine nuclear weapons development effort.

The West German parliamentary investigation also unearthed evidence that Brazil had transferred technology from its safeguarded, FRG-sponsored civil nuclear program to the unsafeguarded parallel program by reasssigning personnel from the former to the latter after they had acquired important skills and technical data. According to a West German Foreign Office report of October 26, 1987, some 20 percent of the Brazilian staff that had been trained in Germany under the 1975 agreement had "wandered" from the safeguarded program into "other areas," and West German officials discussed their concerns with their Brazilian counterparts about three specific individuals, who were known to have been transferred from the safeguarded to the unsafeguarded program.[62]

Moreover, owing to a lack of cooperation on the part of West German suppliers, and particularly on the part of Brazil, the IAEA was apparently unable to keep track of equipment and blueprints that Brazil received from Germany for the safeguarded Resende enrichment plant and that it had pledged to use only in safeguarded installations. Indeed, according to a January 1987 German Foreign Office memorandum, Brazil had informed the IAEA of "no single hardware delivery" to the Resende plant in 10 years.[63] This would have left Brazil free to transfer such items quietly to its unsafeguarded enrichment program or to duplicate them for use there.

Although the Resende plant uses the jet nozzle technique, which differs considerably from the gas centrifuge technology used at Ipero, both plants process highly corrosive uranium hexafluoride gas in "cascades" consisting of thousands of individual enrichment units. Thus, some skills gained at the safeguarded Resende plant would be of great value in operating the unsafeguarded Ipero facility. The January 1987 Foreign Office memorandum described monitoring deficiencies that had occurred prior to that date. Issued eight months before Sarney's September 1987 announcement that Brazil had mastered the enrichment process, the report's findings suggest that such clandestine technology transfers might have played a significant role in Brazil's supposedly indigenous advances and in the development of the Ipero plant, which was commissioned in April 1988. The reorganization of August 1988, which placed both enrichment programs under the CNEN, triggered additional concerns in the German Foreign Office that jet-nozzle enrichment work at Resende "could be integrated into" the parallel program, in the words of a subsequent Foreign Office report.[64] Again, the improper exploitation of German-trained personnel for the parallel program and the efforts to bypass technology controls by failing to cooperate with the IAEA appear inconsistent with a nuclear program limited to strictly peaceful objectives.[65]

One more addition to the parallel program deserves mention in this regard. In November 1988, scientists working in the Navy-supervised program at IPEN, in São Paulo, started a zero-power research reactor, known as MB-01, intended to permit calculations needed to build a larger prototype nuclear power reactor in the 60-megawatt range.[66] The new reactor required 466 pounds (212 kilograms) of 4-percent enriched uranium as fuel. The source of this material was not disclosed, but production of enriched material in this quantity appears to have been beyond the capacity of the Ipero facility at the time.[67] This strongly suggests that Brazil's military had engaged in another clandestine nuclear transaction to obtain the material, although Admiral Pinheiro da Silva claimed that the fuel was indigenously produced.[68]

The mid-1987 disclosures in West Germany concerning Brazil's parallel program and the possible diversion of German technology led to calls in Bonn for the termination of further cooperation with Brazil under the 1975 agreement. After a review of the matter during the fall, however, the West German government of Chancellor Helmut Kohl extended the agreement for an additional five years in February 1990. German officials stated that they were satisified that Brazil was "pledged to the exclusively peaceful use of nuclear energy" and that sufficient measures were in place to address the question of technology diversion.[69]

Missile program. As Brazil pursued these advances in its nuclear capabilities, it also proceeded with work on a number of powerful launch vehicles. Over the past decade and a half, Brazil has launched a series of sounding rockets with increasing payloads and ranges. These have subsequently been modified into surface-to-surface missiles, which have been offered for sale on the international market.[70] The largest Brazilian missile currently available is the SS-60 (based on the Sonda III sounding rocket), which can carry a 330-pound (150 kilogram) payload about 40 miles. However, Brazil has had three successful test launches of the two-stage Sonda IV sounding rocket, the most recent in April 1989. The Sonda IV can lift a 1,100-pound (500 kilogram) payload to an altitude of 200 miles, making it potentially suitable as a nuclear delivery system.[71]

By the mid-1990s, this rocket is expected to be converted into a 375-mile-range military version, known as the Orbita MB/EE-600, for sale to foreign customers.[72] Flight tests of another Brazilian missile, the Avibras SS-300, comparable in range and payload to the Scud-B, with a 190-mile range and a 2,200-pound (1,000 kilogram) payload, were expected in 1989 but did not take place.[73]

Brazil has collaborated with other countries in developing some of these systems. Libya was reported to be contributing to the financing of the Orbita system, and Iraq, to the Avibras missile.[74] These programs will undoubtedly be delayed, however, if they are pursued at all, because of the financial difficulties of some of the companies involved; in early 1990, one of the partners in Orbita was near insolvency, and Avibras had filed for bankruptcy.[75] These problems are also likely to set back efforts by Iraq and Brazil to collaborate on the development of satellites and missiles, pursuant to their discussions in March 1989.[76]

The government-backed Sonda IV project is also experiencing technical difficulties. Brazil's space program had benefitted over the years from extensive official cooperation with the United States, Canada, the European Space Agency, and the Soviet Union, and a number of Western European firms are said to have assisted in the development of components for the Sonda IV.[77] However, the establishment of the Missile Technology Control Regime (MTCR) in 1987 by Canada, France, Great Britain, Italy, Japan, the United States, and West Germany has drastically curtailed such assistance. In particular, the regime has prevented Brazil from obtaining guidance technology it needs to proceed with the Sonda IV, according to Brazilian officials.[78] Brazil may be continuing to receive substantial assistance from France for its larger satellite launch vehicle, the VLS, however. Brazil successfully tested a prototype of the VLS in May 1989. France has offered to supply Brazil the liquid-fuel Viking rocket engine for the project, a step the United States has considered a violation of the MTCR. France had argued that the transfer was permissible since it was for a peaceful space program, which MTCR guidelines state are not to be inhibited by the regime's restrictions.[79] Nonetheless, France appears to have suspended the sale.

Under a 1986 agreement with the People's Republic of China, Brazil is also reportedly to receive advanced technology on liquid rocket fuels and rocket guidance systems in return for providing Beijing with solid-fuel technology.[80] In 1989, Avibras and China's Great Wall Industry Corporation entered into a joint venture to provide satellite tracking and launching services (using Chinese launchers).[81]

Nuclear exports. Although Brazil's nuclear export activities have been less extensive than those of Argentina, its capabilities would permit it to export virtually all of the equipment and know-how needed for nuclear weapons. (See Tables 4.1 and 4.2 in Chapter 4.)

In the late 1970s, Brazil sold sizeable quantities of natural uranium to Iraq, which could have contributed importantly to the latter's nuclear weapons effort at the time. During the Iran-Iraq War, moreover, Brazil was a major arms supplier to Baghdad. These links and the two countries' possible collaborative work on missiles could pave the way for weapons-relevant nuclear sales in the future. Brazil's arms sales to Libya similarly could set the stage for Brazilian nuclear transfers to Tripoli. Iran might also be interested in acquiring Brazilian nuclear goods.

Since Brazil is not a party to the Non-Proliferation Treaty, it is not obligated to require recipients to place Brazilian nuclear exports under IAEA safeguards. Moreover, unlike a number of other emerging nuclear suppliers, including, Argentina, China, and South Africa, Brazil has never declared that it will impose this requirement voluntarily, although it has done so in individual agreements for nuclear cooperation. The fact that the CNEN and elements of the Brazilian military demonstrated their disdain for international nuclear controls by mounting clandestine operations to circumvent them adds to concerns that Brazil might engage in irresponsible nuclear sales in the future.

Confidence-building with Argentina. Despite the many disturbing revelations concerning Brazil's unsafeguarded nuclear program during late 1988 and 1989, Brazil and Argentina have continued to exchange visits to each other's nuclear installations and relations have remained cordial. As noted in Chapter 13, during Sarney's visit to Buenos Aires for a summit with Raul Alfonsín in November 1988, he toured the partially completed Ezeiza plutonium extraction plant, apparently the only nuclear facility in Argentina or Brazil that had not yet been visited by scientists or political leaders of the other country.

Confidence-building between the two countries continued after the June 1989 change of government in Buenos Aires. In August of that year, Argentina's new president, Carlos Menem, visited Brazil, where he and Sarney signed a number of agreements, including the extension of a prior accord for joint civilian nuclear research activities and a new agreement for cooperation on space-related projects.[82] By October 1989, moreover, when virtually all of the revelations of suspect Brazilian nuclear activities noted above had been published, Argentine nuclear aides declined to criticize their Brazilian counterparts, choosing instead to stress to reporters the positive aspects of the relationship evident in the increasingly frequent reciprocal nuclear facility visits.[83]

Economic and political crises. Sarney's last year in office was marked by the failure of yet another austerity plan, the return of hyperinflation, and a general loss of public confidence in his ability to govern. In May, labor strikes spurred by the economic crisis were taking place at a rate of 450 per month, and in June, the inflation rate climbed to one percent per day. By early December, inflation for the year was calculated at 1,114 percent. The country's long-awaited elections and the prospect of new leadership appeared to be the crucial safety valves that averted serious social unrest.

Policies under Collor. On December 17, 1989, Fernando Collor de Mello was elected president of Brazil in the country's first direct elections for the office in more than twenty years. Unlike Sarney, Collor possessed a popular mandate and seemed likely to reduce the influence of the Brazilian military, which had played such a prominent behind-the-scenes role in domestic political affairs during Sarney's tenure.

Immediately after taking office on March 15, 1990, Collor decisively seized the reins of power and imposed an unusually tough austerity plan to curb the extreme inflation that had plagued the country throughout the 1980s.[84] He also moved aggressively to gain control of the country's nuclear affairs, naming José Goldemberg, a well-known critic of both Brazil's safeguarded and parallel nuclear programs, as Secretary of Science and Technology.[85] Simultaneously, he ousted Rex Nazareth, the *de facto* father of the latter program, as the chairman of the CNEN and appointed José Luis Santana Carvalho as his successor. Carvalho had served as the Minister for Housing and Urban Affairs under Sarney and had previously been the director of CNEN's office of radiation protection.[86] Collor also replaced the heads of the military services, generally favoring moderates, and took steps to reduce the role of the military within the government.[87]

These personnel changes suggested that a significant alteration in Brazil's nuclear policies might soon follow. Indeed, Collor criticized some elements of Brazil's nuclear program during his campaign, declaring that the nuclear agreement with Germany was "against society's wishes" and that the nuclear program as a whole was "unrealistic" and "needs to be reviewed."[88]

In an interview granted shortly after his appointment, however, Carvalho struck a nationalistic, pro-nuclear-development stance little different from that of his predecessor, declaring that Brazil would "never" accept international supervision—an apparent reference to IAEA safeguards—on that portion of the nuclear program managed by the country's armed forces. Nevertheless, Carvalho stressed that under the Brazilian constitution, all of the country's nuclear activities had to be approved by the Brazilian Congress, and he pledged to facilitate the civilian oversight process.[89] Goldemberg similarly underscored the importance of civilian supervision of the sensitive nuclear field, commenting to the press that "internal control measures" were essential to guarantee that the country did not build nuclear arms.[90] The tone of these comments suggested that the parallel program would receive increased scrutiny by the Collor Administration, and, indeed, within a month of taking office, Collor established an interministerial working group to review all of the country's nuclear activities.[91]

Before Collor took office, there were also indications that he might reconsider Brazil's opposition to joining the Nuclear Non-Proliferation Treaty and to bringing the Treaty of Tlatelolco into force. A number of Collor's advisors appeared to believe that this stance was the product of the outmoded chauvinism of the country's military and that it had caused unnecessary damage to

Brazil's diplomatic relations, especially with the United States.[92] At least one prestigious daily supported this view during the three-month transition period following Collor's election.[93]

Nevertheless, shortly after Collor's inauguration, the country's new Foreign Minister Francisco Rezek declared that Brazil would not adhere to the NPT, arguing that adherence would result in "technological stagnation," presumably because it would require Brazil to place all of its nuclear installations under IAEA safeguards. Although Rezek was silent about the Treaty of Tlatelolco, the fact that it, too, would require full-scope safeguards suggested that no change in Brazil's status under that accord was being contemplated. Goldemberg also rejected the NPT as discriminatory.[94] A visit to Brasilia by U.S. Ambassador-at-Large for Non-Proliferation Richard T. Kennedy on May 23, during which he urged Brazil to ratify the NPT prior to the Non-Proliferation Treaty Review Conference in mid-August 1990, proved fruitless.[95]

Prospects

During the opening months of Collor's tenure, as he was struggling to implement a radical economic reform package and to restructure the military, Collor predictably deferred decisions on major changes in Brazil's nuclear program. Successful efforts to bring inflation under control, however, would greatly enhance his political stature, and it is possible that under these circumstances he might reopen the issue of adhering to the NPT or bringing the Tlatelolco Treaty into effect. Moreover, should South Africa join the NPT during the summer of 1990, as is expected, the development could greatly strengthen the hand of those in Collor's government—and of outside states—urging a similar course for Brazil.

As his appointees gain day-to-day control of the country's nuclear affairs in subsequent months, and as the Brazilian Congress begins to exercise its own prerogatives in this field, it is also likely that the activities in the parallel program will be examined with increasing care. An investigation of the illicit aspects of this program, including the smuggling of equipment from abroad and the use of secret bank accounts, could well lead to new controls on program activities and to possible funding cuts.

Although the Collor Administration had not decided on continuing the exchange of nuclear facility visits with Argentina as of mid-June 1990, there were early indications that it would maintain these links.[96] Indeed, Collor, who was scheduled to visit Buenos Aires in July 1990, may be more inclined to institutionalize these arrangements than his predecessor and to make the mutual visits more scientifically rigorous. With both countries now able to produce 20-percent enriched uranium, such changes could be important to reduce ambiguities that might arise regarding whether weapons-grade material was being produced by one party or the other.[97]

Caribbean Sea
Atlantic Ocean
VENEZUELA
GUYANA
SURINAME
FR. GUIANA
COLOMBIA
Macapa
Manaus
Belem
Amazon River
Fortaleza
Natal
Recife
BRAZIL
Cachimbo
Salvador
PERU
Brasilia
BOLIVIA
Poços de Caldas
Belo Horizonte
São Paulo (IPEN)
Resende
Ipero
Rio de Janeiro
PARAGUAY
Santos
Angra dos Reis
Pacific Ocean
Curitiba
São José dos Campos
Porto Alegre
URUGUAY
ARGENTINA
CHILE
BRAZIL
uranium mining area
0
800
MILES
Atlantic Ocean

Brazil

Power Reactors/Operating or Under Construction

Angra I (light-water/low-enriched uranium, 626 MWe)

- supplier: Westinghouse (U.S.)
- start up: 1982 (commercial operation, 1985)
- fuel source: Initial load, U.S.; current reloads URENCO (W. German, British, Dutch)[a]; Brazilian enriched uranium after 1989 (est.).[b]
- safeguards: yes

Angra II (light-water/low-enriched uranium, 1,300 MWe)

- supplier: Kraftwerk Union (West Germany)
- start up: 1992 (est.)[c]
- fuel source: Initial load, URENCO (W. German, British, Dutch); subsequent loads, Brazilian enriched uranium.[b]
- safeguards: yes

Angra III (light-water/low-enriched uranium, 1,300 MWe)

- supplier: Kraftwerk Union (West Germany)
- start up: 1995 (est.)[c]
- fuel source: Initial load, URENCO (W. German, British, Dutch), subsequent loads, Brazilian enriched uranium.[b]
- safeguards: yes

Uranium Resources/Active Mining Sites/Uranium Mills

- reasonably assured reserves: 192,500 metric tons[d]
- currently active site: Poços de Caldas[d]
- mills in operation: Poços de Caldas, Itataia[d]

Uranium Purification (UO_2)

IPEN, São Paulo

- capacity: more than 10 metric tons per year[e]
- supplier: Brazil, with some W. German equipment.[f]
- start up: 1981 (?)[e]
- safeguards: no

Uranium Conversion (UF_6)

Resende

- capacity: planned 500 metric tons per year (eventually 2000 metric tons per year)[g]
- supplier: P.U.K. (France)[h]
- start up: indefinitely postponed[j]
- safeguards: yes (planned)

IPEN, São Paulo (pilot scale)

- capacity: 90 metric tons UF_6 per yr.[i]
- supplier: Brazil[i]
- start up: 1984[i]
- safeguards: no

IPEN, São Paulo (laboratory scale)

- capacity: 15 metric tons/yr.[i]
- supplier: Brazil (with some foreign assistance)[i]
- start up: 1981-1982[i]
- safeguards: no

Enrichment

Resende

- type: jet-nozzle
- capacity: 5 metric tons of 0.85% enriched uranium planned in 1988[k]
- supplier: Kraftwerk Union (West Germany)[l]
- start up: testing and evaluation to begin in 1990.[d]
- safeguards: yes

Belo Horizonte

- type: jet-nozzle
- capacity: laboratory scale
- supplier: Kraftwerk Union (West Germany)
- start up: 1980[n]
- safeguards: yes

Aramar Research Center, Ipero

- type: ultracentrifuge
- capacity: 400 kilograms of 5% enriched uranium in 1989; small amount of 20% enriched uranium also produced.[k]
- supplier: Brazil
- start up: 1988[z]

IPEN, São Paulo

- type: ultracentrifuge
- capacity: lab scale ("kilogram quantities" of 20% enriched uranium produced)[o]
- supplier: Brazil
- start up: 1986-1987
- safeguards: no

Fuel Fabrication

Resende

- capacity: Planned 100 tons per year for Angra I, II, III[p]
- supplier: West Germany
- start up: 1982 (initial phase)[q]; additional stages, 1990, 1991[m,c]
- safeguards: yes

IPEN, São Paulo (pilot plant)

Reprocessing (Plutonium Extraction)

Resende

- capacity: pilot scale, 10 kg spent fuel per day[r,s]
- supplier: West Germany
- start up: indefinitely postponed[m]
- safeguards: yes[t]

IPEN, São Paulo (laboratory scale)

- capacity: varying estimates[u]
- supplier: Brazil with some West German equipment[f]
- start up: not operating[u]
- safeguards: only when safeguarded fuel is processed[w]

Principal Research Reactors[x,y]

IEAR-1, São Paulo (Pool/highly-enriched uranium,[v] 5 MWt)

- supplier: United States
- start up: 1957
- safeguards: yes

RIEN-1, Rio de Janeiro (Argonaut/medium-enriched uranium, 10 KWt)
• supplier: Brazil
• start up: 1965
• safeguards: yes

Triga-UMG, Belo Horizonte (Triga I/medium-enriched uranium 100 KWt)
• supplier: United States
• start up: 1960
• safeguards: yes

IPEN-Zero Power, São Paulo (light-water/4-percent enriched uranium, 100 w.)
• supplier: Brazil
• start up: 1988
• safeguards: no

Sources and Notes

a. "German Uranium for Angra," *Jornal do Brasil,* translated in *Joint Publication Research Service (JPRS)-Nuclear Development and Proliferation,* January 18, 1982, p. 28.

b. Assis Mendonca, "Germany Denies Having Negotiated Another Process," *O Estado de São Paulo,* translated in *JPRS-TND,* May 13, 1982, p. 21; Charles Thurston, "Brazil Readies Enrichment Plant," *Nuclear Fuel,* December 3, 1984, p. 8. Availability of Brazilian-produced fuel likely to be delayed.

c. *World Nuclear Industry Handbook, 1990* (Sutton, Surrey, U.K.: Nuclear Engineering International, 1990).

d. "Datafile: Brazil," *Nuclear Engineering International* (April 1990): 47.

e. Personal communication with informed U.S. official, 1985.

f. Bernardo Kucinski, "Argentine Boast Spurs Brazil into Race for Bomb," *The Guardian,* reprinted in *JPRS-TND,* February 19, 1982, pp. 6-8.

g. "Researchers Developing Brazil's Know-how on UF-6 Beyond the Plant Now Being Built," *Nuclear Fuel,* September 27, 1982, p. 13.

h. "Uranium Enrichment Plant in Itataia Confirmed," *O Estado de São Paulo,* translated in *JPRS-TND,* January 25, 1982, p. 22.

i. "Fluorine Technology for Enrichment Developed," *Fôhla de São Paulo,* April 23, 1988, translated in *Foreign Broadcast Information Service (FBIS)-Latin America (LAT),* April 27, 1988, p. 35; Mark Hibbs, "Germans Say Brazil Developing Two Production Reactors," *Nucleonics Week,* July 27, 1989, p. 4; "Datafile: Brazil" (see note d).

j. Charles Thurston, "Brazil's First Enrichment Cascade," *Nuclear Fuel,* November 26, 1984, p. 6. "Specifications of IPEN Uranium Enrichment Facility Noted," *Jornal do Brasil,* October 8, 1984. Expansion beyond lab-scale postponed. See note m.

k. "Enriched Uranium Production Goals Previewed," *O Globo,* February 25, 1990, translated in *JPRS-TND,* March 2, 1990, p. 18; "Uranium Enrichment Process Perfected," *O Globo,* April 9, 1989, translated in *FBIS-LAT,* April 11, 1989, p. 36; David Albright, "Bomb Potential for South America," *Bulletin of the Atomic Scientists* (May 1989): 16.

l. "KWU Recommends Development of Jet-Nozzle Enrichment Process Here," *O Estado de São Paulo,* translated in *JPRS-TND,* June 3, 1982, p. 22.

m. Rik Turner, "Brazil's Sarney Defines Emphasis for National Fuel Cycle Development," *Nuclear Fuel,* August 11, 1986, p. 10.

n. Charles Thurston, "Critics See Brazil as Taken For a Ride on Its 'Last Train' to SWU Technology," *Nuclear Fuel,* April 26, 1982, p. 3.

o. Rik Turner, "Brazil's Parallel Program," *Nuclear Fuel,* October 19, 1987, p. 3; "Implications of Mastery of Uranium Enrichment," *Veja,* September 9, 1987, translated in *JPRS-TND,* December 1, 1987, p. 12.

p. "Operation of Fuel Elements Factory in Resende Discussed," *Manchete* (Rio de Janeiro), translated in *JPRS-TND,* December 6, 1982, pp. 18-21.

q. Juan de Onis, "Brazil's Crash Program Slows to Realistic Pace," *International Herald Tribune,* October 1982.

r. William W. Lowrance, "Nuclear Futures For Sale: To Brazil From West Germany, 1975," *International Security* (Fall 1976): 152.

s. "Resende Site of Uranium Reprocessing Plant Confirmed," *O Estado de São Paulo*, translated in *JPRS-TND*, May 10, 1983, p. 21. Later start-up date probable; reportedly little construction to date (personal communication with informed U.S. official.)

t. International storage also required for any plutonium reprocessed from URENCO fuel.

u. "Brazil Says It Now Produces Small Amounts of Plutonium," *Washington Post*, December 18, 1986. Only milligram quantities of plutonium have been produced, if that, according to Brazilian Embassy officials in Washington. See also, Leonard S. Spector, *Nuclear Proliferation Today* (New York: Vintage Books, 1984), pp. 247-252 (possible startup in 1983). Capacity estimates range from minute quantities (scientist Roberto Hukay in "Report on IPEN Reactor," *Jornal do Brasil*, translated in *JPRS-TND*, February 18, 1983, p. 5) to five kg plutonium per year Kucinski, "Argentine Boast Spurs Brazil into Race for the Bomb.") But see, Hibbs, "Germans Say Brazil Developing Two Production Reactors" (see note i), suggesting plant may have operated in mid-1980s, producing small quantities of plutonium.

v. Currently 90-percent enriched uranium, soon to be converted to 20-percent enriched uranium.

w. Currently, only safeguarded fuel is available.

x. James Katz and Onkar Marwah, *Nuclear Power in Developing Countries* (Lexington, Mass.: Lexington Books, 1982), p. 100.

y. International Atomic Energy Agency, *The Annual Report for 1981*, p. 70.

z. "IPEN Head Guarantees Nuclear Project Safety," *O Estado de São Paulo*, September 9, 1987, translated in *FBIS-LAT*, September 11, 1987, p. 11; "Funding Needs, Supplies of Parallel Program Discussed," *Exame*, October 14, 1987, translated in *JPRS-TND*, February 24, 1988, p. 21.

Part V
Africa

The only sub-Saharan African nation posing a significant proliferation risk today is South Africa, whose ability to produce nuclear arms and apparent interest in acquiring them may have led it to build a small nuclear arsenal of perhaps 15 to 25 nuclear devices.

In late 1989, South Africa's new State President, Frederik W. de Klerk appeared to open a new era in South African nuclear policy. De Klerk accelerated negotiations on the country's joining the Nuclear Non-Proliferation Treaty (NPT) and, in early 1990, shut down the uranium enrichment facility at Valindaba that long had been assumed to be the country's source of nuclear-weapon material. By mid-1990, there was increasing optimism that Pretoria would soon ratify the nuclear accord. Despite the difficulties of implementing the treaty's comprehensive inspection requirements in a country presumed to have a stockpile of weapons-grade nuclear material of indeterminate size, South Africa's adherence to the accord would be a singular achievement for international efforts to curb the spread of nuclear arms.

It was anticipated that most of the "front-line" black African states neighboring South Africa would simultaneously join the treaty at the time of Pretoria's ratification. This action would be a valuable demonstration of support for the NPT but would be largely symbolic, inasmuch as no other country in the region has more than the most rudimentary nuclear program.

During the 1980s however, Nigerian leaders repeatedly declared their interest in obtaining nuclear arms, despite the fact that Nigeria is a party to the Non-Proliferation Treaty. In August 1987, for example, former Nigerian Foreign Minister Bolanji Akinyemi, implicitly invoking the nuclear threat from South Africa, called for Nigeria to develop nuclear weapons "to forestall nuclear blackmail" of the black race.[1] Earlier, in April 1986, when asked whether Nigeria should develop nuclear weapons, Akinyemi had declared, "I can't give a direct answer but I'll say that a country the size of Nigeria, with its role and status in the world, cannot rule out any option."[2] Nigerian President Ibrahim Babangida echoed these sentiments in a contemporaneous statement, rejecting the principle of non-proliferation by stating that "the argument about the non-proliferation of nuclear weapons, besides its social overtones, now seems to border on the irresponsibility of [sic] weaker nations."[3]

Notwithstanding these statements, Nigeria does not appear to have taken any steps toward developing nuclear weapons.[4] Indeed, except for South Africa, nuclear programs in sub-Saharan Africa are virtually non-existent. No other state in the region possesses a nuclear power plant, and only Zaire has a research reactor, a one-megawatt, U.S.-supplied unit. Uranium mines and ore-processing mills in Gabon, Niger, and newly independent Namibia are the only other significant nuclear installations. While the long-term potential for further proliferation in Africa remains, it is hard to imagine a second nation obtaining a nuclear weapons capability for decades to come.

15
South Africa

South Africa has had the capability to produce enough highly enriched uranium not subject to international non-proliferation controls to manufacture two to three nuclear weapons annually since 1980 or 1981. In view of past South African activities indicating an intent to develop nuclear arms, it has been widely assumed that the country has used this capability to build up a stockpile of weapons-grade uranium, in effect acquiring a *de facto* nuclear arsenal of 15 to 25 nuclear devices by the end of 1989.

During the past year, however, tensions between South Africa and its neighbors eased significantly, and the country's new State, President Frederik W. de Klerk acted to liberalize South Africa's repressive racial policies (*apartheid*) and to end South Africa's isolation from the international community. These developments set the stage for a major shift in South Africa's nuclear posture. In early 1990, Pretoria voluntarily closed the key facility that had provided it with its presumed stockpile of weapons-grade nuclear material. By mid-year, moreover, South Africa was poised to sign the Nuclear Non-Proliferation Treaty (NPT), a step that would entail its formal renunciation of nuclear arms and the placing of all of its nuclear materials and activities under International Atomic Energy Agency (IAEA) safeguards.

Implementing the treaty in a country suspected of having accumulated a stockpile of weapons-grade uranium would pose a number of challenges. Assuming these can be met, the possibility of South Africa's joining the treaty would be an exciting precedent and would constitute the most significant rollback of nuclear weapons proliferation since the termination of the South Korean and Taiwanese nuclear weapon programs in the mid-1970s.

Background

South Africa's nuclear program dates to the late 1940s, when the country began to supply uranium to the United States and Great Britain for use in their respective nuclear weapon development efforts.[1] During the 1950s, as the country's uranium production grew, it also expanded its nuclear research capabilities, training numerous nuclear specialists in the United States and obtaining the 10-megawatt SAFARI I research reactor under an agreement for cooperation with Washington. The reactor, which is subject to IAEA safeguards, began operating in 1965 and, until the mid-1970s, was fuelled with U.S.-supplied highly enriched uranium. South Africa also participated in the establishment of the IAEA in 1957 and was given a permanent seat on its board of governors, as the country with the "most advanced" nuclear program in its region.

The first public hint that Pretoria might be considering the development of nuclear weapons came in 1965, the year after China became the fifth country to detonate a nuclear device, when Dr. Andries Visser, a member of South African Atomic Energy Board, declared, "We should have the bomb to prevent aggression from loud-mouthed Afro-Asiatic states . . . money is no problem. The capital for such a bomb is available."[2] Later that year Prime Minister H. F. Verwoerd offered a second veiled hint of South Africa's interest in nuclear arms during the inauguration ceremony for the SAFARI I reactor, urging that with respect to nuclear materials, "It is the duty of South Africa to consider not only the military uses of the material, but also to do all in its power to direct its uses to peaceful purposes."[3]

The 1960s were a period of rapid decolonization in Africa, and South African leaders became increasingly concerned about the nation's long-term security, as evidenced by the sixfold increase in South Africa's defense expenditures between 1961 and 1968.[4] These anxieties undoubtedly contributed to Pretoria's decision not to sign the Nuclear Non-Proliferation Treaty when it was opened for signature in 1968. Indeed, in December 1968, General H. J. Martin, then-army chief of staff, was quoted as stating that South Africa was ready to make its own nuclear weapons and was also pursuing the development of ballistic missiles. His statement was quickly repudiated by then-Minister of Defense (later Prime Minister and State President) Pieter W. Botha, but it indicated that some members of the South African defense establishment remained interested in developing nuclear arms.[5]

Uranium enrichment capability; U.S. response. By this point, South Africa had apparently begun to develop the technology for enriching uranium, and, in July 1970, Prime Minister John Vorster announced that the country had mastered a new enrichment process, which was "unique in its concept." The method is believed to be a variation of the German-developed Becker jet-nozzle process, which South Africa has dubbed the "helikon" process. It is widely believed that in the late 1960s and early 1970s Pretoria obtained clandestine assistance from the German firm STEAG (Steinkohlen-Elektrizitats AG) in developing a pilot-scale enrichment plant using this technology at Valindaba.[6] The facility began operating in April 1975, although it did not reach full capacity for a number of years. The pilot plant has never been placed under IAEA safeguards, and most observers assume that it was subsequently used to produce weapons-grade uranium, thereby providing Pretoria with a ready nuclear weapons option.[7]

As the unsafeguarded pilot plant, with its obvious military potential, was nearing completion, the United States appeared relatively unconcerned about South Africa's growing capabilities.[8] In early 1974, for example, the Nixon Administration agreed to an amendment of the U.S.-South Africa agreement for nuclear cooperation to permit expanded nuclear sales to Pretoria. The amendment became effective on June 22, barely a month after India's nuclear detonation reawakened international concerns over the dangers of nuclear proliferation. On July 12, responding to that event, the vice-president of the South African Atomic Energy Board, Dr. Louw Alberts, declared that the country had the technology and resources to produce an atomic bomb.[9] Despite this provocative reminder of South Africa's refusal to renounce nuclear arms, a month later, Washington signed a contract with Pretoria to enrich the uranium South Africa would need to fuel two large power reactors it was planning to build at Koeberg.

Although both the U.S. enriched fuel and the reactors were to be subject to IAEA safeguards, by publicly signing the new agreement and the subsidiary contract, Washington, in effect, declared that it saw nothing amiss in Pretoria's overall conduct of its nuclear affairs. Judging from a September 4, 1974, U.S. Central Intelligence Agency analysis, Washington was well aware at the time of the military potential of the South African nuclear program, but was relatively unalarmed by it. The analysis concluded:

> . . . [South Africa] apparently has developed a technology for enriching uranium that could be used for producing weapons-grade material. South Africa probably would go forward with a nuclear weapons program if it saw a serious threat from its African neighbors beginning to emerge. So serious a threat is highly unlikely in the 1970s.[10]

Whatever the CIA's views, the U.S. Congress began to press for a halt to U.S. nuclear cooperation with South Africa because of growing fears over the country's nuclear potential and because of concerns that cooperation in this sensitive area gave the impression of condoning Pretoria's racial policies. As a result of these efforts, President Gerald R. Ford suspended fuel shipments for the SAFARI I reactor in early 1975. The following year, after the Valindaba pilot plant had begun operating, congressional opposition led to the cancellation of the planned sale of two U.S. nuclear power reactors to South Africa; in late May, France was given the contract for the Koeberg units instead.[11]

Growing insecurity; nuclear test site. Contrary to the CIA's September 1974 projection, during the mid-1970s, South Africa's concerns over its long-term security seriously intensified. In 1975, following the leftist revolution in Portugal the previous year, Mozambique and Angola gained their independence. Mozambique quickly established a Marxist government. In Angola, a civil war ensued, which eventually brought to power the Marxist Popular Movement for the Liberation of Angola, a faction which was heavily supported by Soviet advisers and Cuban troops. Rhodesia, too, was undergoing change that would lead to the advent of a black government there in 1980 (when the country was renamed, "Zimbabwe"). By the end of the decade, South Africa's defense establishment would declare that the country faced the threat of "total onslaught" by Soviet-backed forces in the front-line states. At home, meanwhile, the memory of the 1976 Soweto riots underscored the danger of internal violence against South Africa's Afrikaner regime.

Pretoria responded to these mounting challenges by building up its military power, and defense expenditures doubled between 1975 and 1977.[12] South Africa also pursued a strategy of destabilizing its neighbors by supporting anti-government guerrillas in the Angolan and Mozambican civil conflicts and mounted armed incursions there and in other front-line states, generally targeted at training camps and headquarters of forces of the Namibian independence movement and of the anti-*apartheid* African National Congress, which sought to overthrow the white government in Pretoria.

During this period, Pretoria also hinted that it might be considering the development of nuclear arms. In May 1976, South African Prime Minister John Vorster, when asked whether the country's defenses included a nuclear capability, reiterated that the country possessed the option of developing nuclear arms, stating, "We are only interested in the peaceful applications of nuclear power. But we can enrich uranium, and we have the capability. We did not sign the Nuclear Non-Proliferation Treaty."[13]

Indeed, by 1976, according to former U.S. officials, South Africa had begun to take steps toward developing nuclear weapons, and in August 1977, Soviet and U.S. satellites detected a South African nuclear test site in the Kalahari Desert.[14] Under intense international pressure, Pretoria soon agreed to dismantle the site and pledged not to build or test nuclear arms.[15]

At the time the Kalahari site was discovered, U.S. officials assumed that South Africa had obtained weapons-grade uranium from the pilot-scale plant at Valindaba and that a test might be imminent.[16] Subsequently, however, additional data on the Valindaba plant persuaded U.S. analysts that Pretoria could not have produced enough highly enriched uranium for a test in 1977 and that South African nuclear planners apparently had prepared the test site at that juncture on the basis of over-optimistic projections. Sufficient material for a test was apparently not available until 1980 or 1981. Nonetheless, the Kalahari test site episode provides convincing evidence that in the late 1970s,

Pretoria planned to use its unsafeguarded nuclear capabilities for the development of nuclear explosives.[17]

IAEA, U.N. sanctions. Later in 1977, the United Nations Security Council adopted a mandatory arms embargo against South Africa. The embargo had been under consideration when the Kalahari test site was first exposed, and some observers believe that the crisis over its discovery contributed to the passage of the arms embargo resolution.[18] This setback at the UN followed Pretoria's ouster from its permanent position on the IAEA Board of Governors in June. Both episodes undoubtedly intensified South Africa's sense of isolation.

In 1979, Pretoria suffered another setback at the IAEA, when its credentials were rejected at the organization's General Conference. Rather than risk a repetition of the episode, South Africa has chosen not to attend subsequent general sessions of the agency.

VELA incident. On September 22, 1979, a U.S. VELA satellite passing above the South Atlantic observed a double flash of intense light—the characteristic signature of a nuclear test. A panel of experts convened by President Jimmy Carter determined that the event "probably" was not a nuclear detonation, but the panel's conclusions were disputed by the Defense Intelligence Agency and the Naval Research Laboratory. As a result, the event remains shrouded in uncertainty,[19] although in recent years, the weight of opinion within the U.S. government appears to have moved toward the conclusion that the event was an Israeli test using South African facilities. If so, it is possible that Pretoria obtained data from the event.

Weapon capability obtained. In April 1981, Frederik W. de Klerk, then South Africa's Minister of Mineral and Energy Affairs (and now its State President), announced that South Africa had produced uranium enriched to 45-percent, which it intended to use to fuel the U.S.-supplied SAFARI I research reactor.[20] Since the same enrichment plant could be employed to improve 45-percent enriched uranium to the 90-percent or greater level normally used for weapons, U.S. analysts now date South Africa's arrival as a state capable of producing nuclear arms from 1980 or 1981.[21]

In theory, South Africa could produce approximately 110 pounds (50 kilograms) of highly-enriched uranium per year at the Valindaba pilot-scale plant.[22] Pretoria's steadfast refusal to adhere to the NPT throughout the 1980s or to place the Valindaba enrichment plant under IAEA safeguards has led U.S. officials and other observers to assume that the country has, in fact, stockpiled weapons-grade uranium from Valindaba as part of a nuclear weapons program, although the size of this stockpile can only be surmised from published sources, and it is not known whether Pretoria has fabricated complete nuclear devices.[23]

U.S. nuclear embargo; European and Chinese fuel sales. Pretoria's refusal to place its entire nuclear program under IAEA safeguards led to the formal termination in 1980 of U.S. nuclear reactor and fuel exports under the terms of the 1978 Nuclear Non-Proliferation Act.[24] As a result, South Africa became ineligible to receive the U.S. low-enriched uranium fuel that it had ordered in 1974 for the two French-supplied nuclear power reactors at Koeberg. These were scheduled to come on line in the early 1980s, posing a difficult problem for Pretoria. Washington, however, did not exploit the leverage inherent in this situation in order to press South Africa to accept added safeguards on its nuclear program.

In 1981, the newly inaugurated Reagan Administration, concerned about Soviet inroads in the region, attempted to rebuild ties with Pretoria, which had been badly frayed during President Jimmy Carter's years in office. Reagan adopted a new, non-confrontational strategy, known as "constructive engagement," in dealing with South Africa on a wide range of issues, including the ending of *apartheid*.[25] In the nuclear field, consistent with this approach, Washington promised the government of Prime Minister P. W. Botha that it would authorize nuclear exports to Pretoria that were not expressly prohibited by law, in the hopes that this would permit a dialogue on other aspects of Pretoria's nuclear program and eventually lead to additional restraints.[26]

In keeping with this stance, the Reagan Administration, rather than seeking to persuade other nuclear suppliers to embargo nuclear fuel sales for the Koeberg reactors until South Africa accepted added safeguards at Valindaba, chose instead to turn a blind eye when two U.S. uranium brokering firms arranged for South Africa to purchase a significant portion of the needed fuel from European sellers.[27] Washington also authorized American companies to provide technical services for the Koeberg reactors and granted export licenses for a number of dual-use commodities destined for the installations.[28] The European fuel, like the Koeberg reactors, was subject to IAEA safeguards.

Concurrently, Pretoria is also believed to have obtained 60 tons of low-enriched uranium not subject to IAEA safeguards from the People's Republic of China.[29] This material could have been used in the Koeberg units, but could also have served (in hexafluoride form) as feed for the Valindaba pilot-scale enrichment plant in lieu of natural uranium, potentially increasing the facility's output of unsafeguarded weapons-grade uranium at least three-fold.

1984: Export restraint. Washington's flexible approach appeared to have a modest success in early 1984, when the Botha government declared that its own nuclear exports would be made only under strict controls, comparable to those adopted by the Nuclear Suppliers Group, and when it agreed to open negotiations with the IAEA on the application of the agency's safeguards to a second, semi-industrial-scale uranium enrichment plant that it had begun to build at Valindaba. However, the Botha government explicitly rejected calls to place the smaller, pilot-scale enrichment plant at the site under comparable controls, thereby retaining the country's nuclear weapons option.[30]

Also in 1984, South Africa's strategic environment improved somewhat when Pretoria negotiated a cease-fire agreement with Angola in March, and a similar accord with Mozambique in April. Under the agreements, the black-governed states agreed to cease providing sanctuary to anti-South African guerrillas, while Pretoria agreed to stop supporting anti-government rebels in both countries. The accords soon broke down, however, as Pretoria resumed its efforts to keep its neighbors off balance through a series of military incursions and renewed support for rebel forces. Within South Africa, meanwhile, the adoption of a new constitution in September 1984, which granted limited new powers to Indian and mixed race groups, while continuing to deny even these rights to the country's black majority, triggered widespread unrest in black urban "township" areas. This, in turn, led to increased repression and violence by the government.

1985: Repression spurs new sanctions. Through mid-1985, the Western nuclear supplier states other than the United States had been reluctant to use sanctions against Pretoria as a means for gaining its acceptance of added safeguards. In the summer of 1985, however, the country's nuclear program became a target of pressure from this quarter after P. W. Botha, whose title changed to State President under the new constitution, declared a state of emergency, giving South African security forces broad new powers to crack down on black protesters. In July, for example, France imposed an embargo on new nuclear trade arrangements with South Africa in response to the new repression.[31]

In September, at the insistence of Congress and against his preference, President Reagan also imposed a series of additional trade sanctions against South Africa by executive order. These sanctions included an embargo on all classes of nuclear exports to Pretoria, including those still permitted under the Non-Proliferation Act.[32] Like the French nuclear trade sanctions, the U.S. embargo was imposed exclusively as an expression of American disapproval of *apartheid*, however, and not in direct response to Pretoria's nuclear activities.[33]

Pressures on the South African government continued to mount during late 1985 and early 1986, despite Botha's announcement of a series of carefully limited reforms of *apartheid*, as the British Commonwealth nations, the European Community, and the United States pressed Botha to begin substantive negotiations on a transition to majority rule. Botha's response to these efforts throughout the spring and summer of 1986 was one of bitter defiance—and sometimes outright provocation—as he attempted to demonstrate that South Africa's domestic policies could not be dictated by other states.

Botha's intransigence was met, however, with decreased tolerance by his opponents overseas. In August, six Commonwealth states voted to impose sanctions on South Africa and, one month later, the European Community did as well, although these sanctions appeared slightly less harsh.[34] Then, in October, the U.S. Congress overrode President Reagan's veto and enacted the Comprehensive Anti-Apartheid Act.

The new U.S. law prohibited further imports of South African uranium ore and oxide and codified the ban in the Reagan executive order on all classes of U.S. nuclear exports to South Africa. Unlike the executive order, however, the legislation specified that the *export* ban would be lifted if South Africa ratified the NPT or placed all of its nuclear activities under IAEA safeguards. This provision was intended to give Pretoria an incentive to take this step.[35]

NPT offer. As noted above, in mid-1984, South Africa had attempted to put off its IAEA critics by suggesting that it would accept IAEA safeguards on the semi-commercial-scale uranium enrichment plant it was building at Valindaba. Negotiations with the IAEA on the details of a safeguards agreement for the facility proceeded for two years, but were broken off in 1986 because of Pretoria's insistence on conditions that were known to be contrary to long-standing IAEA safeguards practice.[36] The resulting impasse precipitated a crisis within the agency, where it was expected that, at the urging of a group of African states led by Nigeria, the IAEA General Conference would vote to expel South Africa from the organization at its 1987 meeting.

As the General Conference was convening in Vienna, State President Botha announced that South Africa "hopes that it will soon be able to sign the NPT and has decided to open discussions with others to this end."[37] The declaration led the Soviet Union—which had backed the African bloc campaigning for such a move at an IAEA Board of Governors meeting in June—to withdraw its support for immediate expulsion, and the conference adopted a compromise that postponed a vote on the issue until its September 1988 meeting.[38]

Growing regional tensions. By late 1987, South Africa had come under increasing international pressure on several fronts. In addition to international economic sanctions and threats of expulsion from the IAEA, South African troops were also heavily engaged in Angola in support of National Union for the Total Independence of Angola (UNITA) rebel forces. Some clashes, according to South African military spokesmen, pitted South African forces against Cuban troops and Soviet advisers fighting alongside Angolan government forces.[39] During October 1987, South African troops assisted UNITA in a decisive battle to repel an Angolan government offensive supported by Cuban mechanized infantry, tank units, and MiG-23 aircraft. Referring to the episode in November, South African Defense Minister Magnus Malan declared that the defeat of UNITA would have brought South Africa "to the brink of the abyss" eventually leading to Communist domination of the region.[40] In a subsequent press conference he stated, "The reality is that if the Cubans and Russians should destroy UNITA, there is no guarantee for us they will stop in the southeast of Angola."[41]

In light of such concerns, through early 1988, elements of South Africa's leadership undoubtedly saw the country's nuclear capabilities as a valuable deterrent against such possible Soviet and Cuban aggression. South Africa's continuing efforts to destabilize Mozambique and to maintain pressure on Zambia, Botswana, and Zimbabwe through sporadic raids on anti-South Africa guerrilla bases provided further evidence of Pretoria's continuing security anxieties.[42]

Political and Technical Developments

Capability declared. Though South Africa's nuclear policies would begin to change significantly in the latter part of 1989, in mid-1988, there was little reason to anticipate such a shift. Despite State President P. W. Botha's statement in September 1987 that South Africa hoped to sign the Non-Proliferation Treaty within a year, the matter remained at an impasse. Not only did Pretoria decline to join the treaty, but in the intervening months, two South African officials openly declared that the country possessed the capability to manufacture nuclear weapons.

In July 1988, Foreign Ministry spokesman Eric de Montille denied that South Africa possessed twenty nuclear weapons, as asserted in an American press report, but went on to say that any country that maintained an "enrichment program the size of South Africa's would naturally possess a nuclear weapons capability."[43] Similarly, in August, South African Foreign Minister Roelof (Pik) Botha, when asked in Vienna whether his country had nuclear weapons declared, "We have the capability to make one if we want to," offering no further elaboration on the point.[44] Though these comments merely confirmed what was already widely understood about the country's nuclear capabilities, they suggested that Pretoria had little interest in gaining international approval by moderating its nuclear stance.

At the 1988 IAEA General Conference, South Africa again temporized, as it had in 1987. To deflect continued pressure for its expulsion from the organization, it declared in a diplomatic note that signing the NPT remained its "objective and desire." Pretoria also stated, however, that before taking this step, it wished to determine whether anti-*apartheid* nuclear trade restrictions imposed by the United States and other Western countries would be lifted after it became a party to the treaty. While offering little of substance, the declaration was sufficient to allow the United States, Britain, and the Soviet Union to argue successfully that the bid to expel South Africa from the IAEA should be deferred for another year.[45]

New enrichment plant. The start-up of the second uranium enrichment plant at Valindaba in August 1988 added to concerns about South Africa's nuclear capabilities. The new plant was a semi-commercial-scale facility, which South African officials declared would be used to produce low-enriched uranium fuel for the country's two IAEA-safeguarded nuclear power reactors at Koeberg.[46] Like the pilot-scale plant, the new installation used the helikon jet-nozzle enrichment process. Through the end of 1989, the facility operated at an unspecified level below its full capacity of 82.5 short tons (75 metric tons) of 3- to 4-percent enriched uranium per year.[47] IAEA safeguards have not been applied to the facility. Thus, despite official declarations of the intended use of the facility, Pretoria remained free to employ it in support of a nuclear weapons program.

In theory, a plant of the size of the new Valindaba unit could be configured to produce 3,300 pounds (1,500 kilograms) of weapons-grade uranium annually—enough for 60 to 100 nuclear weapons.[48] Even if configured to produce low-enriched uranium at the 3- to 4-percent enrichment level, the facility could nonetheless be used to produce weapons-grade material, enriched to 90 percent or more, by extending the time that uranium remained in the plant's thousands of jet-nozzle enrichment units and by recycling partially upgraded material through the plant for a second time. In addition, low-enriched uranium produced in the plant could be upgraded in the older pilot-scale facility. Indeed, by starting with low-enriched material rather than natural uranium, the older plant's output of weapons-grade uranium could be increased at least three-fold, from about 110 pounds (50 kilograms) of weapons-grade uranium to 330 pounds (150 kilograms) of the material, enough for as many as nine nuclear devices.

Regional settlements. As Pretoria expanded its nuclear capabilities and rebuffed the extension of IAEA safeguards on its key nuclear plants, its overall security environment began to improve dramatically. In December 1988, South Africa signed a series of important undertakings with its neighbors, collectively known as the Angola-Namibia Agreement, which largely eliminated the external threats to the country. Often exaggerated by South Africa's right-wing leadership in the past, these threats may nonetheless have been an important motive for the country's bid to acquire a nuclear weapons capability. The Angola-Namibia Agreement was midwifed by the United States and had the backing of the Soviet Union. The latter had supplied arms and economic aid to Pretoria's regional adversaries, but by the late 1980s was seeking to reduce its military commitments in the region.

Under the regional pact, South Africa agreed to withdraw from Angola and undertook not to intervene there or in Mozambique on behalf of the rebel groups that it had covertly supported throughout the 1980s. In addition, Pretoria agreed to grant independence to Namibia, the former German colony, which it had controlled since World War I. In return, Cuba agreed to withdraw its 50,000 troops from Angola over a 27-month period. Angola, in turn, undertook to remove guerrilla training bases of the anti-*apartheid* African National Congress and to prevent guerrillas fighting to end South African rule in neighboring Namibia from interfering in the independence process.[49]

The net result of the agreement was to reduce greatly the external military threat to South Africa. Indeed, in December 1989, following the initial withdrawals of Cuban forces from Angola and elections for a constituent assembly in Namibia, Pretoria halved the mandatory service period for the country's military conscripts and announced major cuts in military spending.[50]

Domestic liberalization. In January 1989, shortly after the Angola-Namibia agreement was concluded, P. W. Botha suffered a debilitating stroke, a development that led to a major shift in the country's leadership. In February, Education Minister F. W. de Klerk, a far more pragmatic figure, replaced Botha as head of the National Party. When Botha was forced to resign as State President in August, de Klerk was appointed to replace him by a party caucus. The following month, de Klerk was elected to a full term in that office in nation-wide balloting—from which South Africa's blacks were excluded—ushering in a new era of politics in the country.

Immediately upon taking office, de Klerk began to implement a number of unprecedented reforms to ease the country's policy of racial separation and to expand the civil rights of blacks, although he continued to oppose steps that might lead to black majority rule. In September, he ended the ban on peaceful protests against *apartheid*, which had been imposed under the state of emergency that had been renewed annually since June 1986; a series of massive anti-government rallies followed. The next month, he released Walter Sisulu and six other long-imprisoned leaders of the outlawed African National Congress (ANC) and the Pan-Africanist Congress.

In late November, de Klerk dismantled the State Security Management System. This network of committees, comprising officials at various levels of government, was dominated by the police and the military and had been responsible for suppressing the township uprisings that had begun in 1984 and continued into 1987. At the same time, de Klerk abolished the National Security Council, a body dominated by intelligence, police, and military officials that had increasingly displaced the policymaking role of the cabinet.[51]

Continuing his drive to open up the country's political system, on February 2, 1990, de Klerk formally legalized three leading anti-*apartheid* organizations, the ANC, the Pan-Africanist Congress, and the South African Communist Party; the activities of all three groups had been prohibited for decades. Nine days later, in an act of enormous political consequence, he released Nelson Mandela, the jailed chief of the ANC and *de facto* leader of the anti-*apartheid* movement, after 27 years of incarceration. Mandela's release created the conditions necessary to open a serious dialogue between the nation's blacks and the government on the country's future.

These extraordinary changes in South African domestic affairs and the formal granting of independence to Namibia on March 21, 1990, did much to restore South Africa's international respectability. On February 10, for example, anticipating Mandela's imminent release, President George Bush invited de Klerk to Washington, the first such invitation to be extended to a South African leader in many years. (Nine days earlier, Bush had invited Mandela to the White House, once the latter regained his freedom.) On February 21, similarly, after Mandela had been freed, Great Britain ended a ban on new investments in South Africa, breaking ranks with the other members of the European Community. Following the Namibian independence celebration, moreover, U.S. Secretary of State James Baker met with de Klerk in Cape Town, the first visit by a U.S. Secretary of State to South Africa since 1978. And, in May, de Klerk visited nine Western European capitals, many of which had shunned his predecessor.

On June 7, de Klerk took yet another step toward political liberalization by lifting the repressive state of emergency in three of South Africa's four provinces.[52]

NPT adherence reexamined. Against this background of improved regional security, domestic political liberalization, and increased international acceptability, de Klerk began what appeared to be a major shift in South Africa's nuclear policies. At the September 1989 IAEA General Conference, after de Klerk had become State President, the Soviet Union's representative to the body told journalists that he had "received assurances" that South Africa remained interested in joining the NPT and that he was "optimistic" Pretoria would ultimately take this step. Although South Africa's ambassador to the IAEA advised the organization only that the country was considering adherence in a "serious manner," growing hopes for progress on the issue under the new de Klerk government led to an agreement to postpone for one more year consideration of whether to expel South Africa from the organization.[53]

In December, South African Foreign Minister Pik Botha met in Vienna with U.S., Soviet, and British representatives to discuss the country's adherence to the treaty. Unlike previous talks dating from 1987, which had yielded very little, Botha reported that "progress" had been made at the December 1989 session and that the participants would meet again in early 1990 on the matter.[54] In January 1990, a well-regarded South African newsletter reported that the country was "prepared" to sign the treaty and in March, American, Soviet, and British diplomats in Vienna were quoted as stating that Pretoria was indeed poised to take this step.[55] By May, U.S. officials, noting that there had been additional diplomatic contacts on the matter, said they considered the country's NPT adherence to be imminent, and in a May 21 interview, de Klerk himself stated that "we are moving toward a situation where we are making progress on how to handle this problem in the future."[56] As of June 1990, it was widely assumed that South Africa would join the treaty before the mid-August NPT Review Conference so that it would be able to participate in the session and prolong the favorable international attention that would no doubt follow its action.

Several front-line states, including Mozambique, Angola, and Zimbabwe, were expected to adhere to the treaty simultaneously with Pretoria. Since none has a nuclear program, the step is apparently intended to be a political gesture, expressing satisfaction with South Africa's renunciation of nuclear arms.

South Africa's possible reasons for reexamining its twenty-two year opposition to the NPT can only be surmised. Presumably, the easing of tensions with its neighbors has eliminated any military rationale for acquiring nuclear arms, even as symbols of strength, while Pretoria's desire to gain international acceptance offered a strong incentive to join the pact. De Klerk may also hope that adherence will lead to the lifting of restrictions on nuclear trade with South Africa that have been imposed by the United States and a number of other advanced nations, in particular restrictions on the importation of South African uranium and on the sale of additional nuclear power plants to Pretoria.[57]

Pilot enrichment plant closed. In January 1990, South African nuclear officials announced that the country would shut down the older, pilot-scale uranium enrichment plant at Valindaba, the source of its presumed stockpile of weapons-grade uranium.[58] Operations at the plant ceased in early February, according to U.S. and South African officials.[59] Although South African officials stated that the step was taken as a cost-cutting measure, it was seen by U.S. diplomats as a further indication that Pretoria intended to renounce nuclear arms and join the NPT.

The facility is one of less than a dozen installations outside the industrialized world that are known to be capable of producing weapons-grade nuclear material free from IAEA safeguards. Moreover, for more than a decade, South Africa had refused to restrict activities at the facility in any fashion, despite pressure from the United States and other concerned governments. Thus de Klerk's decision to close the plant was an event of considerable importance.

Unfortunately, it has not precluded the country's continued production of unsafeguarded weapons-grade uranium, since the material could also be produced in the new Valindaba plant. As of mid-1990, there was no reason to believe that the new facility was being used for this purpose. Indeed, South African nuclear aides announced in May that they had successfully produced a re-load for one of the Koeberg power reactors, indicating that at least part of the output of the new enrichment facility was in the form of low-enriched uranium, rather than weapons-grade material.[60]

Starting inventory ambiguities. If Pretoria does accede to the NPT, the secrecy that has surrounded its nuclear activities in the past will make it difficult to ensure that all of the enriched uranium it has produced in previous years at the two Valindaba enrichment plants is properly accounted for and placed under IAEA safeguards. This matter will have to be resolved in order to establish a starting inventory for the application of safeguards on the country's entire nuclear program. Of particular concern is the stockpile of weapons-grade uranium that Pretoria is presumed to have accumulated since 1980 or 1981 using the pilot-scale Valindaba enrichment facility and, possibly, the new larger plant there. Unless South Africa can show persuasively that it has made all of this material available for inspection, fears will remain that it has retained a secret cache of bomb-grade uranium—or of nuclear weapons themselves.

A key indicator of how much weapons-grade material South Africa has produced is the capacity of the pilot-scale plant at Valindaba. Under IAEA rules, however, IAEA inspectors would be entitled to visit the plant—and to assess its capacity first-hand—only if South Africa still had nuclear materials there. These materials may no longer be present now that the plant has been shut down. Unless Pretoria voluntarily permitted the IAEA to visit the unit, the IAEA would have little choice but to rely on South Africa's unverifiable declarations as to the plant's capabilities, leaving major questions as to the accuracy of the organization's initial inventory.

Even if the IAEA were given access to the unit, significant uncertainties would still remain, because the plant's operating history would be difficult to verify. As a practical matter, the IAEA would have to rely on South African assertions to establish when the plant began operating, whether it operated at full capacity throughout its lifetime, whether it produced weapons-grade uranium or material of lower enrichment, and whether it used natural uranium or low-enriched uranium as feedstock. All of these factors would affect how much weapons-grade material South Africa might have produced over the years and would introduce doubts as to whether South Africa was tendering all such material for inspection.[61] Similar problems would be posed in establishing how the new, larger Valindaba enrichment plant had been used since its start-up in late 1988.

Published estimates state that using natural uranium as feed, the smaller plant can produce 110 pounds (50 kilograms) of weapons-grade uranium annually, enough for two or three weapons. This can provide a benchmark for evaluating the inventory South Africa presents to the IAEA. Assuming that the small plant started producing this material in late 1980 or early 1981 and continued to do so through January 1990, South Africa would have to account for the output of some nine years of operation. This could be as much as approximately 990 pounds (450 kilograms) of the material, possibly enough for 30 nuclear devices, assuming 33 pounds (15 kilograms) of material per device. As a practical matter, however, the plant probably was not able to sustain operations at full capacity throughout this entire period. (The estimate used earlier, that South Africa's current stockpile of weapons-grade uranium is probably sufficient for only 15 to 25 weapons, is based on this conservative assumption.)

One important check against South African deception is that Pretoria would have to assume that American and Soviet intelligence agencies had been monitoring its nuclear activities for many years and would have to expect that if it withheld weapons-grade material from the IAEA, this discrepancy could ultimately be discovered and publicly exposed. Such exposure would vitiate the diplomatic benefits that de Klerk presumably would be seeking by joining the NPT and would cast South Africa as a nuclear renegade once again.

In the end, whether South Africa's adherence to the treaty will reap the diplomatic dividends that de Klerk is apparently seeking will depend on whether the country is perceived to be acting in good faith to meet its safeguards obligations under the accord. Meeting this challenge will be difficult, but if Pretoria is able to convince the United States and the Soviet Union on this score, others nations may be prepared to accept their reassurances on this issue.[62]

Stockpiling, withdrawal. Assuming that South Africa accedes to the NPT and that an initial inventory is satisfactorily established, Pretoria still could retain a ready nuclear weapons option by stockpiling weapons-grade uranium under IAEA safeguards. If it subsequently abrogated the NPT or withdrew from it on 90 days' notice, as permitted under Article X if its "supreme interests" are threatened, the stockpiled material could be transformed into nuclear arms in a matter of weeks, if not days, assuming the proper preparations had been made.

South Africa has never claimed to need weapons-grade uranium in its peaceful nuclear program, inasmuch as its power reactors operate on 3- to 4-percent enriched uranium and its SAFARI I research reactor now uses 45-percent enriched fuel. Thus, its retention of uranium enriched to 90 percent or more would be highly provocative. Indeed, disclosure at the time of the original IAEA inventory that it possessed such material would itself be extremely embarrassing, since it would provide strong circumstantial evidence that the country had developed a *de facto* nuclear weapons capability during the 1980s, despite numerous denials. Possibly, de Klerk's nuclear aides will arrange for any such material to be diluted to the 45-percent level or lower to avoid controversies on these issues.

Missile development; alleged collaboration with Israel. In parallel with its nuclear program, South Africa has pursued the development of a powerful rocket booster, the Arniston, with the potential to serve as a nuclear-capable ballistic missile. The booster was tested on July 5, 1989, at the Overberg test range, but officials offered no further details about the rocket. As discussed in Chapter 9, a lengthy investigation by NBC television news concluded that the rocket had been built in collaboration with Israel and that the missile had a range of 900 miles, suggesting it was a version of the Israeli Jericho Follow-on missile.[63] The report alleged that Israel had provided missile technology to South Africa in return for a supply of enriched uranium for nuclear weapons and access to long-range missile test facilities on the South African mainland at Overberg and on the Prince Edward Islands in the Antarctic.

Both Israel and South Africa denied the charges in the NBC report, although some Israeli statements left open the possibility that collaboration in the missile field might be continuing under contracts signed by the two states some years ago.[64] Moreover, as the story unfolded, additional press reports, quoting unnamed U.S. officials, essentially confirmed the existence of joint missile activities between the two states.[65] There appeared to be no additional support, however, for the view that Israel had received a nuclear *quid pro quo* for providing missile technology to Pretoria, and even NBC News began to downplay this aspect of the story.

The missile test took place prior to de Klerk's appointment as State President in August, and it is not clear whether he supports the program and is willing to continue to fund it, despite the cuts he has ordered in South Africa's military budget. South Africa's apparent readiness to join the NPT, as noted earlier, undoubtedly reflects in part de Klerk's conclusion that the country's improved security environment makes nuclear weapons unnecessary. Presumably, this assessment would also make the acquisition of medium-range missiles superfluous. Thus, the project's future remains uncertain.

It has been alleged for many years that Pretoria has significantly enhanced its nuclear weapon capabilities through extensive cooperation with Israel, another technologically advanced state that, like South Africa, has been diplomatically isolated and surrounded by hostile neighbors. According to various reports, which have never been fully substantiated, cooperation between the two countries dates back to the 1950s and has involved sales of South African uranium to Israel to fuel the Dimona reactor and extensive scientific exchanges, including Israeli assistance in developing South Africa's now decommissioned SAFARI II research reactor in the late 1960s.[66]

As noted earlier there has also been speculation that the double flash observed by a U.S. satellite in the South Atlantic in September 1979 was a joint Israeli-South African nuclear test.[67]

It is difficult to assess the validity of these allegations given the secrecy that surrounds the nuclear programs of both countries. Still, there seems to be little question that Israel and South Africa have close political ties and that they have cooperated extensively in the area of conventional armaments. By one estimate, Israeli sales of conventional weapons to South Africa have amounted to between $400 and $800 million annually.[68] South Africa's most advanced fighter, the Cheetah, is a close copy of the Israeli Kfir and is believed to have been developed with Israeli assistance.[69] The two countries were also said to be collaborating in the development of Israel's advanced Lavi fighter; when Israel cancelled the plane in 1987, there were reports that South Africa was attempting to recruit 600 Israeli technicians who had worked on the project, although Pretoria publicly denied this.[70]

In March 1987, responding to charges that its arms transfers to Pretoria violated the 1977 U.N. arms embargo, Israel agreed not to sign new military contracts with South Africa and to let existing contracts expire. In reporting to Congress on arms trade between the two countries, President Reagan stated that prior to the March decision, "Israel appears to have sold military systems and sub-systems and provided technical assistance on a regular basis."[71] Although the allegations of nuclear links between the two nations remain to be substantiated, they would be consistent with this pattern of bilateral security collaboration and should be taken seriously.

Prospects

As of mid-1990, there was growing optimism that the military threat posed by South Africa's nuclear program might soon be greatly reduced. Under de Klerk, South Africa's commitment to regional accommodation and internal liberalization appeared to be gaining momentum and, much like the revolutionary changes in Eastern Europe, seemed irreversible. Joining the Non-Proliferation Treaty could be the next milestone on this odyssey.

Nonetheless, the possibility of a conservative backlash cannot be ruled out. Though it remains far too weak to challenge de Klerk's parliamentary majority, the power of the Conservative Party has grown in recent elections. Extra-parliamentary neo-Nazi extremist groups have also gained strength, and de Klerk's dismantling of the security apparatus that effectively governed the country for much of the 1980s created further disaffection on the political right.[72] Under these circumstances, de Klerk could well decide to consolidate his accomplishments and defer major new initiatives, such as renouncing the country's nuclear option by joining the NPT.[73]

If South Africa joins the pact, however, the development would have far-reaching significance. Since the late 1970s, six countries with substantial unsafeguarded nuclear activities—Argentina, Brazil, Israel, India, Pakistan, and South Africa—have been the focus of international concern over the spread of nuclear arms, and more recently, North Korea has been added to this list. To remove any of these states from this group of so-called "problem countries" would be a singular achievement. Not only would it reduce the threat of nuclear conflict in one region of the globe, but it would also demonstrate that the decision to acquire a nuclear weapons option can be reversed under the right combination of regional, domestic, and international conditions.

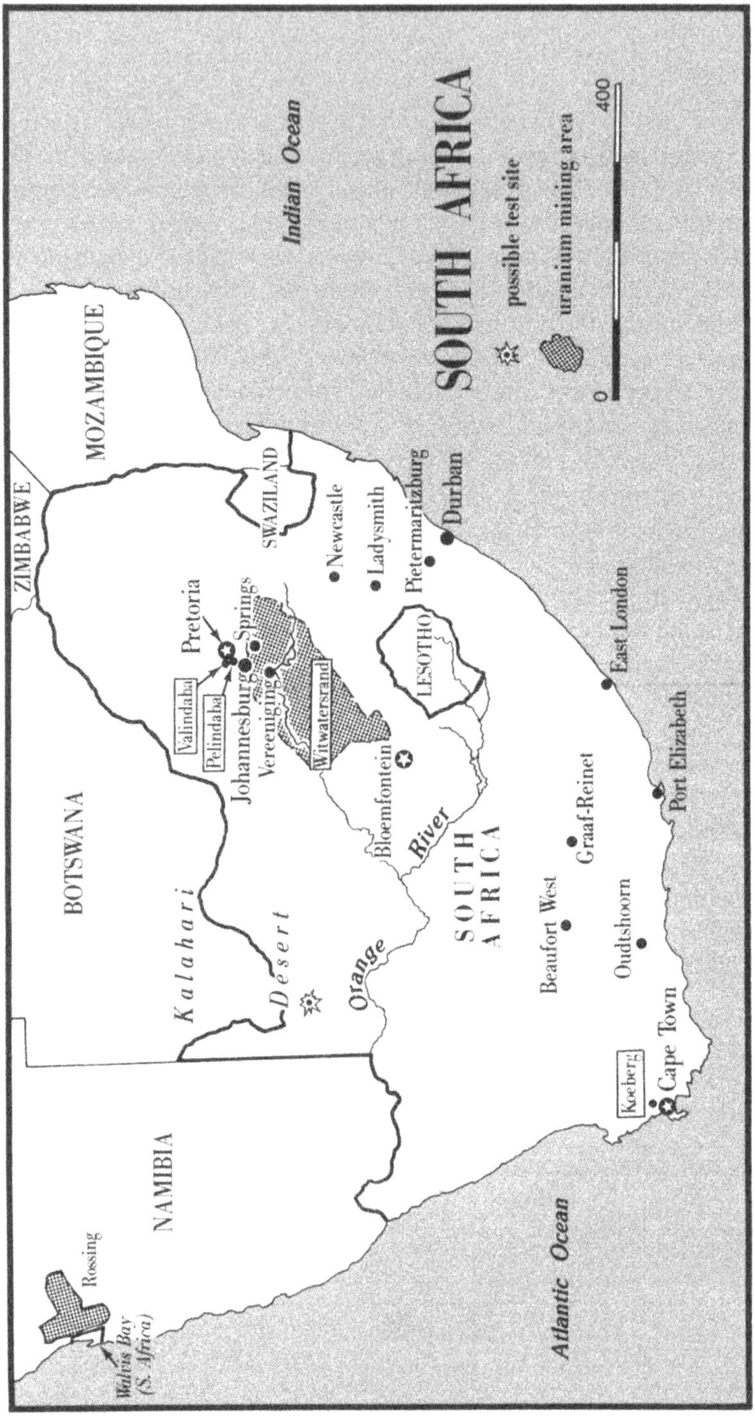

SOUTH AFRICA
possible test site
uranium mining area
0
400
Indian Ocean
MOZAMBIQUE
ZIMBABWE
SWAZILAND
BOTSWANA
NAMIBIA
Kalahari
Desert
Orange
River
Valindaba
Pelindaba
Pretoria
Johannesburg
Springs
Vereeniging
Witwatersrand
Newcastle
Ladysmith
Pietermaritzburg
Durban
LESOTHO
Bloemfontein
East London
Port Elizabeth
Graaf-Reinet
Beaufort West
Oudtshoorn
SOUTH AFRICA
Koeberg
Cape Town
Rossing
Walvis Bay (S. Africa)
Atlantic Ocean

South Africa

Power Reactors/Operating or Under Construction

Koeberg I (light-water/low-enriched uranium, 922 MWe)

• supplier:	Framatome (France)
• start up:	1984[a]
• fuel source:	Switzerland, France, W. Germany, Belgium, China for initial loads; S. Africa after 1988.[b]
• safeguards:	yes

Koeberg II (light-water/low-enriched uranium, 922 MWe)

• supplier:	Framatome (France)
• start up:	1985
• fuel source:	Switzerland, France, W. Germany, Belgium, China for initial loads; S. Africa after 1988.[b]
• safeguards:	yes

Uranium Resources/Active Mining Sites/Uranium Mills

• reasonably assured reserves:	385,700 metric tons[c]
• currently active sites:	Witwatersrand Basin, Palabora
• mills in operation:	Buffelsfontein Chemwes East Rand Gold and Uranium Harmony (3 mills) Hartebeestfontein Joint Metallurgical Scheme Randfontein St. Helena-Beisa Vaal Reefs (3 mills) Western Areas West Driefontein Palabora[d]

Uranium Conversion (UF_6)

Valindaba

• capacity:	Sufficient for Valindaba pilot-scale enrichment plant (presumed).
• supplier:	Great Britain (?)[e]
• start up:	prior to 1975
• safeguards:	no

Valindaba

• capacity:	700 metric tons/yr[d]
• supplier:	South Africa
• start-up:	1986[d,j]
• safeguards:	no

Reprocessing (Plutonium Extraction)

Pelindaba[n]

• capacity:	hot cell complex; size unknown; able to handle Koeberg, SAFARI fuel[n]
• supplier:	South Africa (?)
• start-up:	1987?[o]
• safeguards:	only when safeguarded spent fuel present[n]

Enrichment

Valindaba (pilot plant)[g,n]

- type: stationary-wall centrifuge/jet-nozzle[f]
- capacity: 110 lbs. (50 kg.) high-enriched uranium per year[g]
- supplier: South Africa (some equipment reportedly supplied by U.S., West German, French, and Swiss firms)[h]
- start up: 1975; full operation in 1977[i] (sufficient material for nuclear weapon probably not available until 1980-1981); shut down in February 1990.[p]
- safeguards: no

Valindaba (semi-commercial plant)

- type: stationary-wall centrifuge/jet-nozzle[f]
- capacity: 50 metric tons of low-enriched uranium per year[g]
- supplier: South Africa
- start up: 1988[j]
- safeguards: no

Fuel Fabrication

Elprod (Pelindaba)[n]

- capacity: sufficient to fuel SAFARI I reactor and partial fueling of Koeberg reactors.[p]
- supplier: South Africa (?)
- start-up: 1981[n]
- safeguards: no

Research Reactors

Safari I, Pelindaba (light-water/high-enriched uranium, 20 MWt)

- supplier: Allis Chalmers Corporation (U.S.)
- start up: 1965
- fuel source: United States;[k] after 1981, South Africa.
- safeguards: yes

Pelinduna Zero (Safari II) (heavy-water/low-enriched uranium, 0 MWt)

- supplier: South Africa
- start up: 1967[l] (now decommissioned)
- fuel source: United States
- heavy water: United States
- safeguards: yes

Sources and Notes

a. "Koeberg Ups Power," *Rand Daily Mail, Joint Publication Research Service-Nuclear Development and Proliferation (JPRS-TND)*, April 17, 1984, p. 36.

b. See discussion in text accompanying notes 27-29.

c. Reserves at less than $130 per kg; *Uranium Resources, Production, and Demand,* A Joint Report of the OECD Nuclear Energy Agency and the International Atomic Energy Agency (Paris: OECD, 1986), pp. 299-305.

d. *World Nuclear Industry Handbook 1990* (Sutton, Surrey, U.K.: Nuclear Engineering International, 1990), pp. 50-51.

e. J. E. Spence, "The Republic of South Africa: Proliferation and the Politics of 'Outward Movement,' " in Lawrence and Larus, eds., *Nuclear Proliferation Phase II.* (Lawrence, Kans.: University Press of Kansas), p. 217.

f. Possibly based on West German jet-nozzle process, degree of similarity uncertain.

g. Department of Political and Security Council Affairs, United Nations Centre for Disarmament, Report of the Secretary General, *South Africa's Plan and Capability in the Nuclear Field,* Report

A/35/402 (New York: United Nations, 1981), p. 22, updated in U.N. documents A/39/470 (September 12, 1984) and A/42/649 (October 16, 1987).

h. C. Raja Mohan, "Atomic Teeth to Apartheid: South Africa and Nuclear Weapons," in K. Subrahmanyam, *Nuclear Myths and Realities* (ABC, New Delhi, 1981), p. 123.

i. David Fishlock, "The South African Nuclear Weapons Scare," prepared for Congressional Research Service, December 1977, p. CRS-5; communications with informed U.S. officials.

j. "Nuclear Self-Sufficiency Expected in 1988," Johannesburg Domestic Service 1600 GMT, May 26, 1987, reprinted in *Foreign Broadcast Information Service-Africa*, May 28, 1987, p. U-11.

k. U.S. exports ceased in 1975.

l. J. E. Spence, "South Africa: the Nuclear Option," *African Affairs*, October 1981, p. 441. The reactor is presently decommissioned; Robert Jaster, "Politics and the Afrikaner Bomb," *Orbis* (Winter 1984): 827.

m. "South Africa," *Nucleonics Week*, March 21, 1985, p. 15; "Koeberg Reaches Full Capacity," *The Citizen, JPRS-TND*, July 16, 1984.

n. Jozef Goldblat, ed., *Non-Proliferation: The Why and the Wherefore*, (Philadelphia: Taylor and Francis, 1985), p. 315; *Report by the Board of Governors and the Director General to the General Conference, South Africa's Nuclear Capabilities*, GC (XXVIII)/724, September 24, 1984, Annex 1, Attachment 2, p. 2. All of South Africa's spent fuel is currently under IAEA safeguards.

o. "Implementation of the Declaration on the Denuclearization of Africa," Report of the Secretary General, U.N. document A/42/649, October 16, 1987 (IAEA safeguards formalized, September 1, 1987).

p. Michael Knapik and Ann MacLachlan, "Closing of Pilot Enrichment Plant Seen as Sign that South Africa Will Sign NPT," *Nuclear Fuel*, May 14,1990, p. 1

Part VI
The Non-Proliferation Regime

16
Controls and Safeguards

The nuclear non-proliferation regime—a constellation of international treaties, institutions, codes, and bilateral nuclear trade arrangements—is a major restraint on the spread of nuclear arms. Elements of the regime have been discussed throughout this book, but several important developments deserve to be covered separately in the brief analyses that follow. Between late 1988 and mid-1990, the regime was strengthened in some notable respects, but a number of troubling developments have underscored its limitations and indicated the need for better vigilance in enforcing some of its guidelines and standards.

The Non-Proliferation Treaty (NPT) and the 1990 Review Conference

New Signatories. The last edition of this volume reported that in April 1988, Saudi Arabia had communicated its intent to sign the Non-Proliferation Treaty. That country did, in fact, accede to the treaty on October 3, 1988. Since then, three other states have also joined the pact: Bahrain, on November 3, 1988; Qatar on April 3, 1989; and Kuwait, on November 17, 1989. This brought the total number of parties to 140, including three nuclear-weapon states, the United States, the Soviet Union, and Great Britain.

As this book went to press, there was considerable optimism that South Africa would also join the NPT, a development that would be a major step forward for international non-proliferation efforts. In addition, it was anticipated that most of the "front-line" black African states neighboring South Africa, including Angola, Mozambique, and Zimbabwe, would simultaneously join the treaty at the time of Pretoria's ratification. If these accessions take place, it would be reasonable to expect that newly independent Namibia will also follow suit.

The 1990 NPT Review Conference. Every five years since the NPT entered into force in 1970, the treaty's signatories have convened a conference "in order to review the operation of [the] Treaty with a view to assuring that the purposes of the Preamble and the provisions of the Treaty are being realized," as specified in Article VII.3 of the pact. The fourth of these Review Conferences—and the last before the treaty must be extended in 1995—is scheduled to be held between August 20 and September 14, 1990, in Geneva.[1] This conference likely will set the stage in tenor and in substance for discussions in 1995, adding to the overall importance of the 1990 review.

Several areas are likely to prove particularly controversial at the 1990 session.[2] If not resolved through compromise, disputes on these matters could lead various parties to insist on the inclusion of language in the conference's "Final Document" that is unacceptable to others. This would block a consensus on the adoption of the document, which will also contain the broad statement of support for the treaty that the United States and many other parties hope will be endorsed unanimously by the gathering.

Article VI and pressures for more rapid progress in superpower arms control. In all three previous Review Conferences, Article VI of the NPT, which requires nuclear-weapon state (NWS) parties to pursue an end to the arms race and nuclear disarmament, emerged as a major topic of debate. In particular, the neutral and non-aligned non-nuclear-weapon state (NNWS) signatories, which often vote as a bloc, have asserted that the superpowers have not lived up to their obligations under this provision.

Substantial progress in American-Soviet arms control efforts since 1985 may reduce the intensity the controversy on this issue in 1990. This period has seen the implementation of the U.S.-Soviet Treaty on Intermediate Range Nuclear Forces, eliminating an entire class of U.S. and Soviet weapons, and significant steps towards the signing of a Strategic Arms Reduction Treaty (START). In addition, it was expected that after Moscow and Washington agreed on improved verification arrangements in June 1990, the 1976 Peaceful Nuclear Explosion Treaty and the 1974 Threshold Test Ban Treaty might soon be ratified, offering modest progress toward limitations on nuclear testing. (The treaties, which though unratifed, have been observed by both sides, prohibit underground nuclear tests with yields in excess of 150 kilotons.) Important advances have also been made towards a treaty reducing the conventional forces of the Warsaw Pact and the North Atlantic Treaty Organization and towards a global ban on chemical weapons, with the United States and Soviet Union entering into a separate preliminary agreement to destroy a significant proportion of their existing chemical weapon stockpiles.

Nevertheless, the neutral and non-aligned NNWS's at the Conference are likely to maintain pressure for further arms control measures, especially for a comprehensive ban on nuclear testing, an initiative that has long been the top item on their disarmament agenda but which has been opposed by the United

States.[3] This group may also call for a cut-off of the production of weapons-grade nuclear materials and for the extension of negative security assurances by the nuclear weapon states, *i.e.*, guarantees to non-nuclear weapon states that they will not be subjected to a nuclear attack.

Israel and South Africa. Israel and South Africa long have drawn criticism from their respective neighbors and traditional adversaries—the Arab states in the case of Israel, and the African bloc in the case of South Africa—for their nuclear activities in light of their continued refusal to accede to the Non-Proliferation Treaty. The record of criticism in arenas such as the annual General Conferences of the International Atomic Energy Agency (IAEA) and the 1985 NPT Review Conference, combined with press disclosures in late 1989 alleging extensive Israeli-South African collaboration on the development of a nuclear-capable missile,[4] make it virtually certain that these two countries and their unsafeguarded nuclear programs will be controversial in 1990.

Of the two, the Israel is likely to be the target of the most severe criticism, stirred by such events as the Palestinian uprising in the Israeli-occupied West Bank and Gaza Strip. This criticism may take the form of condemnatory floor statements to the conference by Israel's antagonists or proposals to include language denouncing that country in the Final Document. At the January 1989 Paris Conference on Chemical Weapons, the Arab states demonstrated their readiness to raise the issue of Israel's nuclear capability in international fora by refusing to support restrictions on the spread of chemical weapons unless Israel placed its nuclear program under international inspection.[5]

South Africa presumably would avert criticism at the session if it acceded to the treaty beforehand. Even if it did not, the country's antagonists might moderate their denunciations of Pretoria because of expectations that it might soon take this step. South Africa's recent liberalization of its racial policies and its opening of a political dialogue with its black citizens are additional factors that could lead its traditional antagonists to moderate their demands at the Conference.

Iran, Iraq, North Korea. At the 1985 Review Conference, an intense dispute erupted between Iran and Iraq, then in the midst of their eight-year war, over the latter's bombing of Iran's incomplete nuclear power reactors at Bushehr. While that conflict has ended, relations between the two states remain embittered, and it is possible that Iran will seek to embarrass its longtime adversary at the conference by declaring that Iraq's efforts to acquire components for nuclear weapon triggers in early 1990 constituted a violation of the NPT. (See Chapter 11.)

South Korea and other interested states can also be expected to express concerns over North Korea's refusal to accept IAEA inspections on all of its nuclear activities as required by its ratification of the treaty in 1985. (See Chapter 8.) It is possible, though unlikely given the continuing negotiations on this issue, that North Korea will be accused of violating the pact.

French, Chinese participation. The conference will confront these and other controversies that may impede its adoption of a final consensus statement supporting the treaty. A more positive development, however, is that France and the People's Republic of China, the two remaining declared nuclear-weapon states that have not signed the pact, have agreed to participate in the meeting as observers. This implicit show of support for the treaty, which may signal that the two states are giving serious consideration to joining it, represents a tacit change of course for both states and could contribute to an atmosphere conducive to consensus at the session.

The International Atomic Energy Agency

Safeguards activities. Each year the IAEA prepares a confidential internal evaluation of its safeguards activities, known as the Safeguards Implementation Report (SIR). The agency issues a summary of that report following the meeting of its Board of Governors in June of each year. In June 1990, however, it released an unusually abbreviated summary (see below), apparently because the Board was considering the release of the entire report for the first time but had not completed its deliberations on the matter. The agency's 1988 SIR—the most recent whose contents have been disclosed in any detail by the organization—recorded that, as in previous years, the Secretariat of the IAEA "did not detect any anomaly which would indicate the diversion of a significant amount of safeguarded nuclear material—or the misuse of facilities, equipment or non-nuclear material subject to safeguards under certain agreements—for the manufacture of any nuclear weapon, or for any other military purpose, or for the manufacture of any other nuclear explosive device, or for purposes unknown."[6]

Table 16.1 summarizes relevant information concerning safeguards agreements, inspection-related activity, and materials under safeguards for the years 1987 and 1988.[7] The figures show a clear increase in the amount of material under safeguards and a slight increase in the number of relevant agreements and countries covered. The total number of inspections decreased slightly.

Additional safeguards developments of importance noted in the IAEA report covering 1988 included selected details on new safeguards agreements. The People's Republic of China, in conjunction with Brazil, agreed in January 1988 to let the IAEA safeguard a joint Brazilian-PRC effort to manufacture reactors fueled by thorium. The safeguards would apply to work on, and the security of, reactors built by the two countries. The agreement further restricts the export to third countries of the materials exchanged between them as a result of cooperation in this area, to reduce the risks of proliferation.[8] The willingness of Brazil and the PRC, two countries outside the NPT, to submit this project to IAEA safeguards is a valuable show of restraint by two emerg-

Table 16.1
IAEA Safeguards Activities and Coverage

	1987	1988
Number of countries with safeguards agreements	97	99
Total number of safeguards agreements in force	166	168
Number of installations covered by inspections	631	—
Location of inspections	52 NNWS 4 NWS	53? NNWS 4 NWS
Number of inspections	2133	2128
Approximate number of discrepancies	290	—
Number of major facilities at which inspection goals were attained for the entire facility	214	248
Number of plutonium and uranium samples analyzed	1340	1170
Analytic results of sample analysis	3600	3040
Detachment and verification of seals applied to nuclear material or to Agency safeguards equipment	12,500	15,500
Separated plutonium outside reactor cores	8.8t	10.9t
Recycled plutonium in fuel elements in reactor cores	0.6t	0.5t
Plutonium contained in irradiated fuel	224.2t	254.4t
High-enriched uranium (uranium enriched to more than 20%)	12.2t	13.1t
Low-enriched uranium	29,252t	31,704t
Source material (uranium oxide)	50,867t	54,514t

Source: IAEA, *Annual Report for 1988*, GC(XXXIII)/873

ing nuclear supplier countries that are not required by any international undertaking to place nuclear exports or imports under the IAEA system. (See Chapter 4.)

In March 1988, Albania—not an NPT signatory—agreed to place all of its nuclear material and facilities under IAEA safeguards. In September 1988, India agreed to place under safeguards the two Soviet-supplied VVER-1000 reactors to be built at Koodankulam in Tamil Nadu, as specified in the agreement between the two countries which covers the transfer. Also in September 1988, China signed a voluntary offer agreement that would subject some of its civilian nuclear facilities to safeguards; China thus joined all of the other declared nuclear weapon state in granting such access.[9] The resulting safeguards agreement entered into force on September 18, 1989.[10]

In more recent developments, Vietnam signed a safeguards agreement in September 1989 at the 33rd General Conference of the IAEA;[11] China agreed in November 1989 to apply agency safeguards to a nuclear power reactor that it intends to sell to Pakistan;[12] and, in November 1989, Pakistan inaugurated a safeguarded 27-kilowatt research reactor that was built in collaboration with the PRC.[13]

Last, the IAEA report noted that Japan had cooperated extensively with the IAEA in designing the Rokkasho-Mura plutonium extraction facility in order to facilitate the effective application of agency safeguards when the plant is completed. Japan contributed $250 thousand to the effort. (Critics believe, however, that the processing and accumulation of plutonium for Japan's civilian nuclear energy program, even if effectively safeguarded to preclude diversion of the material to military purposes by the Japanese government, would still pose a serious risk. They are concerned that terrorists might be able to obtain the material, either by stealth or by direct assault on the installations involved. Since opponents of the Japanese program argue that plutonium use cannot be justified on economic grounds, they believe Japan's activities expose the international community to unnecessary risks.)

In June 1990, the IAEA issued a press release commenting on the 1989 SIR. The agency noted:

> The SIR records that in 1989, as in previous years, the IAEA did not detect any event which would indicate the diversion of a significant amount of safeguarded nuclear material—or the misuse of facilities, equipment or non-nuclear materials subject to safeguards—for the manufacture of any nuclear weapon, or for any other military purpose, or for the manufactiure of any other nuclear explosive device or for purposes unknown.

The press release noted that 922 installations were under safeguards or contained safeguarded material.

A series of developments, which took place or were revealed during 1989-1990 and which are discussed elsewhere in this volume, have confronted the agency with a number of serious challenges. These include the failure of North Korea to sign a comprehensive safeguards agreement with the agency; the agency's difficulties in monitoring West Germany's transfers of nuclear technology to Brazil; the repeated violations of IAEA rules on transfers of heavy water to India; and the violation of German export controls that would otherwise have required the application of IAEA safeguards to certain nuclear equipment delivered to Pakistan. The agency's press release made no mention of these matters.

Additional Approaches to Controlling the Spread of Nuclear Weapons

In addition to the Non-Proliferation Treaty and International Atomic Energy Agency safeguards, a number of additional mechanisms have contributed to controlling the spread of nuclear weapons. These mechanisms vary greatly in their emphasis, scope of application, and potential achievements.

Nuclear Supplier Controls. Zangger Committee trigger lists and London Nuclear Supplier Guidelines (see Appendix F) have combined with the basic provisions of the NPT (i.e., Articles I, II, III and IV) to create a web of national and international constraints on the diffusion of technology and equipment that could support nuclear weapons development programs. These regimes have strengthened international non-proliferation efforts, for example, by facilitating the harmonization of national export controls. However, during the 1980s they experienced a number of serious failures and setbacks because of the successful nuclear smuggling operations of several emerging nuclear states. These are discussed in detail in Chapter 4.

Full-scope safeguards. The United States, Australia, Canada, and Sweden have gone beyond the basic requirements of these export control systems by prohibiting major nuclear exports to countries that have not placed *all* of their nuclear activities under IAEA oversight. Since the late 1970s, the United States has sought, with some success, to persuade the other advanced supplier countries to make this requirement for "full-scope" safeguards a condition for major new nuclear supply commitments.

As noted in Chapter 6, however, the Soviet Union's 1988 decision to sell India two 1,000-megawatt power reactors broke an eight-year *de facto* moratorium on major new nuclear sales to states that had not accepted full-scope safeguards. China then followed suit, by agreeing in November 1989 to sell a 300-megawatt nuclear power plant to Pakistan without requiring full-scope safeguards, and in February 1990, France agreed to permit the sale of a 900-megawatt power reactor to Pakistan under similar terms. (See Chapter 7.) All of the reactors and any plutonium produced through their use will be under IAEA safeguards, but both India and Pakistan will remain free to advance their nuclear weapon capabilities using their unsafeguarded facilities.

On the positive side, Japan announced in 1989 that it would adopt the U.S. position of requiring full-scope safeguards as a condition for major new nuclear supply commitments and it appeared that West Germany was also prepared to follow this policy. [14]

The Missile Technology Control Regime (MTCR). Chapter 3 describes three potential links between ballistic missile proliferation and the spread of nuclear weapons: missiles provide an unparalleled nuclear-weapon-delivery capability; by exposing the cities of *de facto* nuclear states, like Israel, to devastating non-nuclear attacks they increase the risk of nuclear retaliation; and, by providing a potent non-nuclear retaliatory capability, they can enable a would-be nuclear-weapon state to deter an adversary from striking key nuclear facilities during the period when the nuclear aspirant is attempting to develop a nuclear capability of its own. Thus, the uncontrolled spread of ballistic missile capabilities can increase the risk of nuclear weapons being used and enhance the ability of some states to acquire such armaments.

As described in Chapter 3, in 1987, Canada, France, Italy, Japan, the United Kingdom, the United States, and West Germany created the Missile Technology Control Regime (MTCR) in order to slow missile proliferation by restraining the transfer of technology, components, and complete systems to Third World missile programs. The regime consists of parallel export controls adopted by the participating nations. These controls are designed to slow the development of missiles that are capable of hitting a target the size of a small city from a distance of 190 miles (300 kilometers) or further and which are able to carry a warhead weighing 1,100 pounds (500 kilograms) or more.[15]

Since late 1988, several developments have considerably strengthened the regime. The most important is the decision by the Soviet Union to abide by the regime's export standards (except with respect to transfers of Scud missiles to Afghanistan)[16] and China's pledge to the United States that it will not transfer medium-range missiles to the Middle East.[17] In addition, Spain joined the regime in 1989; Sweden and Switzerland agreed to abide by its export requirements that year; and in May 1990, Belgium, Luxembourg, and the Netherlands joined the group.[18]

The system has also had a success in halting at least one missile system of concern, the Argentinian-Egyptian-Iraqi Condor II project. (See Chapter 13.) It also appears that a controversy between France and the United States involving France's proposed sale of the Viking liquid-fuel rocket engine to Brazil has been resolved, with France agreeing to withhold the technology at issue. (See Chapter 14.) Advances by Iraq, which test fired the first stage of its Al-Abid space launch vehicle; India, which test fired the intermediate-range Agni; Israel, which tested its intermediate-range Jericho II Follow-on missile and launched two space satellites; South Africa, which tested the intermediate-range Arniston; and Pakistan, which tested the short-range Hatf I and Hatf II, indicate, however, that the spread of nuclear-capable missiles in the developing world is continuing at a disturbing rate. (See Tables in Chapter 3.)

Regional Confidence-Building Measures

Confidence-building measures (CBMs) are commitments undertaken by rival states to increase the transparency of certain potentially threatening activities (usually military operations, such as exercises) in order to reduce mutual suspicion and tensions. Although the adoption of nuclear-related CBMs by emerging nuclear powers falls short of iron-clad non-proliferation restrictions, such measures can help to reduce the risk of a nuclear arms race where nuclear weapon capabilities are suspected to be under development and, by reducing the risk of conflict, can diminish the possibility of nuclear use where such weapons already have been acquired. Nuclear CBMs in Latin America and South Asia have become an increasingly important element of efforts to reduce nuclear dangers in these regions.

Brazil-Argentina: Mutual visits to nuclear facilities. In 1985, Argentina and Brazil agreed in principle to open up their nuclear facilities to mutual inspection.[19]

While actual inspections have yet to be agreed to in fact, the two countries have undertaken a series of *ad hoc* visits to each other's nuclear installations by politicians and technicians to lessen mutual suspicion. The goals of increased confidence and cooperation in the nuclear sphere have been strengthened by the Declaration on Joint Nuclear Policy[20] and by the Ezeiza Nuclear Statement[21] both signed by the presidents of the two countries in 1988, as well as by a series of agreements of bilateral nuclear trade and research. Teams have visited all nuclear facilities in both states. (See Chapters 13 and 14.)

The visits have not expanded into full-fledged inspections, in large part due to Brazilian opposition to such a move. Without inspections, neither country can be certain of the output or the quantities of enriched uranium being produced at the other's unsafeguarded enrichment plant. Nonetheless, the mutual visits have made a significant contribution to the abatement of suspicions in each state concerning the other's nuclear activities. Their continuation appears likely: Argentinian President Carlos Menem has supported them, and Brazil's new President, Fernando Collor de Mello, is expected to do the same.

India-Pakistan: No attack pledge. In 1985, then-Indian Prime Minister Rajiv Gandhi and then-Pakistani President Mohammed Zia ul-Haq reached an oral agreement to refrain from attacking each other's nuclear facilities. This agreement, which was codified in written form by Gandhi and Pakistan's new Prime Minister Benazir Bhutto on December 31, 1988, states that the countries "will refrain from undertaking, encouraging or participating indirectly or directly in any action aimed at causing the destruction or damage to any such [nuclear] installations or facilities in the other country,"[22] including nuclear power and research facilities, as well as uranium enrichment plants and other nuclear-related installations. The agreement calls for each country to provide the other with a list of nuclear facilities that are to be covered by the no-attack pledge; the lists are to be updated annually. Both countries ratified the agreement in 1989; however neither the instruments of ratification nor the lists of covered facilities have been exchanged.

The fate of the accord, in light of challenges to Bhutto's authority in Pakistan and the election of V.P. Singh as India's new Prime Minister in November 1989, remains to be seen. A key question is whether Pakistan will include on its facility list nuclear installations—such as a uranium hexafluoride plant at Dera Ghazi Khan—the existence of which has been well-documented internationally, but which the Pakistani government has never acknowledged. If the accord is fully implemented, it could be an important step in opening a nuclear dialogue between the two South Asian rivals.

Nuclear Free Zones. Nuclear or Nuclear-Weapon Free Zones (NFZs or NWFZs) have been relied upon to prevent further proliferation of nuclear weapons.[23] Several zones have been created to date, but only two of them cover wide expanses of inhabited areas: the Latin American Nuclear Weapon Free Zone, established by the Treaty of Tlatelolco, and the South Pacific Nuclear Free Zone (SPNFZ), established by the Treaty of Rarotonga. The Antarctic Treaty (1961), the Outer Space Treaty[24] (1967), and the Seabed Arms Control Treaty[25] (1972) also define particular areas that are to be kept free of nuclear weapons.

It can be argued that each of these so-called zones was created with the ultimate goal of impeding further vertical proliferation by the nuclear-weapon states, *i.e.*, preventing the further spread of nuclear weapons by the declared nuclear powers: the first moves for the creation of the zone in Latin America came on the heels of the Cuban Missile Crisis, when the USSR attempted to deploy missiles in that Caribbean country; support for SPNFZ drew largely from opposition to the French nuclear testing program, which is located in French Polynesia, as well as from fears arising from the presence of U.S. nuclear-powered and nuclear-capable ships in the region; and the Antarctic, Outer Space, and Seabed treaties each sought to prevent the introduction of international conflict and nuclear weapons into uninhabited areas that hitherto had been free of them.[26]

NFZs can also create a politically practicable mechanism for preventing horizontal proliferation by obliging countries located within the zone to foreswear the acquisition, development, manufacture and/or deployment of nuclear weapons.

The Latin American Nuclear Weapon Free Zone. The Treaty for the Prohibition of Nuclear Weapons in Latin America, also known as the Treaty of Tlatelolco, created the first nuclear weapons free zone in a populated region. (See Annex E for a full description.) Despite rumors that Argentina might consider ratification following the inauguration of President Carlos Menem in June 1989, Foreign Minister Domingo Cavallo ruled out ratification of the Treaty of Tlatelolco, as well as the NPT, in September 1989. Argentine objections to the Treaty of Tlatelolco include its control system (i.e., safeguards administered by the IAEA), which, it is believed, could compromise Argentina's industrial secrets.[27]

Although Brazil has ratified the Tlatelolco Treaty, it has refused to waive the treaty's provision which stipulates that it will enter into force only with the adherence of all states in the region specified (Article 28). The continued refusal of Argentina and Cuba to ratify the treaty means that it is binding on neither Argentina nor Brazil, the only countries in Latin America with sizeable nuclear programs, leaving them free to operate nuclear facilities that are not under IAEA supervision.

The South Pacific Nuclear Free Zone. The Treaty of Rarotonga, creating the South Pacific Nuclear Free Zone, came into effect in December 1986. The Treaty bans the manufacture, acquisition, possession, control, and stationing of nuclear weapons, and seeks to prevent the testing of them in the South Pacific.[28] None of the regional parties appears to be a potential proliferant, and each has signed the NPT.

As stated above, one of the major objectives of the Treaty was to bring the French nuclear testing program in the region to an end. Protocol 3 would prohibit such tests, but France has refused to sign the document. Despite careful tailoring of the treaty and its three accompanying protocols to accommodate U.S. interests in the region, the Reagan Administration refused to sign any of the three protocols to the treaty, citing its "global security responsibilities" as the predominant reason for abstention. While the administration did not elaborate publicly on this position, a number of factors may have led to this stance.

First, Washington may have wanted to avoid a confrontation with the French, an important—and historically uncompromising—ally. As revealed by Richard Ullman in a 1989 *Foreign Policy* article, the two countries have secretly engaged in cooperation relating to nuclear weapons and strategy over the years. U.S. signature to the SPNFZ Protocols, particularly Protocol 3 prohibiting testing in the zone, which could have been directed only against the French nuclear testing program, could have injured these important defense ties.[29] Second, American officials may have feared that nuclear free zones in other areas, where the U.S. military presence is more extensive than in the South Pacific, could be encouraged by a U.S. endorsement of SPNFZ. Third, unlike the Treaty of Tlatelolco which served to exclude Soviet nuclear weapons from America's "back yard," no commensurate strategic advantage would have accrued to the U.S. from signing the SPNFZ protocols, although this might have improved U.S. relations with the states in the region. Finally, a primary regional concern for the U.S. at the time of the treaty's inception was New Zealand Prime Minister David Lange's policy of excluding nuclear-capable and nuclear-powered ships from that country's ports. The SPNFZ Treaty left the question of whether or not to permit port calls open to determination by each individual party to the treaty. U.S. protocol signature conceivably could have been perceived as indirect acceptance of New Zealand's policy, a situation that would have been in direct opposition to the US position on the New Zealand ban.

The Bush Administration is not known to be reconsidering the U.S. position. The Solomon Islands ratified the Treaty in 1989.

(Efforts to promote nuclear weapon free zones in the Middle East and other regions have not advanced in recent years.)

***

Despite certain promising developments during 1989 and 1990 in the field of nuclear non-proliferation, a number of disturbing trends in this area point to the need for continued vigilance and intensified international efforts to curb the spread of nuclear weapons. Even as the risk of nuclear confrontation between the superpowers has diminished, for example, concerns have increased that conflict may erupt between the two *de facto* nuclear powers in South Asia, a conflict in which nuclear weapons would inevitably play a role. Similarly, as hopes have been raised that South Africa may join the NPT and thereby reduce the number of states outside the international non-proliferation regime, anxieties have increased over the nuclear intentions of North Korea and Iraq and over the commitments of these two NPT parties to the treaty. In addition, the demise of the Condor II missile program, attributed in part to the Missile Technology Control Regime, and Chinese and Soviet pledges to restrict missile transfers have been counterbalanced by the significant advances in the missile capabilities of Israel, Iraq, India, and Pakistan. Argentina and Brazil have undertaken important confidence-building measures through their exchanges of visits to nuclear facilities, but fears of unconventional warfare continue to mount in the Middle East, as evidenced by Iraqi President Saddam Hussein's threats to destroy half of Israel with chemical weapons, if attacked, and by Israel's threats to respond to such an act a hundred times over.

While those concerned about nuclear proliferation may take heart from the positive developments, such successes must not eclipse the very real dangers that remain, as a number of regional powers pursue their nuclear ambitions. With this dichotomy in mind, the international community must pursue its efforts to slow the spread of nuclear weapons with renewed vigor.

Notes

Chapter 1: The Silent Spread of the Bomb

1. "Testimony of Rear Admiral Thomas A. Brooks, Director of Naval Intelligence," Subcommittee on Seapower, Strategic, and Critical Materials of the Committee on Armed Services, U.S. House of Representatives, 101st Cong., 1st Sess., February 22, 1989 (mimeo).
2. All three countries are parties to the Nuclear Non-Proliferation Treaty (NPT), which requires them to place all of their nuclear plants under International Atomic Energy Agency (IAEA) inspection, once nuclear materials have been introduced. The nuclear weapons efforts of Iraq and Iran have not yet reached this point, however. North Korea, though it signed the NPT in 1985 and has since completed at least one nuclear-weapons-relevant facility, had not executed its inspection agreement with the IAEA as of June 1990.
3. These events have included Iran's holding more than 50 American diplomats in captivity for over a year in 1979-81 and the initiation of U.S. naval patrols in 1987-88 to protect shipping in the Persian Gulf from Iranian depredations. The latter action, which effectively allied the United States with Iraq in the Persian Gulf War, led to a series of brief engagements between U.S. and Iranian forces and to the accidental shooting down by an American Aegis-class cruiser of an Iranian civil airliner. More recent incidents, such as the continued holding of American hostages by the Iran-backed Party of God in Lebanon and the placing of a bomb aboard Pan Am flight 103 by an Iranian-backed terrorist group, which destroyed the aircraft over Scotland on December 21, 1988, have further inflamed tensions between the two countries.
4. U.S. Arms Control and Disarmament Agency (ACDA), *International Arms Transfers and World Military Expenditures: 1988* (Washington, D.C.: U.S. ACDA, 1989), p. 18.
5. The confirmation is contained in a formerly classified U.S. Defense Intelligence Agency document, entitled, "Surface-to-Surface Missile Handbook—Free World." The document was released in late 1989 under the Freedom of Information Act to the Natural Resources Defense Council, a Washington-based environmental group, which monitors nuclear issues. "U.S. Document Cites Israeli Missile Progress," *Washington Times*, November 15, 1989. One page of the

document is headed, "Jericho I," and under the column for type of warhead, "nuclear, chemical, or HE [high-explosive]" are listed.

6. Conventionally armed missiles roughly comparable to the Prithvi were used extensively during the Iran-Iraq War and, during 1989, the Afghan government fired hundreds of such missiles against guerrilla forces, backed by the U.S. and Pakistan, in that country.
7. "Testimony of William Webster, Director of Central Intelligence," U.S. Congress, Senate Governmental Affairs Committee, *Hearings on Missile and Nuclear Proliferation*, 101st Cong., 1st Sess., May 18, 1989.
8. The design work for the nuclear device India detonated in 1974 was undertaken at the Bhabha Center, and the plutonium for the device was produced there.
9. Mark Hibbs, "Illegal Export Charges May Spur Tighter German Export Controls," *Nucleonics Week*, January 5, 1989; Mark Hibbs, "German Firms Exported Tritium Purification Plant to Pakistan," *Nuclear Fuel*, February 6, 1989; "Germans Sell Nuke Equipment to Pakistan," *News India*, March 3, 1989; Mark Hibbs, "U.S. Repeatedly Warned Germany on Nuclear Exports to Pakistan," *Nuclear Fuel*, March 6, 1989; Mark Hibbs, "German Firm's Exports Raise Concern About Pakistan's Nuclear Capabilities," *Nuclear Fuel*, March 6, 1989; John J. Fialka and Thomas F. O'Boyle, "West German Firms Admit Supplying Nuclear-Weapons Material to Pakistan," *Wall Street Journal*, April 21, 1989; Mark Hibbs, "Prosecutors Link Tritium Plant to Pakistan Weapons Program," *Nuclear Fuel*, May 1, 1989.
10. "NBC Nightly News," National Broadcasting Company, October 25, 1989.

Chapter 2: American-Soviet Relations

1. The MTCR prohibits transfers of ballistic and cruise missiles and related equipment and technology for systems that are capable of carrying a payload of 1,100 pounds or more to a distance of 190 miles. The Scud-B is able to carry double this payload to almost 190 miles, an overall capability that brings it within these limits, in the view of the U.S. government.
2. See Chapter 15 for a discussion on South Africa.
3. The changing security environment in Europe suggests that the Non-Proliferation Treaty is likely to play an increasingly important role in this region in the years ahead and provides added reason to support the indefinite extension of the pact, on which its parties must vote in 1995.
4. Robert C. Toth, "U.S. Worried by Nuclear Security in Instable Soviet Empire," *Los Angeles Times*, December 15, 1989; William J. Broad, "Specter is Raised of Nuclear Theft," *New York Times*, January 28, 1990.
5. Bill Gertz, "Soviet Rebels Storm an A-Bomb Facility," *Washington Times*, February 6, 1990; Leonard S. Spector, *Going Nuclear* (Cambridge, MA: Ballinger Publishing Company, 1987), Chapter 2.
6. R. Jeffrey Smith, "Soviets Remove Some Nuclear Arms from Areas Marked by Ethnic Strife," *Washington Post,* June 23, 1990.

Chapter 3: Ballistic Missile Proliferation

1. U.S. Defense Intelligence Agency, "Surface-to-Surface Missile Handbook—Free World," released under the Freedom of Information Act to the Natural Resources Defense Council; "U.S. Document Cites Israeli Missile Progress," *Washington Times*, November 15, 1989. One page of the document is headed,

"Jericho I," and under the column for type of warhead, "nuclear, chemical, or HE [high-explosive]" are listed.

2. *Ibid.*
3. David Ottaway, "Egypt Drops Out of Missile Project," *Washington Post*, September 20, 1989; "Defense Minister Says Condor II Project Halted," *Clarin*, April 22, 1990, translated in *Foreign Broadcast Information Service-Latin America*, April 27, 1989, p. 12.
4. Michael Gordon, "Beijing Avoids New Missile Sales Assurances," *New York Times*, March 30, 1990.
5. Geoffrey Kemp, " 'Solving' the Proliferation Problem in the Middle East," in Robert Inman, Joseph Nye, William Perry, and Roger Smith, eds., *New Threats: Responding to the Proliferation of Nuclear, Chemical and Delivery Capabilities in the Third World* (Lanham, Md.: Aspen Strategy Group and University Press of America, 1990).
6. Kenneth Adelman, "Global Threat Soars on Wings of Agni," *Washington Times*, May 29, 1989; Jasjit Singh, "Arms Control and Proliferation of High Technology Weapons in South Asia and Near East: A View from Pakistan," prepared for the Carnegie Endowment Conference on Arms Control and the Proliferation of High Technology Weapons in the Near East and South Asia, Bellagio, Italy, October 23-27, 1989, to be published in forthcoming compendium of conference proceedings.
7. Air Commodore Jasjit Singh, Director of the Institute for Defense Studies and Analysis in New Delhi, for example, has proposed a regional ban on intermediate-range ballistic missiles that would include China, India, Pakistan, Israel, and the Soviet Union.

Chapter 4: New Patterns of Nuclear Trade

1. France, for example, had agreed to sell plutonium plants to Pakistan and South Korea, while West Germany had agreed to transfer both enrichment and plutonium extraction units to Brazil. Eventually the French plutonium plant sales to Pakistan and South Korea were cancelled, and, although the West German-Brazil nuclear contract has remained in force, Brazil has abandoned the project to build the plutonium facility and has built only a small enrichment plant at Resende with West German help. In the late 1970s, both France and West Germany declared that, for the foreseeable future, they would not enter into new commitments to export plutonium extraction plants.
2. Simon Henderson, "Pakistan Builds Second Plant to Enrich Uranium," *Financial Times*, December 11, 1987; Neil A. Lewis, "Reports of Second Pakistani Uranium Plant Weighed," *New York Times*, January 10, 1988. Pakistan denied the allegation that it was building a second enrichment plant.
3. If it is building a new unsafeguarded reactor, as discussed below, however, Pakistan would acquire a source of unsafeguarded material for processing at the New Labs that would not trigger IAEA coverage.
4. For a discussion of these various operations, see Leonard S. Spector, *The Undeclared Bomb* (Cambridge, Mass.: Ballinger Publishing Company, 1988), pp. 136-139; Leonard S. Spector, *Going Nuclear* (Cambridge, Mass.: Ballinger Publishing Company, 1987), pp. 105, 114-116; Leonard S. Spector, *The New Nuclear Nations* (New York: Vintage, 1985), pp. 22-40.
5. "Confessions of Illegal Export of Nuclear Facilities," *Welt am Sonntag*, December 25, 1988, translated in *Joint Publications Research Service (JPRS)-Nuclear*

Development and Proliferation (TND), January 13, 1989, p. 25; Mark Hibbs, "U.S. Repeatedly Warned Germany on Nuclear Exports to Pakistan," *Nuclear Fuel*, March 6, 1989, p. 12.

6. Mark Hibbs, "Illegal Export Charges May Spur Tighter German Export Controls," *Nucleonics Week*, January 5, 1989; Mark Hibbs, "German Firms Exported Tritium Purification Plant to Pakistan," *Nuclear Fuel*, February 6, 1989; "Germans Sell Nuke Equipment to Pakistan," *News India*, March 3, 1989; Hibbs, "U.S. Repeatedly Warned Germany..." (see note 5); Mark Hibbs, "German Firm's Exports Raise Concern About Pakistan's Nuclear Capabilities," *Nuclear Fuel*, March 6, 1989; John J. Fialka and Thomas F. O'Boyle, "West German Firms Admit Supplying Nuclear-Weapons Material to Pakistan," *Wall Street Journal*, April 21, 1989; Mark Hibbs, "Prosecutors Link Tritium Plant to Pakistan Weapons Program," *Nuclear Fuel*, May 1, 1989; "NTG Nuclear Proliferation Case," *Der Spiegel*, November 6, 1989, translated in *JPRS-TND*, November 29, 1989, p. 35.
7. See references in previous note.
8. "Nuclear Exports to Pakistan Reported," *Der Spiegel*, February 20, 1989, reprinted in *Congressional Record*, May 16, 1989, p. S 5441.
9. Gary Milhollin, "Dateline New Delhi: India's Nuclear Cover-Up," *Foreign Policy* (Fall 1986): 161; "Heavy Going for Heavy Water," *Indian Express*, June 5, 1988 (reviewing a report by the Indian Comptroller and Auditor General, which concluded that India's indigenous production of heavy water was very far below planned levels). See also Spector, *The Undeclared Bomb* (see note 4), p. 108. The smaller Cirus reactor, which provided the plutonium for India's 1974 nuclear test, has been operating since the early 1960s. The Cirus reactor originally used U.S. heavy water, which India had pledged would be used exclusively for peaceful purposes. A significant portion of the heavy water in the reactor during the time it was used to produce plutonium for the 1974 test was U.S.-origin material. In 1970, Washington warned New Delhi that its peaceful-use pledge prohibited the use of plutonium produced through the use of U.S. heavy water for any type of nuclear explosives, but India never formally accepted this interpretation. In 1974, India asserted that it had not violated its assurances to Washington because it claimed that the test was a "peaceful nuclear explosion," referring to experimental U.S. and Soviet programs at the time, which explored the use of nuclear explosives for civil excavation and similar non-military projects. Though Washington believed that India had violated its pledges, it declined to pursue the matter because of the ambiguity of India's original undertaking and because of the difficulty of establishing conclusively that U.S. heavy water had remained in the Cirus reactor; India had built a small heavy water plant of its own several years earlier. (India has never claimed to have removed U.S. heavy water from the Cirus reactor, however.)
10. Gary Milhollin was the first investigator to conclude that China was the source of the bulk of the material, but his seminal article on the subject of India's heavy water imports did not offer any detail to substantiate the point. See previous note.
11. Michael Gordon, "Romania Is Reported in Nuclear Deal with India," *New York Times*, April 30, 1990.
12. When Switzerland tightened its transit regulations in 1986, Hempel began to use Amsterdam, Vienna, and ultimately, Monrovia, Liberia, as transshipment points.
13. This paragraph and the discussion below are based on, "Heavy Water Transferred," *Die Tageszeitung*, October 7, 1989, translated in *JPRS-TND*, Novem-

ber 6, 1989, p. 23; Michael Gordon, "Oslo Asserts India Got Atom Material," *New York Times*, November 4, 1988; Sanjoy Hazarika, "Norway Details How Heavy Water Went to India," *New York Times*, May 7, 1989. See also, Mark Hibbs, "Norway Investigating Reports of Heavy Water Diversions," *Nuclear Fuel*, May 30, 1988, p. 9; "Wanted: Bombs for Business: Nuclear Aid for Pakistan and India," ARD [West German Television], 2115, November 3, 1986; "Remarks of Ambassador Roland Timerbaev," Friedmann Conference, Columbia University Law School, March 20, 1987 (transcript provided to author by *Columbia Journal of Transnational Law*). Without mentioning Hempel or that the shipment ultimately went to India, Timerbaev, a Soviet diplomat, stated that a 6.2 metric ton shipment of Soviet heavy water that was diverted to unintended end-uses in 1986 was composed of a number of smaller shipments, each of which was less than 1 metric ton in size and therefore exempt from IAEA controls. Other sources have established that the material went to India.

14. "Ministerial Complicity in Illegal Material Shipments Charged," *Der Spiegel*, October 17, 1988, translated in *JPRS-TND*, December 23, 1988.
15. "Heavy Water Transferred," *Die Tageszeitung* (see note 13).
16. See, e.g., Serge Schmemann, "Heavy-Water Transfer Defended," *New York Times*, May 7, 1988.
17. During 1989, a separate clandestine export transaction of relevance to India's nuclear weapons program was also revealed, namely India's importation in 1984 of 210 pounds of high-purity beryllium from West Germany. Beryllium, a dual-use material, is used in nuclear weapons and in nuclear reactors. Although Indian authorities obtained the necessary German export licenses, the export by the German company, Degussa, had triggered concerns on the part of U.S. authorities, in part because it was made to the Bhabha Atomic Research Center, the presumed home of India's nuclear weapons research program. In addition, a portion of the beryllium originated in the United States and had been exported to Germany on condition that U.S. approval be obtained before the material was reexported. German officials did not obtain U.S. consent to the transfer to India, however, although they did obtain Indian assurances that the material would not be used for nuclear weapons.

    While India acted legally, taking advantage of the pro-export stance of German authorities, the transaction took place out of public view, and in this sense relied on an element of clandestinity. Had the transaction been publicly disclosed, it would surely have stirred considerable controversy, and the need for U.S. approval undoubtedly would have emerged. At the time, there were indications that India was considering a second nuclear detonation. See, Stephen Engelberg, "German Atomic Sale Challenged," *New York Times*, February 1, 1989.

    In March 1990, Degussa agreed to pay $800,000 in fines imposed by the U.S. Commerce Department. In a statement announcing the fine, Under Secretary of Commerce Dennis Kloske declared, "We believe the company knew that re-exporting U.S.-origin nuclear related materials required authorization from Commerce. Authorizations to India and North Korea [the destination of a shipment of U.S.-origin zirconium used to clad uranium fuel] would have been denied." "U.S. Fines Germany's Degussa $800,000 for Illegal Exports to India, North Korea," *Nuclear Fuel*, April 2, 1990, p. 1.
18. Developments with respect to a separate controversy between Norway and Israel over the latter's use of Norwegian-supplied heavy water are discussed in Chapter 9. See also, John J. Fialka, "Israel Plans Sale of Heavy Water Back to Norway," *Wall Street Journal*, April 30, 1990.

19. "CNEA Chairman Makes Announcement," Telam, 2102 GMT, November 18, 1983, translated in *JPRS-TND*, December 12, 1983, p. 7.
20. Richard Kessler, "CNEA Officials Provide More Details About Pilcaniyeu Enrichment Plant," *Nuclear Fuel*, May 30, 1988, pp. 4-5.
21. Judith Miller, "U.S. Is Holding Up Peking Atom Talks," *New York Times*, September 19, 1982; Rob Laufer, "Interview with Malone: Defense of Policy and Assessment of 'Hot Spots,' " *Nucleonics Week*, August 19, 1982, p. 1; Richard Kessler, "Argentina to Tender for Heavy Water After June Purchase from China," *Nucleonics Week*, September 25, 1986, p. 5. These articles noted, among other points, that Argentina had received some 4 to 6 metric tons of Chinese heavy water. However, it appears from subsequent reports that the amount involved was considerably greater, amounting to more than 50 metric tons. "Ministerial Complicity in Illegal Material Shipments Charged," *Der Spiegel* (see note 14).
22. Mark Hibbs, "British Report Details Heavy Water, Uranium Trades by Hempel Firms," *Nucleonics Week*, July 25, 1988, p. 7. U.S. government sources have since confirmed to the author that Chinese uranium hexafluoride was sent to Argentina.
23. *Ibid.*
24. Whether U.S. officials—who claimed to be taken by surprise when the existence of the plant was later revealed—deduced that the country had begun down this path is not known.
25. The agreement was renewed in 1989 in a scaled-back form.
26. Mark Hibbs, "Germans Say Brazil Developing Two Production Reactors," *Nucleonics Week*, July 27, 1989.
27. "Nuclear Program, Enriched Uranium Plans Unveiled, *Jornal do Brasil*, May 22, 1988, translated in *JPRS-TND*, June 16, 1988, p. 1.
28. " 'Tricks' to Gain Nuclear Technology Disclosed," *O Globo*, April 9, 1989, translated in *JPRS-TND*, April 25, 1989, p. 11. Secret bank accounts were apparently used to finance much of this activity. See "Parallel Program's Achievements, CNEN Secret Accounts Viewed," *Veja*, January 14, 1987, translated in *JPRS-TND*, March 4, 1987, p. 11; "Purchase of Nuclear Materials Reported," *Folha de São Paulo*, August 13, 1989, translated in *Foreign Broadcast Information Service (FBIS)-Latin America (LAT)*, August 15, 1989, p. 28.
29. Mark Hibbs, "Kohl Pressed by Opposition, IAEA Over Technology Sent to Brazil," *Nucleonics Week*, July 27, 1989, p. 5; Mark Hibbs, "West Germany to Investigate Charges of Brazilian Technology Diversions," *Nuclear Fuel*, September 4, 1989, p. 6.

    Current and former Brazilian nuclear officials have denied that the country has illegally obtained any items or technology for the parallel program. See "Charges of Smugggling Nuclear Equipment Refuted," *O Estado de São Paulo*, April 19, 1990, translated in *JPRS-TND*, June 12, 1990, p. 6; " 'Hypocrisy' Noted," *O Estado de São Paulo* (editorial), April 19, 1990, translated in *JPRS-TND*, June 12, 1990. For a further discussion of these denials, see Chapter 14, note 65.
30. Richard Burt, "U.S. Says Italy Sells Iraq Atomic Bomb Technology," *New York Times*, March 18, 1980.
31. Ann MacLachlan, "Iraq Nuclear Export Vetoed," *Energy Daily*, October 2, 1980; "Italian Uranium Shuffle to Iraq Raises Concern in Bonn over Euratom Loopholes," *Nuclear Fuel*, May 12, 1980. The Osiraq reactor was destroyed by an Israeli air raid in 1981, in order to preempt an Iraqi nuclear weapons program. Iraq has claimed, however, that the reactor was to be dedicated to peaceful uses only.

32. *Ibid.*, pp. 44-54. The discussion of this episode is based on European press accounts, a review of the prosecutor's dossier in the case, an in-person interview with one defendant, and a telephone interview with a second.
33. One defendant stated in an interview that he discussed the price of the plutonium with a "general of the Iraqi Army in charge of missiles," during a May 1982 meeting in Rome. A May 7, 1982, cable to him from a second member of the smuggling ring quotes the price for the material, apparently in preparation for the meeting. (The price was $60,572,000 or $1.78 million per kilogram.)
34. "Iraq and the Bomb," *MidEast Markets*, December 11, 1989, p. 13.
35. According to the indictment subsequently filed in the case in San Diego, the items were "co-axial, high voltage, low inductance capacitors of the same size, weight, energy storage performance specifications as relating to temperature range, inductance, voltage capacitance, resistance to humidity, vibration, and shock, and altitude performance requirements as those used in the detonation of nuclear warheads." Indictment, *U.S. v. Daghir, et al.*, U.S. District Court for the Southern District of California (San Diego), Criminal Case No. 90-0211-JLI, pp. 2-3.
36. *Ibid.*
37. "Face the Nation," CBS Television Network, April 1, 1990 (statement of Nazir Hamdoon, Deputy Foreign Minister of Iraq).
38. Edward Chen, "U.S. Accuses Iraq in A-Trigger Plot," *Los Angeles Times*, March 30, 1990.
39. Iraq has already used this approach with considerable success to develop chemical weapons and medium-range missiles, activities which have established crucial trading relationships that it appears ready to exploit for its nuclear weapons effort.
40. A sixth country, North Korea, is also taking advantage of loopholes in Western nuclear export controls to advance its nuclear capabilities, although its growing nuclear infrastructure appears to be largely indigenous. Sometime between late 1986 and late 1987, North Korea obtained a shipment of zirconium—a controlled dual-use commodity used as cladding for nuclear reactor fuel—from the West German firm, Degussa. The material involved had been previously exported to Germany from the United States, but Degussa did not obtain U.S. approval for the re-export of the material to North Korea, as required. A U.S. Department of Commerce official commenting on the case declared that the United States would not have approved the export to North Korea had it been notified. U.S. Department of Commerce Press Release BXA-90-10, March 28, 1990.
41. "Bonn Outlaws Work Abroad on 'ABC' Weapons," *This Week in Germany*, June 8, 1990, p. 2; "Bundesrat for Tougher Weapons Law Amendments," *This Week in Germany*, June 29, 1990, p. 2. See also Mark Hibbs, "Bonn Holding Up Tougher Controls on West German Nuclear Exports," *Nucleonics Week*, December 7, 1989, p. 4; Mark Hibbs, "Ruling Conservatives Delay Legislation to Stiffen West German Export Controls," *Nuclear Fuel*, December 11, 1989, p. 3; Simon Henderson, "China May Help Build Pakistan's N-Bomb," *Financial Times*, September 29, 1986; "Pakistan's Atomic Bomb," *Foreign Report*, January 12, 1989, p. 1.

    In a related development, a West German court sentenced Jürgen Hippenstiel-Imhausen, the key figure in the sale of West German equipment for Libya's Rabta chemical weapons plant, to five years imprisonment for tax evasion and export law violations. See Serge Schmemann, "German is Jailed in Sale to Libya," *New York Times*, June 28, 1990.
42. Although China has been a nuclear weapon state since 1964, it did not become a

significant exporter of nuclear goods until the 1980s and is thus considered an "emerging" or "second-tier" supplier.

43. Leslie H.Gelb, "Pakistan Links Peril U.S.-China Nuclear Pact," *New York Times*, June 22, 1984; Leslie H.Gelb, "Peking Said to Balk at Nuclear Pledges," *New York Times*, June 23, 1984; "Pakistan's Atomic Bomb," *Foreign Report*, January 12, 1989, p. 1; Henderson, "China May Help Build Pakistan's N-Bomb" (see note 43).
44. Interviews with knowledgeable U.S. officials; "Heavy Water Transferred," *Die Tageszeitung* (see note 13).
45. See note 34.
46. Farzad Bazoft, "Iran Signs Secret Atom Deal," *Observer*, June 12, 1988; "Nuclear Cooperation with Pakistan Reported," *Al-Watan Al-'Arabi*, translated in *JPRS-TND*, December 23, 1989, p. 11.
47. Although not traditionally viewed as an emerging supplier, North Korea has been included in this section because of its newly apparent nuclear capabilities, its relationships with other pariah states (*e.g.*, Romania under Ceaucescu, and Iran), and its willingness to cooperate with other developing countries in the development of conventional weapons.
48. These include ballistic, ship-to-ship, surface-to-ship, surface-to-air, air-to-air, and anti-tank missiles, multiple-rocket systems and launchers, towed guns and howitzers, and free-rocket-over-ground (FROG) systems.
49. "Iraq and the Bomb" (see note 34).
50. Mark Hibbs, "Prosecutors Link Tritium Plant to Pakistan Weapons Program," *Nuclear Fuel*, May 1, 1989.
51. India's agreements with Afghanistan, Algeria, Argentina, Iraq, Syria, and Yugoslavia were in effect as of 1980, per an unpublished chronology prepared by Tom Graham, cited in William C. Potter, ed., *International Nuclear Trade and Nonproliferation: The Challenge of the Emerging Suppliers* (Lexington, Mass.: Lexington Books, 1990), pp. 155-156.

    Transactions between China and India, China and South Africa, and India and Pakistan were engineered through West German middlemen.

    For more information on the Iraq-China connection, see "Iraq and the Bomb," *MidEast Markets* (see note 34).

    The nuclear cooperation agreement between India and Libya expired in 1984. See Leonard S. Spector, *The New Nuclear Nations* (New York: Vintage Books, 1985), P. 155.

    Data on conventional weapon transfers and cooperation have been culled from *SIPRI Yearbook* (Stockholm: Stockholm International Peace Research Institute, 1981, 1982, 1983, 1984, 1985, 1986, 1988, 1989); Michael Brzoska and Thomas Ohlson, *Arms Transfers to the Third World, 1971-1985* (New York: Oxford University Press, 1987); U.S. Arms Control and Disarmament Agency (ACDA), *World Military Expenditures and Arms Transfers, 1988* (Washington, D.C.: U.S. ACDA, 1989); Douglas J. Murray and Paul R. Viotti, eds., *The Defense Policies of Nations: A Comparative Study, Second Edition* (Baltimore: The Johns Hopkins University Press, 1989); James Everett Katz, ed., *Arms Production in Developing Countries* (Lexington, Mass.: Lexington Books, 1984).
52. This process could be initiated through broad cooperation agreements, such as that between Argentina and Iran, and offers to provide training for the importing country's engineers and technicians. In the years ahead this process could end in the further diffusion of nuclear materials and know-how among yet another tier of potential suppliers.

    A longstanding obstacle to such a course has been the general unavailability of financing in the emerging supplier states to support such projects, and multilateral assistance through the World Bank has not favored the application of "high tech" solutions, such as large nuclear power plants, to meet the energy shortfalls in countries with a limited technological base. However, this attitude conceivably could be revised in the event that countries such as Brazil and Argentina successfully develop small, affordable reactor plants that are suitable for developing economies possessing no feasible energy alternatives. See "Low Power Nuclear Reactor Under Development," Telam, June 7, 1988, 0903 GMT, translated in *FBIS-LAT*, June 10, 1988, p. 20; "Argentine Province Studying 25-MW LWR for Local Use, Exports," *Nucleonics Week*, October 27, 1988, p. 13; "Nuclear Miniplant Project Described," *O Globo*, April 9, 1989, translated in *JPRS-TND*, April 25, 1989, p. 15; "CNEN Director Describes Miniplant Program," *ibid.*

Chapter 5: Nuclear Submarines

1. Thomas B. Cochran, William M. Arkin, Robert S. Norris and Milton M. Hoenig, *Nuclear Weapons Databook Volume II: U.S. Nuclear Warhead Production* (Cambridge, Mass.: Ballinger Publishing Company, 1987), pp. 70-74.
2. The loophole was left open due to Italian and Dutch interest in reserving an option to build nuclear submarines and to Great Britain's desire to ensure that the treaty would not interfere with its acquisition of nuclear materials from the United States for its own submarine propulsion systems. See *The Structure and Content of Agreements Between the Agency and States Required in Connection with the Treaty on the Non-Proliferation of Nuclear Weapons*, INFCIRC/153 (Vienna: International Atomic Energy Agency, 1972), Article 14.
3. "Nuclear Submarine Ready 'Within 5 or 6 Years,' " *Fôlha de São Paulo*, September 12, 1989, p. A-6, translated in *Joint Publications Research Service (JPRS)-Nuclear Development and Proliferation (TND)*, October 6, 1989, p. 12 .
4. S.D. Singh, "More Info on Submarine Deal," *News India*, April 6, 1990; "The Scandal Surfaces," *India Today* (March 15, 1990): 42; Sanjoy Hazarika, "India Charges 6 Over Submarine Kickbacks," *New York Times*, March 7, 1990; "Lawmakers Question Covert Purchases of Submarines," Yonhap, 0108 GMT, September 29, 1989, reprinted in *Foreign Broadcast Information Service (FBIS)-East Asia (EAS)*, September 29, 1990, p. 29; "Contracts for 3 Submarines From FRG Signed," *Korea Herald*, October 7, 1989, reprinted in *FBIS-EAS*, October 19, 1989, p. 36.
5. "India Has 'Sovereign Rights' Over Soviet Nuclear Submarine," *Times of India*, February 20, 1988, reprinted in *JPRS-TND*, April 28, 1988, p. 16.
6. "Need for India to Have Nuclear Submarine Examined," *Hindu,* July 14, 1988, in *JPRS-TND*, September 2, 1988, p. 17; Sanjiv Prakash, "India Emerges," *Defense and Foreign Affairs* (December 1988): 12; "Second 'Charlie I' SSGN for Indian Navy," *Jane's Defence Weekly,* October 29, 1989, p. 893.
7. "India to Return Charlie-Class Submarine to Soviets," *Navy News and Undersea Technology*, November 13, 1989, as cited in "Deadly Submarine," *Bulletin of the Atomic Scientists* (March 1990): 48.
8. "Largely Indigenous Submarine Launched," *India News*, October 6, 1989; "Need for India To Have Nuclear Submarine Examined," *Hindu*, July 14, 1988, reprinted in *JPRS-TND*, September 2, 1989, p. 17.
9. "India To Return Charlie-Class Submarine to Soviets" (see note 7).
10. "Nuclear Negotiations," *Far Eastern Economic Review*, January 19, 1990, p. 6.
11. David Foxwell, "Staying Under Longer: The SSK Closes the Gap," *Military Technology* (November 1988): 61.

12. A. F. Oliva and R. J. Gosling, "The SS*n*: An Affordable Platform for Enhancing Undersea Naval Capability," paper presented at Underseas Defense '88, San Diego, California, October 3-6, 1988.
13. *Jane's Fighting Ships, 1988-89* (London: Jane's Publishing Co., Ltd., 1989), p. 204.
14. "Navy to Install PWR Nuclear Reactor in 1995," *Fôlha de São Paulo*, December 3, 1989, translated in *FBIS-Latin America (LAT)*, December 6, 1989, p. 27; Tonia Malheiros, "Navy Calls for Environmental Impact Report, " *Fôlha de São Paulo*, September 13, 1989, translated in *JPRS-TND*, October 6, 1989, p. 13.
15. Helio Contreiras, "Progress of Submarine Reconstruction Reviewed," *Manchete*, May 27, 1989, translated in *FBIS-LAT*, July 6, 1989, pp. 52-55.
16. "Navy Inaugurates IPEN-MB-01 Research Reactor," *Jornal do Brasil*, November 9,1988, translated in *JPRS-TND*, December 23, 1988, p. 8; Jania Malheiros, "Navy Calls for Environmental Impact Report" (see note 14); "CNEN Authorizes Operation of Aramar Facility," *Jornal do Brasil*, November 11, 1988, in *JPRS-TND*, December 23, 1988, p. 10.
17. Eduardo Italo Pesce, "Brazil's Silent Service," *Naval Institute Proceedings* (March 1989): 63.
18. *Jane's Fighting Ships, 1988-89* (see note 13), pp. 9-10; "Advantages of Building a Nuclear Submarine Cited," *La Prensa*, September 5, 1988, translated in *JPRS-TND*, September 21, 1988, p. 20; Peter Calvert, *The Falklands Crisis* (New York: St. Martin's Press, 1982), p. 51.
19. The reason for the secrecy has not been revealed. Possibly, political and military leaders in Seoul wished to avoid domestic political opposition to the high cost of the deal. See "Lawmakers Question Covert Purchase of Submarines," *Korea Times*, September 23, 1989, reprinted in *FBIS-EAS*, September 19, 1989, p. 29; "Contracts for 3 Submarines From FRG Signed," *Korea Herald*, October 7, 1989, reprinted in *FBIS-EAS*, October 19, 1989, p. 36; "Seoul Orders Submarines from FRG Company," Yonhap, 0626 GMT, October 30, 1989, printed in *FBIS-EAS*, October 31, 1989, p. 35; "South Korea Confirms Submarine Order," *Defense & Foreign Affairs Weekly*, October 30-November 5, 1989, p. 1; "S. Korea Orders Three More Submarines," *Jane's Defence Weekly*, November 11, 1989, p. 1018.
20. A sum of $250 million reportedly was provided for this purpose. See *Jane's Fighting Ships, 1988-89* (see note 13), p. 339. Interestingly, one account of political criticism of the deal alleges that the Defense Ministry "wasted a total of $234 million" by paying above-market prices. See "Contracts for 3 Submarines From FRG Signed" (see note 19).
21. See "German Submarines for Israel," *The Middle East* (October 1989): 24; "Israel Leans on Bonn to Allow SSK Buy," *Jane's Defence Weekly*, September 9, 1989, p. 417; "Contracts for 2 Submarines Signed With FRG," *Ha'Aretz*, August 31, 1989, translated in *FBIS-Near East and South Asia,* August 28, 1989, p. 19; "Israel to Buy 2 Subs From Germany," *Defense News*, July 31, 1989; "...Israel Gets SSK Go-Ahead," *Jane's Defence Weekly*, August 5, 1989, p. 189; "Israel Revives Submarine Plans," *Jane's Defence Weekly*, June 17, 1989, p. 1209. While Israeli military planners have called for a third submarine, funding currently permits the purchase of only two vessels.
22. HDW later returned 21 million Rand, but South Africa retained the blueprints.
23. "West German Firm Fined $800,000," *News India*, April 6, 1990.
24. *Ibid.*
25. Heinz Schulte, "HDW Papers Seized," *Jane's Defence Weekly*, January 20, 1990, p. 98; James P. McWilliams, *ARMSCOR: South Africa's Arms Merchant* (London: Brassey's (UK), 1988), p. 117; "German Submarines for Israel," *The Middle East* (October 1989): 24.

Part II: Asia - Introduction

1. Lincoln Kaye, "Atomic Intentions," *Far Eastern Economic Review*, May 3, 1990, p. 9.
2. Stephen Engelberg and Michael Gordon, "Taipei Halts Work on Secret Plant to Make Nuclear Bomb Ingredient," *New York Times*, March 23, 1988; R. Jeffrey Smith and Don Oberdorfer, "Taiwan to Close Nuclear Reactor," *Washington Post*, March 24, 1988.

   Washington is said to have learned of the clandestine installation from the deputy director of Taiwan's Institute for Nuclear Energy Research, Colonel Chang Hsien-Yi, who disappeared from Taiwan in January 1988, reportedly with the help of the U.S. Central Intelligence Agency (CIA). See references above. A U.S. visit to the site where the clandestine plutonium unit was being built is said to have confirmed Chang's information.

   U.S. officials dismissed the notion that the construction of the plutonium unit might have been a rogue operation undertaken by Taiwanese nuclear aides without higher authorization. (Confidential interview, March 1988.)

   Taiwan has one of the largest nuclear power programs in the developing world, with six operating nuclear power reactors, all supplied by the United States.
3. It is not clear how Taiwan could have bypassed IAEA controls, given the considerable quantity of spent fuel that it would have had to divert in order to obtain a bomb's worth of plutonium. Assuming that the plutonium in each spent fuel rod represented .004 of its total weight, to obtain 11 pounds (5 kilograms) of the material, well over a ton (2,000 pounds or 909 kilograms) of spent fuel would have had to have been diverted for reprocessing. Since the IAEA would have periodically inspected the reactor to determine its output and the whereabouts of all spent fuel and would have used surveillance cameras and seals to keep track of activities between inspections, diverting this much material would not have been easy.
4. State Department Regular Briefing, 12:15 EST, March 23, 1988. Safeguards would be violated if spent fuel were moved from a safeguarded storage area to a unit not covered by the IAEA.
5. In the late 1960s and particularly during the mid-1970s, when the United States was opening friendly relations with mainland China, Taiwan made repeated attempts, some apparently secret, to develop a plutonium separation capability. Because of Taiwan's precarious security situation and the lack of any clear need for this capability as part of its nuclear energy program, U.S. officials came to believe that these efforts were aimed at eventually producing plutonium for nuclear weapons.

   In 1969, for example, Taiwan sought to buy a plutonium separation plant from the United States, but the Nixon Administration denied this request and barred U.S. firms from selling key components for such a facility. See Edward Schumacher, "Taiwan Seen Reprocessing Nuclear Fuel," *Washington Post*, August 29, 1976. Taiwan then turned to other possible suppliers, including Belgium and France, from whom it sought to purchase a large plutonium extraction facility. When the United States intervened, discussions on the sales were broken off. France, however, apparently supplied a number of components useful for such a plant. See Steve Weissman and Herbert Krosney, *The Islamic Bomb* (New York: Times Books, 1981), pp. 152-153.

   From 1974 to 1976, Taiwan tried to negotiate a commercial agreement with Great Britain for the latter to extract plutonium from Taiwanese spent fuel in

England and for the return of the plutonium to Taiwan. Once again, the United States stepped in, this time to insist that any plutonium produced under this arrangement be shipped to the United States instead. See Melinda Liu, "Accounting for the N-Factor," *Far Eastern Economic Review*, December 17, 1976, p. 33.

By 1974, the U.S. CIA had concluded:

> Taipei conducts its small nuclear program with a weapon option in mind, and it will be in a position to fabricate a nuclear device after five years or so. Taipei's role in the world is changing radically, and concern over the possibility of complete isolation is mounting. Its decisions will be much influenced by U.S. policies in two key areas—support for the island's security and attitudes about the possibility of a nuclear-armed Taiwan. Taipei's present course probably is leading it toward development of nuclear weapons.

See CIA, "Prospects for Further Proliferation of Nuclear Weapons," DCINIO 1945/74, Sanitized Copy, September 4, 1974, released to the Natural Resources Defense Council under the Freedom of Information Act.

Despite U.S. interventions, by 1975 Taiwan was constructing a hot cell facility (a laboratory-scale plutonium extraction unit) on its own. The facility was apparently constructed with components purchased from around the world, including some from France. See Schumacher, "Taiwan Seen Reprocessing Nuclear Fuel." Although the facility was capable of separating only small amounts of plutonium, it would have provided Taiwan with valuable experience in this field. In that year, Taiwan asked the United States for permission to separate plutonium from a small amount of spent fuel from a U.S.-supplied research reactor in Taiwan, but the United States did not reply to the inquiry. See Don Oberdorfer, "Taiwan to Curb A-Role," *Washington Post*, September 23, 1976.

In 1976, reports surfaced that Taiwan might have been secretly extracting plutonium from spent fuel at a clandestine plutonium extraction facility of some kind. See Schumacher, "Taiwan Seen Reprocessing Nuclear Fuel;" Oberdorfer, "Taiwan to Curb A-Role." In congressional hearings U.S. officials stated that no plutonium separation had actually taken place and that such a facility had not, in fact, been built. Nonetheless, U.S. suspicions over Taiwan's activities in this area became so intense that the Ford Administration insisted that it cease all activities in the plutonium separation field, dismantle the hot-cell laboratory, and pledge not to engage in any activities related to plutonium extraction. See U.S. Congress, Senate Subcommittee on Arms Control, International Organizations, and Security Agreements, *Hearings on Non-Proliferation Issues*, September 22, 1976 (Washington, D.C.: U.S. Government Printing Office, 1976), pp. 346-371.

As late as 1977, however, Washington remained concerned that Taipei was continuing to engage in activities relevant to nuclear arms, although no details have been publicly revealed. See Milton R. Benjamin, "Taiwan's Nuclear Plans Concern U.S. Officials," *Washington Post*, December 20, 1978.

In addition to its diplomatic initiatives, from 1975 to 1978, the United States reportedly sought to thwart Taiwan's nuclear ambitions by mounting a covert operation, which led to the theft of its weapons plans and files, and by attempting to choke off its efforts to purchase sensitive items. See Bob Woodward, *Veil: The Secret Wars of the CIA* (New York: Simon and Schuster, 1987), p. 170.

6. Benjamin, "Taiwan's Nuclear Plans Concern U.S. Officials" (see note 5).
7. Because individual power-reactor spent fuel elements are much larger than those used in the Lung Tan research reactor, are much more radioactive, and are stored at electric power plants rather than at a classified research center, safeguarding the material against diversion is easier than safeguarding the Lung

Tan spent fuel. In addition, since power reactor fuel typically remains in a reactor far longer than research reactor fuel, its plutonium contains a larger proportion of the isotope Pu-240, which makes the material less desirable for use in nuclear weapons.

8. Judith Miller, "U.S. Is Holding Up Peking Atom Talks," *New York Times*, September 19, 1982.
9. Kenneth Adelman, Director, Arms Control and Disarmament Agency, "Non-Proliferation Assessment Statement for the Peaceful Nuclear Cooperation Agreement between the United States and China," July 19, 1985.
10. Leslie H. Gelb, "Pakistan Link Perils U.S.-China Nuclear Pact," *New York Times*, June 22, 1984.
11. Adelman, "Non-Proliferation Assessment Statement," p. I-5 (see note 9). The details of the U.S.-Chinese understanding on the matter were contained in a classified note submitted by the Reagan Administration to Congress along with the agreement. The document is said to contain a U.S.-prepared summary of oral understandings reached between U.S. and Chinese negotiators. Apparently China did not sign the document but did not object to its being used by U.S. officials to characterize China's position.
12. Daniel Southerland, "U.S., China Sign Accords," *Washington Post*, August 28, 1985. It appears that some unsafeguarded transfers of Chinese nuclear materials to Argentina and possibly South Africa continued into 1985, although these were supposedly based on pre-existing contracts. See Patrick Tyler, "Hill Panels Add Conditions to U.S.-China Nuclear Pact," *Washington Post*, November 14, 1985; Richard Kessler, "Premier Zhao Ziyang Appeared to Confirm that China . . .," *Nucleonics Week*, November 14, 1985; Richard Kessler, "Argentina Officials Deny Any Purchases of Chinese Heavy Water," *Nucleonics Week*, October 31, 1985. Chinese sales of unsafeguarded heavy water to India (through a West German middleman) also apparently continued until 1987. See "Heavy Water Transferred," *Die Tageszeitung*, October 7, 1989, translated in *Joint Publications Research Service-Nuclear Development and Proliferation*, November 6, 1989, p. 23.
13. Atomic Energy Act of 1954, Sections 123 and 130 (1985).
14. S. J. Res. 238 (99th Cong., 1st Sess.) (1985).
15. Confidential interview, June 1990.

Chapter 6: India

1. "Shadow of an Indian H-Bomb," *Foreign Report*, December 13, 1984, p. 1; "India: Fixed-Wing Nuclear Delivery by Mid-1989," *Defense and Foreign Affairs Weekly*, October 3-9, 1988, p. 1.
2. This historical summary of the Indian nuclear program is based on a more detailed review found in Leonard S. Spector, *Nuclear Proliferation Today* (New York: Vintage Books, 1984), pp. 23-48.
3. At the time, the United States and the Soviet Union had programs to explore the use of nuclear devices for such peaceful purposes.
4. See, *e.g.*, G. G. Mirchandani, *India's Nuclear Dilemma* (New Delhi: Popular Book Services, 1968), p. 35.
5. India chose this characterization to deflect international criticism and also to support the claim that the test did not violate agreements with Canada and the United States. The former had transferred the Cirus reactor to India, where plutonium for the 1974 test was produced, and the United States had provided the heavy water needed to permit the facility to operate. India's agreements

with both states required that it use the facility, the heavy water, and any plutonium produced through their use only for peaceful purposes. India claimed that its nuclear detonation met this standard, but Ottawa and Washington had previously made clear that they understood the restrictions in their respective agreements with India to preclude the use of the transferred items and produced plutonium for *any* type of nuclear explosive, on the ground that peaceful nuclear explosives and nuclear weapons were essentially the same.

Canada subsequently ended all nuclear cooperation with India because of this dispute. The United States continued, however, to supply low-enriched uranium fuel for the two-reactor, U.S.-supplied Tarapur nuclear power station. Washington obtained a clarification from India that all plutonium produced from U.S.-origin fuel (the only fuel that would be used in the reactors) would either be stored or used in the reactors themselves, as a substitute for low-enriched uranium fuel, thereby precluding the use of such plutonium for any type of nuclear explosive. Washington obtained no further assurances regarding the Cirus reactor, where U.S.-origin heavy water was apparently still in use.

The continued sale of fuel for the Tarapur reactors proved highly controversial in the United States and led to a major dispute with India in the late 1970s and early 1980s. Under the 1978 Nuclear Non-Proliferation Act (NNPA), after March 1980, U.S. exports of nuclear fuel were prohibited to countries that had not accepted IAEA safeguards on all of their nuclear activities, but India claimed that its prior nuclear agreement with the United States bound Washington to supply the fuel for the Tarapur units; New Delhi threatened to remove plutonium-bearing, U.S.-origin spent fuel at Tarapur from IAEA safeguards if the United States did not comply. President Jimmy Carter waived the NNPA prohibition to permit one fuel shipment in 1980. In 1982, the Reagan Administration ended the controversy by arranging for France to replace the United States as the supplier of fuel for the units. See, Spector, *Nuclear Proliferation Today* (see note 2), pp. 31-37, 46-48.

6. Jerrold F. Elkin and Andrew Ritzael, "India," in Douglas J. Murray and Paul R. Viotti, eds., *The Defense Policies of Nations: A Comparative Study, Second Edition* (Baltimore: The Johns Hopkins University Press, 1989), p. 523.
7. Steve Weissman and Herbert Krosney, *The Islamic Bomb* (New York: Times Books, 1981), pp. 43-46. The authors' information comes from two participants in the key meeting, held in Multan, at which Bhutto announced that Pakistan would develop nuclear weapons.
8. Michael T. Kaufman, "India Gives Warning of Atom-Arms Race," *New York Times*, August 16, 1979. Other top officials of his government made similar statements during this period.
9. "Gandhi Says National Interest May Require Nuclear Blasts," *Washington Post*, March 14, 1980; Judith Miller, "Cranston Says India and Pakistan Are Preparing for Nuclear Testing," *New York Times*, April 28, 1981; Roger Gale, "India to Press 'Start Button' on Nuclear Fuel Reprocessing Plant," *Energy Daily*, February 5, 1981; Yogi Aggarwal, "India Makes Another Bomb," *Sunday Observer* (Bombay), August 30, 1981 (government had one atomic bomb ready in April, could prepare another one in weeks); K. Subrahmanyam, "Bomb—The Only Answer," *Times of India*, April 26, 1981, p. 1; Milton R. Benjamin, "U.S. Is Delaying Nuclear Exports to India," *Washington Post*, June 23, 1983.
10. This apparently was the view of the Pakistani Army. See Stephen P. Cohen,

The Pakistani Army (Berkeley: University of California Press, 1984), p. 153; "Shadow of an Indian H-Bomb," *Foreign Report* (see note 1). In a 1985 interview with the principal author of this book, K. Subrahmanyam, then the director of the Indian Institute for Defence Studies and Analyses, indicated that he assumed Mrs. Gandhi had pursued such activities because not to have done so would have been "irresponsible."

Adding to New Delhi's concerns over India's strategic environment at the time was yet another Chinese nuclear advance, its first test in 1980 of an intercontinental ballistic missile. See Chong-Pin Lin, *China's Nuclear Weapons Strategy: Tradition Within Evolution* (Lexington, Mass.: Lexington Books, 1988), p. 67.

11. See chart at end of this chapter.
12. These reactors were built domestically and commissioned in 1983 and 1985, respectively. India declined to place them under IAEA safeguards, thus preserving the option to use the plutonium they produced for nuclear weapons.
13. Commissioned in 1985, Dhruva is one of the largest civilian research reactors in the developing world. See discussion later in this chapter.
14. The Trombay plant reopened in late 1983 or early 1984. It is designed to extract plutonium from metallic uranium fuel, which is used in the unsafeguarded Cirus and Dhruva research reactors. For a further discussion of constraints on the use of plutonium produced in the Cirus reactor, see notes 5 and 27.
15. The Tarapur plant, which is much larger than that at Trombay, began operating at full capacity in 1982. It possesses the capability of extracting plutonium—a procedure known as "reprocessing"—both from metallic fuel and from the oxide fuels used in India's nuclear power reactors. Since the early 1980s, it has been reprocessing power-reactor fuel from the safeguarded Rajasthan reactors and from the unsafeguarded Madras I reactor. Fuel from Madras II may also have been reprocessed there. IAEA inspectors are present when fuel from the safeguarded Rajasthan reactors is processed to ensure that plutonium from them is not diverted to military purposes. However, India has not permitted the IAEA to oversee the reprocessing of the fuel from the Madras I plant (or from the Madras II reactor, if indeed material from this reactor has been processed there).
16. Rick Atkinson, "3 Pakistanis Indicted on A-Arms Charges," *Washington Post*, July 17, 1984; Leslie Gelb, "Peking Said to Balk at Atom Pledges," *New York Times*, June 23, 1984.
17. The accounts were based on a U.S. intelligence briefing to a congressional committee. The report was originally carried by ABC News. See Don Oberdorfer, "U.S. Sees India-Pakistan Rifts Not As Signals of Imminent War," *Washington Post*, September 15, 1984; Don Oberdorfer, "Pakistan Concerned about Attack on Atomic Plants," *Washington Post*, October 12, 1984. India was reportedly considering such a raid as early as 1982. See Milton R. Benjamin, "India Said to Eye Raid on Pakistani A-Plants," *Washington Post*, December 20, 1982.
18. "Paper Assesses Pakistani Umbrella Proposal," *Indian Express*, October 15, 1984, reprinted in *Foreign Broadcast Information Service (FBIS)-South Asia*, October 29, 1984, p. E-3; William K. Stevens, "India Worried by U.S. Links to Pakistan," *New York Times*, October 21, 1984.
19. "People's Expectations Are Scary," *India Today*, February 15, 1985, p. 8; "Gandhi on Nuclear Weapons, Sri Lanka, Punjab," Delhi Domestic Service, 0240 GMT, January 31, 1985, reprinted in *FBIS-South Asia*, January 31, 1985, p. E-1; John Elliott, "We've Got Five Years," *Financial Times*, April 4, 1985; "Defense Ministry Security Report Raps Pakistan," Delhi Domestic Service,

1530 GMT, April 16, 1985, reprinted in *FBIS-South Asia*, April 17, 1985, p. G-1; "India to Review Nuclear Policy, Gandhi Says," *Washington Post*, May 5, 1985; "Congress-I Party Meets in New Delhi," *FBIS-South Asia*, May 6, 1985.

20. "*Le Monde* Interview," *Le Monde*, June 4, 1985, translated in *FBIS-South Asia*, June 5, 1985, p. E-1.
21. John Scali, "Good Morning America," American Broadcasting Company, July 11, 1985.
22. Rone Tempest, "India Starts Up New Reactor That Can Make Arms-Grade Plutonium," *Los Angeles Times*, August 9, 1985; "New Delhi Says It Can Make Plutonium from Indian Fuel," *New York Times*, August 9, 1985; Swaminathan S. Aiyar, "Dhruva Reactor an N-Bomb Spinner," *Indian Express*, August 10, 1985. Earlier official descriptions of the reactor, in contrast, had stressed its use for medical isotope production and engineering research. See "New Reactor Capable of Producing Plutonium," Agence France Presse, 0905 GMT, August 8, 1985, reprinted in *FBIS-South Asia*, August 9, 1985, p. E-1.

    Dhruva's capacity would enable it to produce the substantial quantity of 55 pounds (25 kilograms) of plutonium annually. (Fifty-five pounds of plutonium, once it is separated from the spent fuel at Dhruva, would be sufficient to produce three to five nuclear weapons, based on the standard assumption that 11 to 17.6 pounds (5 to 8 kilograms) of the material are needed per device.)
23. Mohamed Aftab, "Pakistan Concern Over India's Nuclear Move," *Financial Times*, August 8, 1985; "India Can Make 30 N-Bombs a Year," *Pakistan Times Overseas Weekly*, August 18, 1985; "Karachi Raps India's Nuclear Plans, 'Propaganda,' " Karachi Domestic Service, 1715 GMT, August 11, 1985, reprinted in *FBIS-South Asia*, August 13, 1985, p. F-1; "*The Muslim* on India's Nuclear Capabilities," *Muslim*, August 13, 1985, reprinted in *FBIS-South Asia*, August 21, 1985, p. F-1.
24. "Report on Kahuta Denied," Karachi Domestic Service, 1703 GMT, September 30, 1985, reprinted in *Joint Publications Research Service (JPRS)-Nuclear Development and Proliferation (TND)*, October 21, 1985, p. 37; "Minister Talks About India's Alleged Attack Plans," Karachi Domestic Service, 1500 GMT, October 16, 1985, reprinted in *JPRS-TND*, November 4, 1985, p. 42; "Plans to Attack Pakistan Nuclear Complex Denied," Delhi Domestic Service, 1540 GMT, November 6, 1985, reprinted in *FBIS-South Asia*, November 6, 1985, p. E-1.
25. This information was provided by a senior Indian official during a June 1986 interview with the principal author of this book in Bombay. See also Vyvyan Tenorio, "India's Supply of Unsafeguarded Pu Grows As Reprocessing of MAPS Fuel Begins," *Nuclear Fuel*, August 11, 1986, p. 3. Ironically, the Dhruva reactor, whose plutonium production capabilities had been highlighted at its inauguration, experienced severe operating problems that would not be corrected for many months.
26. The plutonium was extracted at the Tarapur plutonium plant from spent fuel removed from the unsafeguarded Madras I nuclear power reactor, some 1,000 miles away. Since there was no urgent need for the material in India's peaceful nuclear program, the decision to go to the expense and risk of transporting the highly radioactive material such a great distance suggests that the material was sought for other purposes.
27. The restricted plutonium had come from the Rajasthan I and II nuclear power reactors and the Cirus research reactor. The former units are subject to IAEA safeguards, which would cover any plutonium produced in their spent fuel.

(This means IAEA inspectors periodically conduct on-site inspections to verify that the material is being used only for non-military purposes.)

The 1956 Indo-Canadian agreement covering the operation of the Cirus reactor prohibited the use of plutonium produced therein for non-peaceful purposes. This reactor, however, is not under IAEA safeguards, and no other measures to verify India's pledge have been implemented. As noted earlier, India used plutonium produced in the Cirus reactor for its 1974 test, claiming, contrary to the Canadian view, that "peaceful nuclear explosions" were permissible under the Indo-Canadian agreement. India has never asserted the right to use Cirus-origin plutonium for nuclear weapons, as such, however. Nonetheless, because Canada has not taken steps to ensure that Cirus-origin plutonium is not used for this purpose, as a practical matter, India could use it for nuclear weapons without being detected and may well have earmarked the bulk of Cirus-origin plutonium for this purpose. See note 61.

28. The issue of bilateral treaties to establish a foundation for better relations has a long and tortuous history, with proposals having been tabled in 1959, 1965, 1979, and 1981. Among other issues in these negotiations have been India's demand that Pakistan agree not to allow any foreign bases on its soil and that it agree to the resolution of outstanding disputes through bilateral negotiations. Pakistan, though stating it has no intention of allowing foreign bases, has objected to India's attempt to dictate its foreign policy; Pakistan has also been unwilling to give up recourse to multilateral fora, including the United Nations, for settling the Kashmir dispute. For a description of previous proposals, see Leonard S. Spector, *The Undeclared Bomb* (Cambridge, Mass.: Ballinger Publishing Company, 1988), p. 351 (note 21).
29. Gandhi initiated the no-attack proposal to lay to rest speculation that India was planning a strike against Pakistani facilities. See note 17. The status of the no-attack agreement is discussed below (see note 91 and accompanying text).
30. "Back From the Brink of War," *Asiaweek*, February 15, 1987, p. 11; "Back from the Edge," *India Today*, February 28, 1987, pp. 24-25; "On India's Border, a Huge Mock War," *New York Times*, March 6, 1987.
31. On events in the Siachen Glacier region, see "Indians and Pakistanis Clash in Kashmir," *New York Times*, October 1, 1987; on tensions in Punjab, see "India Direct Rule Imposed in Punjab," *New York Times*, May 12, 1987, and "President's Rule in Punjab," *Statesman Weekly*, May 16, 1987; "The Obstacles Remain," *Far Eastern Economic Review*, March 12, 1987, pp. 35-36; and on Pakistani accusations of Indian terrorism, see "Riot Victims Blame Unseen, Uncontrolled Forces," *New York Times*, December 20, 1986; "Roots of the Rift Run Deep," *Far Eastern Economic Review*, January 1, 1987, pp. 16-18; and "Pakistan's Premier Points at India in Fatal Bombings," *Washington Post*, July 17, 1987.
32. Bob Woodward, "Pakistan Reported Near Atom Arms Production," *Washington Post*, November 4, 1986.
33. "Official: India Has Nuclear Weapon Capability," United Press International, AM cycle, November 4, 1986; Ivan Fera and Kannan Srinivasan, "Keeping the Nuclear Option Open: What It Really Means," *Economic and Political Weekly*, December 6, 1986, p. 2119.
34. Ramanna did not explain India's rationale behind pursuing this technology, but three explanations appear the most likely. First, Gandhi and Ramanna may have sought to silence domestic critics who feared that, despite the country's generally more advanced nuclear capabilities, India would fall behind Pakistan in the nuclear technology race in South Asia. Second, India may have

wanted an enrichment capability as part of its reported research program related to the development of thermonuclear weapons. See "Shadow of an Indian H-Bomb," *Foreign Report*, June 27, 1987. Third, India may have sought a uranium enrichment capability to produce fuel for nuclear-powered submarines.

Although India's announcement of its enrichment breakthrough came as a response to the report of Pakistan's progress in the field, India's work on the technology had obviously been initiated a number of years earlier and, it appears, would have progressed whether or not Pakistan achieved weapons-grade levels of enrichment. In this sense, the new enrichment capability—and many other elements of India's burgeoning nuclear program that were initiated prior to the advent of a serious nuclear challenge from Pakistan—should not be considered exclusively as a response to Pakistani nuclear advances but also as a manifestation of India's broader ambitions to enhance its technological capabilities and overall strategic potential. Nonetheless, the centrality of the Pakistani threat to Indian nuclear policy has been manifest for the past decade.

35. Although there is some reason to believe that he was misquoted, Khan was attributed with saying that Pakistan had already built nuclear weapons. While he later denied that he had made the statement, the interview elicited a strong reaction in India nonetheless. See Kuldip Nayar, "We Have the A-Bomb, Says Pakistan's Dr. Strangelove," *Observer*, March 1, 1987; *India Today*, cover story, March 31, 1987. For an analysis raising questions about the authenticity of the interview, see Spector, *The Undeclared Bomb* (see note 28), pp. 91-94.

President Zia, in an interview with *Time* magazine, stated that "Pakistan has the capability of building the Bomb whenever it wishes." Zia stressed, however, that Pakistan did not intend to manufacture nuclear arms and asserted that it had not yet enriched uranium to weapons grade. See "Knocking at the Nuclear Door," *Time*, March 30, 1987, p. 42.

36. See, *e.g.*, Tom Diaz, "Gandhi Warns U.S. Is 'Soft' on 'Islamic Bomb,' Threat," *Washington Times*, April 27, 1987; "India Announces Review of Nuclear Policy," Reuters, April 27, 1987; Amalendu Das Gupta, "Challenge and Response," *Statesman Weekly*, May 16, 1987, p. 12; "Natwar Singh Discusses 'Nuclear Option,' " Delhi Domestic Service, 1230 GMT, October 18, 1987, reprinted in *FBIS-Near East and South Asia (NES)*, October 19, 1987, p. 51; "India Won't Tolerate Pak Nuclear Bomb: Gandhi," *Indian Observer*, November 3, 1987.

37. The controversies concerning U.S. aid to Pakistan are discussed in detail in Chapter 7.

38. "Pakistan Proposes Nuclear Test Ban," *New York Times*, September 14, 1987. The proposal had first been broached with Gandhi in June.

39. "Address by His Excellency Mr. Mohammad Khan Junejo, Prime Minister of the Islamic Republic of Pakistan at the Forty-Second Session of the United Nations General Assembly," September 24, 1987 (text provided by the Embassy of Pakistan), p. 11. Previous Pakistani proposals included a number of comprehensive bilateral nuclear arms control options, such as mutual ratification of the Nuclear Non-Proliferation Treaty (NPT), mutual acceptance of IAEA safeguards on all nuclear facilities, the establishment of a South Asian nuclear-weapon-free zone, comprehensive bilateral nuclear inspections, and a joint declaration renouncing nuclear weapons.

India had rejected these proposals for a variety of reasons. It has long opposed the NPT and comprehensive IAEA safeguards on the grounds that they are discriminatory instruments of the nuclear weapon powers designed to legitimize the latters' nuclear status while freezing others in permanent inferiority. Citing the potential threat from China's nuclear weapons, India has also opposed the establishment of a South Asian nuclear-weapon-free zone unless China is included in the arrangement, a condition China is certain to reject. New Delhi has also turned down comprehensive bilateral nuclear inspections with Pakistan, arguing that this (like ratification of the NPT and acceptance of comprehensive IAEA safeguards) would foreclose its nuclear option while leaving a nuclear-armed China on its border.

Perhaps most important is India's objection to all of these measures on the ground that they treat India and Pakistan as relative equals, despite India's far greater size and economic and military power—although this has not stopped India from pursuing the bilateral no-attack-on-nuclear-facilities agreement. Indian analysts have also complained that even if some form of comprehensive inspections were adopted, India could never be confident that Pakistan had not previously hidden away some nuclear weapons material, thereby retaining a secret nuclear weapons option. (India has said little about a possible joint declaration renouncing nuclear arms.)

Since many of India's objections to these proposals have a long history, there is some question as to whether Pakistan has advanced them in good faith. India, however, has refused to put Pakistan to the test by agreeing to discuss any of these initiatives, despite the possibility that some of New Delhi's concerns might be resolvable through negotiation. For example, mutual inspection regimes of limited duration could ease pressures toward nuclear arming between India and Pakistan in the short term, without permanently foreclosing India's nuclear option vis-à-vis China. Similarly, even if neither side could be sure that the other had not previously hidden away some nuclear weapons material—and Pakistan would have as much reason for concern as India in this regard—prospective limits on the production or use of such material would still be a valuable arms control measure.

40. Most observers believe that neither is likely to test a nuclear device because of the risks of triggering similar moves by the other party and of inviting sanctions by the United States and other countries. U.S. sanctions could include the termination of U.S. foreign aid and Export-Import Bank credits (Foreign Assistance Act, section 670; Export-Import Bank Act, section 2(b)(4)), and reduced financing from multilateral lending institutions (International Financial Institutions Act, section 701). In addition, under the Nuclear Suppliers Group Guidelines, the nuclear supplier countries have pledged to consult on a possible embargo of all nuclear transfers to non-nuclear-weapon states that conduct nuclear explosions. (Section 129 of the U.S. Nuclear Non-Proliferation Act also specifies that U.S. nuclear cooperation would be cut off in such an event, but this has already essentially ended in the case of India and Pakistan because of their refusal to accept IAEA safeguards on some of their nuclear facilities. See Nuclear Non-Proliferation Act, section 128.)

41. "Pakistan Proposes Nuclear Test Ban" (see note 38); "India Not in Favor of Regional Disarmament," *News India*, November 14, 1986.

42. "Gandhi's Talks with U.S. Leaders Reported," Delhi Domestic Service, 0240 GMT, October 21, 1987, reprinted in *FBIS-NES*, October 21, 1987, p. 57; "India-Pakistan Talks Urged," *Washington Post*, October 21, 1987; Neil Lewis, "Better Ties Seen in Visit by Gandhi," *New York Times*, October 21, 1987.

At the time, because of the quirks of the congressional appropriations process, aid for Pakistan had, in fact, been suspended. In 1981, Congress had waived the Symington Amendment (prohibiting aid to countries that import uranium enrichment technology while refusing IAEA safuguards on all of their nuclear activities) for six years through a special statutory provision. That period expired on September 30, 1987, the end of the fiscal year. Legislation to renew aid to Pakistan—and to waive the Symington Amendment for an additional period—had not yet been enacted, however, because it was to be included in an omnibus, government-wide spending bill, known as a "continuing resolution." That legislation had been delayed because of a number of major disputes between Congress and the Reagan Administration over budgetary priorities.

43. P.L. 100-202, section 557, amending Foreign Assistance Act, section 620E(d).
44. These concerns also led Congress to accept President Reagan's waiver in early 1988 of a U.S. law which would have required the termination of U.S. assistance to Pakistan because of its nuclear smuggling activities in the United States. Foreign Assistance Act, section 670(a).
45. Foreign Assistance Act, section 620E(e).
46. Moreover, Administration sources, in off-the-record interviews, confirmed the advanced status of the Pakistani program as characterized in a March 1988 U.S. press report. (Interviews with the principal author of this book, March-June 1988.) The article stated that the Pakistani weapons weighed only 400 pounds (180 kilograms) and that the nuclear weapons India had designed, presumably without outside help, weighed "under a ton" (2,000 pounds or 909 kilograms). The first U.S. weapons, by contrast, weighed approximately 10,000 pounds (4,545 kilograms). The article made it clear that Pakistan could rapidly deploy nuclear weapons in any future conflict. See Hedrick Smith, "A Bomb Ticks in Pakistan," *New York Times Magazine*, March 6, 1988, p. 38.
47. " 'Constant Vigil' on Pakistani Nuclear Issues," Delhi Domestic Service, 0830 GMT, March 17, 1988, reprinted in *FBIS-NES*, March 17, 1988, p. 50; "Gandhi Comments on Nuclear Issue, Afghanistan," Delhi Domestic Service, 0240 GMT, April 12, 1988, reprinted in *FBIS-NES*, April 12, 1988.
48. India's Department of Atomic Energy stated explicitly in 1987 that "plans for reprocessing [extracting plutonium from] spent fuel from the new reactor, Dhruva, are being prepared." See Department of Atomic Energy, *Annual Report: 1986-1987* (Government of India, Department of Atomic Energy, 1987), p. 42.
49. Steven R. Weisman, "India's Nuclear Energy Policy Raises New Doubts on Arms," *New York Times*, May 7, 1988. This claim was reiterated in a July 1988 interview with the principal author of this book, when Srinivasan declared that India was separating plutonium from Madras I spent fuel as part of this effort. The material, he said, would be used in the country's first large-scale breeder reactor, for which design work had already begun. It seems unlikely, however, that India would undertake the considerable cost of transporting Madras I spent fuel more than 1,000 miles and of extracting its plutonium solely for the purpose of starting to accumulate fuel for a project for which, as of mid-1990, ground had not been broken and which could not conceivably come on-line until after the year 2000. Indian nuclear aides, it may be noted, adopted a similar public rationale in 1966, when they first began extracting plutonium from Cirus reactor spent fuel, claiming that the plutonium was needed for projects that could be realized only in the distant future. In fact, as noted earlier, Cirus-origin plutonium formed the core of the nuclear device India tested in 1974. India's experimental breeder reactor, which also uses Cirus-origin plutonium as fuel, did not begin operating until 1985.
50. "Testimony of Deputy Assistant Secretary of State Robert A. Peck," *Hearings on the Export of Nuclear Materials*, Subcommittee on Asian and Pacific Af-

fairs, Committee on Foreign Affairs, U.S. House of Representatives, 100th Cong., 2nd Sess., February 18, 1988 (mimeo); confidential interviews, April-May 1988.

51. Richard Sale, "India Said to Be Building 20 Nuclear Weapons a Year," United Press International, April 25, 1988, PM cycle. This timetable would have placed initial production at approximately one year after India first began extracting plutonium from Madras I spent fuel and about the time of the first Western press reports that Pakistan was producing weapons-grade uranium; presumably information concerning Pakistan's uranium enrichment activity had been available to U.S.—and possibly to Indian—intelligence sources a number of months earlier.

    In an account published in March 1988, Sale quoted U.S. intelligence sources as stating that India had assembled a "handful of highly sophisticated low-yield atomic bombs" and that there was strong evidence that it had developed a "nuclear warhead for use on a surface-to-surface missile with a 200-mile range." See Richard Sale, "Exclusive: India Said to Upgrade Nuclear Arsenal," United Press International, March 19, 1988, AM cycle.

    See also, "Shadow of an Indian H-bomb," *Foreign Report* (see note 34). This report states, "India's claim that it could make nuclear weapons but has not yet decided to do so is a myth, American officials have told *Foreign Report*. The bombs exist and need only to be tested to become operational they say."

52. Notwithstanding these irritations stemming from Washington's support for Islamabad, India-U.S. relations had improved steadily. Apart from Washington's increased sensitivity to Indian concerns over U.S. arms sales to Pakistan, the most visible evidence of this was increased U.S. willingness to transfer advanced civilian and military technology for such projects as India's Light Combat Aircraft development program.

53. For a more recent article on Indian attitudes on some of these issues, see, for example, "Indian Policy Analysts Wary of Emerging Pak-Iran Connection," *News India*, February 19, 1990, p. 6.

54. Concerning the regional military balance, see, *e.g.*, Ashley J. Tellis, "The Air Balance in the Indian Subcontinent: Trends, Constants, and Contexts," *Defense Analysis* (December 1986): 263; Brian Cloughley, "And Whose 'Zone of Peace' Is It Going to Be, Anyway?" *Far Eastern Economic Review*, April 7, 1988, p. 25; "A Sea Change" and "Hard Choice," *India Today*, December 31, 1987, p. 72.

55. "Soviet Union Leases Nuclear Submarine to Navy," Delhi Domestic Service, 1530 GMT, January 5, 1988, reprinted in *FBIS-NES*, January 6, 1988. See Chapter 5 for a discussion of India's submarine acquisition plans and activities, as well as for a broader discussion of the potential proliferation challenges of nuclear submarines.

56. Sanjoy Hazarika, "India Tests Its Own Surface-to-Surface Missile," *New York Times*, February 26, 1988.

57. James Clad, "A Calming Influence," *Far Eastern Economic Review*, March 15, 1990.

58. Like the Madras I and II nuclear power plants, the Narora facility is continuously fueled. Normally, the refueling of nuclear power reactors occurs only when the uranium's fuel value has been exhausted. The resulting byproduct under these conditions is known as "high burn-up" spent fuel. Plutonium produced in high burn-up spent fuel is usable for nuclear weapons, but would not be preferred by weapons designers because it contains more of the isotope Plutonium-240 than is desirable; the presence of this isotope can lead to uncer-

tainty in the yield of nuclear weapons by causing them to detonate imperfectly.

The continuous refueling capabilities of the Madras and Narora reactors, however, permit Indian nuclear planners to avoid this problem by refueling the reactors at a faster pace. This process yields "low burn-up" spent fuel, in which the plutonium is ideally suited for nuclear weapons. This practice would be inefficient from one point of view, in that the power plants would use more uranium fuel than normal, which would fractionally increase the cost of the electricity they produce. By another standard, however, producing weapons-grade uranium in electric power reactors is quite efficient, since it would make unnecessary the construction and operation of military reactors of equivalent size; in effect, most of the costs of producing plutonium would be offset by the revenues from the sale of electricity.

The military utility of India's unsafeguarded nuclear power reactors has been recognized by one of India's leading strategic analysts, K. Subrahmanyam. See "Commentator Scores U.S. Stand on Pakistan Bomb," *Hindu*, March 2, 1990, reprinted in *JPRS-TND*, May 15, 1990, p. 12.

59. The plutonium could be extracted at India's plutonium extraction facility at Tarapur. See note 14.
60. The capacity of the Tarapur plutonium extraction plant is roughly sufficient to handle the annual spent fuel discharges from Madras I, Madras II, and Narora I.
61. Plutonium from the Canadian-origin Cirus reactor may be used only for peaceful purposes under the terms of the Indo-Canadian agreement governing the transfer of the facility. There is some question as to whether India accepts the restrictions in the Cirus agreement today. See notes 5 and 27.

    Since the facility is not under IAEA safeguards, Canada has no means for identifying the plutonium produced in it. Ottawa could, however, insist that a portion of India's plutonium stocks be specially earmarked as limited to peaceful uses, and set aside for periodic inspection, an approach it followed in the 1960s but has since abandoned. See Gordon H. E. Sims, *A History of the Atomic Energy Control Board* (Hull, Québec, Canada: Canadian Government Publishing Center, 1981), p. 193.
62. Vyvyan Tenorio, "India's Supply of Unsafeguarded Pu Grows as Reprocessing of MAPS Fuel Begins," *Nuclear Fuel*, August 11, 1986, p. 3; *Department of Atomic Energy, Annual Report 1988-1989* (Bombay: Government of India, 1989), p. 3.26.
63. Steven R. Weisman, "India's Nuclear Energy Policy Raises New Doubts on Arms," *New York Times*, May 7, 1988; "AEC Chairman Plans for More Nuclear Power Plants," *Times of India*, July 19, 1988, reprinted in *JPRS-TND*, September 21, 1988; "Atomic Energy Official on Nuclear Plants," *Hindustan Times*, July 24, 1988, reprinted in *FBIS-NES*, August 11, 1988, p. 61. Also, see note 49.
64. See, *e.g.*, "Kalpakkam Nuclear Power Plant Set for Production," *Times of India*, October 12, 1989, reprinted in *JPRS-TND*, November 29, 1989, p. 25.
65. "MAPS Both Back on Line at Half Power," *Nuclear Engineering International* (September 1989): 4; Guha Prasad, "Powerless Performance," *India Today*, June 30, 1989, p. 86.
66. For a similar estimate, see David Albright and Tom Zamora, "India, Pakistan's Nuclear Weapons: All the Pieces in Place," *Bulletin of the Atomic Scientists* (June 1989): 20. The estimate in the text is substantially lower than past estimates by the principal author of this volume. For comparison, see Spector, *The Undeclared Bomb* (see note 28), pp. 92-93.

67. This is consistent with assessments by Albright and Zamora (see note 66) and comparable to the low estimates made by the principal author of this volume in the past.
68. The design work for the nuclear device India detonated in 1974 was undertaken at the Bhabha Center, and the plutonium for the device was produced there.
69. "Testimony of William Webster, Director of Central Intelligence," Hearings on Missile and Nuclear Proliferation, Committee on Governmental Affairs, U. S. Senate, 101st Cong., 1st Sess. May 18, 1989.
70. *Department of Atomic Energy, Annual Report 1988-1989* (Bombay: Government of India, 1989), p. 3.3.
71. Gary Milhollin, "Dateline New Delhi: India's Nuclear Cover-Up," *Foreign Policy* (Fall 1986): 161.
72. "Minister Discloses Plans for Fast Breeder Reactors," *Hindu*, August 12, 1988, reprinted in *JPRS-TND*, October 11, 1988, p. 11 (declaration by by Minister of State for Science and Technology, K. R. Narayanan); "Atomic Energy Official on Nuclear Projects," *Hindustan Times*, July 24, 1988, reprinted in *JPRS-TND*, September 2, 1988, p. 14 (declaration by Atomic Energy Commission Chairman M. R. Srinivasan); "Illegal Import of Norwegian Heavy Water Denied," Delhi Domestic Service, 1530 GMT, November 2, 1988, reprinted in *JPRS-TND*, November 28, 1988, p. 5 (declaration by Foreign Ministry spokesman); Sanjoy Hazarika, "Norway Details How Heavy Water Went to India," *New York Times*, May 7, 1989 (quoting AEC Chairman Srinivasan as stating, "We are one of the most important producers of heavy water in the world and we do not have any imports except from the Soviet Union").
73. Michael Gordon, "Oslo Asserts India Got Atom Material," *New York Times*, November 4, 1988; Hazarika, "Norway Details How Heavy Water Went to India" (see note 72).
74. Michael Gordon, "Romania Is Reported in Nuclear Deal with India," *New York Times*, April 30, 1990.
75. "Heavy Water Transferred," *Die Tageszeitung*, October 7, 1989, translated in *JPRS-TND*, November 6, 1989, p. 23; "Wanted: Bombs for Business: Nuclear Aid for Pakistan and India," ARD [West German Television], 2115, November 3, 1986.
76. "Heavy Going for Heavy Water," *Indian Express*, June 5, 1988; "Audit Agency Report Puts Heavy Water Claims in Doubt," *Patriot*, May 18, 1988, reprinted in *JPRS-TND*, July 13, 1988, p. 17; "Program May Not Be Achieved," *Sunday Statesman*, May 22, 1988, reprinted in *JPRS-TND*, July 13, 1988, p. 18.
77. "Funding of Atomic Energy Department Projects," *Hindu*, March 4, 1988, reprinted in *JPRS-TND*, May 5, 1988, p. 12; Rita Manchanda, "Heavy Water Drought," *Far Eastern Economic Review*, August 31, 1989, p. 18.
78. See Chapter 4, note 17.
79. *Department of Atomic Energy, Annual Report 1988-1989* (Bombay: Government of India, 1989), p. 3.17.
80. See, for example, "Gandhi: India Has No Plans to Build Nuclear Weapons," *Statesman*, June 6, 1988, reprinted in *JPRS-TND*, August 10, 1988, p. 13; Badruddin, "Hypocrisy and Duplicity Behind Pakistan's Nuclear Program," *Mainstream*, April 14, 1990, p. 19.
81. Dilip Bobb, "Chariot of Fire," *India Today*, June 15 1989.
82. "Defense Ministry Adviser on 'Prithvi' Test Firing," Delhi Doordarshan Television Network, 1600 GMT, October 25, 1989, reprinted in *JPRS-TND*, October 26, 1989, p. 32.

83. "Rajiv Gandhi in China: Breaching the Wall," *India Today*, January 15, 1989, p. 20.
84. "Official Notes PRC Concern Over Indian Missile," *Dawn*, July 26, 1989, reprinted in *FBIS-NES*, August 11, 1989, p. 55; " 'Deep Concern' Over Indian Missile Expressed," Islamabad Domestic Service, 1500 GMT, May 11, 1989, translated in *FBIS-NES*, May 15, 1989.

    India's close ties to the Soviet Union have undoubtedly contributed to India's willingness to forego the development of a full-fledged nuclear deterrent against China. In 1971, New Delhi and Moscow signed a Treaty of Peace and Friendship, which India sought with a view to ensuring Soviet support for India in the event of another conflict with either China or its ally, Pakistan. In the ensuing 1971 Indo-Pakistani War, many observers believe China did not intervene directly on behalf of Pakistan because of the threat of Soviet involvement in support of India. Soviet economic aid and concessionary arms sales to India since that time have bolstered bilateral ties. In 1988, as Moscow began an effort to improve relations with Beijing, it took pains to reassure India of continued support, agreeing to sell it two nuclear power reactors and to lease it a nuclear-powered submarine, two controversial transactions discussed earlier.

    Today, however, with Moscow reducing its global commitments as it focuses on problems in Eastern Europe and at home, India could become fearful about the reliability of its Soviet ties. This factor could lead New Delhi to reconsider its need for a nuclear deterrent against China.
85. Statement by Prime Minister Rajiv Gandhi, "Agni Launch," May 22, 1989 (text provided by the Embassy of India).
86. "India Prepares New Agni Tests, Production," *Defense and Foreign Affairs Weekly*, February 26-March 4, 1990, p. 1.

    Indian officials have portrayed the Agni as a system based entirely on domestically developed technology. A persuasive case has been made, however, that India relied heavily on foreign technical assistance in advancing the project, much of it provided to India's space program. Spector, *The Undeclared Bomb* (see note 28), pp. 38-39; Gary Milhollin, "India's Missiles: With a Little Help From Our Friends," *Bulletin of the Atomic Scientists* (November 1989): 31.
87. See *e.g.*, Pushpindar Singh, "India's Agni Success Poses New Problems," *Jane's Defence Weekly*, June 3, 1989, p. 1052; K.K. Sharma, "Keeping Track as India Reaches for the Sky," *Financial Times*, August 25, 1989.
88. Saudi Arabia does possess intermediate-range CSS-2 missiles, which are assumed to be conventionally armed; however these missiles were purchased from China, which had built them origninally as a nuclear system, and were not developed by Saudi Arabia indigenously. Although they are much shorter range systems, Iraq's 325-mile range Al-Husayn and 600-mile range Al-Abas missiles were adaptations of the Soviet-origin Scud-B missile, a system which was also developed for use with nuclear warheads. Iraq is developing the 1,200-mile range Tammouz-1—also on the basis of Scud technology—but is also known to be working on nuclear arms and, undoubtedly, is hoping to equip the new missiles with nuclear warheads.
89. This concept gained prominence in a highly publicized study issued by the Commission on Integrated Long-Term Strategy. See *Discriminate Deterrence, Report of the Commission on Integrated Long-Term Strategy* (Washington: U.S. Department of Defense, 1988).
90. The Agni, with its advanced guidance and 1,500-mile range, could easily have a unit cost in excess of $10 million, particularly if the costs of research and

development are included. This suggests that even if India deploys only 100 Agni missiles, total costs could run to $1 billion.

For comparison, Israel's 400-mile range Improved Jericho missiles are believed to have cost $1.5 million each in 1971, which would be close to $4 million today. Iraq's less accurate 375-mile range Al-Husayn, similarly, is said to cost about $2 million each. The system closest to the Agni in terms of capability is the Chinese CSS-2, purchased by Saudi Arabia in 1988. The latter is said to have paid at least $1 billion for no more than 60 missiles, which would come to at least $16 million per missile. See W. Seth Carus, "Trends and Implications of Missile Proliferation," paper presented at International Studies Association Convention, Washington, D.C., April 13, 1990, pp. 31-34.

91. "Pakistan and India Sign Pact," *Washington Post*, January 1, 1989.
92. These include the Dera Ghazi Khan uranium hexafluoride plant, a tritium purification and storage facility said to be about 100 miles south of Rawalpindi, and a possible uranium enrichment plant at Golra. See chart at end of Chapter 7.
93. "India/Pakistan: Nuclear Non-Attack Pact Proceeds," *Nucleonics Week*, January 18, 1990, p. 16.
94. Bilateral talks, for example, appeared to promise an end of skirmishing between the two states on the Siachen Glacier, along an undemarcated portion of their common border. See Ramindar Singh, "Breaking the Ice," *India Today*, July 15, 1989, p. 79. In addition, India credited Ms. Bhutto with reducing assistance to Sikh separatists in India's Punjab region, a longstanding source of considerable bilateral friction.
95. Zalmay Khalilzad, "The United States in South Asia," *Current History* (December 1989): 417.
96. Washington had maintained high levels of aid to Pakistan throughout the 1980s because of the presence of Soviet troops in Afghanistan.
97. "Heading for a Crisis," *India Today*, February 28, 1989; "Build-up Back-Lash," *Far Eastern Economic Review*, July 27, 1989, p. 18.
98. "Trouble Between Neighbors," *Asiaweek*, April 7, 1989; "India's Widening Embrace," *Asiaweek*, September 1, 1989.
99. "India 'Should Produce' Nuclear Weapons," *Defense and Foreign Affairs Weekly*, October 2-8, 1989, p. 2.
100. "Reportage on Appointments of Ramanna, Iyengar," *JPRS-TND*, March 16, 1990, pp. 16-19 (reprinting various Indian press reports).
101. "Pakistan Welcomes Indian Offer of Nuclear Dialogue," *Far Eastern Economic Review*, December 14, 1989, p. 16 (citing a statement by Singh of December 5).
102. "Stresses Foreign Policy, Defense," Agence France Presse (Hong Kong), 1040 GMT, December 15, 1989, reprinted in *FBIS-NES*, December 15, 1989, p. 48; "V.P. Singh on Pakistani Nuclear Weapons, Options," Delhi Domestic Service, 1230 GMT, February 20, 1990, reprinted in *FBIS-NES*, February 21, 1990, p. 48 ("India will have to review its peaceful nuclear policy if Pakistan manufactures nuclear weapons").
103. David Housego, "India Warns Pakistan Over Kashmir," *Financial Times*, January 29, 1990; Paul Kreisberg, "Meltdown in Kashmir," *Newsday*, May 6, 1990.
104. "Turning Bitter," *India Today*, February 15, 1990; "Growing Fears of Clash Over Kashmir," *Financial Times*, January 30, 1990; Al Kamen, "U.S. Voices Concern Over Kashmir," *Washington Post*, April 19, 1990.

105. *Wall Street Journal*, April 11, 1990; "Singh Says Pakistan 'Encouraging Insurgency,' " Agence France Presse, 1525 GMT April 10, 1990, reprinted in *FBIS-NES*, April 11, 1990, p. 47.
106. "Separatists Seized in Kashmir," *New York Times*, April 16, 1990.
107. Barbara Crossette, "India's Growing Peril: Kashmir and Punjab Separatism," *New York Times*, April 17, 1990; "PM Donates $4.8 Million to Kashmir Fundamentalists," *News India*, March 23, 1990; "Beg Praises Kashmir Militants," *News India*, March 30, 1990.
108. Barbara Crossette, "India Rejects Talks with Pakistan Over Kashmir," *New York Times*, June 9, 1990.
109. Nevertheless, there was considerable concern in both capitals that the border clashes could escalate to wider conflict. Border tensions remained relatively high into the early 1970s.
110. "If Pushed Beyond a Point by Pakistan, We Will Retaliate," *India Today*, April 30, 1990, p. 74.

Chapter 7: Pakistan

1. Steve Weissman and Herbert Krosney, *The Islamic Bomb* (New York: Times Books, 1981), pp. 43-46.

   A detailed history, published in 1987, of the early stages of Pakistan's nuclear development, from 1953 through 1971, makes a persuasive case that during this period the program was devoted exclusively to energy production and peaceful research and that Pakistan's leadership did not take steps to exploit its military potential. See Ashok Kapur, *Pakistan's Nuclear Development* (London: Croom Helm, 1987), Chapters 3 and 4. The study also makes a convincing case that Zulfikar Ali Bhutto, Pakistan's energy minister during the mid-1960s, made only a nominal contribution in building the program, contrary to Bhutto's subsequent assertions that he played a leading role in Pakistani nuclear affairs during this period.
2. For a detailed history of Pakistani nuclear policy under Bhutto from 1971-1977, see Kapur, *Pakistan's Nuclear Development* (see note 1), Chapter 6.
3. Leonard S. Spector, *Nuclear Proliferation Today* (New York: Vintage Books, 1984) pp. 74-81. Negotiations on the plutonium plant began in 1973, prior to the Indian test, and a contract for it was signed in October 1974, some six months afterward.
4. *Ibid*. Pakistan's continuing interest in plutonium extraction is discussed below.
5. Zulfikar Ali Bhutto, *If I Am Assassinated . . .* (New Delhi: Vikas, 1979), p.138.
6. "Report of the Inter-Ministerial Working Party Responsible for Investigating the 'Khan Affair,' " Foreign Ministry of the Netherlands, October 1979 (mimeo, English version).
7. A 1979 Dutch government investigation concluded, "It is reasonable to assume that through Dr. Khan, Pakistan has been able to obtain possession of essential gas centrifuge know-how." See *ibid*. Dr. Khan has repeatedly denied that he engaged in any illegal conduct, pointing out that no criminal charges were ever brought against him for the activities alleged in the Dutch government report. See, *e.g.*, A.Q. Khan, "Letter to the Editor," *International Herald Tribune*, July 1, 1986. Dutch officials state Khan was not prosecuted for his activities in 1974-1975 because it was impossible to prove what information he may have taken. A Dutch prosecution for Khan's subsequent attempt to solicit allegedly classified information from two former colleagues was dropped after

Khan won an appeal on the procedural ground that he had not been properly served with relevant court documents. Interviews by the principal author of this volume, Amsterdam and The Hague, May 1985.

In 1986 a number of additional details concerning the early stages of Pakistan's uranium enrichment program were disclosed in an article in a prominent Pakistani defense publication, apparently based on authoritative Pakistani sources. The account in the text is based on that article. See Hafiz R. Khan, "The Kahuta Story," *Defence Journal* (August 1986): 11-13. Small portions of "The Kahuta Story" appear to have been taken verbatim from an article by A. Q. Khan that appeared in a Karachi paper on August 1, 1986. See "Khan Narrates Uranium Enrichment Experience," *Dawn*, August 1, 1986, reprinted in *Joint Publications Research Service (JPRS)-Nuclear Development and Proliferation (TND)*, September 2, 1986, p. 76.

8. See chart at end of chapter.
9. Leonard S. Spector, *The New Nuclear Nations* (New York: Vintage Books, 1985), p. 31. Migule was given a $10,000 fine and a six-month suspended sentence for his smuggling activities.
10. "Testimony of Assistant Secretary of State Thomas R. Pickering," *Hearings on Nuclear Proliferation: The Situation in India and Pakistan*, Subcommittee on Energy, Nuclear Proliferation, and Federal Services, Committee on Governmental Affairs, U.S. Senate, 96th Cong., 1st Sess., May 1, 1979 (Washington, D.C.: U.S. Government Printing Office, 1979) p. 10.
11. Interview with former Carter Administration senior intelligence official.
12. Spector, *Nuclear Proliferation Today* (see note 3), pp. 91-93; Foreign Assistance Act of 1961, Section 670(b)(2) (1981).
13. "Scientist Affirms Pakistan Capable of Uranium Enrichment, Weapons Production," *Nawa-I-Waqt*, February 10, 1984, translated in *JPRS-TND*, March 5, 1984, p.32.
14. "Zia Chastises Western Media for Accounts of Khan's Remarks on Weapons Capability," *Nuclear Fuel*, February 27, 1984, p. 11.
15. Senator Alan Cranston, "Nuclear Proliferation and U.S. National Security Interests," *Congressional Record*, June 21, 1984, p. S 7901. Indeed, in early April 1984, the Reagan Administration publicly opposed legislation that would have cut off U.S. aid unless the President certified that such activities had ceased. Reagan aides claimed that they could not make the necessary certification and that the aid cutoff, which the legislation would set in motion, was unacceptable, given the continuing Soviet presence in Afghanistan. See *United States Security Interests in South Asia: A Staff Report,* prepared for the U.S. Senate Foreign Relations Committee (Washington, D.C.: Government Printing Office, April 1984).
16. Seymour M. Hersh, "Pakistani in U.S. Sought to Ship A-Bomb Trigger," *New York Times*, February 25, 1985.
17. *Ibid*. Apparently, the United States had evidence as early as 1981 that Pakistan was working on nuclear-weapon triggering mechanisms. See Barry Schweid, "Turks Ship U.S. A-Tools to Pakistan," *Washington Post*, June 28, 1981.
18. Leslie H.Gelb, "Pakistan Links Peril U.S.-China Nuclear Pact," *New York Times*, June 22, 1984; Leslie H. Gelb, "Peking Said to Balk at Nuclear Pledges," *New York Times*, June 23, 1984.
19. Gelb, "Pakistan Links Peril U.S.-China Nuclear Pact" (see note 18); Gelb, "Peking Said to Balk at Nuclear Pledges" (see note 18). Chinese assistance apparently ceased temporarily in late 1984, and the administration submitted the pact for congressional review in July 1985. The agreement entered into

force in December of that year. See "U.S.and China Exchange Notes," *Nuclear Engineering International* (February 1986): 4.

To permit the Nuclear Regulatory Commission to issue nuclear export licenses under the pact, however, the President was required by separate legislation to certify that China was not assisting other states to develop nuclear weapons; that the provisions of the accord would be "effective" in ensuring that U.S. nuclear exports were used exclusively for peaceful purposes; and that the United States would retain certain controls over China's extraction of plutonium from spent fuel produced from U.S.-supplied fuel or in a U.S.-supplied reactor. Neither the Reagan nor the Bush Administration has been able to make these findings, and U.S. officials have stated privately that secret Chinese nuclear aid to Pakistan has continued.

20. Hedrick Smith, "A Bomb Ticks in Pakistan," *New York Times Magazine*, March 6, 1988, p. 38; David Ignatius, "U.S. Pressuring Pakistan to Abandon Controversial Nuclear-Arms Program," *Wall Street Journal*, October 25, 1984; "Shadow of an Indian H-Bomb," *Foreign Report*, December 13, 1984; Don Oberdorfer, "Pakistani Spurns Soviets' Afghan Pullout Plan," *Washington Post*, July 18, 1986. According to Smith, the Reagan Administration considered explicitly threatening in the letter that the United States would cut off aid if the 5-percent enrichment level were exceeded but rejected this option in favor of the more ambiguous "grave consequences" formulation.
21. It is generally believed that reliable first-generation nuclear weapons can be built without testing. Indeed, the bomb dropped on Hiroshima was of a type that had never been tested. If Pakistan, as reported, had obtained a previously tested nuclear weapon design from China, it would have added reason to be confident in any untested devices it might build. See note 25 and accompanying text.
22. Smith, "A Bomb Ticks in Pakistan" (see note 20); Oberdorfer, "Pakistani Spurns Soviet's Afghan Plan" (see note 20). In off-the-record interviews with the principal author of this volume in 1988, Reagan Administration sources confirmed that Pakistan gave explicit assurances on the five-percent enrichment cap. According to one report, the original pledges were made during a mid-November 1984 visit to Washington by Pakistani Foreign Minister Yaqub Ali Khan. See "Shadow of an Indian H-Bomb," *Foreign Report* (see note 20).

    In early 1985, moreover, President Zia appeared to allude to the five percent enrichment cap when he announced that the Kahuta plant had been able to produce a small amount of enriched uranium but stressed that "it is less than five percent. . . ." See Mary Ann Weaver, "Pakistan 'Over the Hump'—Zia Interview," *Christian Science Monitor*, March 1, 1985; "A Step Nearer the Bomb . . .," *Observer* (London), February 24, 1985.
23. Smith, "A Bomb Ticks in Pakistan" (see note 20).
24. "Pakistan to Get U.S. Air-to-Air Missiles," *Financial Times*, March 14, 1985; Dusko Doder, "Gorbachev Warns on Afghan War," *Washington Post*, March 16, 1985; Joanne Omang, "Lawmakers Exercise Foreign Aid Initiative," *Washington Post*, March 21, 1985.
25. John Scali, "Good Morning America," American Broadcasting Company, July 11, 1985. In nuclear weapons of the kind Pakistan is believed to be making, a hollow sphere of highly enriched uranium is surrounded by conventional high explosives that have been specially shaped so that when the conventional explosive is detonated, it crashes inward (or "implodes") symmetrically, instantaneously compressing the highly enriched uranium core to form a solid, explosive "critical mass" of the material. Additional hollow spheres of metal,

such as natural uranium and beryllium, are typically placed between the core and the high explosives. These smash into the core during the implosion and reflect neutrons back into it to improve the efficiency of the nuclear reaction. The parts of the weapon that surround the core are known as the "triggering package." Any nation developing nuclear arms would want to master the implosion technique before building or testing a nuclear device in order to gain confidence that it would work as planned. India detonated similar dummy nuclear weapons in preparing for its 1974 test. See "Reports, Comment on Pakistan Nuclear Development," *Times of India*, July 14, 1986, reprinted in *JPRS-TND*, September 23, 1985, p. 63.

26. Foreign Assistance Act of 1961, section 620E(e).
27. See Foreign Assistance Act of 1961, section 670(a) (1985).
28. Smith, "A Bomb Ticks in Pakistan" (see note 20). Administration officials have subsequently confirmed this account in off-the-record interviews.
29. "Pakistan Persists," *Foreign Report*, March 27, 1986; U.S. Department of State, Press Guidance, "Pakistan: Post-FY 1987 Assistance," March 24, 1986.
30. During 1985, Pakistan continued its clandestine efforts to acquire equipment for its nuclear program. Major operations included successful attempts to obtain high-speed flash X-ray machines, used to photograph detonations of dummy nuclear weapon triggering packages and about 1,900 pounds (880 kilograms) of specially hardened maraging steel in bars identical to the size needed for uranium enrichment centrifuge rotors. Pakistan also attempted unsuccessfully to acquire several hundred tons of reactor-grade graphite in the United States, Great Britain, France, and West Germany. In theory, this material could be used to build a natural-uranium-fueled reactor to produce plutonium or tritium, used in advanced nuclear weapons. See Simon Henderson, "U.S. Halts High-Tech Camera Sale to Pakistan," *Financial Times*, August 17, 1985; David Albright, "Pakistan's Bomb-Making Capability," *Bulletin of the Atomic Scientists* (June 1987): 30; Christer Larssen and Jan Melin, "Third World Countries Buy Swedish Nuclear Weapons Technology," *Ny Teknik*, May 2, 1986, p. 1, translated in *JPRS-TND*, July 30, 1986, p. 1; Egmont Koch and Simon Henderson, "Auf dunken Wegen zur Atommacht" ("Secret Routes to the Bomb"), *Stern*, April 30, 1986, p. 52; "Pakistan Persists," *Foreign Report* (see note 29).
31. Bob Woodward and Don Oberdorfer, "Pakistan A-Project Upsets Superpowers," *Washington Post*, July 15, 1986; Gerald M. Boyd, "Pakistan Denies Developing Bomb," *New York Times*, July 17, 1986; Don Oberdorfer, "Nuclear Issue Clouds Junejo Visit," *Washington Post*, July 17, 1986; communications with knowledgeable U.S. officials. See also "Pakistan Persists," *Foreign Report* (see note 29).
32. Smith, "A Bomb Ticks in Pakistan" (see note 20); Bob Woodward, "Pakistan Reported Near Atom Arms Production," *Washington Post*, November 4, 1986.
33. Woodward, "Pakistan Reported Near Atom Arms Production" (see note 32).
34. See note 26 and accompanying text.
35. Another factor that probably contributed to Pakistan's desire to acquire nuclear arms was the Soviet occupation of Afghanistan. This undoubtedly heightened Pakistani security concerns and increased fears of Soviet-Indian collaboration during any future Indo-Pakistani war. Pakistan's relatively meager nuclear capability and limited delivery systems would have been unlikely, in themselves, to deter aggression by Soviet or Soviet-backed Afghan forces. Nevertheless, Pakistani strategists may have hoped that the country's posses-

sion of nuclear weapons would have made Moscow more cautious about intervening militarily through Afghanistan in a future Indo-Pakistani conflict, perhaps out of fear that the heightened stakes might lead to a confrontation with the United States.

36. For a discussion of Zulfikar Ali Bhutto's push for regional non-proliferation agreements, adopted almost immediately after he took the decision to develop a Pakistani nuclear bomb, see Kapur, *Pakistan's Nuclear Development*, pp. 155, 165-166 (see note 1).
37. The improvement took place despite the fact that during the fall of 1985, a series of unusually outspoken calls by Pakistani political figures for the development of nuclear weapons exacerbated tensions on the nuclear front. See "Panel Supports Pakistan Becoming a Nuclear Power," *Nawa-I-Waqt*, November 15, 1985, translated in *JPRS-TND*, February 7, 1986; "Leader Urges 'Full Use of Nuclear Capabilities,' " *Muslim*, December 2, 1985, reprinted in *JPRS-TND*, January 9, 1986, p. 70.

    The August revelation of Pakistan's attempted purchase of flash X-ray machines, used in testing nuclear-weapon triggering packages, was a further irritant. See note 30.
38. The domestic political scene was made more complicated in Pakistan by President Zia's termination of martial law on December 30, 1985, a step which significantly increased the power of Prime Minister Junejo and the nation's parliament, elected in non-party balloting eleven months earlier.
39. "Pakistan Persists," *Foreign Report* (see note 29).
40. "They told us that Pakistan could never produce the bomb and they doubted my capabilities," Khan was reported as saying, "but now they know we have done it. . . . America knows it. What the CIA has been saying about our possessing the bomb is correct and so is the speculation in some foreign newspapers." See Kuldip Nayar, "India Forcing Us to Go Nuclear: A. Q. Khan," *Muslim*, March 1, 1987; Kuldip Nayar, "We Have the A-Bomb, Says Pakistan's 'Dr. Strangelove,' " *Observer* (London), March 1, 1987. The piece also appeared in a New Delhi weekly. See Steven R. Weisman, "Report of Pakistani A-Bomb Causes Stir in the Region," *New York Times*, March 2, 1987.
41. Embassy of Pakistan, Press Release, "Dr. A. Q. Khan Denies Allegation in London *Observer* Article," February 28, 1987 (sic).
42. "Abdul Qadir Khan Sets 'Record Straight' on Interview," *Muslim*, March 2, 1987, reprinted in *JPRS-TND*, April 8, 1987, p. 35.
43. Mushahid Hussein, a Pakistani editor, claimed to have been present at the meeting and appeared to validate the contents of the interview in an editorial. See "Bomb Controversy," *Muslim*, March 3, 1987. Days after this editorial appeared, apparently under pressure from the Pakistani government, Hussein resigned from his paper, and declared that Khan, in fact, had been misquoted.
44. At least six substantial segments of the Kuldip Nayar interview appear to be near verbatim repetitions of passages from Khan's *Dawn* article, which appeared on August 1, 1986. For a comparison of the language in the two articles, see Leonard S. Spector, *The Undeclared Bomb* (Cambridge, Mass.: Ballinger Publishing Company, 1988), pp. 372-374 (note 47). The common language found in the Nayar interview and in Khan's *Dawn* article was first brought to the attention of the principal author of this book by Nazir Zaidi, a Washington-based journalist.

    Nayar defended the authenticity of the interview in a column published in December 1987, noting that the Press Council of Britain had issued an opinion on the matter. The opinion, he stated, had concluded, "It is agreed by the parties that a meeting and a conversation took place between the complainant, Dr. A. Q. Khan, and the writer of the article, Kuldip Nayar. Dr. Khan strongly denies quotations attributed to him and the interview reported in the article. On the evidence before it, however, the Press Council has not been satisfied that the interview did not take place or that the quotations were untrue. . . ." See "A. Q. Khan Interview: Journalist Reiterates Position," *Sunday*, December 20-26, 1987, reprinted in *JPRS-TND*, February 25, 1988, p. 23.

45. Some observers speculated that Khan's comments had been approved by higher authorities, possibly as a warning to New Delhi. The interview had taken place on January 28 at a time when a conflict with India loomed as a serious possibility, but by March, this threat had receded, which, according to this hypothesis, would explain the government's subsequent disclaimer. Other analysts suggested that Khan had been indiscreet and caused considerable embarrassment to the Zia government, which had consistently maintained that Pakistan's nuclear activities were entirely peaceful. This view assumed that a split had developed within Pakistan's nuclear establishment between hardliners, who wanted to declare Pakistan's nuclear capabilities and present India and the United States with a *fait accompli*, and more cautious elements, who preferred a policy of ambiguity.

    For a detailed discussion of the controversy, see Nayan Chanda, "Yes, We Have No Bomb," *Far Eastern Economic Review*, March 12, 1987; John Elliot, "Net Closes In on Pakistan's Nuclear Capabilities," *Financial Times*, March 7, 1987.

46. "Knocking at the Nuclear Door," *Time*, March 30, 1987, p. 42.
47. Arnaud de Borchgrave, "Pakistan's Zia Insists Solarz is Voice of India in Congress," *Washington Times*, November 16, 1987.
48. Egmont R. Koch and Simon Henderson, "Nennen wir ihn einfach 'Kotari' (Let's Just Call Him 'Kotari')" *Stern*, April 29, 1987, pp. 198-201; Mark Hibbs, "German Firm Implicated in Trying to Aid Pakistan Enrichment Effort," *Nucleonics Week*, May 4, 1987; Mark Hibbs, "Attempt Revealed to Export to Pakistan HEU-Capable Components," *Nucleonics Week*, May 7, 1987.
49. Simon Henderson, "Cologne Company Raided Over Supply of Nuclear Equipment Plans to Pakistan," *Financial Times*, April 30, 1987.
50. James M. Markham, "Bonn Checks Report of Smuggling of Atomic Technology to Pakistan," *New York Times*, May 5, 1987.
51. "Pakistan Spurns Nuclear Inspection," *Washington Post*, August 4, 1987; Michael R. Gordon, "Pakistan Rejects Atomic Inspection," *New York Times*, August 6, 1987.
52. Steven R. Weisman, "Pakistani Parties Denounce the U.S.," *New York Times*, September 3, 1987. The Pakistani parliament was generally subordinate to Zia, and Junejo had been hand-picked by Zia to serve as prime minister. Thus Junejo's proposal undoubtedly had received Zia's approval and was not a personal initiative. When Junejo began to operate with increasing independence in 1988, Zia dismissed him and dissolved parliament.
53. "Pakistan Proposes Nuclear Test Ban," *New York Times*, September 14, 1987; "India Not in Favor of Regional Disarmament," *News India*, November 14, 1986. Perhaps to underscore the sincerity of its offer, in December 1987, Pakistan ratified the Partial Test Ban Treaty, prohibiting nuclear tests in the atmosphere, in outer space, or under water. Pakistan had signed the accord in 1963. The action brought Pakistan in line with India, which had ratified the pact upon signature (also in 1963).
54. Michael R. Gordon, "Congress Delays New Pakistan Aid Amid Nuclear

Rift," *New York Times*, September 30, 1987.

55. Michael R. Gordon, "U.S. Official Urges New Aid to Pakistan," *New York Times*, October 23, 1987; Robin Pauley, "Pakistan 'Producing High Grade Uranium,' " *Financial Times*, October 26, 1987.
56. Simon Henderson, "Pakistan Builds Second Plant to Enrich Uranium," *Financial Times*, December 11, 1987; Neil A. Lewis, "Reports of Pakistan Uranium Plant Weighed," *New York Times*, January 10, 1988; "Pakistan Denies New Enrichment Plant," *Nuclear Engineering International* (February 1988): 7.

Pervez was convicted of illegally attempting to export a 154-pound (70-kilogram) bar of beryllium from the United States to Pakistan. The export of beryllium is controlled because of its use as a neutron reflector to enhance the yield of nuclear weapons. Pervez was also convicted of three counts of giving false statements on export license applications concerning the end use of the twenty-five tons of maraging steel he was seeking; he had claimed the steel was to be used for "high-speed turbines and compressors." He was found not guilty, however, of trying to export the material illegally. On this and on two other counts of the eight-count indictment the jury believed that the government had improperly entrapped him.

Pervez had been the subject of a ten-month undercover investigation involving U.S. and Canadian authorities which began after Carpenter Technology Corporation, a steel-maker based near Reading, Pennsylvania, informed the U.S. government in October 1986 of an inquiry its Toronto agent had received from Pervez for maraging steel. Carpenter had earlier alerted the government after it received an order for the same metal in May of 1985 directly from the Pakistani Embassy in London. See Mark Hosenball and James Adams, "A-Bomb Plot is Linked to Embassy," *Sunday Times* (London), July 26, 1987.

Testimony at Pervez's trial established that the maraging steel—a specialized alloy made from nickel, cobalt, and molybdenum with an unusually high tensile strength—is used principally to make uranium enrichment centrifuges of a West German design. The centrifuges are used by the British-Dutch-West German consortium, Urenco; Pakistan is thought to have duplicated them at Kahuta, allegedly by using technology obtained by Dr. A. Q. Khan while he worked at a Urenco plant in the Netherlands. Pervez stipulated that the maraging steel be manufactured as bars 5.9 inches in diameter, a measurement that matches the size of German-designed units, according to scientific witnesses called by the government. These witnesses also stated that 35,000 pounds of the metal would allow Pakistan to build enough centrifuges to produce one nuclear device per year, apparently requiring 55 pounds (25 kilograms) of highly enriched uranium. Pervez had sought 50,000 pounds of the special steel and had put in a bid for eleven additional 50,000-pound shipments of the metal, to be supplied every four months. See Andy McCord, "Trial Witnesses Link Pervez to Pakistan Nuclear Plans," *India Abroad,* December 18, 1987; Mike Holderness, "Special Steel Evidence of Pakistan's Nuclear Plans," *New Scientist*, July 30, 1987; Smith, "A Bomb Ticks in Pakistan" (see note 20).

Testimony by government experts also established that the only realistic use for the beryllium at issue in the case was in the manufacture of nuclear weapons. See "Testimony of John J. Keeney, Deputy Assistant Attorney General, Criminal Division," Subcommittee on Asia and Pacific Affairs, Committee on Foreign Affairs, U.S. House of Representatives, 100th Cong., 2nd Sess., February 18, 1988, p. 5 (mimeo).

Among the evidence linking Pervez to the Pakistani government were entries in his notebooks (that had been seized by Canadian authorities) identifying his project as "atom" and "military" and noting that "my expert is procurement manager for nuclear plant."

On February 10, 1988, Pervez was sentenced to five years' imprisonment and an additional term of five years' probation—the longest prison term yet imposed for attempted nuclear smuggling to an emerging nuclear state.

Pervez's conviction was overturned on appeal in February 1990, on the grounds that the question of entrapment had not been properly considered in his trial in view of a subsequent Supreme Court ruling in an unrelated case. Rather than face a retrial on the charges, Pervez pled "no contest" (*nolo contendere*) and had his sentence commuted to the time he had already served, approximately three years in all.

57. Under the Solarz Amendment, aid is prohibited for any country that the president determines has illegally exported, or attempted to export illegally from the United States, equipment, materials, or technology to be "used in the manufacture of a nuclear explosive device." (The law then provides that the prohibition may be waived if the president determines that an aid cut-off would be seriously prejudicial to U.S. non-proliferation interests or would otherwise jeopardize the common defense and security.) In mid-January 1988, President Reagan determined that Pakistan had violated this stipulation, but then went on to waive the aid ban.

    In testifying on the president's decisions, Deputy Assistant Secretary of State Robert Peck made clear that the provision had been triggered by the activities of Arshad Pervez, who was determined to be working on behalf of the government of Pakistan, contrary to repeated denials by senior Pakistani officials during the summer and fall of 1987. See, *e.g.*, Richard Weintraub, "Pakistan Denies Connection to Any Nuclear-Export Plot," *Washington Post*, July 22, 1987. In addition, he stated that the maraging steel Pervez was seeking to acquire "would have been used in Pakistan's uranium enrichment plant, and *therefore* would have been 'used in the manufacture of a nuclear explosive device.' " See "Testimony of Deputy Assistant Secretary of State Robert A. Peck," *Hearings on the Export of Nuclear Materials*, Subcommittee on Asian and Pacific Affairs, the Committee on Foreign Affairs, U.S. House of Representatives, 100th Cong., 2nd Sess., February 18, 1988, (mimeo) (emphasis added).
58. Smith, "A Bomb Ticks in Pakistan" (see note 20). With respect to weapons that India had designed (presumably without outside help), the article stated that these weighed "under a ton."
59. As a measure of how far the program had progressed toward the actual fabrication of weapons, Smith noted that Pakistan had transformed the highly enriched uranium hexafluoride into uranium metal—the form of the element used in nuclear devices—and that some Reagan Administration sources believed that Pakistan had already machined the metal into nuclear-weapon cores.
60. These sources also validated the article's description of the timing of U.S.-Pakistani interactions, especially the failure in October 1985 of President Reagan to challenge President Zia on Pakistan's enrichment of uranium to more than five percent in violation of pledges given to Washington.
61. There is no question that Pakistani analysts perceive India's stockpiling of unsafeguarded plutonium in strategic terms. See, *e.g.*, "Commentary Alleges Indian Bid for Nuclear Hegemony," *Dawn*, October 6, 1987, reprinted in *JPRS-TND*, December 1, 1987, p. 21.
62. Stephen P. Cohen, "Our Bomb and Theirs: Reflections on McGeorge Bundy's

Danger and Survival," mimeographed working paper, undated, p. 1. (Stephen Cohen is a professor of political science and Asian studies at the University of Illinois at Urbana-Champaign.)

63. Elaine Sciolino, "Zia of Pakistan Killed as Blast Downs Plane; U.S. Envoy, 28 Others Die," *New York Times*, August 18, 1989; "Pakistan Points to Sabotage in Zia Crash," *New York Times*, October 17, 1988. In addition to Zia, twenty-eight other people on board were killed, including the U.S. Ambassador and the U.S. military attaché to Pakistan, the head of the Pakistani joint chiefs of military staff, and a number of other senior Pakistani military officers. Although the cause of the crash has never been conclusively established, sabotage of some sort is suspected, usually attributed to the Afghan secret police. India vehemently denied any involvement at the outset, a denial that appears to have been accepted; the tragedy did not lead to a flare-up in Indo-Pakistani tensions.
64. Zia had executed the elder Bhutto in 1979. Zia also imprisoned Benazir Bhutto, as well as her mother and other family members, for several years under harsh conditions.
65. In a report prepared for the U.S. Senate Foreign Relations Committee, for example, committee staff member Peter Galbraith noted that during a visit to the United States in the spring of 1988, Bhutto declared her opposition to a Pakistani nuclear weapon program. Although opposed to Pakistan's signing the NPT without India's doing the same, she noted that just as U.S. intelligence has been able to detect Pakistan's efforts to enrich uranium and obtain weapon components, it would be able to verify a Pakistani pledge not to engage in such activity. See *Nuclear Proliferation in South Asia: Containing the Threat*, Committee on Foreign Relations (Staff Report), U.S. Senate, 100th Cong., 2d Sess. (1988). See also, Peter Adams, "Zia's Strong Stand on Pakistani Nuclear Arms Not Shared by Bhutto Supporters," *Defense News*, September 5, 1988; Ahmed Rashid, "Bhutto Vows to Permit Inspection at Atomic Plant," *Independent*, November 14, 1988; Shahir-ur-Rehman, "Benazir Bhutto Commits Pakistan to Peaceful Nuclear Program," *Nucleonics Week*, December 8, 1988.
66. Foreign Assistance Act of 1961, section 620E(e). The U.S. government fiscal year begins on October 1. Reagan's delayed finding had not had a direct impact on Pakistan, however, because the Reagan Administration interpreted the Pressler Amendment as affecting only new funding, not monies appropriated in the preceding fiscal year, which were already in the "pipeline."
67. "No Nuclear Arms in Arsenal Pakistani Officials State," *Washington Post*, January 29, 1990.
68. The size of the Kahuta plant has not been disclosed. According to one report quoting U.S. officials, by the end of 1987, Pakistan had acquired 220 pounds (100 kilograms) of weapons-grade uranium. Since the plant is thought to have begun producing this material in 1986, this would suggest that the facility had an output of about 110 pounds (50 kilograms) per year. See Smith, "A Bomb Ticks in Pakistan" (see note 20). For the purposes of its regulations, the IAEA assumes that 55 pounds (25 kilograms) of highly enriched uranium is required for a nuclear device, but only 33 pounds (15 kilograms) would be needed if Pakistan employed certain well-known design features to improve the efficiency of its weapons. Using these two alternative figures to establish a range of capabilities, it would appear that the Kahuta plant—as it was configured at the end of 1987—could probably produce enough weapons-grade uranium for two or three devices annually.

Using these figures, by the end of 1988, when Prime Minister Bhutto took office, the Kahuta plant would have been producing weapons-grade material for about three years, and thus Pakistan could have accumulated about 330 pounds (150 kilograms) of material, enough for 6 to 10 weapons. (Given the degree of uncertainty surrounding this estimate, this figure has been rounded off in the text.)

David Albright, a researcher at the Federation of American Scientists, has developed a roughly comparable estimate. He has attempted to size Kahuta's output from reports of the number of uranium enrichment centrifuges said to be operating there and the probable capabilities of this type of unit. In 1987, he estimated that the facility could produce between 46 and 139 pounds (21 and 63 kilograms) of weapons-grade uranium annually. In a subsequent 1989 analysis, he and a colleague estimated the plant's output at between 33 and 165 pounds (25 to 75 kilograms) of such material annually. Using a figure of 44 pounds (20 kilograms) of weapons-grade uranium per device, they estimated that when Ms. Bhutto took office Pakistan possessed enough material for 6 to 8 devices. See David Albright, "Pakistan's Bomb Making Capability," *Bulletin of the Atomic Scientists* (June 1987): 30; David Albright and Tom Zamora, "India, Pakistan's Nuclear Weapons: All the Pieces in Place," *Bulletin of the Atomic Scientists* (June 1989): 20.

If Pakistan has been expanding its enrichment capability, as discussed below, its actual capabilities may be somewhat larger today.

69. Zalmay Khalilzad, "The United States in South Asia," *Current History* (December 1989): 417. This was the first time a knowledgeable U.S. aide with access to classified information on the issue had publicly acknowledged this critical point.

70. Gelb, "Pakistan Links Peril U.S.-China Nuclear Pact" (see note 18); Gelb, "Peking Said to Balk at Nuclear Pledges" (see note 18); "Pakistan's Atomic Bomb," *Foreign Report*, January 12, 1989, p. 1; Simon Henderson, "China May Help Build Pakistan's N-Bomb," *Financial Times*, September 29, 1986. According to Gelb, the design Pakistan obtained was that of the device tested in China's fourth nuclear test. This test involved the firing of a ballistic missile with a live nuclear warhead over Chinese territory to the Lop Nor test site. This would mean that Pakistan obtained a design suitable for adaptation as a missile warhead.

In an interview with the principal author of this volume, a former Reagan Administration official confirmed that China had, indeed, supplied the design of a nuclear device to Pakistan, although he offered no other details.

It has also been reported that during the early 1980s, China supplied Pakistan with enough weapons-grade uranium for two nuclear devices. See "Pakistan's Atomic Bomb." See also, Simon Henderson, "Why Pakistan May Not Need to Test a Nuclear Device," *Financial Times*, August 14, 1984.

China has denied charges that it has assisted the Pakistani nuclear weapons effort. See "China Denies Reports It Helped Pak in Nuke Weapons Production," *News India*, January 27, 1989.

71. John Scali, "Good Morning America," American Broadcasting Company, July 11, 1985; "Reports, Comment on Pakistan Nuclear Development," *Times of India*, July 14, 1986, reprinted in *JPRS-TND*, September 23, 1985, p. 63.

72. According to the Los Alamos National Laboratory, hemispherical metal shells are "among the essential parts of a nuclear bomb." See "Machining Hemispherical Shells," *Research Highlights: 1988* (Los Alamos, Calif.: Los Alamos National Laboratory). In the early 1980s, Pakistan had sought to obtain 13-

inch hemispherical shells in Western Europe; though it reportedly had little success, it apparently acquired the ability to manufacture the weapon components itself. See, "Zia's Hollow Spheres," *Foreign Report*, February 12, 1982; Henderson, "Why Pakistan May Not Need to Test a Nuclear Device" (see note 70). During the mid-1980s, Islamabad also obtained special West German milling machines and "hot isostatic presses," useful in machining weapons-grade uranium into nuclear-weapons components. See "Nuclear Exports to Pakistan Reported," *Der Spiegel*, February 20, 1989, reprinted in *Congressional Record*, May 16, 1989, p. S 5441.

73. The State Department interpretation is contained in a March 5, 1987, letter to a subcommittee of the House Foreign Affairs Committee, quoting the Department's Office of the Legal Advisor. The letter states:

> A state may possess a nuclear explosive device, and yet maintain it in an unassembled form for safety reasons or to maintain effective command and control over its use or for other purposes. The fact that a state does not have an assembled device would not, therefore, necessarily mean that it does not possess a device under the statutory standard.

See *Hearings on Foreign Assistance Legislation FY 1988/1989*, Subcommittee on Asian and Pacific Affairs, Committee on Foreign Affairs, U.S. House of Representatives, 100th Cong., 1st Sess., March 5, 1987 (Washington: U.S. Government Printing Office, 1987), pp. 487-488. In the 1950s the United States routinely stored its nuclear weapons in separate parts for safety reasons. See, Senator John H. Glenn, "Pakistan's Dispossessed Bomb," *Congressional Record*, November 16, 1989, p. S 15880.

74. Smith, "A Bomb Ticks in Pakistan" (see note 20); Nayan Chanda, "Nuclear No-No," *Far Eastern Economic Review*, June 22, 1989, p. 24.
75. Simon Henderson, "Pakistan Builds Second Plant to Enrich Uranium," *Financial Times*, December 11, 1987; Neil A. Lewis, "Reports of Second Pakistani Uranium Plant Weighed," *New York Times*, January 10, 1988. Pakistan denied the allegation that it was building a second enrichment plant.
76. According to various press accounts, describing the conclusions of a number of official investigations, during the mid-1980s two related West German firms provided Pakistan with test quantities of tritium gas, a tritium purification and storage plant, tritium precursor materials, the design for a reactor that could be used for tritium production, and material and equipment for fabricating fuel for that reactor, including special American-made welding lasers. See Mark Hibbs, "Illegal Export Charges May Spur Tighter German Export Controls," *Nucleonics Week*, January 5, 1989; Mark Hibbs, "German Firms Exported Tritium Purification Plant to Pakistan," *Nuclear Fuel*, February 6, 1989; "Germans Sell Nuke Equipment to Pakistan," *News India*, March 3, 1989; Mark Hibbs, "U.S. Repeatedly Warned Germany on Nuclear Exports to Pakistan," *Nuclear Fuel*, March 6, 1989; Mark Hibbs, "German Firm's Exports Raise Concern About Pakistan's Nuclear Capabilities," *Nuclear Fuel*, March 6, 1989; John J. Fialka and Thomas F. O'Boyle, "West German Firms Admit Supplying Nuclear-Weapons Material to Pakistan," *Wall Street Journal*, April 21, 1989; Mark Hibbs, "Prosecutors Link Tritium Plant to Pakistan Weapons Program," *Nuclear Fuel*, May 1, 1989; "NTG Nuclear Proliferation Case," *Der Spiegel*, November 6, 1989, translated in *JPRS-TND*, November 29, 1989, p. 35.
77. Hibbs, "U.S. Repeatedly Warned Germany" (see note 76); Hibbs, "German Firm's Exports Raise Concern" (see note 76).

It was reported that Pakistan initially planned to produce tritium by irradiating lithium targets surreptitiously in a U.S.-supplied research reactor at the Pakistan Institute for Nuclear Science and Technology (PINSTECH). This would violate Pakistan's agreement covering the reactor, which precludes its use for any non-peaceful purpose.

78. The plant was thought not to be operating because it has not been placed under IAEA safeguards. All of Pakistan's spent fuel has been produced in reactors that are subject to IAEA inspection. Spent fuel from these reactors could not be moved legally to the plutonium extraction plant unless it, too, were placed under the IAEA system, along with any plutonium produced from the inspected spent fuel, steps Pakistan has refused to take. Spent fuel from the new reactor, however, would not be under IAEA restrictions, and Pakistan would thus be free to use plutonium extracted from such spent fuel for nuclear arms.
79. See note 77.
80. See note 77.
81. Pool-type research reactors normally use uranium enriched to 20 percent or more as fuel and ordinary, or "light," water as a moderator. (See Glossary.) In theory, Pakistan could use enriched uranium from Kahuta for the former purpose. This would allow it to produce tritium by irradiating targets of lithium; Pakistan has no other facility in which it can produce tritium legally. (As noted above, the alternative of using the U.S.-supplied research reactor at the Pakistan Institute of Nuclear Science and Technology would violate Pakistan's agreement covering the reactor, which precludes its use for any non-peaceful purpose. See note 77.) Although the new reactor could also be used to produce plutonium, it is not clear that this would be an efficient mechanism for increasing Pakistan's stockpile of weapons-grade nuclear material, since it would probably use up enriched uranium at a greater rate than it would produce plutonium.
82. "Surface-to-Surface Missiles Successfully Tested," Islamabad Domestic Service, 1500 GMT, February 11, 1989, translated in *JPRS-TND*, March 20, 1989, p. 23.
83. "Pakistan in Missile Build Claim," *Jane's Defence Weekly*, February 18, 1989, p. 267. The Missile Technology Control Regime (MTCR), described in Chapter 3, presumes that any missile with such a payload can carry a nuclear weapon manufactured by a developing country, which presumably would not be as light as one built by a more advanced nuclear power. According to one report, Pakistan's nuclear device, which is said to be based on a Chinese design, weighs only 400 pounds. See Smith, "A Bomb Ticks in Pakistan" (see note 20).

    The MTCR seeks to limit the spread of missiles that, like the Hatf II, are capable of carrying such a 1,100-pound payload 186 miles (300 kilometers) or more.
84. Iraq adapted its Scud-B missiles in this way during the Iran-Iraq War to enable them to reach Tehran. See Chapter 11.
85. U.S. Arms Control and Disarmament Agency (ACDA), *World Military Expenditures and Arms Transfers, 1988* (Washington, D.C.: U.S. ACDA, 1989), p. 19.
86. Mushahid Hussain, "Pakistan 'Responding' to Change," *Jane's Defence Weekly*, October 14, 1989, p. 779. China is also said to have developed a missile in this range, the M-9, and it is possible that China might sell this or other systems to its regional ally. It has pledged to the United States only that it will not sell medium-range missiles to countries in the Middle East.
87. The Army appears to have been responsible for funding and arranging the construction of the key installations needed in the country's nuclear weapons

effort. The facilities were apparently designed and operated by a team of nuclear scientists, headed by Drs. Abdul Qadeer Khan and Munir Khan, who have acquired considerable bureaucratic power in their own right. The Army, which also manufactures conventional munitions and other military hardware, also appears to have been directly responsible for undertaking certain aspects of the nuclear weapons program, such as the fabrication and testing of key non-nuclear components. Some of this Army work is said to take place at the Pakistani Ordnance Factory in Wah. See Khan, "The Kahuta Story" (see note 7); "Pakistan Bomb Secrets," *Foreign Report*, November 11, 1983.

As noted earlier, A. Q. Khan had been in charge of Pakistan's effort to develop a uranium enrichment capability, the key to its nuclear weapons program. Under Zia, he had become something of a national hero because of his success, a factor which undoubtedly gave him considerable power within governmental circles responsible for nuclear policy. Munir Khan, the chairman of the Pakistani Atomic Energy Commission (PAEC) and a prominent figure at the IAEA, had charge of the official, publicly acknowledged elements of the Pakistani nuclear program, but undoubtedly wielded influence with respect to its undeclared elements, as well. One report states that the PAEC's Directorate of Technical Development is responsible for design work on nuclear arms. See "Pakistan's Atomic Bomb," *Foreign Report*, January 12, 1989.

88. In 1977, Zia had taken power in a military coup and had governed the country as a military dictatorship until 1985, when he granted limited authority to a parliament under a hand-picked prime minister, while retaining full control over Pakistan's foreign policy—and nuclear affairs. In early 1988, Zia had dissolved the parliament, promising that constitutionally mandated free elections would follow but repeatedly postponing them. After his death, however, the elections were scheduled, ultimately bringing Bhutto to office.
89. On these and related issues concerning Pakistani security affairs and the domestic political scene, see Mahnaz Ispahani, *Pakistan: Dimensions of Insecurity*, Adelphi Paper 246 (London: International Institute for Strategic Studies, Winter 1989/1990); Rodney W. Jones, "Pakistan After Afghanistan," *Washington Quarterly* (Summer 1989): 65.
90. "U.S. Approval Doubtful for Pak's F-16s," *News India*, February 10, 1989.

Over fiscal years 1982 through 1987, Pakistan received $3.2 billion in aid from the United States. Half of these funds were for economic assistance, half for security assistance. The Reagan Administration planned a second phase of assistance to Pakistan, totalling $4.02 billion for fiscal years 1988 to 1993, with the proportion of economic assistance increasing to 57 percent. However, U.S. budget pressures reduced aid disbursements from annual targets of $670 million in both FY 1988 and FY 1989, to actual payments of only $611 million and $576 million, respectively. Despite these aid shortfalls, Pakistan continued to rank among the top five annual recipients of U.S. foreign aid. See Thomas P. Thornton, *Pakistan and the United States*, Media Briefing (New York: Asia Society, June 1989), p. 12; U.S. Department of State, *Congressional Presentation for Security Assistance Programs, Fiscal Year 1990* (Washington, DC: U.S. Government Printing Office, 1989), pp. 223-225.

91. "U.S. Approval Doubtful for Pak's F-16s" (see note 90).
92. "Testimony of Philip E. Jones," Subcommittee on Asian and Pacific Affairs, Committee on Foreign Affairs, U.S. House of Representatives, 101st Cong., 1st Sess., March 8, 1989; "Testimony of Paul H. Kreisberg," *ibid.*; Thornton, *Pakistan and the United States* (see note 90), pp. 12-13.
93. "Testimony of William Webster, Director, U.S. Central Intelligence Agency,"

Committee on Governmental Affairs, U.S. Senate, 101st Cong., 1st Sess., May 18, 1989, p. 23.

94. Stephen Engelberg, "C.I.A. Chief Wary of Pakistani Nuclear Program," *New York Times*, May 19, 1989.

95. David B. Ottaway, "Addressing Congress, Bhutto Formally Renounces Nuclear Arms," *Washington Post*, June 8, 1989. For a similar subsequent statement by Bhutto, see "Bhutto Discusses Nuclear Weapons Question," *Al-Ra'y*, July 19, 1989, translated in *JPRS-TND*, August 14, 1989, p. 24.

96. Nayan Chanda, "Nuclear No-No" (see note 74); David B. Ottaway, "U.S. Relieves Pakistan of Pledge Against Enriching Uranium," *Washington Post*, June 15, 1989; William Beecher, "Pakistan Will Halt Nuclear Arms Work to Keep U.S. Aid," *Minneapolis Star-Tribune*, June 14, 1989; Stephen Engelberg, "U.S. Sees Pakistan Seeking an A-Bomb," *New York Times*, June 11, 1989.

97. Michael Gordon, "U.S. Fears That Chinese May Again Sell Missiles," *New York Times*, November 9, 1989.

98. It is possible that the contours of the deal were discussed with General Beg during his trip to Washington in February 1989.

One indicator of Bhutto's increasing authority over security affairs at the time was that in late May she replaced the head of the ISID, Lt. General Hamid Gul, a Zia appointee she suspected of undermining her support. See John Kifner, "Bhutto, in Fateful Move, Ousts a Top General," *New York Times*, June 1, 1989.

99. Khalilzad, "The United States in South Asia" (see note 69).

100. "Ms. Bhutto Says She Favors Non-Nuclear Pact with India," *New York Times*, July 10, 1989; "Bhutto Advisor Rebuts Gandhi on Nuclear Plants," *Saudi Gazette*, July 20, 1989, reprinted in *Foreign Broadcast Information Service (FBIS)-Near East and South Asia (NES)*, July 31, 1989, p. 61.

101. See note 109 and accompanying text.

102. "Bonn Unhappy with U.S. F-16 Deal with Country," *Der Spiegel*, July 24, 1989, translated in reprinted in *FBIS-NES*, August 16, 1989, p. 56. See also, "Pakistan's Atomic Bomb" (see note 87) (also notes progress on weapons casing and fusing); *U.S. News and World Report*, February 2, 1990.

103. Gordon, "U.S. Fears That Chinese May Again Sell Missiles" (see note 97).

104. "Pakistan Bomb Secrets" (see note 70).

105. Khalid Mahmud Arif, "Military Balance and National Security," *Defence Journal* (July 1989): 6.

106. Brigadier Abdul R. Siddiqi (Ret.), "Military Balance and National Security: A Post-Script," *Defence Journal* (July 1989): 3; "General Supports Nuclear Weapons Development," *Jang*, August 10, 1989, translated in *JPRS-TND*, September 18, 1989, p. 25. But see "Commentary Takes Issue with Pro-Bomb Lobby," *Muslim*, August 19, 1989, reprinted in *JPRS-TND*, September 18, 1989, p. 26.

107. "Bhutto Claims Nuclear 'Knowledge' Available," Agence France Press (Hong Kong), 1108 GMT, August 29, 1989, reprinted in *FBIS-NES*, August 29, 1989, p. 47. The Pakistani press promptly noted this shift. One editorial stated:

> The prime minister's remarks here are slightly different from her earlier stand and the categorical statements she made during the tour of the United States that "we neither have any nuclear bomb nor will we make one." Now coming closer to national aspirations, the prime minister has at least made an encouraging declaration that Pakistan might make use of the nuclear know-how in the event of any threat. In this way Pakistan's new elected government has given up its earlier hesitation and given an indication of keeping the nuclear option open.

"Editorial Urges Pakistan to Become Nuclear Power," *Nawa-i-Waqt,* August 31, 1989, translated in *FBIS-NES,* October 6, 1989, p. 22.

108. Hussain, "Pakistan 'Responding' to Change" (see note 86.)
109. *Hearings on the Proposed Sale of F-16s to Pakistan*, Subcommittee on Asian and Pacific Affairs, Committee on Foreign Affairs, U.S. House of Representatives, 101st Cong., 1st Sess., August 2, 1989 (Washington, D.C.: U.S. Government Printing Office, 1989), pp. 35, 40-41.
110. President George Bush to Senator Claiborne Pell, Chairman, Committee on Foreign Relations, U.S. Senate, October 5, 1989.
111. Fiscal Year 1990 Foreign Aid Appropriation, P. L. 101-167, Section 591. The bill was considered and passed in the House and Senate on November 20, 1989, and signed into law on November 31, 1989.
112. "China to Supply PWR in 300-MWe Range," *Nuclear News* (January 1990): 73; Shahid-ur-Rehman Khan, "China Agrees to Supply 300-MW PWR to Pakistan," *Nucleonics Week*, November 23, 1989, p. 1.
113. Steve Coll, "France to Sell Nuclear Plant to Pakistan," *Washington Post*, February 22, 1990; Ann MacLachlan, "France Okays PWR, Reprocessing Plant Breach Talks with Pakistan," *Nucleonics Week*, March 1, 1990, p. 3.
114. Farzad Bazoft, "Iran Signs Secret Atom Deal," *Observer*, June 12, 1988; "Nuclear Cooperation with Pakistan Reported," *Al-Watan Al-'Arabi*, translated in *JPRS-TND*, December 23, 1989, p. 11. One report states that Pakistan is assisting Tehran to build a plutonium producing reactor. See "Washington Whispers," *U.S. News and World Report*, February 12, 1990, p. 20.
115. "Substance to Nuclear Allegations Against Iraq," *MidEast Markets*, December 11, 1989, p. 13.
116. A report in June 1990 stated that images from a U.S. intelligence satellite revealed that Pakistan had sent a heavily armed convoy from a nuclear center to an air base, indicating that Pakistan was preparing nuclear arms for possible use; the story also stated that the United States had determined that Pakistan was refitting some of its F-16s with special racks to carry the weapons. See, James Adams, "Pakistan 'Nuclear War Threat,' " *Sunday Times* (London), May 27, 1990. Bush Administration officials have denied both points in confidential interviews.

Chapter 8: North Korea

1. U.S. Department of State, *A Historical Summary of United States-Korean Relations* (Washington, D.C.: U.S. Government Printing Office, 1962), p. 75.
2. The origins of the Korean War have been the subject of considerable historical debate in recent years. In the United States, the war is generally thought of as an unprovoked attack by the North against the South. Some historians, however, believe the conflict should be thought of more as a civil and revolutionary war in which the North supported nationalist rebels in the South seeking to displace a foreign (U.S.) imposed military dictatorship under Syngman Rhee. See, *e.g.*, Jon Halliday and Bruce Cumings, *Korea: The Unknown War* (New York: Pantheon Books, 1988). The newest insight into the debate was a broadcast by Radio Moscow, in which a Soviet historian adopted the U.S. view of the conflict. See " 'North Korea Ignited Korean War,' " *North Korea News*, April 30, 1990, p. 4.
3. U.S. security obligations arise only upon external armed attack on South Korean territory, thus precluding assistance in the event that the South initiates an attack, as stipulated by a unilateral understanding communicated to South Korea by the U.S. Government. See "Mutual Defense Treaty Between the

United States of America and the Republic of Korea," Vol. 5 UST, p. 2375 (1953).

4. Dean Acheson, *The Korean War* (New York: W. W. Norton & Company, Inc., 1967), p. 83.
5. See, *e.g.*, McGeorge Bundy, *Danger and Survival* (New York: Random House, 1988), pp. 239, 242.
6. John Foster Dulles, "Policy for Security and Peace," *Foreign Affairs* (March 1954): 356, 358-360. Dulles went on to state that "local defense will always be important . . . [but] must be reinforced by the further deterrent of massive retaliatory power."
7. "Warning on Atom Given By Radford," *New York Times*, January 3, 1955.
8. Two detailed, largely overlapping accounts of the early years of the North Korean nuclear program have appeared in the recent past. This section is based on these reports, except as otherwise indicated. See Joseph S. Bermudez, Jr., "N. Korea—Set To Join the Nuclear Club?" *Jane's Defence Weekly*, September 23, 1989, p. 594; "North's Nuclear Capability Assessed," *Hoguk* (July 1989): 119, translated in *Foreign Broadcast Information Service (FBIS)-East Asia (EAS)*, August 3, 1989, p. 23.
9. Bermudez, "N. Korea—Set To Join the Nuclear Club?" (see note 8).
10. Comparison based upon data for reasonably assured, estimated additional, and speculative uranium reserves for these countries. See *Uranium: Resources, Production and Demand* (Paris: Organization for Economic Corporation and Development, 1988).
11. "Military Cooperation With USSR Intensifies," *Naewoe Tongsin*, October 13, 1989, translated in *FBIS-EAS*, January 22, 1990, p. 14.
12. The most authoritative evidence that South Korea had launched a nuclear weapons program is contained in a report of the House International Relations Committee on its 1978 investigation into South Korean attempts to influence U.S. politicians. The report states:

> There are indications that in the early 1970s, some steps were taken which appeared designed to pave the way for an ROK nuclear-weapons program. Specifics on this matter came from a subcommittee staff interview (on February 28, 1978) with a former high-ranking Korean Government official who was a member of the WEC [the secret Weapons Exploitation Committee of senior Korean officials]. He told the subcommittee that the WEC voted unanimously to proceed with the development of nuclear weapons. Subsequently, the Korean Government discussed the purchase of a nuclear fuel reprocessing [plutonium extraction] facility from France and a mixed-oxide fuel reprocessing lab from Belgium. The explosion of an Indian nuclear device in April 1974 using fissionable material produced with the assistance of a Canadian NRX research reactor led to greater caution by nuclear technology suppliers, however, and the Belgians and the Canadians withdrew offers for certain technology. Negotiations between the ROK and France continued for some time over a reprocessing facility. Ultimately, it appears that by some time in 1975, any ROK nuclear weapons program had been canceled and the negotiations for purchase of a fuel reprocessing facility also ended.

See Subcommittee on International Organizations, Committee on International Relations, U.S. House of Representatives, *Investigation of Korean-American Relations*, 95th Cong., 2nd Sess., 1978, p. 80.

13. Robert Gillette, "U.S. Squelched Apparent S. Korea A-Bomb Drive," *Los Angeles Times*, November 4, 1978.
14. This finding is detailed in a study of South Korean nuclear activities, based on interviews with former U.S. officials. See Peter Hayes, "South Korea," unpublished paper delivered at the Conference on Emerging Nuclear Suppliers, Bellagio, Italy, June 1988. Hayes also quotes an unnamed former U.S. ambas-

sador to South Korea as stating that Seoul never dismantled the facilities set up to pursue the weapons option, but merely stopped operating them. Some senior civilian South Korean officials, according to the study, still harbor an interest in acquiring a nuclear weapons option.

In 1984-1985, South Korea and Canada held discussions on Canada's extracting plutonium from spent South Korean reactor fuel, apparently for the purpose of recycling the plutonium as a fuel in a new power reactor Canada was to supply Seoul. According to one account, South Korea was in fact seeking to obtain detailed technical data on plutonium extraction. Intervention by the United States, however, led Canada to end the talks. See, T. Shorrock and O. Gadacz, "U.S. Intervention in Nuclear Fuel Project Still a Mystery," *Business Korea* (April 1985): 47.

15. John W. Finney, "U.S. Says Russians Now Have Deployed 60 MIRV Missiles," *New York Times*, June 21, 1975; International Institute for Strategic Studies (IISS), *Strategic Survey, 1975* (London: IISS, 1976), p. 98. Later that year, however, Schlesinger was reported as saying that the "massive conventional firepower" of the United States would be brought to bear against the North in the event of an invasion. When questioned on the possibility of nuclear use, Schlesinger demurred, noting instead only the conventional aspect of American forces. See "Schlesinger Vows Defense of Korea," *New York Times*, September 2, 1975.
16. "Nuclear Weapons Use Discussed by Carter," *New York Times*, May 30, 1977.
17. Richard Burt, "New Study Raises U.S. Estimate of North Korean Army Strength," *New York Times*, January 4, 1979; Hedrick Smith, "Carter Pressed Brezhnev at Vienna to Allow U-2 Flights Over Turkey," *New York Times*, June 22, 1979; "Joint Chiefs Ask Delay of Korea Pullout to '81," *New York Times*, July 19, 1979; Richard Burt, "U.S. Halts Pullout of Combat G.I.'s From South Korea Till at Least '81," *New York Times*, July 21, 1979.
18. An indigenous program is consistent with the principle of self-reliance—or *Juche*—which the Kim Il Sung government has embraced as a doctrine of national development.
19. John R. LaMarsh, "On the Construction of Plutonium-Producing Reactors by Small and/or Developing Nations," in *Nuclear Proliferation Factbook*, Joint Committee Print, Committee on International Relations, U.S. House of Representatives, and Committee on Governmental Affairs, U.S. Senate, September 23, 1977, p. 533.
20. India's Cirus reactor, used to produce the plutonium for the country's 1974 nuclear test, has a power rating of 40 megawatts. Similarly, although it has since been expanded, Israel's Dimona reactor was rated at 26 megawatts when it apparently began to produce plutonium for Israel's nuclear weapons in the mid-1960s.
21. "North's Nuclear Capability Assessed," *Hoguk* (see note 8).
22. LaMarsh, "On the Construction of Plutonium-Producing Reactors by Small and/or Developing Nations" (see note 19), p. 552.
23. North Korea is believed to have a "uranium processing facility," apparently a mill, northwest of the research complex at Yongbyon, capable of treating 660 pounds (300 kilograms) of uranium ore per day. See "Military Cooperation With USSR Intensifies," *Naewoe Tongsin* (see note 11).
24. Even when its military superiority over the South was unquestioned, the North often claimed that it was threatened with aggression from this quarter.
25. Burt, "U.S. Halts Pullout of Combat G.I.'s" (see note 17).
26. Masao Okonogi, "The Ideology and Political Leadership of Kim Jong Il," in Masao Okonogi, ed., *North Korea at the Crossroads* (Tokyo: Japan Institute of

International Affairs, 1988), pp. 1-7.

27. Do Young Chang, "Inter-Korea Relations and the Major Powers," *Asian Affairs* (Summer 1983): 27; Maggie Ford, "Two Koreas To Allow Divided Families To Meet," *Financial Times*, October 17, 1989; Maggie Ford, "Seoul and Pyongyang Edge a Little Closer," *Financial Times*, November 22, 1989.
28. "Seoul Reports Sinking North Korean Boat," *New York Times*, August 5, 1983. A second ship was able to escape. See "Seoul Says Spy Ship Was Forced to Flee," *New York Times*, August 7, 1983. While such an effort could certainly be consistent with the North's generally aggressive activity against the South, the primary purpose could have been to induce a nuclear accident at the reactor. The political fallout that likely would have resulted from such an "accident" could have generated new fears concerning nuclear weapons, and thus intensified pressures on the U.S. to remove its military forces from the South.
29. Following revelations of North Korean responsibility for the KAL bombing, Japan imposed sanctions on that country, and the U.S. halted efforts to open diplomatic discussions with Pyongyang.

    The Rangoon bombing and the destruction of the KAL flight, although consistent with past Northern efforts to intimidate and undermine South Korea, extended the political conflict beyond the Peninsula's boundaries for the first time and involved unprecedented levels of overt violence. Both acts are thought to have been carried out under the direction of Kim Il Sung's son and heir apparent, Kim Jong Il, suggesting that the hostility of North-South relations is unlikely to abate if the younger Kim succeeds his father.
30. It has also been alleged that Moscow has provided assistance for North Korea's Scud-B missile modernization program. See Janne E. Nolan, "U.S. Options for Countering the Proliferation of Ballistic Missiles: An Assessment of Possible Arms Control Measures," in Robert Shuey, *Missile Proliferation: A Discussion of U.S. Objectives and Policy Options* (Washington, D.C.: Congressional Research Service, February 21, 1990), p. 39; Hideshi Takesada, "Military Trends in North Korea," in Masao Okonogi, ed., *North Korea at the Crossroads* (see note 26), pp. 79-80.
31. International Atomic Energy Agency, "Operational Projects: Full Project Status Report by Project," March 4, 1990.
32. See "U.S. Fines Germany's Degussa $800,000 for Illegal Exports to India, North Korea," *Nuclear Fuel*, April 2, 1990, p. 1; personal interview with Department of Commerce official, April 27, 1990. Zirconium is not essential for encasing uranium fuel in the Yongbyon reactor; aluminum can also be used. Another incident, although not involving nuclear technology or material, revealed the existence of a defense procurement connection between North Korea and West German businessmen. In 1985, the U.S. Department of Commerce revealed that it had evidence that, over a two-year period, a West German businessman had engineered the sale of approximately 82 Hughes 500 helicopters—a type deployed by South Korean forces—to North Korea, in violation of U.S. export laws. The export documents filed with the Department of Commerce fraudulently indicated that the helicopters were bound for West Germany. See Clyde H. Farnsworth, "U.S. Investigates Helicopter Sales," *New York Times*, February 3, 1985.
33. Seth Carus and Joseph S. Bermudez, Jr., "The North Korean 'Scud B' Programme," *Jane's Soviet Intelligence Review* (April 1989): p. 177. The normal selling price for the Scud-B is said to be approximately $1 million per missile (interview with Joseph S. Bermudez, Jr.).
34. Construction activity was reported as early as February 1989. See John McBeth,

Nayan Chanda, and Shada Islam, "Nuclear Jitters," *Far Eastern Economic Review*, February 2, 1989, p. 15.

35. "Testimony of Under Secretary of Defense for Policy, Paul Wolfowitz," Committee on Armed Services, U.S. Senate, 101st Cong., 2nd Sess., April 19, 1990 (hereafter, Wolfowitz 1990 Testimony). Earlier in April, Assistant Secretary of Defense Henry Rowen had made a similar public statement. See "Nuclear N. Korea?" *Washington Post*, April 13, 1990. (Although the article paraphrases Rowen as stating that North Korea had completed the plutonium plant, Rowen has advised the authors that this is incorrect and that the Department of Defense had evidence only that the plant was under construction.)
36. Wolfowitz 1990 Testimony (see note 35).
37. Under IAEA practice, uranium mines and the mills where uranium concentrate ("yellowcake") is produced are not placed under IAEA inspection, since these parts of the nuclear fuel cycle are deemed to be too many steps away from nuclear weapons to warrant such oversight. Uranium concentrate production figures, however, are to be reported to the agency to help it maintain an overall nuclear materials inventory for the state in question.
38. IAEA Document, INFCIRC/252, November 14, 1977.
39. Article III, paragraph 4 of the NPT states, "Negotiation of such agreements shall commence within 180 days from the original entry into force of this Treaty. For States depositing their instruments of ratification or accession after the 180-day period, negotiation of such agreements shall commence not later than the date of such deposit. Such agreements shall enter into force not later than eighteen months after the date of initiation of negotiations."
40. It is also clear, however, that the United States opposes the North Korean position in part because it wishes to avoid the restrictions on the deployment or use of its own nuclear forces in the region that would be required by the declarations sought by Pyongyang. If Washington were forced to declare explicitly that it had no nuclear weapons in South Korea or that it would not use such weapons in the event of a Northern attack, U.S. officials fear that the deterrent value of the U.S. presence in the region would be reduced. Indeed, when New Zealand, a U.S. ally, banned U.S. nuclear-powered and nuclear-weapon-capable ships from its waters in 1985, Washington withdrew its decades-long alliance commitment to Wellington rather than accept restrictions on U.S. nuclear operations or openly declare that U.S. ships did not carry nuclear weapons. Thus it is highly unlikely that Washington would accept the attempt by Pyongyang, an adversary of the U.S., to constrain U.S. nuclear policy in Korea, unless, possibly, this were part of a broader negotiation involving both the North and the South to reduce military forces in the region.

    Some analysts have suggested that Washington could offer North Korea the assurances it is seeking without harming the deterrent value of U.S. forces and that the United States should do so if this *quid pro quo* would help ensure that Pyongyang abandoned its bid for nuclear weapons. These observers argue that, as a practical matter, the United States is highly unlikely to use nuclear weapons in the defense of South Korea, and that, in any event, it could maintain nuclear deterrence with nuclear cruise missiles aboard U.S. ships in the region. This perspective, however, ignores the need for the United States to maintain a consistent, global policy on nuclear arms deployments and overlooks the widespread concern of U.S. officials that exceptions in one theater could lead to the unravelling of U.S. policy elsewhere.
41. For a detailed discussion of the North's 1987 and 1988 arms control proposals and U.S. attitudes towards them, see *Dialogue with North Korea: Report of a Seminar on "Tension Reduction in Korea"* (Washington, D.C.: Carnegie Endowment

for International Peace, 1989), pp. 5-14 and Appendix B. See also, "Pyongyang Offers New Disarmament Proposals," *People's Korea*, June 16, 1990.

42. In a speech on October 23, 1989, for example, Secretary of State James A. Baker, III, reviewing emerging threats to peace in the developing world, stated that "nuclear proliferation, notably North Korea's reactor program, remains dangerous." See "Address and Q&A Session by the Honorable James A. Baker, III, Secretary of State, Before the Commonwealth Club, San Francisco, California, October 23, 1989," Department of State PR. No. 197, October 23, 1989, p. 8. Shortly thereafter, on November 2, Under Secretary of Defense for Policy, Paul Wolfowitz, publicly referred to "signs that North Korea may be in the process of developing a nuclear weapons capability." See "Remarks Prepared for Delivery by the Honorable Paul Wolfowitz, Under Secretary of Defense for Policy, University of Pittsburgh, Thursday, November 2, 1989," U.S. Department of Defense, OASD(PA) News Release No. 498-89, pp. 3-4. Richard H. Solomon, Assistant Secretary of State for East Asian and Pacific Affairs, similarly stated in February 1990, "We are deeply concerned about Pyongyang's continuing unsafeguarded nuclear activity. . . . We view Pyongyang's acceptance of an IAEA safeguards regime as an important indicator of North Korean intentions." See Richard H. Solomon, "Sustaining the Dynamic Balance in East Asia and the Pacific," *Current Policy*, No. 1255, February 1990.
43. Wolfowitz 1990 Testimony (see note 35), Federal News Service Transcript, p. B-4-11,1.
44. "Joint Soviet-American Statement," Released by TASS, Moscow, reprinted by Federal News Service, February 10, 1990, p. SB-6-2, 1-2. See also "Is North Korea Going Nuclear?" *Newsweek*, February 26, 1990, p. 8.
45. "DPRK Urged to Allow Nuclear Inspections," Kyodo, 1408 GMT, February 9, 1990, reprinted in *Joint Publication Research Service (JPRS)-Nuclear Development and Proliferation (TND)*, March 2, 1990, p. 3. See also, "Japan Suspects DPRK of Developing Nuclear Arms," *Tokyo Shimbun*, November 11, 1989 (evening edition) (Foreign Ministry source quoted as stating that "if North Korea has started manufacturing nuclear weapons, this would have a grave impact on the political and military situation on the Korean Peninsula; Japan could not leave the situation unchecked").
46. During the mid-1970s, for example, France had agreed to sell South Korea a plutonium extraction plant to permit it to recycle plutonium as fuel in its conventional reactors in lieu of fresh, low-enriched uranium. Washington ultimately persuaded the two countries to cancel the sale, however, citing the proliferation risks involved. The United States also acted during this period to dissuade Taiwan from obtaining a plutonium plant from France.
47. "Testimony of Rear Admiral Thomas A. Brooks, Director of Naval Intelligence," Subcommittee on Seapower, Strategic, and Critical Materials of the Armed Services Committee, U.S. House of Representatives, 101st Cong., 1st Sess., February 22, 1989 (hereafter, "Brooks 1989 Testimony").
48. *Ibid.*; William Safire, "Those Chinese Missiles," *New York Times*, February 23, 1989; Louise Lief, "Will Baghdad Get the Bomb?" *U.S. News and World Report*, April 9, 1990, p. 34; Bermudez and Carus, "North Korean 'Scud B' Programme" (see note 33).
49. Brooks 1989 Testimony (see note 47).
50. South Korea has also manufactured surface-to-surface missiles. One system, the NHK-1, with a range of 110 miles, is said to be based on the U.S. Nike-Hercules. A second missile is said to have a range of 160 miles. See W. Seth Carus, "Trends and Implications of Missile Proliferation," unpublished paper

delivered at the International Studies Association Convention, April 13, 1990. South Korea is not believed to possess chemical weapons and is not thought to be developing nuclear weapons at this time.

51. To be sure, Kim could probably launch a raid against Seoul with manned bombers, but the Scud alternative would avoid the risk of losing costly and difficult-to-replace aircraft and pilots.
52. Bennett Ramberg, *Nuclear Power Plants as Weapons for the Enemy: An Unrecognized Military Peril* (Berkeley: University of California Press, 1984), pp. 19-55.
53. Bermudez and Carus, "The North Korean 'Scud-B' Programme" (see note 33), p. 180; Ramberg, *Nuclear Power Plants as Weapons for the Enemy* (see note 52), p. 92. The four power reactors are Uljin 1 and 2, and Yeongkwang 1 and 2.
54. Bermudez and Carus, "The North Korean 'Scud B' Programme" (see note 33). In June 1990, it was reported that the missile was being prepared for its first test launch. See Bill Gertz, "North Korea's Ballistic Missiles Have U.S. Worried," *Washington Times*, June 4, 1990.
55. At present, six power reactors are under construction or in planning. A report that is intended to be the basis of the ROK's future nuclear power strategy states that construction of some 50 nuclear units will be required in the first quarter of the 21st century. See "Datafile: Korea," *Nuclear Engineering International* (April 1989): 67-72; "Korean Nuclear Plans Revealed," *Nuclear Engineering International* (January 1990): 7.
56. The South's nuclear power plants could also, of course, be the target of Northern air strikes. However, the South has extensive defenses against such attacks. Such defenses would not be effective against Northern missiles, and for this reason they pose a unique threat for Seoul.
57. "Tear Down Barrier, Start Exchanges in All Fields," *Korea Times*, July 7, 1988.
58. "Seoul and Moscow Set Up Partial Ties," *New York Times*, December 9, 1989; Peter Maass, "Seoul, Moscow Agree on Goal of Normalizing Ties," *The Washington Post*, March 31, 1990.
59. In the final 1989 agreement to exchange family visits, for example, Roh accepted every one of the North's proposals on the matter and most of its suggestions on related topics. Similarly, one of North Korea's longstanding demands that has acted as a barrier to North-South dialogue has been that the ROK change its National Security Law, which makes contact with the North by South Koreans a criminal offense. Work is now under way to revise that law, although prospects for major changes remain cloudy. See Maggie Ford, "Seoul and Pyongyang Edge a Little Closer" (see note 27).
60. "USSR's Arbatov Discusses Policy Toward DPRK," *Chungang Ilbo*, September 12, 1989, translated in *FBIS-EAS*, September 16, 1989, p. 26. Arbatov, Director of the Research Institute for the United States and Canada, made this statement at a conference in Seoul in September 1989. Arbatov's remarks may have been directed primarily to a U.S. audience, and his presence may have signalled a desire to pursue a final settlement of one of the few remaining irritants in relations between East and West.
61. "Now that the Party is Over," *AsiaWeek*, July 28, 1989, p. 22. Western press reports suggest that Kim Jong Il is currently in charge of much of the day-to-day government activity in the North and that the full transition is imminent. See "Kim Jong Il Assumes a Higher Leadership Profile," *Far Eastern Economic Review*, February 22, 1990; "Rumors on Ideological Struggle in North Korea Reported," *Dong-A Ilbo*, August 23, 1989, translated in *FBIS-EAS*, August 25, 1989, pp. 25-26. When the General Secretary of the Chinese Communist Party

visited Pyongyang in March 1990, he was greeted by both Kims, an unusual occurrence in a system that has striven to maintain a clear division of labor between father and son, possibly indicating that the transition is actually in progress.

62. "Differing Views on Kim Chong-il's Succession," *The Korea Herald*, February 16, 1990, reprinted in *FBIS-EAS*, February 16, 1990, p. 37; "British Analyst in Seoul Sees DPRK Changes," Yonhap, 0232 GMT, April 14, 1990, reprinted in *FBIS-EAS*, April 16, 1990, p. 21; Nicholas D. Kristof, "Great Leader to Dear Leader," *New York Times Magazine*, August 20, 1989, p. 73. Senior Chinese and Soviet officials both privately acknowledged to foreign visitors in recent months that they expect such a power struggle.
63. "North Korean Inaugural," *Washington Post*, May 25, 1990.
64. Kim outlined his strategy, which he called "south-south cooperation," in a 1988 New Year's address, stating:

    If the peoples of developing countries and oppressed nations are to achieve complete national liberation and build a new independent society, they must continue to intensify their anti-imperialist national-liberation struggle and fight to abolish the unfair, outdated international economic order and establish a new fair one. At the moment, the people of developing countries should make a joint effort to effect full-scale south-south cooperation.

    See "Kim Il Song Gives New Year's Address," KCNA, 0834 GMT, January 1, 1988, reprinted in *FBIS-EAS*, January 4, 1988, p. 6.
65. "Events in January 1989," *Vantage Point* (February 1989): 25.
66. "Soviet N-Plant Sale to DPRK Contradicts Worsening Relations," *Defense and Foreign Affairs Weekly*, March 12-18, 1990, p. 6.
67. In December 1988, the Iranian Minister of Mines and Metals flew to Pyongyang; during his visit he met with top Northern leaders, including the new premier of the Administrative Council (or Cabinet), as well as Kim Il Sung himself. See "Events in December 1988," *Vantage Point* (January 1989): p. 30. Shortly after this visit ended, North Korean Deputy Premier-Foreign Minister Kim Yong Nam and External Affairs Minister Chong Song Nam visited Iran, where the economic, scientific, and technical cooperation agreement between the two countries was signed. See *ibid*, p. 31; "Events in January 1989," *Vantage Point* (February 1989): 25. See also "North Korean Economic Minister Meets Officials," Tehran Domestic Service, 0330 GMT, March 15, 1990, reprinted in *FBIS-NES*, March 21, 1990, p. 55 (referring to a bilateral agreement on cooperation in "mines and metals").
68. "North Said to Conclude Weapons Treaty with Iran," Seoul Domestic Service, 2200 GMT, December 8, 1989, translated in *FBIS-EAS*, December 11, 1989, p. 11.

    During an earlier visit, Iranian President Seyyed Ali Khamenei suggested the general nature of cooperation between the two countries when he stated,

    All the aggressors of the world must know that if they risk a military gamble against the revolution and revolutionary peoples, they will only sustain an ignominious defeat. . . .

    Further, Khamenei echoed Kim Il Sung's January 1988 call for "south-south cooperation" when he went on to proclaim,

    [The great powers] do not want to see our development and progress but are willing to assist us on condition that we surrender our independence to them. Hence, we attach particular importance to cooperation in the third world, particularly to cooperation among the peoples and governments of such countries as the Islamic Republic of Iran and the Democratic People's Republic of Korea which are standing on the forefront against imperialism and dominationism.

See "Iranian President Gives Speech," KCNA, 1535 GMT, May 16, 1989, reprinted in *FBIS-EAS*, May 17, 1989, p. 16.

69. "Castro Supports Kim Il-song Reunification Plan," KCNA, 0433 GMT, February 10, 1990, reprinted in *FBIS-EAS*, February 12, 1990, p. 12.

At a banquet hosted by the Cuban ambassador to North Korea the same month, North Korean Premier Yong Hyong-muk appeared to reciprocate, declaring that the two countries would cooperate to further socialism, and that great gains had been made by both in spite of confrontational relations with the "U.S. imperialists." " 'North Korea Will Cooperate with Cuba to Strengthen Socialist System'—Premier Yong Hyong-muk Vows," *North Korea News*, March 5, 1990, p. 3.

70. Interestingly, when the North Korean atomic energy delegation returned home, it was met by the Soviet *chargé d'affaires ad interim*, suggesting Soviet awareness of—and possibly tacit support for—Cuban-North Korean nuclear ties. "Nuclear Power Delegation Returns From Cuba," Pyongyang Domestic Service, 1300 GMT, February 9, 1990, translated in *FBIS-EAS*, February 15, 1990.

Irrespective of Soviet intentions, its sale of nuclear power plants to Cuba and proposed sales to North Korea and Iran will provide an element of commonality in the nuclear programs of these countries that will encourage nuclear cooperation among them. See "USSR/Iran: Iranians Say Soviets Will Aid in Nuclear Program," *Nucleonics Week*, March 16, 1990.

71. "*Spiegel* Reports Missile Aid to Romania," *Der Spiegel*, May 8, 1989, pp. 166-168, translated in *JPRS-TND*, May 23, 1989, p. 42; Henry Kamm, "Rumania Makes Threats, Hungary Says," *New York Times*, July 11, 1989.

72. James Adams, "Russians Sound Nuclear Warning on North Korea," *Sunday Times* (London), June 17, 1990. While not commenting on the alleged transfer of nuclear materials to North Korea, U.S. officials privately dismissed as far-fetched another assertion in the article, that the North might obtain nuclear weapons in six months.

73. U.S. Defense Intelligence Agency, *Terrorist Group Profiles* (Washington, D.C.: U.S. Government Printing Office, 1988), p. 114.

74. Brooks 1989 Testimony (see note 47), Federal News Service transcript, p. 48.

75. Peter Maass, "Seoul Agrees to Discuss Troop Cut," *Washington Post*, February 16, 1990.

76. See J. Bennett Johnston and Dale Bumpers, "An Overly Costly 'Trip Wire' in Korea," *New York Times*, July 21, 1989; "Don't Worry, For Now," *The Economist*, February 3, 1990, p. 33; "Cut and Share," *Far Eastern Economic Review*, February 15, 1990, p. 32; "Foreign Bases Up for Closure," *Congressional Quarterly*, February 3, 1990, p. 342.

77. "North Korean Plans Delayed," *Nuclear Engineering International* (February 1990): 16.

Part III: Middle East - Introduction

1. Mohammed El-Sayed Selim, "Egypt," in James E. Katz and Onkar S. Marwah, eds., *Nuclear Power in Developing Countries* (Lexington, Mass.: Lexington Books, 1982), pp. 138-139.
2. "The Trail to Crazy Island," *Wall Street Journal*, March 16, 1984.
3. Joseph S. Bermudez, Jr. and W. Seth Carus, "The North Korean 'Scud B' Programme," *Jane's Soviet Intelligence Review* (April 1989): 177; U.S. Arms Control and Disarmament Agency (ACDA), *International Arms Transfers and World Military Expenditures: 1988* (Washington, D.C.: U.S. ACDA, 1989), p. 18.
4. "Nuclear Equipment Sold to Iraq by FRG Company," *Der Spiegel*, December 18, 1989: Mark Hibbs, "German Firm Implicated in Trying To Aid Pakistan Enrichment Effort," *Nucleonics Week*, May 4, 1987, p. 1; Farzad Bazoft, "Iran Signs Secret Atom Deal," *Observer*, June 12, 1988.

   Concerns have also been raised that Pakistan might assist Saudi Arabia to develop a nuclear warhead for the intermediate-range CSS-2 missiles that it acquired from China in early 1988. Saudi Arabia is thought to have provided some of the financing for Pakistan's nuclear weapons development program and might hope to capitalize on this tie.

   In October 1988, however, Saudi Arabia acceded to the Non-Proliferation Treaty, a step that will reaffirm and formalize its pledges not to develop or acquire nuclear arms, even though the treaty will not provide mechanisms for verifying the status of the new Saudi missiles, specifically.
5. "Israel Bids India Join Attack on Pakistan," *Israeli Foreign Affairs* (April 1987); Shyam Bhatia, "Israelis Plot A-Plant Raid on Pakistan," *Observer*, March 27, 1988. Earlier rumors such as these led then-Indian Prime Minister Rajiv Gandhi and then-Pakistani President Mohammad Zia ul-Haq to agree verbally during a December 1985 summit in New Delhi not to attack each other's nuclear installations. The understanding was subsequently formalized in a written accord, which both countries had ratified by the end of 1989, although they had not yet exchanged lists of installations covered by the agreement.
6. This point is discussed at greater length in the chapter on Israel, below.
7. See, *e.g.* Caryle Murphy, "Syria Urged to Stress Defense," *Washington Post*, November 11, 1989; John P. Hannah, "Moscow Puts the Brakes on Syria," *New York Times*, November 28, 1989.
8. "War of Liberation," *New York Review of Books*, November 22, 1984, p. 36; Neil Roland, "Soviets Reportedly Offer Nuclear Help to Syria," UPI, November 28, 1985, citing an interview with Syrian Defense Minister Mustafa Tlas in *Al-Ittihad*, October 4, 1985, translated in *Foreign Broadcast Information Service (FBIS)-Middle East and Africa*, October 7, 1985.

   A Soviet spokesman claimed Tlas's statement was untrue, but other Soviet officials reportedly told Western visitors that if Israel were to attack Syria itself—even if only with conventional armaments—Moscow would assist its ally with military force, including tactical nuclear weapons if necessary. Ze'ev Schiff, "Dealing with Syria," *Foreign Policy* (Summer 1984):94.

   See also "Defense Minister Interview on Parity with Israel," *Al-Majallah*, December 11-17, 1985, translated in *FBIS-Middle East and Africa*, December 12, 1985, p. H-1 (Tlas stated, "We are capable of acquiring a nuclear option in order to achieve strategic balance with the Zionist enemy"); "USSR Reportedly To Train Syrians on Nuclear Weapons," KUNA 1315 GMT, October 12, 1985, translated in *Joint Publication Research Service (JPRS)-Nuclear Development and Proliferation*

(TND), November 25, 1985, p. 66 (citing a Lebanese magazine, *Al-Nahar Al-Arabi Wa Al-Dawli)*; "Secret Nuclear Weapons Talks with USSR Reported," Salalah (Syrian) Domestic Service, 1600 GMT, March 21, 1986, translated in *FBIS-Middle East and Africa*, March 25, 1986, p. H-1.

Chapter 9: Israel

1. "Revealed: The Secrets of Israel's Nuclear Arsenal," *Sunday Times* (London), October 5, 1986.
2. Frank Barnaby, *The Invisible Bomb* (London: I. B. Taurus, 1989), p. 25.
3. See note 82 and accompanying text.
4. Stephens Broening, "Israel Could Build Missile to Hit Soviets, U.S. Thinks," *Baltimore Sun*, November 23, 1988.
5. Based on Vanunu's testimony and his photos, Taylor told the *Sunday Times* investigative team:

   > Assuming that the photographs were taken at Dimona, the models of the interior components of a nuclear weapon are genuine and Vanunu's identification of the nature and purpose of the various facilities and of the materials corresponding to the "models" are generally correct.
   >
   > The information obtained from Vanunu's statements and photographs as presented to me are entirely consistent with a present Israeli capacity to produce at least five to ten nuclear weapons a year that are significantly smaller, lighter, and more efficient than the first types of nuclear weapons developed by the U.S., U.S.S.R., U.K., France, and China.

   See "Revealed: The Secrets of Israel's Nuclear Arsenal" (see note 1). Taylor, however, did not speak directly to Vanunu.

   Barnaby, in contrast, had the opportunity to debrief Vanunu extensively. He told the *Sunday Times* that he found Vanunu's testimony to be "totally convincing." See *ibid.*
6. Thomas L. Friedman, "Israeli Suspect Flashes a Hint He Was Abducted," *New York Times*, December 23, 1986; "How Israeli Agents Snatched Vanunu," *Sunday Times* (London), August 9, 1987; "Riddle of Vanunu Ship," *Sunday Times* (London), August 16, 1987; "Revealed: The Woman From Mossad," *Sunday Times* (London), February 21, 1988. The Israeli government has never explained how it obtained custody of Vanunu.
7. Glenn Frankel, "Israel Convicts Vanunu of Treason for Divulging A-Secrets to Paper," *Washington Post*, March 25, 1988. Vanunu has appealed his conviction.
8. Explaining his ability to roam about the building where he worked and to take his 60 photos, Vanunu stated that perimeter security at the facility was relatively lax, permitting him to smuggle in a camera, and that as a control room technician, he had the run of the building so that he could respond to irregularities that showed up on his monitors.

   There has been speculation that the Israeli government deliberately planted Vanunu to give its nuclear deterrent more credibility, without actually acknowledging that it existed. This scenario appears rather far-fetched in light of Vanunu's detention and conviction—unless it is assumed that these developments, too, are part of an elaborate Israeli ploy.
9. The following material concerning the first decades of Israel's nuclear program is condensed from a more extensive review of this subject in Leonard S. Spector, *The Undeclared Bomb* (Cambridge, Mass.: Ballinger Publishing Company, 1988), pp. 165-187.

10. Pierre Péan, *Les Deux Bombes* (Paris: Fayard, 1981). The book relies on extensive documentation and interviews with the former French officials and industrialists who had assisted the nuclear Israeli effort.
11. Péan, *Les Deux Bombes*, Chapters V, VIII (see note 10).
12. "Israel: Jericho II and the Nuclear Arsenal," *Defense and Foreign Affairs*, May 9, 1985, pp. 1-2.

    In 1974, the missile's existence was confirmed in a CIA analysis of Israel's nuclear capabilities, which was inadvertently released in 1978. See note 79 below.
13. Péan, *Les Deux Bombes*, Chapter VI (see note 10).
14. The EL-3 reactor used slightly enriched uranium fuel, moderated by heavy water and cooled by air. The Dimona facility, however, is thought to be both cooled and moderated by heavy water and to use natural uranium fuel. See *ibid.*
15. *Ibid.* Péan also states that the cooling ducts were three times as large as needed for a 24-megawatt reactor.
16. *Ibid.*, Chapter VII.
17. Steve Weissman and Herbert Krosney, *The Islamic Bomb* (New York: Times Books, 1981), p. 114.
18. "France Admits It Gave Israel A-Bomb," *Sunday Times* (London), October 12, 1986.
19. Charles De Gaulle, *Memoirs of Hope: Renewal and Endeavor* (New York: Simon and Schuster, 1971), p. 266.

    In his 1986 interview with the *Sunday Times*, however, Francis Perrin appears to contradict this statement—and Péan's understanding—by stating that "General De Gaulle stopped the collaboration on nuclear weapons with Israel in 1959, although he agreed to supply the plutonium plant because we had a reciprocal agreement and we thought it would be good for France to have this possibility of working with Israel." See "France Admits It Gave Israel A-Bomb," *Sunday Times* (London) (see note 18). It is thus possible that De Gaulle, notwithstanding his official utterances, knowingly permitted work on the unit to continue through the end of 1960.
20. John W. Finney, "U.S. Hears Israel Moves Toward A-Bomb Potential," *New York Times*, December 19, 1960; John W. Finney, "U.S. Misled At First on Israeli Reactor," *New York Times*, December 20, 1960. Péan states that the U.S. Central Intelligence Agency got wind of the construction at Dimona in 1958, but that it did not make a public issue of the matter because Israel insisted that the reactor was intended only for peaceful research and promised to accept inspections by American experts after construction was completed. He states that the affair became public only after it was leaked to the press. There is strong evidence, however, that Israel, in fact, deliberately hid the reactor from Washington. See *Executive Sessions of the Senate Foreign Relations Committee* (Historical Series), Vol. XIII, Pt. 1, 87th Congress, 1st Session, 1961 (1984), pp. 7-8. In a 1988 communication with the principal author of this book, a former U.S. nuclear aide familiar with this period of U.S.-Israeli nuclear relations stated he was convinced that Washington did not learn of the Dimona reactor until 1960 and was genuinely surprised by the development.
21. Dana Adams Schmidt, "Israel Assures U.S. on Reactor," *New York Times*, December 22, 1960. On December 21, 1960, Ben Gurion addressed the Israeli Knesset, disclosing the reactor and pledging it would be devoted to peaceful purposes.
22. Alvin Shuster, "Israel Satisfies U.S. on Use of Reactor," *New York Times*, December 23, 1960; Faud Jabber, *Israel and Nuclear Weapons* (London: Chatto and Windus, 1971), p. 35.

23. See Glossary.
24. Bertrand Goldschmidt, *Le Défi Atomique* (Paris: Fayard, 1980), pp. 205-206.
25. Jabber, *Israel and Nuclear Weapons*, p. 89 (see note 22); "CBS Evening News with Walter Cronkite," CBS Television Network, February 21, 1980.
26. Communication with former U.S. official, 1986 (Argentina supplied fifteen tons); Richard Kessler, "Argentine Officials Deny Rumors of [Recent] Nuclear Trade with Israel," *Nucleonics Week*, May 29, 1986, p. 7; Jabber, *Israel and Nuclear Weapons* p. 89 (see note 22).

    In 1960, Israel had obtained a small, 5-megawatt research reactor from the United States, but this required enriched uranium as fuel, then available exclusively from the United States.
27. The details of Norway's sale of heavy water to Israel and of the non-proliferation controls that Norway imposed on the transaction were unearthed and brought to public attention by Gary Milhollin, a professor of law at the University of Wisconsin and director of the Wisconsin Arms Control Project. See John H. Cushman, Jr., "U.S. Sent Liquid to Israel for Reactor," *New York Times*, November 10, 1986; Charles R. Babcock, "Norway Eyes Israel's Use of Nuclear Ingredient," *Washington Post*, November 10, 1986; Gary Milhollin, "Heavy Water Cheaters," *Foreign Policy* (Winter 1987-1988): 100.
28. Press Release, Norwegian Department of Foreign Affairs, October 10, 1987.

    The press release also notes that Norway authorized a 1.1 ton (1.0 metric ton) shipment of heavy water to Israel in July 1970, but refused a contemporaneous Israeli request for 4.4 tons (4.0 metric tons) of the material. (Text courtesy of Gary Milhollin.)
29. Milhollin, "Heavy Water Cheaters," pp. 103-105 (see note 27).
30. Babcock, "Norway Eyes Israel's Use of Nuclear Ingredient" (see note 27); Wolf Blitzer, "Norwegian Heavy Water Said Used for Dimona Plutonium," *Jerusalem Post*, November 19, 1987.

    Péan states that Israel also received several tons of heavy water from France in a single shipment in 1960. If that transfer actually took place, it would have violated retransfer restrictions in France's nuclear supply agreements with the United States and Norway, which until then had provided France with all of its heavy water but which are not known to have authorized its reexport. See Milhollin, "Heavy Water Cheaters," p. 107 (see note 27).

    The United States also supplied 4.3 tons (3.9 metric tons) of U.S. heavy water to Israel, but according to a former senior U.S. nuclear aide, the material was not used in Dimona; the material was covered initially by bilateral U.S. safeguards, and is now under International Atomic Energy Agency inspection.

    Israel may also have been able to produce small quantities of heavy water on its own. By November 1954, Israel already had built a pilot plant for the production of heavy water. See statement of Abba Eban, Israeli Ambassador to the United Nations, November 15, 1954, *Official Records of the General Assembly*, Ninth Session, First Committee, 716th meeting, pp. 335-337.

    It should be noted that the largest proportion of heavy water required to operate a heavy water reactor is needed for the initial inventory of new reactors. Make-up requirements are very small in comparison.
31. Péan, *Les Deux Bombes*, Chapter VII (see note 10).
32. Most other sources place the start-up of the reactor at late 1963 or early 1964 and have generally given the start-up date of the plutonium extraction plant as 1969 or 1970. See, *e.g.*, "How Israel Got the Bomb," *Time*, April 12, 1976. U.S. intelligence sources speculate that the plutonium unit was going through a

shake-down period prior to the 1967 Arab-Israeli War and that significant production did not get under way until at least 1968 or 1969.

33. Barnaby, *The Invisible Bomb*, p. 31 (see note 2).
34. The visits were initiated in 1964 and conducted annually through 1969. See John W. Finney, "Israel Permits U.S. to Inspect Atomic Reactor," *New York Times*, March 14, 1965 (noting visits in 1964 and 1965); John W. Finney, "U.S. Again Assured on Negev Reactor," *New York Times*, June 28, 1966; John W. Finney, "Israel Could Make Atom Arms in 3 or 4 Years, U.S. Aides Say," *New York Times*, July 6, 1966; Hedrick Smith, "U.S. Assumes the Israelis Have A-Bomb or its Parts," *New York Times*, July 18, 1970 (describing visits in 1968 and 1969 and noting that American analysts thought Israel had obtained plutonium for nuclear weapons by having it extracted from Dimona-reactor spent fuel in France).
35. Smith, "U.S. Assumes the Israelis Have A-Bomb or its Parts" (see note 34).
36. In 1976, a major review of the Israeli nuclear program in *Time* magazine stated that Israel had stopped work on the plutonium plant in 1964 and that construction was not restarted until early 1968, nearly a year after the Six Day Arab-Israeli War of June 1967. See "How Israel Got the Bomb," *Time* (see note 32). See also Pringle and Spigelman, *The Nuclear Barons* (New York: Holt, Rinehart and Winston, 1981), p. 296. Under this hypothesis, no improper activities would have been occurring at the site until near the end of the U.S. visits, when in fact the U.S. visitors were becoming suspicious. The evidence of Péan and Vanunu suggests that a stronger—though not conclusive—case can now be made that the plant was completed during the mid-1960s. Significant plutonium production may not have begun until 1968 or 1969, however.
37. In fact, considerably more material was missing from the facility, a total of nearly 600 pounds (about 270 kilograms), but the U.S. Atomic Energy Commission concluded that all but 220 pounds (100 kilograms) of the loss could be explained by process losses, leaving the remainder "unaccounted for." If the larger amount were diverted to Israel, it would have provided enough material for 10 to 18 weapons.
38. "Near Armageddon: The Spread of Nuclear Weapons in the Middle East," American Broadcasting Company, April 27, 1981, transcript pp. 13-14. See also, "C.I.A. Concluded, Ex-Aide Says, That Israel Made Nuclear Arms," *New York Times*, May 2, 1981.
39. Despite extensive U.S. government investigations, in 1980, the Nuclear Regulatory Commission concluded that there was no "concrete information concerning the final disposition or location of any material which may be missing from this facility." See U.S. Nuclear Regulatory Commission, *Safeguards Summary Event List*, NUREG-0525, rev. 3, December 1980.

    Glenn Seaborg, the chairman of the Atomic Energy Commission (whose functions are now exercised by the Nuclear Regulatory Commission and the Department of Energy) at the time the material was determined to be missing, has stated, "It is all but certain . . . that the material was lost during complicated chemical procedures that, to save money, were conducted with inadequate care." See Glenn T. Seaborg, with Benjamin S. Loeb, *Stemming the Tide* (Lexington, Mass.: Lexington Books, 1987), p. 259.
40. EURATOM, the European Community nuclear agency, has confirmed that the yellowcake was diverted during the transit between Antwerp and Genoa; the only outstanding question is whether Israel was the country responsible, and there appears to be little doubt on this score. When European Community officials first confirmed the loss of the material in 1977, for example, press

reports quoted unnamed American and European intelligence sources as being convinced that the material was taken to Israel to fuel the Dimona reactor. See, *e.g.*, William Drozdiak, "Uranium Loss is Admitted," *Washington Post*, May 3, 1977. Various internal U.S. government memoranda on the affair, released under the Freedom of Information Act to the Natural Resources Defense Council and provided to the authors, repeatedly cite Israel as the presumed perpetrator. In answers to questions posed by a subcommittee of the House International Relations Committee, for example, U.S. nuclear aides in 1976 went so far as to volunteer how long the material might last if used to fuel the Dimona reactor.

The affair was first brought to public attention by Paul Leventhal, a former staff member of the Senate Governmental Operations Committee and now president of the Nuclear Control Institute, Washington, D.C. The episode is sometimes referred to as the "*Scheersberg*" incident, after the ship involved or as the "plumbat" affair, because the drums containing the yellowcake were, reportedly, marked with this label (a Latin reference to lead). For a comprehensive investigation of the incident, see Elaine Davenport, Paul Eddy, and Peter Gillman, *The Plumbat Affair* (London: Deutsch, 1978). Israel has denied that it was responsible.

41. Z. Ketzinel, "Uranium Sources, Production, and Demand in Israel," *Proceedings of the Fourth International Conference on the Peaceful Uses of Atomic Energy*, Geneva, September 6-16, 1971, vol. 8, pp. 113-119, as cited in *Israeli Nuclear Armament*, Report of the Secretary General, A/36/431, September 18, 1981. p. 11.
42. Barnaby, *The Invisible Bomb* (see note 2), p. 36.
43. Smith, "U.S. Assumes Israelis Have A-Bomb or Its Parts" (see note 34); "How Israel Got the Bomb," *Time* (see note 32).
44. Interviews, fall 1987.

The possibility of a superpower nuclear confrontation also developed during the conflict. On the night of October 24-25, the United States placed its nuclear forces on alert. This was a response to a note from Soviet General Secretary Leonid Brezhnev, which threatened to introduce Soviet combat troops into Egypt if Israel, which had surrounded the Egyptian Third Army Corps, did not comply with a U.N. ceasefire. U.S. intelligence had confirmed that Moscow was readying its airborne forces for such a mission. Marvin Kalb and Bernard Kalb, *Kissinger* (Boston: Little Brown and Company, 1974), pp. 489-496. Contributing to American concerns was the possibility that the Soviet Union was also about to deploy tactical nuclear weapons in Egypt. See Barry M. Blechman and Douglas M. Hart, "The Political Utility of Nuclear Weapons," *International Security* (Summer 1982): 137.

From these accounts, it was not Israeli nuclear activity on October 8 that triggered the U.S.-Soviet crisis, as is sometime reported, but the risk that weeks later Israel might decimate Egyptian forces by conventional means. The U.S. alert ended on October 26, after the threat of Soviet intervention eased.

45. In September 1974, for example, a CIA analysis (which was inadvertently released under the Freedom of Information Act to the Natural Resources Defense Council) concluded that "Israel has already produced nuclear weapons." See Central Intelligence Agency, "Prospects for Further Proliferation of Nuclear Weapons," DCI NIO 1945/74, September 4, 1974.

In December 1974, Israeli President Ephraim Katzir indirectly confirmed this assessment in an interview stating, "It has always been our intention to develop the nuclear potential. We now have that potential." See *The Guard-*

ian, December 3, 1974.

Similarly, in early 1976, Carl Duckett, a senior CIA official, disclosed in a secret briefing to the Nuclear Regulatory Commission that Israel was making nuclear weapons from plutonium produced at Dimona. See U.S. Nuclear Regulatory Commission, *Inquiry into the Testimony of the Executive Director for Operations*, Volume III (Interviews), February 1978, p. 178.

At another supposedly off-the-record briefing to members of the American Institute of Aeronautics and Astronautics in March 1976, Duckett estimated that the Israeli nuclear arsenal contained between 10 and 20 weapons as powerful as the bomb used on Hiroshima. See Arthur Kranish, "CIA: Israel Has 10-20 A-Weapons," *Washington Post*, March 15, 1976.

46. See, *e.g.*, *Israeli Nuclear Armament* (see note 41), p. 12.
47. "America's Nuclear Pledge to Israel," *Foreign Report*, January 21, 1981 (Israel has 200 weapons, can produce 8 per year); "Israel Said to Deploy Nuclear Missiles," *Aerospace Daily*, May 1, 1985 (Israel might possess as many as 200 weapons); Anthony Cordesman, interview on "NBC Nightly News," National Broadcasting Company, July 30, 1985 (Israel has at least 100 nuclear weapons and perhaps as many as 140).
48. "Californian Indicted in Export of Triggers to Israel," *New York Times*, May 17, 1985; John M. Goshko, "L.A. Man Indicted in Export of Potential Nuclear Bomb Component to Israel," *Washington Post*, May 17, 1985. A krytron releases the precisely timed electrical pulse that simultaneously detonates the high explosive "lenses" that surround the weapon's nuclear core.
49. John M. Goshko, "Israel Got U.S.-Made Devices," *Washington Post*, May 14, 1985; John M. Goshko, "U.S. Asks to Inspect Israeli Atom Sites," *Washington Post*, May 15, 1985.
50. *Hearings on Developments in the Middle East, July 1985*, Subcommittee on Europe and the Middle East, Committee on Foreign Affairs, U.S. House of Representatives, 99th Cong., 1st Sess., July 24, 1985, p. 72. See generally, Leonard S. Spector, *The New Nuclear Nations* (New York: Vintage Books, 1985), pp. 41-44.
51. Interview with a knowledgeable U.S. official.
52. Charles R. Babcock, "Computer Expert Used Firm to Feed Israel Technology," *Washington Post*, October 31, 1986; Richard Sale and Geoffrey Aronson, "Exporting Nuclear Triggers: The Strange Case of Richard Smyth," *Middle East Report* (May-June 1987): 40.

    Israel also appears to have by-passed international controls in obtaining about 52 tons (47 metric tons) of depleted uranium from Luxembourg in April 1984. See Spector, *The New Nuclear Nations* (see note 50), pp. 141-143. The material—uranium whose fissionable isotope, uranium-235, has been partially culled—is a by-product of the uranium enrichment process. It cannot be used for nuclear weapons, but can be irradiated in a reactor to produce plutonium. Depleted uranium is also used for armor-piercing shells, and it appears that this was the use Israel had in mind.
53. Sale and Aronson, "Exporting Nuclear Triggers: The Strange Case of Richard Smyth" (see note 52).
54. "France Admits It Gave Israel A-Bomb," *Sunday Times* (London), October 12, 1986; Weissman and Krosney, *The Islamic Bomb* (see note 17), p. 114.
55. "Israel: The Covert Connection," *Frontline*, Public Broadcasting System, May 16, 1989.
56. Jane Hunter, "Israel, South Africa, and the Bomb," *Middle East Report* (November-December 1986): 13 (citing transcript of BBC 2 TV, "Newsnight,"

2250 GMT, July 11, 1985); Sale and Aronson, "Exporting Nuclear Triggers: The Strange Case of Richard Smyth" (see note 52).

57. Sale and Aronson, "Exporting Nuclear Triggers: The Strange Case of Richard Smyth" (see note 52).
58. U.S. General Accounting Office, *Nuclear Proliferation: Major Weaknesses in Foreign Visitor Controls at Weapons Laboratories*, GAO/RCED-89-31 (October 1988), pp. 25-26. Among other conclusions, the GAO noted that

> four foreign visitors from Israel visited [Lawrence] Livermore [National Laboratory] for activities associated with lasers, including those used in weapons-related applications, . . . and three foreign nationals from . . . Israel visited Los Alamos [National Laboratory] to discuss laser isotope separation. . . .

The latter technique can be used to enrich uranium to weapons grade. The report also states that:

> six foreign nationals from Israel and India visited or were assigned to various laboratories to conduct research on or to discuss theoretical astrophysics. DOE [Department of Energy] documents state that astrophysics is closely tied to understanding nuclear weapons phenomena and nuclear detonations.
>
> One foreign national from Israel assigned to Los Alamos collaborated in research related to high explosives. This research involved predicting the initiation and detonation behavior of a high-explosive specifically used in nuclear weapons applications. Further, four other visitors from Israel . . . were involved in research related to explosives and shock waves.

59. Aharon Levran and Zeev Eytan, *The Middle East Military Balance: 1987-1988* (Boulder, Colo.: Westview Press, 1988), pp. 133-229.
60. Yohanan Ramati, "Israel and Nuclear Deterrence," *Global Affairs* (Spring 1988, Vol III, No.2).
61. "Minister Denies London Report on Nuclear Weapons," Jerusalem Domestic Service, September 28, 1986, reprinted in *Foreign Broadcast Information Service (FBIS)-Near East and Africa (NEA)*, September 29, 1986, p. I-7; "Herzog Meets Lange, Discusses Nuclear Arms," Jerusalem Domestic Service, reprinted in *FBIS-NEA*, November 12, 1986, p. I-4. See also "Rabin Discusses Nuclear Weapons, Transfer Policy," *Yedi'ot Aharonot*, November 12, 1987, translated in *Joint Publications Research Service (JPRS)-Nuclear Developments (TND)*, December 1, 1987, p. 20.
62. "Intelligence Head on Arab Arms," Jerusalem Domestic Service, 0505 GMT, December 24, 1986, translated in *FBIS-Near East and South Asia (NES)*, December 24, 1986.
63. Gideon Raphael, "Syria and Israel: Too Near the Brink," *Washington Post*, January 11, 1987.
64. Mary Curtius, "Israel Needs More US Aid to Spur Economy and Immigration, Economist Says," *Christian Science Monitor*, January 20, 1987.
65. Barnaby, *The Invisible Bomb* (see note 2), p. 31.
66. Barnaby, *The Invisible Bomb* (see note 2), p. 28. This position was also taken by the *Sunday Times*, which suggested that the reactor had "an ingenious cooling system" that disguised its output. The *Sunday Times* offered no further details. Barnaby, however, hypothesized that the upgrade could have been accomplished by modifying the reactor to cool it with heavy water rather than with carbon dioxide (the coolant thought to have been used in the plant as originally supplied by France) and by increasing the density of uranium fuel rods in the reactor's core.

67. Presumably the U.S. aides are referring to evidence from the U.S. visits of the 1960s and to subsequent satellite photos.
68. Assuming the enlargement occured, a total of 825 pounds (375 kilograms) of plutonium could have been produced: 110 pounds (50 kilograms) during intermittent operation from 1962 to 1972 at 40 megawatts, and 715 pounds (323 kilograms) during the 17 years at 70 megawatts between 1973 and the end of 1989.
69. "U.S. Document Cites Israeli Missile Progress," *Washington Times*, November 15, 1989.
70. The U.S. Department of Defense has been far more cautious in describing Israeli nuclear capabilities publicly. Nonetheless, in the 1989 edition of its annual survey, *Soviet Military Power*, the Defense Department for the first time noted, in discussing military developments in the Third World, that Israel, India, Pakistan, and other regional states "are working hard to develop" nuclear weapons. U.S. Department of Defense, *Soviet Military Power* (Washington, D.C.: U.S. Government Printing Office, 1989), p. 124.
71. Barnaby, *The Invisible Bomb* (see note 2), p. 25. In 1974, two Israeli scientists filed patents in West Germany describing a laser enrichment process, but there had been no publicly available evidence since that time that Israel had, in fact, begun to operate an enrichment unit using this, or any other, technology. See, Robert Gillett, "Uranium Enrichment: Rumors of Israeli Progress with Lasers," *Science*, March 22, 1974, p. 15.
72. A set of reporters' notes prepared by the team from the *Sunday Times* that broke the Vanunu story also indicates that he told them that Israel had an enrichment program, corroborating Barnaby's account of Vanunu's comments. The notes, however, provide no details as to the size or operating characteristics of the Israeli enrichment effort.
73. "Testimony of William Webster, Director of Central Intelligence," *Hearings on Missile and Nuclear Proliferation*, Senate Governmental Affairs Committee, U.S. Senate, 101st Cong., 1st Sess., May 18, 1989 (mimeo).
74. Boosting would also permit a designer to use less of these materials to achieve the same yield as that which could be obtained without boosting. Thus boosting would allow more weapons of a particular yield to be produced from a given amount of plutonium or highly enriched uranium—in effect permitting Israel to increase the number of devices in its arsenal. In addition, reducing the amount of nuclear material in the core also reduces the amount of high explosives needed to detonate the device, permitting weapons with smaller physical dimensions to be built. Such miniaturization could facilitate the manufacture of missile warheads and nuclear artillery shells.

    However, as the quantity of plutonium or highly enriched uranium in a device is decreased, the design requirements for the device become increasingly stringent. Inasmuch as Israeli weapon designers cannot count on being able to conduct full-scale nuclear tests to prove out their designs, it is likely that they would favor more conservative designs that did not attempt to take full advantage of boosting to minimize the amount of nuclear weapons material used.
75. A kiloton is the equivalent of 1,000 tons of TNT.
76. See Appendix B. A megaton is the equivalent of 1 million tons of TNT.
77. Lithium-6 is an isotope of lithium that occurs naturally in a concentration of 7.42 percent. Vanunu described in considerable detail how between 1982 and 1984, Israel built and operated a plant for increasing the concentration of lithium-6, or "enriching" lithium in this isotope. Barnaby states (relying on Vanunu) that Israel had built a pilot plant for producing lithium-6 in 1977.

Tritium can be produced by irradiating lithium-6 in a reactor and then processing the irradiated material to draw off the tritium and separate it from other gases, such as helium, that are also created by irradiation. Tritium is also created in heavy water that undergoes irradiation. Thus, before Israel began using lithium-6 to produce tritium, it may have obtained tritium by removing it from the heavy water used in the Dimona reactor, although this is said to be a very difficult process to master.

78. Barnaby, *The Invisible Bomb* (see note 2), p. 25.
79. Inadvertently released under the Freedom of Information Act to the Natural Resources Defense Council, the analysis states:

> We believe that Israel has already produced nuclear weapons. Our judgment is based on Israeli acquisition of large quantities of uranium, partly by clandestine means; the ambiguous nature of Israeli efforts in the field of uranium enrichment, and Israel's large investment in a costly missile system designed to accommodate nuclear warheads. We do not expect the Israelis to provide confirmation of widespread suspicions of their capability, either by nuclear testing or by threats of use, short of a grave threat to the nation's existence. Future emphasis is likely to be on improving weapon designs, manufacturing missiles more capable in terms of distance and accuracy than the existing 260-mile Jericho, and acquiring or perfecting weapons for aircraft delivery.

See Central Intelligence Agency, "Prospects for Further Proliferation of Nuclear Weapons," DCI NIO 1945/74, September 4, 1974.

80. "Israel Said to Deploy Jericho Missile," *Aerospace Daily*, May 1, 1985; "Nuclear Efforts of Israel, Pakistan Prompts Meeting of U.S. Group," *Aerospace Daily*, May 7, 1985.
81. Richard Sale, interviewed on "NBC Nightly News," National Broadcasting Company, July 30, 1985.
82. Elaine Sciolino, "Documents Detail Israeli Missile Deal with Shah," *New York Times*, April 1, 1986. The 750-kilogram payload would encompass the nuclear device itself, the reentry vehicle, and other elements.

The deployment of the missile was also confirmed to the principal author of this book by several current and former U.S. officials during 1986 and 1987.

83. U.S. Arms Control and Disarmament Agency (ACDA), *International Arms Transfers and World Military Expenditures: 1988* (Washington, D.C.: U.S. ACDA, 1989), p. 18.
84. *Ibid.*; "Testimony of Rear Admiral Thomas A. Brooks, Director of Naval Intelligence," *Hearings on Intelligence Issues*, Subcommittee on Seapower, Strategic, and Critical Materials, Committee on Armed Services, U.S. House of Representatives, 101st Cong., 1st Sess., February 22, 1989.

Multiple test launchings of the system had been reported by the end of 1989. See Thomas W. Netter, "Israel Reported to Test New Longer-Range Missile," *New York Times*, July 22, 1987; Thomas L. Friedman, "Soviets Caution Israel Against a New Missile," *New York Times*, July 29, 1987; "Israel in Second Secret Test of Jericho IRBM," *Jane's Defence Weekly*, November 19, 1988, p. 1258; David B. Ottaway, "Israel Reported to Test Controversial Missile," *Washington Post*, September 16, 1989; "Moscow Concern at 'Israel Ballistic Missile Launch,' " *Financial Times*, September 16-17, 1989. According to one report, by May 1987 Israel had tested the missile some ten times. See "Israel's Jericho Missile," *MidEast Markets*, November 23, 1987, p. 11.

The longest range to which the missile had been tested by the end of 1989 was 800 miles, and the missile's range is sometimes described as being 800 miles in the press, even though ACDA and Admiral Brooks used the 900-mile figure. Some confusion may be caused by the fact that 800 nautical miles are the same as 900 statute miles, and Admiral Brooks in his testimony used the nautical mile terminology.

85. U.S. Arms Control and Disarmament Agency, *International Arms Transfers and World Military Expenditures: 1988* (see note 83), p. 18.
86. See also, Stephens Broening, "Israel Could Build Missiles to Hit Soviets, U.S. Thinks," *Baltimore Sun*, November 23, 1988.

    There has been some confusion over the names used in the media to designate these various Israeli missiles; both the Improved Jericho and the Jericho Follow-ons have sometimes been referred to as the Jericho II, and the Jericho Follow-ons have sometimes been called the Jericho IIb. For clarity, the terms, Jericho, Improved Jericho, and Jericho Follow-ons are employed here.
87. Glenn Frankel, "Israel Launches Its First Satellite Into Orbit," *Washington Post*, September 20, 1988.
88. Andrew Whitley, "Israel Presses on With Its Spy-in-Space Satellite Program," *Financial Times*, October 18, 1988. Israeli sources indicated that the Offeq 1 satellite was intended to pave the way for the future launch of a photo-reconnaissance intelligence satellite.
89. Steven E. Gray, "Israeli Missile Capabilities: A Few Numbers to Think About" (Livermore, Calif: Lawrence Livermore National Laboratories, Z Division), October 7, 1988 (mimeo).
90. Janne Nolan and Albert Wheelon, "Third World Missile Proliferation," in Robert Inman, Joseph Nye, William Perry, and Roger Smith, eds., *New Threats: Responding to the Proliferation of Nuclear, Chemical and Delivery Capabilities in the Third World* (Lanham, Md.: Aspen Strategy Group and University Press of America, 1990). This analysis also assumed that the first two stages of the Shavit would be used to launch a guidance/warhead payload.
91. "Official Communique Issued," Jerusalem Domestic Service, 1500 GMT, April 3, 1990, translated in *FBIS-NES*, April 4, 1990, p. 27; "Further on Technical Details," Jerusalem Domestic Service, 0400 GMT, April 4, 1990, printed in *FBIS-NES*, April 4, 1990, p. 27. The launch of Offeq II came one day after Iraqi President Saddam Hussein declared that he would "destroy half of Israel" with chemical weapons, if Israel attacked any of Iraq's nuclear facilities. The Israeli launch, though it was apparently planned months earlier, served as a timely reminder of Israel's own potent deterrent capabilities.
92. This possibility has been mentioned in the press. See, for example, "Israel's Shavit Booster Believed to Be Version of Jericho-2 Ballistic Missile," *Aviation Week and Space Technology*, October 3, 1988, p. 28.
93. Adding to the confusion on this point is that Israel has acknowledged conducting four tests of the Shavit II before the Offeq 1 launch—two of the rocket's first stage and two of two stages. All were fired into the Mediterranean, which is also the site of the tests of the Jericho Follow-on, at least one test of which took place in June 1987. See "Israel Presses on With Its Spy-in-Space Satellite Program," *Financial Times*, October 18, 1988.
94. In several radio broadcasts after a test in July 1987, for example, the Soviet Union warned Israel against deploying the system. See Thomas L. Friedman, "Soviet Cautions Israel Against a New Missile," *New York Times*, July 29, 1987. Moscow also expressed grave concerns about the system in 1989. See discussion in text.
95. "Moscow Concern at 'Israel Ballistic Missile Launch,' " *Financial Times* (see note 84).
96. "NBC Nightly News," National Broadcasting Company, October 25, 1989.
97. Other Middle East states are also reportedly seeking coastal test sites in Africa

for long-range missiles. See Michael R. Gordon, "U.S. Fears Iraq is Seeking Long-Range Missile Site," *New York Times*, April 24, 1990.

98. "NBC Sunrise," National Broadcasting Company, October 26, 1989; "Officials Continue to Deny South African Links," Jerusalem Domestic Service, 0500 GMT, October 30, 1989, reprinted in *FBIS-NES*, October 30, 1989, p. 36; "Pik Botha Denies Israeli Missile Links," SAPA, 1114 GMT, October 27, 1989, reprinted in *FBIS-Africa*, October 30, 1989, p. 18.

Some Israeli denials contained an element of ambiguity, however. In a statement to NBC, for example, a spokesman for Israeli Prime Minister Yitzhak Shamir stated, "We absolutely deny that there is any link between Israel and South Africa in the nuclear field. And therefore there is no truth in this [NBC] story that was published yesterday night. Our cabinet decided in March 1987 that there is going to be no new contract between Israel and South Africa, and Israel scrupulously adheres to and reenforces this decision of the cabinet." (The cabinet decision was taken in response to U.S. legislation that could lead to the termination of U.S. military aid to countries that engage in military sales to South Africa in violation of the 1977 United Nations arms embargo of the latter.) The Israeli spokesman's statement did not explicitly rule out, however, that the two countries were collaborating in the area of missile technology (as distinct from nuclear technology) under contracts signed prior to March 1987. See also, Michael Gordon, "U.S. Says Data Suggest Israel Aids South Africa on Missile," *New York Times*, October 27, 1989.

99. Michael Gordon, "U.S. Data Suggest . . ." (see note 98); David B. Ottaway and R. Jeffrey Smith, "U.S. Knew of 2 Nations' Missile Work," *Washington Post*, October 27, 1989.

100. Transcript, "White House Background Briefing on the Visit of Prime Minister Yitzhak Shamir of Israel, November 15, 1989," Federal News Service, pp. 1, 7-1, 8-1, 8-2; see also, "Israeli-S. African A-Project Would Be 'Taboo,' Bush Says," *Washington Post*, October 29, 1989.

101. See, e.g., interview with Prime Minister Yitzhak Shamir, "MacNeil/Lehrer Newshour," Public Broadcasting System, November 15, 1989.

102. "Israeli Missiles for Pretoria: No," *New York Times* (editorial), November 15, 1989. On the supercomputer controversy, see Senator John Glenn, "Supercomputers and Super Bombs," *Congressional Record*, October 31, 1989, p. S 14382; Gary Milhollin, "Why Are We Helping the Third World to Go Nuclear?" *Washington Post*, April 1, 1990.

103. Since 1988, Israel has received such U.S. know-how in conjunction with a joint project to develop the "Arrow" anti-tactical-ballistic missile. Although the U.S. technology transferred for the Arrow is intended for a defensive system, it undoubtedly can be used by Israel to enhance its offensive missile capabilities, as well. For this reason, the U.S. transfers have been criticized as contrary to the objectives of the U.S.-sponsored Missile Technology Control Regime—the set of parallel export controls adopted by seven Western countries in 1987 to curb the spread of ballistic missiles in the developing world. Moreover, in its investigative reports, NBC News also stated that Israel had transferred technology to Pretoria from Israel's cancelled Lavi advanced jet fighter project; much of the advanced electronics for this project had been supplied to Israel from the United States. U.S. spokesmen stated that Washington had "no indication" that Israel had transferred U.S missile technology to South Africa and "no information to confirm" that Israel had given the latter U.S.-origin technology supplied for the Lavi. See David B. Ottaway, "State Dept. Sees No Israel-South Africa Transfer," *Washington Post*, October 28, 1989; Tony Walker, "Israel Coy About South African

Connection," *Financial Times*, October 31, 1989.

Concerning the gyroscope episode, see Jeff Gerth, "5 Charged in Plot to Export Arms," *New York Times*, November 16, 1989.

104. Why Norway did not implement its verification rights is not known, but given published accounts of American suspicions about Israel's nuclear weapons program dating back to at least 1970, there is reason to suspect that Norway, like France in the 1950s, was sympathetic to Israel's desire for a nuclear deterrent and deliberately chose not to pursue its inspection rights. See, *e.g.*, Smith, "U.S. Assumes Israelis Have A-Bomb or Its Parts" (see note 34).
105. See notes 27 and 28.
106. Benny Morris, "Inspection of Norwegian Heavy Water Barred," *Jerusalem Post*, April 16, 1987; Michael Gordon, "Norway Says Israel Resists a Nuclear Check," *New York Times*, May 26, 1987; "Norway Bid for Heavy Water Supervision Denied," Jerusalem Domestic Service, 2100 GMT, October 3, 1987, translated in *FBIS-NES*, October 6, 1987, p. 20; "Norway to Renew Heavy Water Inspection Request," *Nucleonics Week*, July 2, 1987, p. 13; E. A. Wayne, "Israel Accused Anew of Nuclear Violations," *Christian Science Monitor*, December 2, 1987; "Israel: Norwegian Heavy Water Use to be Probed," *Nucleonics Week*, December 10, 1987, p. 15.
107. Andrew Whitely, "Israel to Permit N-Plant Inspections," *Financial Times*, April 21, 1988; "Norway and Israel Resolve Dispute Over Heavy Water," *International Herald Tribune*, June 11-12, 1988.
108. John J. Fialka, "Norway Rejects a Proposed Agreement with Israel on Nuclear-Plant Coolant," *Wall Street Journal*, October 11, 1988; Hugh Carnegy, "Israel, Norway to Reopen Nuclear Inspection Debate," *Financial Times*, April 4, 1989; "Heavy Water Pact with Norway, 'Scrapped,'" *Jerusalem Post*, April 4, 1989, reprinted in *JPRS-TND*, April 25, 1989, p. 20; "Israel/Norway: Heavy Water Inspection Debate Reopened," *Nucleonics Week*, April 6, 1989, p. 10.
109. "Israel," *Nucleonics Week*, May 3, 1990, p. 15.; John J. Fialka, "Israel Plans Sale of Heavy Water Back to Norway," *Wall Street Journal*, April 30, 1990.
110. The large-scale production of plutonium at Dimona in the mid-1970s described by Vanunu, for example, was presumably authorized by Prime Minister Yitzhak Rabin of the Labor Party. His successor, Likud Party leader Menachem Begin, who served as Prime Minister from 1977 to 1983, apparently authorized the continued build-up of Israeli plutonium stocks, along with the development of the lithium deuteride production capability to be used for advanced nuclear weapons. Begin would also have been responsible for deploying the Improved Jericho and, possibly, for early work on the Jericho Follow-ons. Begin was succeeded by Likud Prime Minister Yitzhak Shamir who governed through September 1984 and apparently continued these policies. Shamir was followed by Labor Party leader Shimon Peres—said to be the father of the Israeli nuclear program because of his negotiations with France to obtain crucial nuclear assistance in the 1950s—who served through October 1986, as part of a Labor-Likud coalition government. During this period, plutonium production at Dimona continued and, judging by Vanunu's photos, Israel was manufacturing advanced nuclear weapons using lithium deuteride components. Work on the Jericho Follow-ons would also have been undertaken. In October 1986, Shamir returned to the post of Prime Minister, which he has held since. Among other developments during his tenure as the head of a Labor-Likud coalition government have been the tests of the Jericho Follow-ons and the launch of Israel's first two space satellites.

In addition to this apparent continuity in Israel's nuclear program, as long as the Likud Party retains the prime ministership, the so-called "Begin Doctrine" of preventing other Middle East states from acquiring nuclear arms—based on former Prime Minister Menachem Begin's decision to destroy the Osiraq reactor in 1981—is also likely to remain a serious element of Israeli strategic doctrine. The Osiraq raid was said to have been opposed by Labor Party leader Shimon Peres, however. Possibly, if Labor regains control of the government, a reconsideration of the Begin Doctrine might follow.

111. "Netanyahu Calls for Israel-Iraq Summit in Cairo," *Jeruslem Post*, September 26, 1989; "Strategic Understanding with Iraq Sought," *Davar*, November 23, 1989, translated in *FBIS-NES*, November 24, 1989, pp. 32-33.
112. For an overview of changing political alignments among the Arab states, see Mark Rodenbeck and Victor Mallen, "Shifting Sands of Politics Create New Arab Hopes," *Financial Times*, December 29, 1989.
113. Caryle Murphy, "Syria Urged to Stress Defense," *Washington Post*, November 11, 1989.
114. Glenn Frankel, "Iraq Said Developing A-Weapons," *Washington Post*, March 31, 1989; George C. Wilson, "Iraq Downs a Visiting Egyptian Jet," *Washington Post*, April 28, 1989; Levran and Eytan, *The Middle East Military Balance 1987-1988* (see note 59), p. 295-305.
115. "Iraq Threatens to Use Chemical Weapons Against Israeli Attack," *Financial Times*, April 3, 1990. See also, "Paper Warns Against Zionist 'Flagrant Threats,'" Baghdad INA, 0545 GMT, September 18, 1988, translated in *FBIS-NES*, September 19, 1988, p. 18.
116. The setbacks to the Iraqi program often attributed to Israel included an April 1979 explosion that destroyed the Osiraq reactor's core structures while they were awaiting shipment to Iraq in a warehouse in Seine-sur-Mer, France; the June 1980 murder in Paris of Dr. Yahya el-Meshad, an Egyptian nuclear scientist working for the Iraqi Atomic Energy Commission; and a series of bombings and bomb threats against a number of French and Italian companies supplying nuclear equipment for the Iraqi nuclear program. See Leonard S. Spector, *Nuclear Proliferation Today* (New York: Vintage Books, 1984), pp. 175-178 (also discusses Israeli diplomatic efforts to slow the Iraqi program).
117. James M. Dorsey, "U.S. Fears Iraq's Chemical Weaponry," *Washington Times*, July 24, 1989; Simon Henderson, "West Blocks Advance of Condor II Missile," *Financial Times*, August 31, 1989; Hugh Carnegy and Victor Mallet, "Launch of Iraqi Space Rocket Fuels Fears in Israel," *Financial Times*, December 20, 1989.
118. Michael Gordon, with Stephen Engelberg, "Iran is Expanding Chemical Stocks Used in Poison Gas," *New York Times*, January 29, 1989; "Defense Industries Manufacture New Missile," IRNA, 1535 GMT, April 17, 1989, reprinted in *JPRS-TND*, May 5, 1989, p. 20; Simon Henderson, "West Blocks Advance of Condor II Missile" (see note 117); Stephen Engelberg with Michael Gordon, "Germans Accused of Helping Libya Build Nerve Gas Plant," *New York Times*, January 1, 1989; Patrick Tyler and Nora Boustany, "Moscow Pursues Diplomacy But Syria Gets Soviet Arms," *Washington Post*, April 25, 1989.
119. Stephen Engelberg, "Iraq Said To Study Biological Arms," *New York Times*, January 18, 1989; "Iraq Has Developed Biological Weapons, Says Israeli Official," *Financial Times*, January 19, 1989.
120. See, *e.g.*, "Sources View Iraqi Missiles as 'Direct' Threat," *Hadashot*, August 5, 1987, translated in *FBIS-NES*, August 5, 1987, p. L-1; Elaine Ruth Fletcher, "Iraqi Missile Poses Threat to Israel," *Jerusalem Post*, August 15, 1987; Ra-

phael, "Syria and Israel: Too Near the Brink" (see note 63).

121. On Israeli concerns about Syrian chemical missile capabilities, see, Raphael, "Syria and Israel, Too Near the Brink" (see note 63); Ian Black, "Israel Tries to Counter Arab Nerve Gas Threat," *Manchester Guardian Weekly*, December 21, 1987.
122. "Testimony of General J.R. Deane, Jr., Director of Research and Development, U.S. Army," *Hearings on the 1975 Defense Authorization*, Committee on Armed Services, U.S. Senate, 101st Cong., 1st Sess., May 7, 1974; Edward Mortimer, "Israel Hints It Keeps Chemical Weapons as Defensive Measure," *Financial Times*, January 10, 1989.
123. Kenneth Kaplan, "Concern Over Arab Chemical, Germ Arms," *Jerusalem Post*, January 28, 1989. For similar earlier statements by Rabin using the "100 times over" formula, see, "Rabin on Saudi Missiles, Other Issues," *Yedi'ot Aharonot*, April 1, 1988, translated in *FBIS-NES*, April 6, 1988, p. 29; "Shomron Discusses Unrest, Saudi Missiles, Syria," Jerusalem Domestic Service, 0400 GMT, April 9, 1988, translated in *FBIS-NES*, April 11, 1988, p. 31; Joel Brinkley, "Israel Fears That Peace Plan in the Persian Gulf Will Unleash Bitter Foe," *New York Times*, July 24, 1988; Yossi Melman and Dan Raviv, "Why Nobody Will Will Win the Next War in the Middle East," *Washington Post*, November 6, 1988.
124. James M. Markham, "Arabs Link Curbs on Gas and A-Arms," *New York Times*, January 9, 1989. The article quotes Iraqi Foreign Minister Tariq Aziz as stating that there was " 'no doubt whatsoever' that it [Israel] had introduced nuclear weapons into the Middle East." (See note 123.)

Chapter 10: Libya

1. The story has been reported by Mohamed Heikel, confidant of then-Egyptian President Gamal Abdul Nasser, whose assistance Khadafi sought in approaching China. See Mohamed Heikel, *The Road to Ramadan* (New York: Quadrangle/The New York Times Book Co., 1975), pp. 76-77. The story was later confirmed by Mohammed al-Mougariaf, the former Libyan ambassador to India. See American Broadcasting Company, "News Closeup," Monday, April 27, 1981, "Near Armageddon: The Spread of Nuclear Weapons in the Middle East," Transcript, p. 43. The story is widely accepted by U.S. non-proliferation officials.
2. Steve Weissman and Herbert Krosney, *The Islamic Bomb* (New York: Times Books, 1981), p. 60. Among other elements of the arrangement, according to knowledgeable Pakistani and Libyan officials interviewed by Western journalists, Libya specifically requested training in the operation of hot cells, the radiologically shielded units in which plutonium can be extracted from spent reactor fuel.
3. Weissman and Krosney, *The Islamic Bomb* (see note 2), pp. 211-212; "Niger Says It Sold Uranium to Libya, Use for Nuclear Weaponry Feared," *Washington Star*, April 14, 1981. Niger has since agreed to make future yellowcake sales exclusively to NPT parties with valid safeguards agreements. Niger reportedly believed this standard was satisfied in the case of Libya between 1978 and 1980 because it was a party to the NPT.
4. U.S. Congress, Office of Technology Assessment, *Technology Transfer to the Middle East* (Washington, D.C.: U.S. Government Printing Office, 1985), pp. 385-386. This analysis states that Libya is known to have purchased 788 tons of uranium from Niger and may have purchased about 2,000 tons. See also, "Ura-

nium Exports for First Six Months Given," Agence France Presse, 0859 GMT, August 27, 1981, translated in *Joint Publications Research Service (JPRS)-Nuclear Development and Proliferation (TND)*, September 24, 1981, p. 17, which gives slightly lower figures.

5. "Libya Presses India on Nuclear Technology," *Nuclear Engineering International* (November 1979): 7.
6. *Ibid.* See also "Libya Charged with Using Oil to Obtain Nuclear Technology," *Business Times* (Kuala Lumpur), August 30, 1979 (quoting from stories in the *Times of India* and the *Hindu*), reprinted in *JPRS-TND*, October 31, 1979, p. 7.
7. "Indo-Libyan Nuclear Accord," *The Telegraph* (Calcutta), August 8, 1984, reprinted in *JPRS-TND*, September 24, 1984, p. 69. India's stance provides evidence of the caution exhibited by some "second tier" nuclear-supplier countries in sharing nuclear technology.
8. Congressional Research Service, *Analysis of Six Issues About Nuclear Capabilities of India, Iraq, Libya and Pakistan*, report prepared for the Subcommittee on Arms Control, Oceans, International Operations, and Environment of the Senate Foreign Relations Committee, 1982, p. 11; Joseph V.R. Micallef, "A Nuclear Bomb for Libya?" *Bulletin of the Atomic Scientists* (August-September 1981): 14.
9. Senate Foreign Relations Committee, *Hearings on the Israeli Air Strike*, June 18, 19, and 25, 1981, 97th Cong., 1st Sess. (Washington, D.C.: U.S. Government Printing Office, 1981) p. 50; Claudia Wright, "Libya's Nuclear Program," *The Middle East* (February 1982): 47.
10. In 1976, Libya also signed a nuclear cooperation agreement with France for the purchase of a 600-megawatt nuclear power plant, but France later barred the deal because of concerns over Libyan-Pakistani cooperation. See "France Is To Build Libyan Atomic Plant," *New York Times*, March 23, 1976; Office of Technology Assessment, *Technology Transfer to the Middle East* (see note 4), p. 385; Thomas O'Toole, "Libya Said to Buy Soviet A-Power Plant," *Washington Post*, December 12, 1977.
11. "Envoy Says Soviets May Build Nuclear Plant in Egypt," Associated Press, May 10, 1986, AM cycle; "Libya Abandons Plans for First Unit," *Nuclear Engineering International* (April 1986): 6; "On-Again, Off-Again Libyan Nuclear Power Plant Surfaced Once More," *Nucleonics Week*, March 31, 1983, p. 11; Ann MacLachlan, "Libyans Are Seeking Broad International Cooperation in Nuclear Area," *Nucleonics Week*, September 27, 1984, p. 1.
12. "U.S. Bans Libyans from Some Studies," *Washington Post*, March 11, 1983.
13. Guy Duplat, "Possible Breakthrough in Belgian Nuclear Cooperation with Arab Countries," *Le Soir*, August 8, 1981, translated in *JPRS-TND*, September 8, 1981, p. 2; "US Presses Belgians to Cut Libya Ties," *Nuclear Fuel*, April 26, 1982, p. 11.
14. Some U.S. officials feared that Libya could use uranium metal as fuel in a crude natural uranium/graphite reactor to produce plutonium. A crude reprocessing plant for extracting the plutonium might be within the range of Libyan capabilities, especially with assistance from Pakistan or, possibly, Argentina.

    Uranium hexafluoride might serve as feedstock if Libya were to build an enrichment plant under its supposed arrangement with Pakistan, or if the material could be sent to Pakistan for enrichment to weapons grade.

    The Belgonucléaire facility might not have been subject to IAEA inspections despite Libya's adherence to the NPT, since IAEA rules may exempt such plants. The issue had, apparently, never been decided by the agency.
15. "Belgium, Libya Reportedly Ready to Sign Nuclear Pact," *Wall Street Journal*,

May 18, 1984; Ann MacLachlan, "Belgians Fail To Get Guarantees from Neighbors on Libyan Aid," *Nucleonics Week,* December 6, 1984, p. 7.

16. Fred Hiatt, "Belgium Urged to Reject Pact With Libyans," *Washington Post*, October 9, 1984.
17. MacLachlan, "Belgians Fail to Get Guarantees from Neighbors on Libyan Aid" (see note 15); "Government Stopping Libyan Deal," *Nuclear News* (January 1985); interviews with Western European officials, May 1985.
18. "Libya Aided Argentina in War," *New York Times*, May 14, 1984.
19. "Brazilian Military Concern Over Argentine Talks with Libya," *Correio Braziliense*, May 24, 1983, translated in *JPRS-TND*, June 30, 1983, p. 8; unpublished interview with Carlos Castro Madero, former head of the Argentine Atomic Energy Commission, by John Cooley, ABC News, Fall 1984. The events suggest the possibility that Tripoli was seeking a nuclear *quid pro quo* for its earlier military assistance.
20. "Uranium Mining Aid to Libya," *Correio Braziliense*, November 8, 1984, translated in *JPRS-TND*, January 14, 1985, p. 42; "Negotiations Begin on Building Warships for Libya," Agence France Presse, 1624 GMT, October 5, 1984, translated in *Foreign Broadcast Information Service (FBIS)-Latin America (LAT)*, October 9, 1984, p. D-1; "Libya Seeks Military Cooperation," *O Estado de São Paulo*, December 23, 1984, translated in *FBIS-LAT*, December 26, 1984, p. D-1.
21. Mac Margolis, "Brazil Plans to Resume Weapons Sales to Libya," *Washington Post*, January 28, 1988. Brazil had suspended sales to Libya in 1983 after discovering that four Libyan cargo planes that had landed for refueling in Manaus carried a clandestine East Bloc arms shipment destined for Nicaragua.
22. This episode is described in detail in Joseph C. Goulden, *The Death Merchant* (New York: Bantam Books, 1985), pp. 289-295; see also, Peter Maas, *Manhunt* (New York: Random House, 1986), pp. 176-177, 179-182. Both appear to rely in part on the testimony of John Heath at Wilson's 1983 trial in Houston, Texas, for illegal export of explosives. Heath was an employee of Wilson who was present at the meetings when the nuclear arms were discussed with Libyan officials. Neither Goulden nor Maas had access, however, to the documents describing the nuclear deal offered the Libyans—including illustrations of the nuclear devices that were purportedly being offered—which were obtained by the principal author of this book in 1986 under the Freedom of Information Act and are discussed in detail in Leonard S. Spector, *Going Nuclear* (Cambridge, Mass.: Ballinger Publishing Company, 1987), pp. 150-159.

    The nuclear scam was reportedly initiated in February or March 1981, when Wilson was contacted by a Belgian associate, Armand Donnay, who claimed to have access to enough "fissionable material" to build a nuclear weapon. It was decided that a formal proposal would be made to Wilson's contacts in Libya, and soon afterwards, Donnay went there with a stack of documents containing the details of the deal.

    Heath was present at the first of the meetings with the Libyans at which Wilson and Donnay discussed the sale of nuclear weapons. According to Heath's testimony at Wilson's 1983 trial, the meeting was held with top Libyan nuclear officials—the "Libyan nuclear committee," in Heath's words—and Wilson and the others were introduced to the group by a senior military aide. See Transcript of testimony of John Heath, January 29 and 31 and February 3, 1983, pp. 77, 89-90, 94-96, *U.S. v. Wilson*, Cr. H-82-139, Federal District Court, Southern District of Texas. The Libyans quickly rejected Donnay's presentation as incomplete. One account—apparently based on Heath's recollections—states that the Libyans wanted to know the quality of the fissionable material Donnay had to offer. When he told them it was "twenty percent" —presumably twenty-percent enriched uranium, unsuitable for weapons—the group's chairman stated, "That's no good for us. We need eighty percent. If you can't do better than this, we have no business." See Joseph Goulden, *The Death Merchant*, p. 291.

Donnay then offered to rework and expand his offer and one or more additional meetings were held with the Libyans to discuss the nuclear weapons deal. The Libyans rejected the revised proposal, however, because of its incompleteness and technical defects. See Heath Testimony, pp. 74, 77.

According to Heath's testimony, Wilson was motivated by simple greed in pursuing the deal. See Heath Testimony, p. 71. In an interview conducted in the fall of 1986, Wilson challenged this version. While confirming that meetings were held with Libyan officials to discuss the nuclear sales, he claimed that Heath "was the one that was really pushing it" because "he wanted to make a commission out of the thing." Indeed, Wilson asserted that he "killed the deal" when interviewed in 1986. See Alan Berlow, "All Things Considered," *National Public Radio*, August 14, 1986, 6:30 PM. The jury in Wilson's trial, however, found Heath a credible witness, and U.S. law enforcement officials have rejected Wilson's account of his role.

23. "Al-Qadhdhafi Lectures University Students," Tripoli Television Service, 1958 GMT, June 21, 1987, translated in *FBIS-Near East and South Asia (NES)*, June 26, 1987.
24. "Al-Qadhdhafi Discusses Upcoming Amman Summit," Tripoli Television Service, 2032 GMT, November 2, 1987, translated in *FBIS-NES*, November 3, 1987, p. 19.
25. Mark Hibbs, "Argentine Officials Optimistic on Futures of Atucha-2, Exports," *Nucleonics Week*, October 12, 1989, p. 11.
26. David B. Ottaway, "CIA Director Says Libya is Building A Large Chemical Weapons Plant," *International Herald Tribune*, October 27, 1988; Stephen Engelberg with Michael Gordon, "Germans Accused of Helping Libya Build Nerve Gas Plant," *New York Times*, January 1, 1989. The matter became a major source of tension between Washington and Bonn in early 1989 and was the subject of innumerable press stories and analyses. In June 1990, the key West German figure who provided assistance to Libya in building the plant, Jürgen Hippelstiel-Imhausen, was convicted of export law violations and tax evasion and sentenced to five years' imprisonment.
27. "Qaddafi—Chemical Weapons Revisited," *Focus on Libya* (Washington, D.C.: PEMCON, Ltd., December 1989): 2.
28. Jennifer Parmelee, "Libya Said to Offer to Dismantle Plant If It Is Given New One," *Washington Post*, May 3, 1990.
29. E. A. Wayne, "Libya Seeks Chemical Weapons in War Against Chad, US Charges," *Christian Science Monitor*, January 5, 1987; "Chad Accuses Libyans of Using Napalm, Gas," *Washington Post*, December 23, 1987.
30. See references in previous note and "The Year in Missiles," *Defense and Foreign Affairs* (March 1988): 22.
31. U.S. Arms Control and Disarmament Agency (ACDA), *World Military Expenditures and Arms Transfers: 1988* (Washington, D.C.: U.S. ACDA, 1989), p. 18.
32. Quoted in "Libyan Weaponry: A Strategic Assessment," *Focus on Libya* (Washington, D.C.: PEMCON, Ltd., September 1989): 3. During the late 1970s and early 1980s, OTRAG briefly set up test facilities in Zaire and Libya to develop an unsophisticated launch vehicle with a 200-mile range. See Aaron

Karp, "Ballistic Missiles in the Third World," *International Security* (Winter 1984/5): 166.

33. "Technology Transfer Reported," *Sunday Correspondent*, October 8, 1989, reprinted in *JPRS-TND*, October 26, 1989, p. 43; "Libya Making ICBM's with W. German Aid," *Washington Times*, October 9, 1989 (citing a story in the London *Sunday Telegraph*); "UK Paper on FRG Assistance for Missile," *Sunday Correspondent*, October 15, 1989, reprinted in *FBIS-NES*, October 16, 1989, p. 15; "Overheard: Libya," *MidEast Markets*, November 13, 1989, p. 3.
34. U.S. Arms Control and Disarmament Agency, *International Arms Transfers and World Military Expenditures: 1988* (See note 31), p. 18.
35. On April 15, 1986, in retaliation for the U.S. air strike on several targets in Libya the previous day, Colonel Khadafi launched a Scud-B attack against the U.S. Coast Guard base on the Italian island of Lampedusa. The missile (or missiles) fell into the sea several hundred yards short of the U.S. base, however. (The United States had launched the air raid against Libya as punishment for its alleged involvement in a terrorist bombing of a Berlin discotheque frequented by American servicemen.)
36. John Barham, "Brazil Ignores US Protest Over Arms for Libya," *Sunday Times* (London), January 31, 1988; "Libya Said Financing Missile Construction," Rio De Janeiro Rede Globo Television, 2200 GMT, February 3, 1988, translated in *FBIS-LAT*, February 4, 1988, p. 32; "Missiles to Libya," *Washington Times*, February 22, 1988.
37. Stephen Engelberg with Bernard E. Trainor, "Soviets Sold Libya Advanced Bomber, U.S. Official Says," *New York Times*, April 5, 1989; John England, "W. Germans Accused of Helping Libya Extend Range of Bombers," *Washington Times*, January 16, 1989; Alan Cowell, "Libyans Are Reported to Find a German Technician's Body," *New York Times*, September 7, 1989; "Further Reaction to Sale of Bombers to Libya," Jerusalem Domestic Service, 0500 GMT, April 8, 1989, reprinted in *FBIS-NES*, April 10, 1989.
38. Testimony of William Webster, Director of Central Intelligence, *Hearings on the Proliferation of Ballistic Missiles*, Committee on Governmental Affairs, U.S. Senate, 101st Cong., 1st Sess., May 18, 1989.
39. "Any Old Excuse Will Do," *The Middle East* (August 1989): 19.
40. "Gaddafi's Garbled Message," *The Middle East* (December 1989): 14; "Gadhafi: Libya Quit Funding Terror Groups," *Washington Post*, October 27, 1989.
41. Jennifer Parmelee, "Libya Seen Cutting Support to Rebel Groups," *Washington Post*, September 5, 1989; "Gadhafi: Libya Quit Funding Terror Groups," *Washington Post* (see note 40); Alan Cowell, "Few Gains for Qaddafi in Drive to End Isolation," *New York Times*, September 9, 1989.
42. "Washington Stays Cautious," *The Middle East* (January 1990): 29.
43. "Al-Qadhdhafi Calls for Arab Nuclear ICBM's," Tripoli Television Service, 2106 GMT, April 19, 1990, translated in *JPRS-TND*, April 29, 1990, p. 34.

Chapter 11: Iraq

1. Paul Power, "The Baghdad Raid: A Retrospect and Prospect," *Third World Quarterly* (July 1986): 860; Jed C. Snyder, "The Road to Osiraq: Baghdad's Quest for the Bomb," *The Middle East Journal* (Autumn 1983): 565; U.S. Congress, Office of Technology Assessment, *Technology Transfers to the Middle East* (Washington, D.C.: U.S. Government Printing Office, 1984), p. 388; Lucien Vandenbroucke, "The Israeli Strike Against Osiraq," *Air University Re-*

view (September-October 1984): 35; Leonard S. Spector, *Nuclear Proliferation Today* (New York: Vintage Books, 1984), pp.166-183.

2. David Ignatius, "Iraq's 13-Year Search for Deadly Chemicals," *Washington Post*, September 25, 1988.
3. According to a recent book on the subject, Iraq was not seeking to purchase a large heavy water power reactor, but a 40- megawatt heavy water research reactor prototype, known as Cirene. See Shyam Bhatia, *Nuclear Rivals in the Middle East* (London: Routledge, 1988), p. 85. A heavy water reactor of this size also would be an efficient plutonium producer, producing about one or two weapons' worth of the material annually.
4. Jed C. Snyder, "The Non-Proliferation Regime: Managing the Impending Crisis," *Journal of Strategic Studies* (December 1985): 11.
5. See references in note 1 and citations therein. Another controversial aspect of the Osiraq project was the fact that it was to be fueled with highly enriched uranium, material directly usable for nuclear weapons. France originally proposed selling Iraq 158 pounds (72 kilograms) of the material, enough for several weapons. When, under international pressure, France proposed a substitute fuel that could not be used for nuclear arms, Iraq objected, insisting that only the fuel France originally promised would permit a full range of research activities at Osiraq. This insistence, too, has been taken as an indication that Iraq saw the reactor and its fuel as means for advancing toward the bomb. Ultimately, France agreed to provide highly enriched material, but only in smaller quantities so that once Osiraq began operating, Iraq would not have on hand at any time enough readily available weapons-grade uranium for a single weapon, assumed to require 55 pounds (25 kilograms) of the material under IAEA guidelines. (Only 55 pounds [25 kilograms] of fresh fuel would be in Iraq at any time, half of which would be in Osiraq and thus highly radioactive, and the other half of which was to be lightly irradiated in the adjacent Isis reactor.) This left irradiation of uranium targets in Osiraq (to be subsequently reprocessed in the hot cells or an expanded reprocessing unit) as the major proliferation risk posed by the reactor.
6. Ann MacLachlan, "Iraq Nuclear Export Vetoed," *The Energy Daily*, October 2, 1980, p. 1; Leonard S. Spector, *The New Nuclear Nations* (New York: Vintage, 1985), pp. 165-166. This episode is also highlighted in Snyder, "The Road to Osiraq" (see note 1). Depleted uranium, a leftover of the uranium enrichment process, is uranium from which the uranium-235 isotope has been culled; if the remaining material—uranium-238—is irradiated in a reactor, a portion is transmuted into plutonium.
7. It should be noted that under the NPT, Iraq had the obligation to open the Italian hot cells to periodic IAEA inspections only after it had introduced nuclear materials there; until Iraq declared that nuclear material had been introduced, IAEA inspectors had the right to check that design information submitted to the agency was accurate, but no on-going authority to inspect these units or another Italian laboratory capable of fabricating uranium specimens that could have been inserted into Osiraq. Although Iraq had not yet declared the labs to the agency at the time of the Israeli raid, using the facilities to produce plutonium clandestinely would still have required circumventing the tight safeguards on Osiraq itself. See Jed C. Snyder, "The Non-Proliferation Regime" (see note 4) and references in note 8.
8. H. Gruemm, "Safeguards and Tamuz: Setting the Record Straight," *IAEA Bulletin* (December 1981): p. 1; Christopher Herzog, "Correspondence to the Editor," *International Security* (Spring 1983): 196. Iraq could have pursued the

manufacture and testing of non-nuclear atomic weapon components in secret. Regarding the subsequently abandoned Swedish nuclear weapons program of the late 1950s and the 1960s, see Spector, *The New Nuclear Nations* (see note 6), pp. 65-77 .

9. The setbacks to the Iraqi program included an April 1979 explosion that destroyed the Osiraq reactor's core structures while they were awaiting shipment to Iraq in a warehouse in Seine-sur-Mer, France; the June 1980 murder in Paris of Dr. Yahya el-Meshad, an Egyptian nuclear scientist working for the Iraqi Atomic Energy Commission; and a series of bombings and bomb threats against a number of French and Italian companies supplying nuclear equipment for the Iraqi nuclear program. Some observers attribute these acts to Israel. See Spector, *Nuclear Proliferation Today* (see note 1), pp. 175-178; reference also discusses Israeli diplomatic efforts to slow the Iraqi program.

   In a new book on Israeli nuclear affairs, British arms control specialist, Frank Barnaby has stated that the destruction of the Osiraq reactor may have been aided by the detonation of explosives that had been placed *inside* the reactor by Israeli agents:

   > It is said that the explosives had been placed at critical points in the reactor core by Israeli agents—perhaps French technicians working on the reactor and sympathetic to, or paid by, Israel. The explosives could have been actually built into the structure while the reactor was being constructed.
   >
   > The explosive charges were apparently set off by a radio signal from Israel. There are reports that such a radio signal was picked up at the crucial time by monitors in Jordan. If this is so, it would certainly account for the complete destruction of the reactor, even though the concrete containment vessel was in place.

   See Frank Barnaby, *The Invisible Bomb* (London: I.B. Tauris & Co., 1989), p. 89. No additional reports have come to light supporting this view.

10. Shlomo Nakdimon, *First Strike* (New York: Summit Books, 1987); Bob Woodward, *Veil: The Secret Wars of the CIA 1981-1987* (New York: Simon and Schuster, 1987), p. 161; Bob Woodward, "Probes of Iran Deals Extend to Roles of CIA, Director," *Washington Post*, November 25, 1987.

    In his book, Nakdimon, who served as spokesman for Israeli Prime Minister Menachem Begin, describes Begin's concerns over Iraq's nuclear program, his lack of confidence in IAEA and French controls, and the extensive, but unsuccessful diplomatic efforts he mounted to slow the Iraqi nuclear effort. Citing, among other evidence, a December 1980 memo that U.S. ambassador to Israel Samuel Lewis discussed with Begin, the book notes that senior American officials had concluded that the Osiraq reactor would allow Iraq to fabricate nuclear arms by the mid-1980s and that they saw Iraq's large-scale purchases of natural uranium as evidence of its intent to use the facility to produce plutonium. U.S. aides, however, appeared to take greater comfort from French restrictions and IAEA controls than their Israeli counterparts. See pp. 136-137, 144, 176-177. Concerning the importance of U.S. aircraft, see pp. 110, 124, 168, 181 and 189.

11. Through a series of interim measures at the 1983 and 1984 General Conferences, a final decision on sanctions against Israel had been postponed to succeeding years, enabling the United States to return as a full participant in IAEA activities in mid-1983. See Spector, *The New Nuclear Nations* (see note 6), pp. 143-146. A motion at the 1986 General Conference for sanctions against Israel—made only days before a major press story indicating that Israel had a more substantial nuclear arsenal than previously assumed—was withdrawn after it became clear it would be defeated. Anti-Israel measures were no more successful at the 1987 session.

12. *Ibid*, pp. 44-54. The discussion of this episode is based on European press accounts, a review of the prosecutor's dossier in the case, an in-person interview with one defendant, and a telephone interview with a second.
13. One defendant stated in an interview that he discussed the price of the plutonium with a "general of the Iraqi Army in charge of missiles," during a May 1982 meeting in Rome. A May 7, 1982, cable to him from a second member of the smuggling ring quotes the price for the material, apparently in preparation for the meeting. (The price was $60,572,000 or $1.78 million per kilogram.)
14. Iraq first used these arms against Iranian forces in 1983 and would continue to do so with increasing abandon through the remainder of the conflict.
15. Aharon Levran and Zeev Eytan, *The Middle East Military Balance 1987-1988* (Boulder, Colo.: Westview Press, 1988), p. 232.
16. "IRNA Commentary Urges CW War Prevention," IRNA, 1102 GMT, September 2, 1988, reprinted in *Foreign Broadcast Information Service (FBIS)-Near East and South Asia (NES)*, September 7, 1988, p. 53.

    Iraq's use of chemical weapons was confirmed on a number of occasions by United Nations fact-finding teams. See "Iraq Chemical War Cited in U.N.," *New York Times*, May 14, 1987; Elaine Sciolino, "Iraq Cited on Chemical Weapons," *New York Times*, March 15, 1986; Michael Berlin, "U.N. Team Says Chemical Agents Used in Gulf War," *Washington Post*, March 27, 1984.
17. "Iran Says Iraq Used Nerve Gas," *International Herald Tribune*, March 21, 1988; Patrick Tyler, "Poison Gas Attack Kills Hundreds," *Washington Post*, March 24, 1988.
18. Jonathan C. Randal, "Kurds Tell of Mass Gassings," *Washington Post*, September 7, 1988; Robert Pear, "U.S. Says It Monitored Iraqi Messages on Gas," *New York Times*, September 15, 1988. Iraq denied using chemical weapons during these operations. See Patrick E. Tyler, "Iraqi Official Says Kurds Fleeing Army Operation," *Washington Post*, September 4, 1988.
19. "Testimony of William Webster, Director of Central Intelligence," Committee on Governmental Affairs, U.S. Senate, May 18, 1989; Patrick E. Tyler, "Both Iraq and Iran Gassed Kurds, U.S. Analysts Say," *Washington Post*, May 3, 1990. Iranian allegations concerning Iraq's use of chemical weapons are summarized in "IRNA Commentary Urges CW War Prevention," IRNA, 1102 GMT, September 2, 1988, reprinted in *FBIS-NES*, September 7, 1988, p. 53. Iranian use of poison gas was first alleged in 1987. See Berlin, "U.N. Team" (see note 16).
20. Iraq ratified the Protocol with two reservations: (1) that it would be bound by the Protocol only vis-à-vis other signatories and (2) that it would cease to be bound with respect to any enemy state whose armed forces do not observe the Protocol.

    Iraq was the first to use chemical weapons in the Gulf War. Since, at the time, Iran was a party to the Protocol and was observing its terms, Iraq's action constituted a violation of the convention. Similarly, since Iran did not reserve the right to use these weapons under any circumstances, it, too, violated the Protocol when it later responded in kind to Iraq's use of chemical arms.
21. See David Ottaway, "U.S. Decries Iraqi Use of Chemical Weapons," *Washington Post*, March 24, 1988; Julie Johnson, "U.S. Asserts Iraq Used Poison Gas Against the Kurds," *New York Times*, September 9, 1988; Julie Johnson, "U.S. Adamant in Charge Against Iraq," *New York Times*, September 10, 1988. Concerning the various UN investigations, see note 16.
22. "Iraq/Chemical Weapons Use," *Brazilian News Agency*, May 4, 1988, quoted in *Current News* (U.S. Deparatment of Defense daily news summary), May 6,

1988, p. 1; "Iraq Acknowledges Its Use of Gas But Says Iran Introduced It in War," *New York Times*, July 2, 1988.

In September, Iraqi Defense Minister Adnan Khair Allah strongly suggested that Iraq reserved the right to use such arms. Clyde Haberman, "Iraq's Right to Chemical Arms Upheld by Official," *New York Times*, September 16, 1988.

23. In the face of the Reagan Administration's inaction, Congress moved in the summer of 1988 to impose sanctions on Iraq, but this initiative did not result in the enactment of punitive legislation. See Helen Dewar and Don Oberdorfer, "Senate Votes Sanctions Against Iraq," *Washington Post*, September 10, 1988; David Hess, "House Votes Sanctions on Iraq for Chemical Weapon Use," *Philadelphia Inquirer*, September 28, 1988; John Felton, "Iraq Sanctions Yield to Special Interests in Senate," *Congressional Quarterly*, October 19, 1988, p. 3141.

24. For an excellent review of the different phases in the Iran-Iraq ballistic missile war, see W. Seth Carus, "Missiles in the Middle East: A New Threat to Stability," *Policy Focus* (Washington, D.C.: Washington Institute for Near East Policy, Research Memorandum, June 1988): 5-6.

Carus notes that from 1980 to 1985, only Iraq possessed surface-to-surface missiles and used some 150 to 200 against Iranian cities during this period. In May 1985, he states, Iran launched 14 Scud-B missiles obtained from Libya against Iraq and Iraq fired 80 missiles in retaliation. Between June 1986 and February 1988, Iraq launched a limited number of missiles against Iran. Finally, the 52-day "War of the Cities" began on February 29, 1988, with Iraq firing 190 Al-Husayn missiles at Iran and the latter firing 80 Scud-Bs and about 250 domestically produced Oghab missiles back at Iraq.

For a comparable analysis using slightly different figures, see, Aharon Levran and Zeev Eytan, *The Middle East Military Balance 1987-1988*, pp. 213 ff. (see note 15). See also, Steven Zaloga, "Ballistic Missiles in the Third World: Scud and Beyond," *International Defense Review* (November 1988):1423.

25. Interview with Iraqi defense advisor, October 1989.

26. The air strikes took place on March 24, 1984; March 4, 1985; February 12, 1985; November 17, 1987 (two strikes); November 19, 1987; and July 19, 1988.

27. James M. Markham, "Arabs Link Curbs on Gas and A-Arms," *New York Times*, January 9, 1989. The article quotes Iraqi Foreign Minister Tariq Aziz as stating that there "was no doubt whatsoever" that Israel had introduced nuclear weapons into the Middle East.

28. "Iraq/Al-Faw Missile," *Current News*, December 6, 1988, p. 12; "Iraq Now Maintains Five Chemical Weapons Facilities," *Defense and Foreign Affairs Weekly*, August 7-13, 1989, p. 2; "Iraq Completes AWACS Project," *Defense and Foreign Affairs Weekly*, July 24-30, 1989; Stephen Engelberg, "Iraq Said To Study Biological Arms," *New York Times*, January 18, 1989; W. Seth Carus, "Chemical Weapons in the Middle East," *Policy Focus* (Washington, D.C.: Washington Institute for Near East Policy, Research Memorandum, December 1988): 15. The last reference states that Iraq is "believed to have an operational biological warfare capability. It probably has a program to make toxins and should be assumed [to be] researching advanced biological agents."

29. "Testimony of Rear Admiral Thomas A. Brooks, Director of Naval Intelligence," Subcommittee on Seapower, Strategic, and Critical Materials, Committee on Armed Services, U.S. House of Representatives, 101st Cong., 1st Sess., February 22, 1989 (mimeo).

30. Glenn Frankel, "Iraq Said Developing A-Weapons," *Washington Post*, March 31, 1989; "Substance to Nuclear Allegations Against Iraq," *MidEast Markets*, December 11, 1989, p. 13; "Barak on War with Syria, Iraqi Nuclear Threat,"

Jerusalem Domestic Service, translated in *FBIS-NES*, April 6, 1989, p. 29.
31. Glenn Frankel, "Iraq Said Developing A-Weapons" (see note 30).
32. Interviews with the principal author of this volume, spring 1989; "Testimony of Reginald Bartholomew, Under Secretary of State for Security Assistance, Science, and Technology," Committee on Governmental Affairs, U.S. Senate, 101st Cong., 1st Sess., May 18, 1989.
33. "Reports on Nuclear Chemical Warheads Denied," Baghdad INA, July 3, 1989, 0745 GMT, translated in *FBIS-NES*, July 3, 1989; "Iraq Denounces A-Bomb Report," *Washington Post*, April 4, 1989; "Iraq Denies Making Nuclear Warheads," *Washington Times*, July 4, 1989.
34. Steve Mills, "U.S. Seizes Local Firm's Pumps Bound for Iraq," *Rochester Democrat and Chronicle*, March 28, 1989; David B. Ottaway, "U.S. Bars Export of Pumps to Iraq," *Washington Post*, May 5, 1989.
35. Interview with Western diplomat, mid-1989; Patrick E. Tyler, "Iraq Nuclear Program Stirs Debate," *Washington Post*, June 4, 1989.
36. "Nuclear Equipment Sold to Iraq by FRG Company," *Der Spiegel*, December 18, 1989. URENCO produces non-weapons grade uranium for use as commercial nuclear power reactor fuel. Centrifuges can, however, be linked in longer than usual arrays (known as "cascades") to produce higher enrichment levels.

    Pakistan is thought to have based its centrifuge enrichment plant at Kahuta on a Dutch URENCO centrifuge design, which Pakistan is believed to have obtained through industrial espionage in the mid-1970s. Pakistan also gained access to blueprints for portions of a second URENCO plant in the early 1980s from a West German URENCO subcontractor. See Mark Hibbs, "German Firm Implicated in Trying To Aid Pakistan Enrichment Effort," *Nucleonics Week*, May 4, 1987, p. 1.
37. "Iraq and the Bomb," *MidEast Markets*, December 11, 1989, p. 130.
38. The report also drew attention to Iraq's hydrogen fluoride plant at a phosphate production facility in Al-Qaim, and noted that hydrogen fluoride is needed to produce uranium hexafluoride, the gaseous form of uranium that is processed in an enrichment plant.
39. According to the indictment subsequently filed in the case in San Diego, the items were "co-axial, high voltage, low inductance capacitors of the same size, weight, energy storage performance specifications as relating to temperature range, inductance, voltage capacitance, resistance to humidity, vibration, and shock, and altitude performance requirements as those used in the detonation of nuclear warheads." Indictment, *U.S. v. Daghir, et al.*, U.S. District Court for the Southern District of California (San Diego), Criminal Case No. 90-0211-JLI, pp. 2-3.
40. *Ibid.*
41. "Face the Nation," CBS Television Network, April 1, 1990 (statement of Nazir Hamdoon, Deputy Foreign Minister of Iraq).
42. Edward Chen, "U.S. Accuses Iraq in A-Trigger Plot," *Los Angeles Times*, March 30, 1990.
43. W. Seth Carus, "Chemical Weapons in the Middle East," *Policy Focus* (Washington, D.C.: Washington Institute for Near East Policy, Research Memorandum, December 1988): 4-5; "Iraq Now Maintains Five Chemical Weapons Facilities," *Defense and Foreign Affairs Weekly*, August 7-13, 1989, p. 2; "Iraq Sows the Seeds of Catastrophe," *South* (May 1987): 63; James M. Dorsey, "U.S. Fears Iraq's Chemical Weaponry," *Washington Times*, July 24, 1989.
44. "Firms Involved in Iraq Missile Production," *Der Spiegel*, March 27, 1989, translated in *Joint Publications Research Service (JPRS)-Nuclear Proliferation*

and Development (TND), April 10, 1989, p. 33; "*Profil* Views Involvement in Iraq Arms Plant," *Profil*, April 24, 1989, translated in *JPRS-TND*, May 5, 1989, p. 34; "Austrian Firms Helped Build Iraqi Missile Complex," *Defense and Foreign Affairs Weekly*, May 8-14, 1989, p. 6; "*Spiegel* Reports Missile Aid to Romania," *Der Spiegel*, May 8, 1989, translated in *JPRS-TND*, May 23, 1989, p. 42; "Big Explosion in Iraq," *MidEast Markets*, September 18, 1989, p. 11; "Iraq Bid to Buy Condor Kit Stopped," *MidEast Markets*, October 16, 1989.

The Condor II project was begun in the early 1980s and built upon the short-range Condor I Argentine rocket. An Egyptian effort to smuggle key components out of the United States was exposed in 1988, and under U.S. pressure, Egypt apparently abandoned the project the following year. Related Argentine activities also ceased by early 1990, presumably at U.S. insistence. At the time, Argentina was seeking U.S. aid in addressing a continuing financial crisis. By this point, it appeared that Iraq, which had become the most technically advanced of the three partners in the missile field, was pursuing the Condor II project alone. See "Secret Egypt-Iraq Accord Collapses," *MidEast Markets*, June 12, 1989, p. 10; "Big Explosion in Iraq . . ." *MidEast Markets* (see above); David B. Ottaway, "Egypt Drops Out of Missile Project," *Washington Post*, September 20, 1989; "Defense Minister Says Condor II Project Halted," *Clarin*, April 22,1990, translated in *FBIS-Latin America (LAT)*, April 27, 1989, p. 12.

45. Michael Gordon, "U.S. Confirms Iraq Has Launched Rocket that Can Carry Satellites," *New York Times*, December 9, 1989. The flight test and new missiles are discussed further below.
46. "The Secrets of Samarra," BBC *Panorama*, 2130 GMT, October 27, 1986; John J. Fialka, "Western Industry Sells Third World the Means to Produce Poison Gas," *Wall Street Journal*, September 16, 1988; David Ignatius, "Iraq's 13-Year Search for Deadly Chemicals," (see note 2); Gary Thatcher and Tim Aeppel, "The Trail to Samarra: How Iraq Got the Materials to Make Chemical Weapons," *Christian Science Monitor*, December 13, 1988.

    Regarding the help of Western firms in advancing Iraq's missile capabilities, see references in note 44. In addition, some sources have alleged that Iraq received assistance from North Korean and/or East German technicians in modifying Soviet-origin Scud-B missiles.
47. "Testimony of William Webster, Director of Central Intelligence" (see note 19). Webster indicated that a number of countries are advancing their nuclear efforts by using the clandestine purchasing networks that were originally set up to assist chemical weapons and ballistic missile programs.
48. " 'Careful Monitoring' of Iraqi Reactor Promised," *Jerusalem Post*, March 28, 1989, reprinted in *JPRS-TND*, April 10, 1989, p. 26; Glenn Frankel, "Iraq Said Developing A-Weapons" (see note 30); "France Denies Negotiating on Iraqi Reactor," Jerusalem Domestic Service, 0900 GMT, May 7, 1989, translated in *FBIS-NES*, May 8, 1989; "Threat to Iraqi Reactor Alleged, *Al-Ra'y Al-'Amm*, March 30, 1989, translated in *JPRS-TND*, April 10, 1989; "Aziz on Tammuz Nuclear Reactor, Syrian Borders," Paris Radio Monte Carlo, 1200 GMT, July 10, 1989, translated in *FBIS-NES*, July 11, 1989, p. 19.

    Some nuclear contacts with France continued during the Gulf War, however, with a French nuclear team visiting Baghdad in early 1987.
49. Atomic Energy Commission of Iraq, *Annual Report for 1988* (Baghdad: Iraq Atomic Energy Commission, 1989), p. 94. Research was also conducted on purifying uranium.
50. Glenn Frankel, "Iraq Said Developing A-Weapons" (see note 30).
51. Interviews with IAEA officials, March 1989; *PPNN Newsbrief*, July 1989, p. 9

(published by the Programme for Promoting Nuclear Non-Proliferation, New York, NY).

52. See the introduction to Part II, on Asia, and Chapter 7, on Pakistan, elsewhere in this book.
53. "Nuclear Program, Enriched Uranium Plans Unveiled," *Jornal do Brasil*, May 22, 1988, translated in *JPRS-TND*, June 16, 1988, p. 1.
54. "Testimony of William Webster, Director of Central Intelligence" (see note 19). At the time of the accord, Brazil intended to become a commercial supplier of enriched uranium, using facilities to be built under its mammoth 1975 nuclear trade agreement with West Germany. These facilities were never constructed, however, but Brazil developed a separate enrichment capability on its own. Although this is currently far too small to permit commercial-scale exports of low-enriched material for nuclear power reactors, exports of smaller but militarily significant quantities of highly enriched material may soon be possible. Presumably Brazil will want to reserve such material for its own uses, including its program for the development of nuclear-propelled submarines, but it might be prepared to share enrichment technology with others.
55. "Joint Space Program Negotiated with Iraq," *Fôlha de São Paulo*, March 19, 1989, translated in *FBIS-LAT*, May 10, 1989 p. 43; "Development of Missiles with Brazil Planned," *O Globo*, March 19, 1989, translated in *JPRS-TND*, April 10, 1989, p. 25; "Iraq to Purchase INPE Satellite Technology," *Fôlha de São Paulo*, March 5, 1989, translated in *FBIS-LAT*, March 7, 1989, p. 40; "Orbita to Export 'Smart' Missiles to Iraq," *O Globo*, December 20, 1988, translated in *FBIS-LAT*, January 24, 1989, p. 39; "Assessing Iraq's Missile Technology," *Jane's Defence Weekly*, December 23, 1989, p. 1371.
56. Testimony of Rear Admiral Thomas A. Brooks (see note 29); "Iraq/Chemical Weapons," *Current News*, May 6, 1988, p. 13 (quoting the Brazilian News Agency); Thalif Deen, "Rebuilding Iraq's Armed Forces," *Jane's Defence Weekly*, April 22, 1989.
57. Richard Kessler, "Argentine Officials Deny Rumors of Nuclear Trade with Israel," *Nucleonics Week*, May 29, 1986, p. 6.
58. Mark Hibbs, "Argentine Officials Optimistic on Futures of Atucha-2, Exports," *Nucleonics Week*, October 12, 1989, p. 11.
59. "Industry Minister Discusses Military Production," *Al-Akhbar*, May 7, 1989, translated in *FBIS-NES*, May 9, 1989, p. 18; Giovanni de Briganti, "Baghdad Exhibit Shows Increased Egyptian-Iraqi Coordination," *Defense News*, May 8, 1989.
60. See note 9. Egypt denied reports in 1989 that it was assisting Iraq to rebuild the Osiraq reactor. "Rebuilding of Iraqi Reactor Denied," MENA, 0745 GMT, March 20, 1989, translated in *JPRS-TND*, March 28, 1989, p. 17.
61. "Arab Atomic Energy Commission Formed," *Nuclear Engineering International* (December 1988): 5.
62. North Korea is another potential source of militarily relevant nuclear know-how and equipment. It may have supplied arms to Baghdad during the Gulf War and is said by some reports to have assisted Iraq in extending the range of its Soviet Scud missiles. See Steven Zaloga, "Ballistic Missiles in the Third World: Scud and Beyond" (see note 24).
63. As a party to the Non-Proliferation Treaty, the Soviet Union is prohibited from making nuclear exports that are not subject to IAEA inspection in the recipient state. This would impose significant restrictions on Iraq's use of any nuclear equipment that it received from the Soviets. Moreover, the latter have never exported uranium enrichment or plutonium extraction technology.

During the early 1980s, however, India was able to obtain a quantity of Soviet heavy water without IAEA safeguards by making the purchase through a West German middleman, Alfred Hempel. Hempel had accumulated the Soviet material in small quantities—for which IAEA inspections are not required—using falsified end-user documents. The Soviet Union claims to have been taken in by this ruse. However, the subterfuge may have been used against it on a number of occasions, at least some of which took place after Moscow had been advised of Hempel's chicanery. Given that the Soviets had a long history of dealing with Hempel, who was their agent for sales of uranium in Western Europe, and that the beneficiary of the heavy water sales was the Soviet Union's most important regional client—a client to whom Moscow made the unprecedented transfer of a nuclear powered submarine in 1988—it is conceivable that Moscow knowingly acquiesced in these transfers. India is presumed to have placed the Soviet-origin heavy water in non-IAEA-inspected nuclear reactors (the Dhruva reactor at the Bhabha Atomic Research Center near Bombay and the Madras Atomic Power Plants) which can now produce plutonium that is free from any non-proliferation controls and thus available for nuclear weapons.

64. "Broadcasts Rocket Launch," INA, 1745 GMT, December 7, 1989, translated in *FBIS-NES*, December 8, 1989, p. 24.
65. Interestingly, the Osiraq reactor was also formally known as Tammuz I.
66. "Iraqi Missile May have Intercontinental Capability," *Defense Daily*, December 12, 1989; Michael Gordon, "U.S. Confirms Iraq Has Launched Rocket That Can Carry Satellites," *New York Times*, December 8, 1989.
67. The first-stage engine was reported to have a thrust to 70 tons, five times that of a single Scud-B. See "Satellite Rocket System Tested," Baghdad Voice of the Masses, 1136 GMT, December 7, 1989, translated in *FBIS-NES*, December 8, 1989, p. 23.
68. "Iraq Reports Second Long-Range Missile Test," *Defense Daily*, December 15, 1989, p. 408.
69. "Iraqi Rocket Called 'Serious Matter,' " *Ma'ariv*, December 8, 1989, translated in *FBIS-NES*, December 12, 1989, p. 27; Hugh Carnegy, "Rabin Tells of Concern Over Iraqi Missiles," *Financial Times*, December 21, 1989.
70. "Iraq Asserts Arabs Must Acquire Atom Arms as a Balance to Israel," *New York Times*, June 24, 1989.
71. Glenn Frankel, "Iraq Said Developing A-Weapons" (see note 30); George C. Wilson, "Iraq Downs a Visiting Egyptian Jet," *Washington Post*, April 28, 1989; Aharon Levran and Zeev Eytan, *The Middle East Military Balance 1987-1988* (see note 15).
72. "Iraq Threatens to Use Chemical Weapons Against Israeli Attack," *Financial Times*, April 3, 1990. See also, "Paper Warns Against Zionist 'Flagrant Threats,' " Baghdad INA, 0545 GMT, September 18, 1988, translated in *FBIS-NES*, September 19, 1988, p. 18.

Chapter 12: Iran

1. Daniel Poneman, *Nuclear Power in the Developing World* (London: George Allen and Unwin, 1982), Chapter 5.
2. Zalmay Khalilzad, *Iran: The Nuclear Option* (Los Angeles: Pan Heuristics, 1977), p.8, quoted in David Blair, *Nuclear Proliferation Prospects for the Middle East and South Asia* (Marina Del Rey: Pan Heuristics, 1981), P. A-17.
3. Daniel Poneman, *Nuclear Power in the Developing World* (see note 1), pp. 89, 92; U.S. Energy Research and Development Administration, *U.S. Nuclear*

Power Export Activities: Final Environmental Impact Statement, ERDA 1542 (April 1976), vol.1, pp. 3-52.

The Shah also entered negotiations to buy a portion of a second enrichment plant planned by another French-led consortium, COREDIF, that was subsequently cancelled.

4. Daniel Poneman, *Nuclear Power in the Developing World* (see note 1), p. 96.
5. Thomas Stauffer, "Ayatollah Rediscovers Nuclear Power, with Kraftwerk Union's Aid," *Energy Daily*, October 2, 1984; "German Team Inspects Stalled Bushehr Plants," *Nuclear News* (June 1984): 95. The West German government has refused to issue export licenses for some 7,000 tons of major nuclear components for the plant. See Richard Kessler, "CNEA Estimates Bushehr-1 Work Worth $100-million to Argentina," *Nucleonics Week*, August 20, 1987, p. 13.
6. Tim Coone and Peter Bruce, "Three-Nation Bid for Iran N-Deal," *Financial Times*, March 9, 1987; Kessler, "CNEA Estimates Bushehr-1 Work Worth $100-Million . . ." (see note 5).

   In November 1987, a technical team from West Germany's reactor inspectorate (Technischer Ueberwachungsverein) examined the facility to determine how much of the equipment that had been delivered in the late-1970s was still serviceable. See Mark Hibbs, "Iraqi Attack on Bushehr Kills West German Nuclear Official," *Nucleonics Week*, November 19, 1987, p.1; Mark Hibbs, "Bushehr Construction Now Remote after Three Air Strikes," *Nucleonics Week*, November 26, 1987, p. 5.
7. Anne MacLachlan, "France, Iran to Keep Talking About Dispute Over EURODIF Loan," *Nuclear Fuel*, January 13, 1986, p. 9.
8. "The Atomic Energy Organization of Iran," Airgram P-750-060-1115, from U.S. Embassy, Tehran, to Department of State, April 15, 1976, released to the Carnegie Endowment for International Peace under the Freedom of Information Act.
9. Martin Bailey and Shyam Bhatia, "Namibia Uranium May Fuel Iran's A-Bomb," *Observer*, May 17, 1987.
10. Richard Kessler, "Panel to Guide Nuclear Technology Sales from Argentina to Algeria," *Nucleonics Week*, May 7, 1987, p. 6.
11. Although Argentina announced in April 1985 that all its exports would be subject to IAEA inspections, it might have deviated from this rule. For some years, Argentina has operated a uranium dioxide production facility that is not under IAEA inspection and thus could have exported the material without the agency's knowledge, while Algeria, since it is not a party to the NPT, would not have been obligated to place the material under IAEA inspection upon its arrival. With the material outside the IAEA system, it could have been secretly transferred to Iran at some later time.
12. "Uranium Processing for Iran," *Mining Journal*, December 25, 1987.
13. "Atomic Energy Official on Exporting Uranium," *Resalat*, September 10, 1989, translated in *Joint Publications Research Service (JPRS)-Nuclear Development and Proliferation (TND)*, October 6, 1989, p. 19.
14. See Akbar Etemad, "Iran," in Harald Müller, ed., *European Non-Proliferation Policy* (Oxford: Oxford University Press, 1987). Etemad was the chairman of the Atomic Energy Organization of Iran during its period of greatest activity under the Shah.
15. Based on interviews with then-current and former U.S. officials, spring 1986. For a detailed review of Iran's nuclear activites under the Shah, see Leonard S. Spector, *Going Nuclear* (Cambridge, Mass.: Ballinger Publishing Co., 1987), pp. 45-47.

16. Etemad, "Iran" (see note 14), p. 9.
17. Shyam Bhatia, *Nuclear Rivals in the Middle East* (London: Routledge, 1988), p. 85.
18. In 1979, Eerkens told the *Los Angeles Times* that he modified the units for export by filling them with carbon dioxide so that they would not emit light at the wavelength thought optimal for uranium enrichment. He noted, however, that by refilling them with chlorine mono-fluoride his Iranian customers could obtain the enrichment wavelength, as specified in his contract. See Robert Gillette, "Iran's Deal with L.A. Firm Widens Nuclear Capability," *Los Angeles Times*, August 22, 1979.

    Although Eerkens claimed that Iran was interested only in producing low-enriched uranium for fueling reactors, rather than in weapons-grade highly enriched uranium, the evidence suggests otherwise. His less-than-candid dealings with U.S. export officials, the use of middlemen to mask Iran's financing of his work, and the scientist's repeated visits to Iran to work on enrichment research despite U.S. government objections, all strongly imply that more was involved than a straightforward arrangement to ensure Iran's future supply of electricity—especially since Iran had already contracted for long-term supplies of low-enriched uranium fuel with several major Western suppliers.

    For a detailed discussion of this episode, see Spector, *Going Nuclear* (see note 15), pp. 47-53.
19. Interview with Dr. Jeffery Eerkens. See also, David Segal, "Iran Speeds Up Nuclear Bomb Development," *Journal of Defense and Diplomacy*, vol. 6, no. 6, 1988, p. 52.
20. Etemad, "Iran" (see note 14), p. 9.
21. "Nuclear Cooperation with Iran Reported," *Al-Watan Al-Arabi*, November 22, 1988, translated in *JPRS-TND*, December 23, 1989 (report states that continuing Iranian laser research was noted at an AEOI conference, September 12-16, 1987, in Tehran); interviews with Dr. Jeffery Eerkens and former Iranian nuclear officials.
22. Etemad, "Iran," p. 9 (see note 14); "Iranian Reactor to Go Critical," *Nuclear Engineering International* (December 1984): 13. (The "reactor" referred to is a subcritical training unit, *i.e.*, one that will not sustain a nuclear chain reaction.)
23. "Iran to Receive Nuclear Technology, Know-How," Noticias Argentinas, 1800 GMT, May 18, 1987, translated in *Foreign Broadcast Information Service (FBIS)-Latin America*, May 19, 1987; "Argentina Confirms Deal for Work on Bushehr," *Nuclear News* (July 1987): 54. See Richard Kessler, "Argentina To Enforce Curbs on Nuclear Trade with Iran," *Nucleonics Week*, March 14, 1987, p.12.

    The Argentinian fuel will be covered by IAEA safeguards, because Iran is an NPT party and because Argentina, though not a party, has announced that it will require IAEA safeguards on all of its nuclear exports. (See note 11.) The IAEA Board of Governors approved an agreement providing inspections of the material in September 1988. See *IAEA Newsbriefs*, October 5, 1988.

    Given these safeguards, the low enrichment level of the fuel, and the small size of the Tehran research reactor, the transaction will not contribute directly to Iran's ability to manufacture nuclear weapons. By permitting Iran to run the unit—which has been operating only intermittently for lack of fresh fuel—Argentina will, however, be assisting Iran to train nuclear technicians, thereby contributing in a general fashion to Iran's overall nuclear capabilities. The deal undercuts U.S. efforts to establish a total embargo on nuclear sales to Tehran as a means of constraining the long-term nuclear weapons potential of the Khomeini regime.

24. "Argentina's Invap to Supply Iran Fuel for Research Reactor," *Nucleonics Week*, May 14, 1989, p. 2.
25. Shyam Bhatia, "Iran's Secret Nuclear Arms Move," *Observer*, March 6, 1988. Iran's only known source of plutonium, however, is the spent fuel from the nuclear reactor at the Tehran Research Center, which contains a very small amount of the material. The spent fuel is subject to IAEA inspection and any attempt to extract its plutonium clandestinely would be a violation of the IAEA system and, in theory, would be easily detectable.
26. David Segal, "Iran Speeds Up Nuclear Bomb Development" (see note 19). Fesharaki's group had concluded that the Shah intended to use Iran's nuclear power program as a means to obtain plutonium for a covert nuclear weapons program, but the specialists considered this approach to be extremely wasteful of resources and time-consuming. They therefore recommended a more direct approach.
27. *Ibid.*; David Segal, "Atomic Ayatollahs," *Washington Post*, April 12, 1987.
   A second energy aide who was part of the Fesharaki group also recounted that he was approached by a figure in the revolutionary government (possibly Beheshti), who made detailed inquiries as to the steps that would be necessary for the country to develop nuclear arms given its existing nuclear infrastructure. See Spector, *Going Nuclear* (see note 15), p. 56.
28. Defense Intelligence Agency document IRI-517-0007-80, from 7602 AINTELG Fort Belvoir, VA, to INER, DIA, Washington, D.C., March 13, 1980, obtained under the Freedom of Information Act. The document, which is heavily redacted, carries the note, "This report contains unprocessed information. Plans and/ or policies should not be evolved or modified solely on the basis of this report."
29. Segal, "Atomic Ayatollahs" (see note 27). Segal took the quoted material from *Nameh Mardom*, a Tudeh party (opposition) newspaper published in Stockholm, Sweden, which claims to have obtained the text from a tape recording that was made by one of those present at Khamenei's address and subsequently smuggled out of the country.
30. The air strikes took place on March 24, 1984; March 4, 1985; February 12, 1985; November 17, 1987 (two strikes); November 19, 1987; and July 19, 1988. See also, "Atomic Energy Official Addresses IAEA Meeting," IRNA, 1830 GMT, September 21, 1988, translated in *FBIS-Near East and South Asia (NES)*, September 23, 1988, p. 56 (thirteen dead). Eleven of the thirteen deaths (including that of a West German technician) occurred as a result of the two attacks on November 17, 1987. See Mark Hibbs, "Iraqi Attack on Bushehr Kills West German Nuclear Official," *Nucleonics Week*, November 19, 1987, p.1; Mark Hibbs, "Bushehr Construction Now Remote after Three Air Strikes, "*Nucleonics Week*, November 26, 1987, p.5. The final Iraqi attack, on July 19, 1988, came one day after Iran had accepted a United Nations-sponsored ceasefire in the Gulf War.
31. Hibbs, "Bushehr Construction Now Remote" (see note 30). Ironically, it was Iraq's delegate who denounced attacks against nuclear installations before a U.N.group in May 1987, declaring that military attacks with conventional weapons against such facilities are "tantamount to the use of radiological weapons because of the release of radioactive material that could result therefrom." See "U.N. Envoy Urges Mid-East Nuclear-Free Zone," Baghdad INA, 1252 GMT, June 4, 1987.
32. Gamini Seneviratne, "IAEA Mission Will Visit Bushehr," *Nucleonics Week*, March 3, 1988, p. 7.
33. "IAEA: Iran Protests Bushehr Attack," *Nucleonics Week*, July 28, 1988 p. 15.
34. Confidential interview, October 1989. Although it would have been at least five

years before the first of the Bushehr reactors could conceivably have started operating—assuming West Germany had been prepared to assist in completing the units—from that time forward, the reactor would have produced significant quantities of plutonium in its spent fuel. In theory, Iran might then have attempted to extract this plutonium, if it had also built a reprocessing capability. All of these activities would have taken place under IAEA safeguards, because of Iran's status as an NPT party, but Tehran could still have accumulated plutonium under IAEA inspection. It could then have quickly converted this material into weapons, if it had chosen to abrogate its NPT commitments or to withdraw from the treaty, as the pact itself permits on 90 days' notice, if a party determines that its "supreme interests" are jeopardized.

In the late 1970s, it appears that Iraq, itself a party to the NPT, was pursuing this very stratagem, using its large Osiraq research reactor as the source of plutonium. (See Chapter 11 on Iraq.) It was this potential that led Israel to attack the Osiraq plant in June 1981. Now, ironically, Iraq may have made the very same calculation that Israel made and may have taken the very same precaution to prevent an adversary from realizing its nuclear ambitions.

35. Patrick E. Tyler, "Both Iraq and Iran Gassed Kurds, U.S. Analysts Say," *Washington Post*, May 3, 1990.
36. Iraq ratified the Protocol with two reservations: (1) that it would be bound by the Protocol only vis-à-vis other signatories and (2) that it would cease to be bound with respect to any enemy state whose armed forces did not observe the Protocol. However, the conditions for Iraqi adherence had not been contravened by Iran when Iraq first used chemical weapons against the latter country. Iraq's action thus constituted a violation of the convention. Similarly, since Iran did not reserve the right to use these weapons under any circumstances, it, too, violated the Protocol when it responded to Iraqi chemical attacks in kind.
37. For an excellent review of the different phases in the Iran-Iraq ballistic missile war, see W. Seth Carus, "Missiles in the Middle East: A New Threat to Stability," *Policy Focus* (Washington, D.C.: Washington Institute for Near East Policy, Research Memorandum, June 1988): 5-6.

    Carus notes that from 1980 to 1985, only Iraq possessed surface-to-surface missiles and used some 150 to 200 of them against Iranian cities during this period. In May 1985, Iran launched 14 Scud-B missiles obtained from Libya against Iraq, and Iraq fired 80 missiles in retaliation. Between June 1986 and February 1988, Iraq launched a limited number of missiles against Iran. Finally, the 52-day "War of the Cities" began on February 29, 1988, with Iraq firing 190 Al-Husayn missiles at Iran and the latter firing 80 Scud-Bs and about 250 domestically produced Oghab missiles back at Iraq.

    For a comparable analysis with slightly different numbers, see, Aharon Levran and Zeev Eytan, *The Middle East Military Balance 1987-1988* (Boulder, Colo.: Westview Press, 1988), pp. 213 ff.
38. "UK Government Blocks Hawk Sale to Iraq . . . During Uneasy Truce on Cease-Fire Line," *Jane's Defence Weekly*, August 5, 1989, p. 196.
39. Michael Gordon, with Stephen Engelberg, "Iran is Expanding Chemical Stocks Used in Poison Gas," *New York Times*, January 29, 1989; James M. Dorsey, "U.S. Fears Iraq's Chemical Weaponry," *Washington Times*, July 24, 1989; "Defense Industries Manufacture New Missile," IRNA, April 17, 1989, 1535 GMT, reprinted in *JPRS-TND*, May 5, 1989, p. 20; Simon Henderson, "West Blocks Advance of Condor II Missile," *Financial Times*, August 31, 1989.
40. "Defense Industries Manufacture New Missile," IRNA (see note 39). Previously, the only missile in the Iranian armory with sufficient range to reach Baghdad

comfortably was the 180-mile Scud-B. Originally, Iran is thought to have obtained a small number of these from Libya and then to have begun purchasing larger quantities from North Korea. The new Iranian missile would increase the number of missiles in the Iranian arsenal that are able to strike Baghdad, and would insulate Iran's supply of such arms against the risk of a possible cut-off in a future conflict.

41. Stephen Engelberg, "Iraq Said to Study Biological Arms," *New York Times*, January 18, 1989; Patrick E. Tyler, "Iraq Nuclear Program Stirs Debate," *Washington Post*, June 4, 1989.
42. W. Seth Carus, "Chemical Weapons in the Middle East," *Policy Focus* (Washington, D.C.: Washington Institute for Near East Policy, Research Memorandum, December 1988): 15. This analysis states that Iran "appears to be developing biological agents in response to Iraq's program."
43. "Hashemi-Rafsanjani Speaks on the Future of the IRGC Iranian Revolutionary Guards Corps," Tehran Domestic Service, 0935 GMT, October 6, 1988, translated in *FBIS-NES*, October 7, 1988, p. 52 (emphasis added).
44. "Testimony of Rear Admiral Thomas A. Brooks, Director of Naval Intelligence," Subcommittee on Seapower, Strategic, and Critical Materials, Committee on Armed Services, U.S. House of Representatives, February 22, 1989 (mimeo).
45. One cryptic April 1989 report in a Kuwaiti newspaper, however, stated that Iran had begun to purchase materials needed for the production of atomic weapons from Japan. The report offered no details and no confirmatory information has come to light. See "Nuclear Weapons Program Reportedly Started," MENA (Cairo), 1340 GMT, April 21, 1989, reprinted in *JPRS-TND*, May 5, 1989, p. 20 (the MENA broadcast quoted the Kuwaiti paper, *Al-Qa-Bas*). Concerning allegations of an Iranian biological weapons program, see note 42.
46. See Chapter 13.
47. "Blix Visits Bushehr; Work Restart Possible?" *Nuclear News* (August 1989):15.
48. Farhang Jahanpour, "Iran After Khomeini," *The World Today*, August 1989, p. 150.
49. Farzad Bazoft, "Iran Signs Secret Atom Deal," *Observer*, June 12, 1988.
50. "Nuclear Cooperation with Pakistan Reported," *Al-Watan Al-'Arabi*, translated in *JPRS-TND*, December 23, 1989, p. 11.
51. Interview with primary author.
52. "Proliferation in Pakistan," *U.S. News and World Report*, February 12, 1990.
53. "Iranian Defense Chief Discusses Military Cooperation," IRNA, 1702 GMT, March 19, 1990, translated in *FBIS-NES*, March 23, 1990, p. 57.
54. See discussion in the introduction to Part II, Asia.
55. "China is Helping Iran to Build Missile Plant, Sources Report," *Baltimore Sun*, March 9, 1989.
56. Joseph S. Bermudez, Jr., and W. Seth Carus, "North Korean 'Scud B' Programme," *Jane's Soviet Intelligence Review* (April 1989): 177.
57. "Defense Cooperation Accord Signed," IRNA, 0826 GMT, December 8, 1989, reprinted in *FBIS-East Asia (EAS)*, December 8, 1989; "North Said to Conclude Weapons Treaty with Iran," Seoul Domestic Service, 2200 GMT, December 8, 1989, translated in *FBIS-EAS*, December 11, 1989. The military ties between the two countries are discussed in greater detail in the chapter of this book on North Korea.
58. Sanjoy Hazarika, "India Says It Sold Iran a Chemical Used in Poison Gas," *New York Times*, July 1, 1989. India has also assisted Iran with post-war reconstruction. See, "Iran Favors Congress (I) in Indian Elections," *Defense and For-*

eign Affairs Weekly, November 20-26, 1989, p. 2.

59. Michael R. Gordon, "Swiss Halt Plans for Plant in Iran," *New York Times*, May 9, 1989; Michael R. Gordon with Stephen Engelberg, "A German Concern Sold Chemicals to Iran, U.S. Says," *New York Times*, June 27, 1989.
60. "Testimony of William Webster, Director of Central Intelligence," Committee on Governmental Affairs, U.S. Senate, 101st Cong., 1st Sess., May 18, 1989 (mimeo).
61. For example, in June 1988, France restored diplomatic relations with Iran and, in October of that year, King Faud of Saudi Arabia ordered that official criticism of Iran be eased.
62. Ed Bianche, "Iran Reported Set to Buy Arms from Soviet Union," *Philadelphia Enquirer*, May 2, 1989. The Soviet Union has publicly denied that it intends to sell arms to Iran, stating that only assistance for reconstruction of the country will be offered. See "Soviet Official Denies Arms Deal," *Al-Ittihad*, May 30, 1989, translated in *FBIS-NES*, June 9, 1989, p. 43.
63. See also Etemad, "Iran" (see note 14), pp. 23-28.

Part IV: Latin America - Introduction

1. Chile has not joined the Non-Proliferation Treaty and, although it has ratified the Tlatelolco Treaty, it is not bound by that accord because it has not waived the treaty's entry-into-force provisions. (See Appendix E.) Its rudimentary nuclear research program is not of proliferation concern, however.

   Cuba, too, is not considered a proliferation threat, even though it has not adhered to the Non-Proliferation Treaty and has neither signed nor ratified the Tlatelolco Treaty. Its fledgling nuclear power program is centered on two Soviet-supplied nuclear power plants currently under construction at Cienfuegos, which will be subject to IAEA safeguards.

   The 1986 nuclear accident at Chernobyl triggered concerns in the United States about the safety of the Cuban plants, some 200 miles from Florida, even though the Cienfuegos reactors are of a different design from the Chernobyl unit. In particular, the Cuban plants are moderated with light (ordinary) water, rather than graphite and possess advanced containment systems, which the Chernobyl facility lacked.

   American safety concerns have resulted in a remarkable level of cooperation between the United States and Cuba. In 1986, both Cuba and the Soviet Union provided details of the Cienfuegos plants' safety features to Washington. Then, in May 1988, following a Cuban inspection of the McGuire nuclear power plant in Cornelius, North Carolina, the chief of the US Interests Section in Havana was allowed to tour the Cienfuegos construction site. In late 1989, officials of the U.S. Nuclear Regulatory Commission also toured the facility and consulted with Cuban regulatory officials. See Jorge I. Dominguez, *To Make a World Safe for Revolution* (Cambridge, Mass.: Harvard University Press, 1989), pp. 58-59, 246; Joseph B. Treaster, "US-Cuba Enmity May be Relaxing," *New York Times*, January 6, 1989; *Nuclear Engineering International* (June 1989): 30.

Chapter 13: Argentina

1. A bloodless military coup on March 24, 1976, ousted Argentina's President Maria Estela Martinez de Peron, wife of two-time president Juan Domingo Peron, and installed a governing junta composed of the commanders of the three armed forces. The coup, led by army General Jorge Rafael Videla, came after months of rampant inflation, right- and left-wing terrorism, and political indecision. General Videla was sworn in as president and served until March 1981, when he was replaced in rapid succession by General Roberto Viola (March-December 1981), General Leopoldo Galtieri (December 1981-July 1982) and General Reynaldo Bignone (July 1982-December 1983).

   This period of military rule was marked by political repression, including suppression of opposition parties and wide-scale detention, torture, and killing of suspected left-wing activists, and by continued economic hardship. After Argentina's defeat in the 1982 Falklands/Malvinas War, General Bignone initiated a transition to civilian rule that culminated in Alfonsín's election.
2. The plants are the West German-supplied Atucha I and the Canadian-supplied Embalse reactors. Both are natural uranium/heavy water reactors.
3. One key commodity still needed from abroad is heavy water, used in Argenna's power reactors. In a controversial 1980 deal, however, the Swiss firm, Sulzer Brothers, with the approval of the Swiss government, agreed to sell Buenos Aires a safeguarded industrial-scale heavy-water production plant, to be built at Arroyito.

   In 1979, Argentina had solicited bids for both the large heavy water plant and for a third nuclear power plant, Atucha II. Canada and West Germany were asked to submit bids for both plants. Canada stated that it would not sell the plants unless Argentina agreed to maintain all of its nuclear installations—current and future—under IAEA safeguards. Under U.S. urging, West Germany agreed to apply the same conditions, apparently because a heavy water plant, rarely sold in international nuclear commerce, was involved. At the last minute, however, Argentina announced it would split the deal, buying the heavy water plant from Switzerland, which had not made these "full-scope" safeguards a condition for the sale, and purchasing the reactor from Germany, which then declared that it, too, would not require comprehensive safeguards, since it was no longer selling the heavy water plant. Canada considered the German action a betrayal of its previous understanding with Bonn. Switzerland and Germany did require, however, that the reactor, the heavy water plant, any plutonium produced through their use, and any facilities based on their technology be placed under safeguards.

Argentina is also developing an experimental heavy water facility indigenously. (See the chart at the end of this chapter.)

4. Argentina's natural uranium/heavy water reactors do not require enriched uranium fuel. Although there may be some small economic benefit to using slightly enriched uranium in them, no other country does so. Argentina's research reactors do require enriched uranium but the Pilcaniyeu plant's capacity, as announced, was more than sixteen times that necessary to serve these small research facilities. Moreover, the decision of Argentina's former leaders not to place the plant under IAEA safeguards meant that if it were configured to produce weapons-grade uranium, Argentina would be free to use the material for nuclear arms.

Commencement of the Pilcaniyeu project coincided with the start of construction of the Ezeiza reprocessing plant, and the two appeared to be a direct response to Brazil's 1975 deal with West Germany, under which it was to receive both enrichment and reprocessing capabilities. (Indeed the individual who then headed the Argentine Nuclear Energy Commission acknowledged to one Carter Administration non-proliferation aide that Argentina could not abandon the Ezeiza plant—the only Argentine plant that had been publicly announced at the time—because of the 1975 Brazilian deal.)

In combination with Argentina's increasing militarism in 1978, as evidenced by its provocation of a near-war with Chile over the Beagle Channel Islands and a major build-up of conventional armaments, these factors strongly suggested that the Pilcaniyeu installation was initiated for the purpose of giving Argentina a nuclear weapons option.

The secret 1978 decision to build the Pilcaniyeu plant may help to explain why Argentina's military government resisted "full-scope" safeguards so vigorously in 1980, when it was negotiating with Canada and West Germany on the purchase of a nuclear power reactor and a heavy-water production plant, although it should be noted that opposition to such constraints predated the decision to build the enrichment facility.

Argentina produces its own uranium and has unsafeguarded facilities for refining it and for converting it to uranium hexafluoride, the form of uranium needed in the enrichment process. (See the chart at the end of this chapter.) At the time the Pilcaniyeu plant project was initiated, however, the indigenous sources of uranium hexafluoride were not yet complete, and it appears that Argentina's military government clandestinely imported uranium hexafluoride not covered by IAEA safeguards from China to permit the start-up of the Pilcaniyeu plant. The secrecy surrounding this transaction, which is discussed in

Chapter 4, is yet another factor suggesting that the Pilcaniyeu plant was part of a covert effort to acquire a nuclear weapons option.

5. Leonard S. Spector, *Nuclear Proliferation Today* (New York: Vintage Books, 1984), pp. 257-263. One well-placed Brazilian official, speaking anonymously in 1986 about the sudden disclosure of the facility, echoed these concerns in a more muted tone: "The people here hadn't expected such a development and felt the Argentines hadn't been quite candid. I believe this has created some uneasiness." See Bradley Graham, "Argentine Leader Proposes Moving Capital to Patagonia," *Washington Post*, April 17, 1986.

   The Pilcaniyeu plant had begun to produce a small quantity of very slightly enriched uranium by mid-1986. Richard Kessler and Mark Hibbs, "Pilcaniyeu Safeguards Offer Tied to Nuclear Transfer by Argentina," *Nuclear Fuel*, February 11, 1988; Richard Kessler, "Argentina's Alfonsín Pledges Funding for Fuel Cycle Projects," *Nucleonics Week*, June 5, 1986. The plant has a capacity of 20,000 separative work units (SWUs) per year (twice that of South Africa's Valindaba pilot enrichment plant); in principle this capacity is enough to produce up to 220 pounds (100 kilograms) of 90 percent-enriched uranium per year, depending on how the facility is laid out. See discussion in text, below.

6. Leonard S. Spector, *The New Nuclear Nations* (New York: Vintage Books, 1985), pp. 182-183.

   Although the Peronists were narrowly defeated by Alfonsín's Radical Civic Union in Argentina's 1983 general elections, they retained a substantial voice in the nation's domestic political debate. Drawing on the coalition put together by Juan Peron, president of Argentina from 1946 to 1955 and again from 1973 until his death the following year, the Peronists achieved political support through a platform of intense nationalism and affiliation with labor. Opposition to Alfonsín's economic policies in 1984, for example, was mounted by the Peronists in conjunction with the General Confederation of Labour (CGT), Argentina's powerful trade union. The November 1985 mid-term congressional elections, however, showed a growing dissatisfaction with the Peronist opposition, especially with the wing of the party most closely associated with union bosses.

7. See, *e.g.*, "Alfonsín Proposes CNEA Reforms," *Nuclear Engineering International* (September 1984): 10. By December 1988, Alfonsín apparently abandoned efforts to enact the legislation. Richard Kessler, "Alfonsín Said to End Plans for Nuclear Program Reform," *Nucleonics Week*, December 1, 1988, p. 5.

8. Richard Kessler, "Argentina, Brazil Agree to Mutual Inspection of Nuclear Facilities," *Nucleonics Week*, March 14, 1985, p. 14. Unlike ratification of the Treaty of Tlatelolco, which might have the appearance of a unilateral concession made at least partly in response to long-standing pressure from the United States and others, implementation of the mutual inspection proposal would have allowed Alfonsín to argue that he had gained something in return for relinquishing Argentina's nuclear privacy.

9. "Joint Declaration on Nuclear Policy by the Governments of Brazil and Argentina," November 29, 1985 (courtesy of the Embassy of Argentina).

   Some important nuclear cooperation had already taken place between the two countries under a 1980 nuclear trade pact, including Argentina's lease of 240 metric tons (265 short tons) of uranium concentrate to Brazil, which Brazil "returned" after its own uranium mining operations were expanded; Argentina's sale to Brazil of 160,000 meters (175,000 yards) of zircalloy tubing for nuclear power reactor fuel elements; and Brazil's contract to manufacture a portion of the reactor vessel for Argentina's Atucha II nuclear power plant, which was delivered in late June 1987. See "Argentina: Atucha-2 Vessel Arrives," *Nucleonics Week*, July 2, 1987.

10. "Alfonsín Wants Nonproliferation," *Jornal do Brasil*, December 1, 1985, translated in *Joint Publications Research Service (JPRS)-Nuclear Developments (TND)*, January 9, 1986, p. 16; Richard Kessler and Jeff Ryser, "Argentina-Brazil Committee to Work Out Details of Joint Inspection," *Nucleonics Week*, December 5, 1986, p. 9.
11. To symbolize the new direction in bilateral relations, at the conclusion of the talks, Alfonsín accompanied Sarney to the Itaipu hydroelectric dam, built by Brazil and Paraguay; Argentina had not joined in the project, which its military government had claimed represented a threat to Argentine security. See Kessler, "Argentine-Brazilian Committee to Work Out Details" (see note 10.)
12. Richard Kessler, "Argentina and Brazil Will Not Provide the IAEA with Information," *Nucleonics Week*, March 2, 1986, p. 9.
13. "Committee Report On Talks with Argentina, Accord with FRG," *O Estado de São Paulo*, April 18, 1986, translated in *JPRS-TND*, June 13, 1986, p. 44; Richard Kessler, "Panel Favors Mutual Inspections of Facilities by Argentina, Brazil," *Nucleonics Week*, May 1, 1986, p. 6.
14. Richard Kessler, "Argentine-Brazil Protocol Promotes Nuclear Accident Accord," *Nucleonics Week*, July 31, 1986, p. 1.
15. Ivo Dawnay, "Brazil and Argentina to Set Seal on Collaboration," *Financial Times*, December 10, 1986; Richard Kessler, "Argentina-Brazil Protocols Detail Broad Nuclear Cooperation," *Nucleonics Week*, December 11, 1986, p. 9. For additional details on cooperation between Brazilian and Argentine nuclear firms, see Richard Kessler, "Argentine-Brazilian Firms Pool Nuclear Design Know-How," *Nucleonics Week*, January 15, 1987. The nuclear protocols provided for joint research efforts on advanced fuels for research reactors, the economic feasibility of breeder reactors, and the production of stable isotopes.
16. Richard House, "Brazil Steps Back from Race to Build Nuclear Weapons," *Washington Post*, August 28, 1986; Richard Kessler, "Electronic Monitors Eyed for Argentine-Brazil Mutual Inspections," *Nucleonics Week*, September 11, 1986.
17. Richard Kessler, "Sarney Visit to Pilcaniyeu Was Key to Reciprocal Inspections," *Nucleonics Week*, July 23, 1987, p. 11.
18. "Joint Argentine-Brazilian Declaration on Nuclear Policy," Viedma, Argentina, July 17, 1987; Michael Gordon, "Brazil and Argentina Start Nuclear Talks," *New York Times*, July 22, 1987; Kessler, "Sarney Visit to Pilcaniyeu Was Key to Reciprocal Visits," (see note 17).
19. "Officials on Navy Nuclear Facilities, Program" *Fôlha de São Paulo*, April 7, 1988, translated in *Foreign Broadcast Information Service (FBIS)-Latin America (LAT)*, April 11, 1988, p. 38.
20. Nonetheless, the important distinction between confidence building and reliable verification must be recognized. See discussion below.
21. Kessler, "Sarney Visit to Pilcaniyeu Was Key to Reciprocal Visits" (see note 17).
22. Richard Kessler, "Brazil Giving Argentina Access to Operating Nuclear Facilities," *Nucleonics Week*, March 24, 1988, p. 9.
23. "Joint Policy Declaration with Argentina," *Fôlha de São Paulo*, April 8, 1988, translated in *FBIS-LAT*, April 11, 1988, p. 36. The number of centrifuges then operating at the Aramar unit was not disclosed, but by this juncture, Brazil claimed to have enriched uranium to 5 percent, and possibly to 8 percent, either at the facility or at IPEN's research unit. See "Officials on Navy Nuclear Facilities, Program," *Fôlha de São Paulo* (see note 19); "Sarney, Alfonsín to Inaugu-

rate Nuclear Plant," *O Globo*, March 2, 1988, translated in *FBIS-LAT*, March 4, 1988, p. 27 (uranium enriched to 8 percent).

24. Richard Kessler, "Costantini Resigns as CNEA Head in Bitter Funding Dispute," *Nucleonics Week*, April 16, 1987, p. 7.
25. "Speech by President Alfonsín at the Central Atomic Energy Day Ceremony, Embalse, Rio Tercero, May 30, 1986," translated by Congressional Research Service (authors' files). Concerning the position of the traditional elements of the Peronist Party, see Richard Kessler, "Argentine Bill Proposes Role for Congress in Nuclear Decisions," *Nucleonics Week*, October 23, 1986, p. 9.
26. "Opposition to Nuclear Plan," *La Nueva Provincia*, June 14, 1987, translated in *JPRS-TND*, August 18, 1987, p. 20. See also, Richard Kessler, "Shakeup Seen for Argentina's CNEA Under New Head Perez Ferreira," *Nucleonics Week*, May 21, 1987, p. 1.

    Alfonsín appointed Emma Perez Ferreira, a respected CNEA physicist, to succeed Costantini.
27. As subsequently published estimates revealed, the start-up dates for both Atucha II and the Swiss-supplied heavy water plant would be postponed for one year, to 1993 and mid-1989, respectively, and cost increases for the latter facility had raised serious questions about its economic viability, even among CNEA officials. Richard Kessler, "Production Delayed for a Year at Argentine Heavy Water Plant," *Nuclear Fuel*, July 13, 1987, p. 13; Richard Kessler, "Atucha-2 to Cost $1 Billion More than Originally Estimated," *Nucleonics Week*, November 19, 1987, p. 6. Alfonsín was said to be seeking to cancel the indigenous heavy water plant at Atucha, and subsequent official reviews of the status of Argentina's principal nuclear projects have not mentioned the unit.
28. The nuclear bureaucracy suffered a further blow to its prestige later in 1987, when Alfonsín appointed a special planning committee to oversee implementation of existing nuclear construction plans. See "Atomic Energy Consultative Council Planned," Telam, 1052 GMT, November 19, 1987, translated in *FBIS-LAT*, November 24, 1987, p. 22.

    Concerning the priority given the Pilcaniyeu facility, see Richard Kessler, "Argentina to Begin Enriching 20% U-235 in Mid-1988," *Nuclear Fuel*, May 18, 1987, p. 3; "CNEA Head Discusses Nuclear Policy Issues," Noticias Argentinas, 1305 GMT, November 24, 1987, translated in *FBIS-LAT*, November 30, 1987, p. 26.
29. In late April 1987, Alfonsín was confronted with military unrest that had been triggered by a mutiny of junior army officers opposed to these prosecutions. Although the uprising ended peacefully after four days, the slow response of the Argentine Army in suppressing the rebellion indicated the depth of sentiment within the military on the issue. In May, stating that the mutiny had brought the country to "the brink of civil war," Alfonsín succumbed to pressure and introduced the legislation noted in the text. See Martin Anderson, "Alfonsín Presents Bill to Absolve Most Officers," *Washington Post*, May 14, 1987. A subsequent mutiny in January 1988, led by the same junior officers, was quickly suppressed by loyal troops, and the mutiny's supporters, apparently confined to the army, were purged, an indication that Alfonsín had satisfied the military's concerns. See Robert Graham, "Alfonsín Battles to Keep House in Order," *Financial Times*, January 29, 1988. Alfonsín would confront yet another insurrection in December 1988, however.
30. Daniel Poneman, "Argentina: The Military Threat Remains," *International Herald Tribune*, January 27, 1988.
31. Alfonsín was constitutionally barred from running for a second term.
32. "Argentina Declares State of Siege Amid Riots Over Economic Steps," *New*

York Times, May 30, 1989; Eugene Robinson, "Bombs Explode as Looting Continues in Argentina," *Washington Post*, May 31, 1989; James Brooke, "Argentina in Chaos as Prices Rise Hourly," *New York Times*, June 2, 1989.

33. "Nuclear Plants Cease Operations for Lack of Funds," *Clarin*, October 17, 1988, translated in *FBIS-LAT*, October 20, 1988, p. 20.
34. David Albright, "Bomb Potential for South America," *Bulletin of the Atomic Scientists* (May 1989): 16.
35. In practice, the material is enriched to a level just below 20 percent, so that, technically speaking, it is still "low-enriched" uranium and thus cannot be used as the core of a nuclear device because it lacks the necessary physical properties. The enriched uranium that Argentina sold to Algeria in 1988, discussed below, was enriched to 19.7 percent, for example. See *Nuclear Engineering International* (December 1988): 5.
36. Albright, "Bomb Potential for South America" (see note 34). As discussed below, Argentina may have a stockpile of 20-percent enriched uranium not subject to IAEA safeguards that it obtained from China in the early 1980s. (See note 58.) If so, another option for producing weapons-grade uranium would be to improve this material in the Pilcaniyeu facility.

    For a variety of technical reasons, in some instances it is not practical to operate a gaseous diffusion enrichment plant to achieve a significantly higher level of enrichment than that for which it has been designed. Obtaining weapons-grade uranium from a commercial-scale enrichment facility—*i.e.*, one many times larger than Pilcaniyeu—that had been designed in order to optimize the production of three-percent enriched uranium might, for example, prove to be infeasible. Fewer obstacles would be presented, however, in obtaining weapons-grade material from a smaller plant like Pilcaniyeu, which apparently has been designed to optimize the production of 20-percent enriched material. See Allan S. Krass, Peter Boskma, Boelie Elzen, and Wim A. Smit, *Uranium Enrichment and Nuclear Weapon Proliferation* (London: Taylor and Francis, 1983).
37. Richard Kessler, "Argentina's CNEA Starts Fabricating Enriched Fuel for PHWR Atucha-1," *Nuclear Fuel*, September 5, 1988, p. 9; Richard Kessler, "Argentina's CNEA Turns to Soviets for Small Supply of Enriched Uranium," *Nuclear Fuel*, November 14, 1988, p. 3.
38. When Argentine nuclear officials announced the existence of the Pilcaniyeu plant in late 1983, they indicated that the facility would shortly have the ability to produce 1,100 pounds (500 kilograms) of 20-percent enriched uranium annually. Operations at this level still remained an unachieved goal at the end of 1988, however. See Richard Kessler, "CNEA Officials Provide More Details About Pilcaniyeu Enrichment Plant," *Nuclear Fuel*, May 30, 1988, p. 4; Kessler, "Argentina's CNEA Turns to Soviets for Small Supply of Enriched Uranium" (see note 37). Details as to the plant's output have not been released, however. In September 1988, a senior Argentine nuclear officials stated that the facility had "75-105 cascades," but he did not describe the output of each of these, or how close this configuration came to achieving the capability of 1,100 pounds of 20-percent enriched uranium per year. (A cascade is a series of enrichment units that have been linked together to permit the continuous incremental improvement of uranium introduced into the system.)
39. "USSR to Enrich Argentine Uranium," *La Nacion*, October 28, 1988, translated in *JPRS-TND*, December 23, 1988, p. 1; Kessler, "Argentina's CNEA Turns to Soviets for Small Supply of Enriched Uranium" (see note 37). Under the arrangement with Moscow, Argentina was to supply natural uranium, which would be enriched in the Soviet Union and returned. The material was to be

delivered in March and September 1989 under IAEA safeguards and bilateral controls prohibiting its reexport without Soviet approval. Thus, in effect, Argentina planned to use the Soviet-enriched material to meet domestic requirements and to export the output from Pilcaniyeu. See note 58, below.

40. Gary Mead, "Alfonsín and Sarney Step Closer to Common Market," *Financial Times*, December 1, 1988; "Alfonsín, Sarney Visit Atomic Center," Telam, 1444 GMT, November 30, 1988, translated in *FBIS-LAT*, December 1, 1988, p. 31; "Ezeiza Declaration—Joint Declaration on Nuclear Policy," November 29, 1988 (courtesy of the Embassy of Argentina).
41. "FRG, Austria Build Rocket Laboratory in Iraq," *Profil*, March 20, 1989, translated in *JPRS-TND*, May 5, 1989, p. 31; "Air Force Chief Comments on Missile Project," DYN, 1140 GMT, October 5, 1989, translated in *JPRS-TND*, October 26, 1989, p. 21. The Condor I would have a range of approximately 60 miles if configured as a missile. See U.S. Arms Control and Disarmament Agency, *World Military Expenditures and Arms Transfers, 1988,* (Washington, D.C.: U.S. Government Printing Office, 1989). At the 1985 Paris Air Show, the Condor I was presented as a multi-purpose rocket, highlighting its military potential. It was test-fired in 1986. See "Pre-Satellite Launching Rocket Tested by Air Force," *La Rioja*, October 14, 1986.
42. *United States v. Helmy*, U.S. District Court for the Eastern District of California (Sacramento), CR. No. S-89-201- RAR, Government's Sentencing Memorandum, pp. 1-2.
43. Robert Gillette, "3d World Missiles Linked to German, Italian Firms," *Los Angeles Times*, February 6, 1989; Simon Henderson and David Goodhart, "Prosecutor Launches Inquiry into MBB's Condor Missile Links," *Financial Times*, April 12, 1989; "*Profil* Views Involvement in Iraq Arms Plant," *Profil*, April 24, 1989, translated in *JPRS-TND*, May 5, 1989, p. 34; "Charges Filed in Missile Parts Sales to Argentina," *L'Unita*, July 30, 1989, translated in *JPRS-TND*, August 14, 1989, p. 41. Concerning the involvement of Italian firms in the project, see "Condor II Scientist Reveals Details of Project," *Europeo*, September 29, 1989, p. 21.
44. "Condor Project Negotiable Demands Discussed," *Somos*, October 4, 1989, translated in *FBIS-LAT*, November 21, 1989, p. 43; "Missile Experts Implicated in Illegal Arms Deal," Rome ANSA, 1033 GMT, July 31, 1989, reprinted in *JPRS-TND*, August 14, 1989, p. 42; "Missile Technology for Iraq Reported," *Stern*, April 6, 1989.
45. Don Oberdorfer, "U.S. Seeks to Curb Argentine Missile Project," *Washington Post*, September 19, 1988; Peter Almond, "Argentine Aide Resists Plea to Stop Missile Project," *Washington Times*, October 13, 1988.
46. "Military Sources on Rocket," Telam, 1525 GMT, September 20, 1988, translated in *FBIS-LAT*, September 21, 1988; "Condor II Development with Egypt to Continue," Noticias Argentinas, 1951 GMT, April 5, 1989, translated in *JPRS-TND*, April 25, 1989, p. 11.
47. Richard Kessler, "Menem Reported Ready to Name Castro Madero CNEA Boss Again," *Nucleonics Week*, January 12, 1989, p. 4; Richard Kessler, "Argentine President-Elect Menem Eyes Nuclear Fuel Cycle Control," *Nucleonics Week*, May 18, 1989, p. 4.
48. Hugh O'Shaughnessy, "People's Choice for New Peron," *Observer* (London), October 16, 1988.
49. "Brazil Pact Widened; Tlatelolco Considered," *Nuclear News* (October 1989): 59. The two countries reduced tariff barriers in September, another indication of their cordial relations. See Ivo Dawnay, "Move on Brazil-Argentine Trade,"

Financial Times, September 13, 1989.

The joint work on space-related projects was an apparent effort to offset the impact of restrictions on technology transfers imposed on both states under the Missile Technology Control Regime.

50. Mark Hibbs, "Argentine Officials Optimistic on Futures of Atucha-2, Exports," *Nucleonics Week*, October 12, 1989, p. 11.
51. Argentina did, however, ratify the Partial Test Ban Treaty, prohibiting nuclear tests in the atmosphere, under water, or in outer space, in July 1986.
52. Richard Kessler, "Argentine Position on NPT, Tlatelolco Unchanged, Says New Foreign Minister," *Nuclear Fuel*, September 4, 1989, p. 7; "Menem Against Signing Nonproliferation Treaty," *Buenos Aires Herald*, November 24, 1989, reprinted in *FBIS-LAT*, November 27, 1989, p. 44; "Argentina Will Not Sign Non-Proliferation Accord," *Defense and Foreign Affairs Weekly*, December 11-17, 1989, p. 5.
53. David B. Ottaway, "Egypt Drops Out of Missile Project," *Washington Post*, September 20, 1989. One factor that had particularly troubled Washington was the discovery that the Egyptian Defense Ministry had mounted a major missile technology smuggling operation related to the project in the United States. See note 42.
54. "Further Comments Reported," Telam, 2028 GMT, October 10, 1989, translated in *JPRS-TND*, October 26, 1989, p. 22 (Air Force Chief of Staff Brigadier José Julia declares that the project "will not be stopped. . . ."); "Air Force Exhibits Technological Developments," Noticias Argentinas, 1705 GMT, November 27, 1989, translated in *FBIS-LAT*, November 30, 1989, p. 43.
55. "Is Condor Kaput?" *U.S. News and World Report*, March 5, 1990, p. 20; "Defense Minister Says Condor II Project Halted," *Clarin*, April 22, 1990, translated in *FBIS-LAT*, April 27, 1989, p. 12; "Menem Says Missile Scrapped Over U.S. Concern," *Clarin*, April 25, 1989, translated in *FBIS-LAT*, April 26, 1990, p. 20.

    Cancellation of the Condor II may also have been related to the restoration of diplomatic ties with Great Britain on February 15, 1990. Relations between the two countries had been broken at the time of the Falklands/Malvinas War in 1982. As part of the reopening of diplomatic links, Great Britain agreed to lift its military exclusion zone around the islands. Although Argentina still claimed sovereignty over them, these developments effectively signaled the end of the military confrontation over the issue and thus may have eliminated the Air Force's rationale for acquiring the Condor II system.
56. "Argentina Announces Delay of Controversial Plant," *Nuclear News* (March 1990): 70. The halting of work on the facility and the suspension of the Condor II project coincided with a spurt of hyper-inflation that necessitated draconian cuts in Argentina's public expenditures.
57. Argentine Atomic Energy Commission Chairman Mondino stated in an interview that Argentina had some twenty bilateral nuclear cooperation agreements as of October 1989. See Hibbs, "Argentine Officials Optimistic on Futures of Atucha-2, Exports" (see note 50).
58. "Alfonsín to Inaugurate Peru's Huarangal Nuclear Complex," *Nucleonics Week*, December 1, 1988, p. 15; "Training Technology Transfer Featured in Exports," *Clarin*, April 2, 1989, translated in *JPRS-TND*, May 23, 1989, p. 10; "Algerian Reactor Opens," *Nuclear Engineering International* (May 1989): 5. Argentina's transfer of radiologically shielded "hot cells" in conjunction with the reactor sale to Peru and Algeria and the transfer of a small plant for fabricating fuel elements in conjunction with the reactor sale to Algeria also appear to be precedent-setting exports for a non-industrialized state.

Argentina contracted to supply 110 pounds (50 kilograms) of 20-percent enriched uranium to Peru and 176 pounds (80 kilograms) of the material to Algeria. See "Argentina Uranium Export," *Telam*, September 13, 1988, reprinted in *Current News*, September 15, 1988, p. 12; "Argentina Agrees to Uranium for Algeria," *Nuclear Engineering International* (December 1988): 5 (material scheduled for delivery in early 1989 to coincide with the completion of the Algerian reactor); Kessler, "Argentina's CNEA Turns to Soviets for Small Supply of Enriched Uranium" (see note 37).

The 10-megawatt reactor that Argentina sold to Peru was inaugurated in December 1988, and the 1-megawatt unit sold to Algeria, in April 1989. There is some question as to whether all of the fuel Argentina promised Peru and Algeria could have been produced at Pilcaniyeu by April 1989, since the Pilcaniyeu facility did not start to produce 20-percent enriched material until the fall of the previous year and was operating well below capacity. See Kessler, "CNEA Officials Provide More Details About Pilcaniyeu Enrichment Plant" (see note 38) (Argentine Atomic Energy Commission Chair Emma Perez Ferreira stated in May 1988 that equipment to produce 20-percent enriched material was "virtually in place," implying that the plant was not yet able to produce this material).

Possibly, Argentina supplied Peru and Algeria only enough material for an initial fueling of their respective reactors, while planning to provide the remainder of the fuel contracted for at a subsequent time. It has also been reported that in the early 1980s, Argentina's military regime obtained 20-percent enriched uranium from the People's Republic of China free from IAEA safeguards. This material may have been used to meet Argentina's late 1988 and early 1989 fuel contract obligations. See Judith Miller, "U.S. Is Holding Up Peking Atom Talks," *New York Times*, September 19, 1982; Rob Laufer, "Interview with Malone: Defense of Policy and Assessment of 'Hot Spots,' " *Nucleonics Week*, August 19, 1982, p. 1.

Argentina had also contracted to sell Iran 255 pounds (116 kilograms) of 20-percent enriched uranium fuel for a research reactor supplied by the United States in the 1970s. The material was to be delivered after Argentina had modified the reactor's core to be able to use such material instead of the more highly enriched fuel it was originally designed to use.

All of these export transactions were to be covered by IAEA safeguards to ensure that transferred materials and equipment, and material produced through their use, were not diverted to military purposes. Argentina also imported 220 pounds (100 kilograms) of safeguarded 20-percent enriched uranium from the Soviet Union in 1980 to use in its domestic research reactors, but presumably this material has been exhausted. (The 330 pounds (150 kilograms) of Soviet-enriched 20-percent material that Argentina contracted for in the fall of 1988 was subject to Soviet re-export controls and presumably was used within Argentina. See note 39).

59. Hibbs, "Argentine Officials Optimistic on Future of Atucha-2" (see note 50).
60. Richard Kessler, "Argentina Expects to Wrap Up U Deal with Algeria Later This Year," *Nuclear Fuel*, October 3, 1988, p. 6.
61. Richard Kessler, "Panel to Guide Nuclear Technology Sales from Argentina to Algeria," *Nucleonics Week*, May 7, 1987, p. 6.
62. Richard Kessler, "Argentina's INVAP to Supply Fuel for Research Reactor," *Nucleonics Week*, May 14, 1987, p. 2.
63. "Training Technology Transfer Featured in Exports," *Clarin* (see note 58); *Programme for Promoting Nuclear Nonproliferation News Brief* (January 1989): 4.
64. Richard Kessler, "Argentina Seeks Nuclear Ties with India and Pakistan," *Nu-*

cleonics Week, April 13, 1989, p. 3. The article notes that an agreement for nuclear cooperation between India and Argentina lapsed in 1979, but that Alfonsín permitted the sale nevertheless.

During his June 1990 interview with the principal author of this book, Alfonsín did not recall the details of Argentine exports to Iran or India and offered no details about Argentine nuclear export policy.

65. *Ibid.* For an overview of Argentine nuclear export activities, see Joel M. Rothblatt, "Argentina's Aggressive Nuclear-Export Policy," *Christian Science Monitor*, June 8, 1989.
66. "Menem Interviewed on Nuclear Policy, Other Issues," *Argentina Nuclear* (No. 19-20, 1989), translated in *JPRS-TND*, March 2, 1990; "Joint Nuclear Project with Turkey Discussed," Buenos Aires Domestic Service, 1600 GMT, translated in *FBIS-LAT*, September 29, 1989, p. 22; Richard Kessler, "Argentina Will Bid on Sale to Egypt of Small Research Reactor," *Nucleonics Week*, November 9, 1989, p. 7; Richard Kessler, "Menem Government Eyes Isotope Production Reactor for Algeria," *Nucleonics Week*, January 4, 1990.
67. "Government Promulgates Law on Nuclear Security," Telam, 1324 GMT, October 10, 1989, translated in *JPRS-TND*, October 26, 1989.
68. Hibbs, "Argentine Officials Optimistic" (see note 50).
69. "Argentina to Push Defense Exports in 'Aggressive' Manner," *Defense and Foreign Affairs Weekly*, February 26-March 4, 1990.
70. "Cucchietti Announces Arms Shipment to Pakistan," Telam, 0008 GMT, November 1, 1989, translated in *FBIS-LAT*, November 3, 1989, p. 22 (sale of 60,000 mortar fuses); "Iranian Ship Departs with Military Cargo," DYN, 2116 GMT, September 3, 1989, translated in *FBIS-LAT*, September 6, 1989, p. 33 (200 tons of rifles, mortars, and ammunition).
71. Shirley Christian, "Argentina's Military Chiefs Warn of Anarchy," *New York Times*, January 2, 1990; Eugene Robinson, "Argentina's Army Receives Domestic Role," *Washington Post*, March 10, 1990.
72. The situation could be greatly complicated if the nuclear submarine programs that Brazil, and possibly Argentina, are pursuing ultimately require weapons-grade uranium fuel. In this case, even if additional verification procedures were implemented, it would be difficult to be certain that weapons-usable material produced to run submarine propulsion reactors was not being diverted for nuclear explosives.

Chapter 14: Brazil

1. See Leonard S. Spector, *Nuclear Proliferation Today* (New York: Vintage Books, 1984), pp. 239-244.
2. "Angra I Inaugurated," *O Estado de São Paulo*, January 18, 1985, translated in *Joint Publications Research Service (JPRS)-Nuclear Proliferation and Development (TND)*, March 12, 1985. The facility first operated in 1982, but underwent an unusually long period of debugging, which delayed commercial operation for three years. The facility has since experienced serious operating problems.
3. Rik Turner, "Brazil's Sarney Defines Emphasis for National Fuel Cycle Development," *Nuclear Fuel*, August 11, 1986, p. 10. Sarney's decision implemented the recommendations of a special commission he had appointed in 1985 shortly after he took office and codified what had been *de facto* policy for a number of years. See "Brazil Commission Opts to Continue Nuclear Program," Reuters, April 18, 1986; Robert Townley, "Brazil Rekindles Reactor Program and Denies Proliferation Plans," *Nucleonics Week*, August 14, 1986, p. 12.

The reprocessing plant was intended to extract plutonium from spent nuclear-power-reactor fuel. Brazil planned to recycle the plutonium by using it in fresh reactor fuel instead of low-enriched uranium. At the time, a world-wide uranium shortage was predicted, and it was assumed that such recycling of plutonium would be essential for the economical operation of Brazil's nuclear power reactors. Today, because of the global down-turn in the growth of nuclear energy, there is a glut of inexpensive enriched uranium on world markets. This has undermined the economic rationale for the costly facility.

Although Brazil cancelled the reprocessing plant, it had already received extensive plutonium extraction technology from West Germany. Under the terms of the 1975 agreement, however, if Brazil were to build a separate plutonium extraction plant using this technology in the future, it would have to place the unit under IAEA safeguards.

As of late 1989, the Angra II reactor was said to be 60 percent complete and scheduled to be on-line in 1993, while little work had been completed on Angra III. See Christina Garcia, "Osart Report Key to Furnas Appeal of Court Order to Keep Angra-1 Shut," *Nucleonics Week*, November 2, 1989.

4. Rik Turner, "Brazil's Nuclebras Wants to Sell Slightly Enriched U [Uranium] to Argentina," *Nucleonics Week*, July 30, 1987. The material was to be enriched only to 0.85 percent uranium-235, fractionally above the naturally occurring rate of the isotope, which is 0.7 percent. Brazil's nuclear power reactors require 3- to 5-percent enriched uranium; nuclear weapons normally use uranium enriched to 90 percent or more.

   For a variety of technical reasons, it appears that adapting the German uranium enrichment plant, as built, to produce weapons-grade material would be infeasible. See Allan S. Krass, Peter Boskma, Boelie Elzen, and Wim A. Smit, *Uranium Enrichment and Nuclear Weapon Proliferation* (London: Taylor and Francis, 1983). Moreover, such a step would violate Brazil's understandings with Germany and would be unlikely to escape detection by the IAEA, whose safeguards are being applied to the facility. The Brazil-West German agreement specifies that Brazil must accept IAEA safeguards on any other jet-nozzle enrichment plant it may build in the future.

5. Brazil has ratified the Treaty of Tlatelolco, but is not bound by its terms because the treaty enters into force only when ratified by all regional states; Argentina and Cuba have not taken this step. Most Latin American states have waived this entry-into-force provision and are therefore bound by the accord. Brazil has declined to do so, however. See Appendix E.

6. According to one account, a 1978 "summit" of representatives of the CNEN, Nuclebras, and the National Intelligence Service first determined that Brazil should pursue nuclear activities in parallel with those under the 1975 West German accord. See "Ten Billion Cruzeiros to be Allocated to Nuclear Research," *Jornal do Brasil*, January 16, 1983, translated in *JPRS-TND*, March 21, 1983, p. 28. Geisel is said to have given his approval shortly before he left office in March 1979, and planning activities began later that year. See "Officials on Navy Nuclear Facilities, Program," *Fôlha de São Paulo*, April 7, 1988, translated in *Foreign Broadcast Information Service (FBIS)-Latin America (LAT)*, April 11, 1988, p. 38; "Do It Yourself Reprocessing," *Economist*, January 24, 1987, p. 80.

   Activities under the Navy's enrichment program commenced in February 1980, and the first enrichment experiment took place in September 1982 at the Institute for Energy and Nuclear Research (IPEN) in São Paulo; two years later nine centrifuges had been linked there in the first experimental "cascade." See "Officials on Navy Nuclear Facilities, Program," just cited.

7. José Sarney gained the presidency following indirect elections that were held in January 1985. The winner of the elections, Tancredo Neves—a long-time opponent of military rule—died before he could take office, and Sarney, his running mate, succeeded to the post. Until a few months before the election Sarney had been the leader of the military government's Social Democrat Party—the party whose candidate, Paulo Maluf, Neves had resoundingly defeated. With his popular mandate thus highly uncertain, Sarney relied heavily on the political support of the military, which, unlike its counterpart in Argentina, had not been tainted by a recent defeat and continued to wield considerable influence.
8. The program is under the control of the Navy and received funding from the Navy, the CNEN, and the National Security Council. See Richard Kessler, "Sarney Visit to Pilcaniyeu Was Key to Reciprocal Inspections," *Nucleonics Week*, July 23, 1987, p. 11; "Officials on Navy Nuclear Facilities, Program," *Fôlha de São Paulo* (see note 6).
9. "IPEN's UF-6 Conversion Technology Said to be Ready for Commercial Use in Brazil," *Nuclear Fuel*, July 28, 1986, p. 5.
10. Brazil has also pursued research and development on a second uranium enrichment technique, laser isotope separation. Work on this is being carried out at IPEN and at the Brazilian Air Force's Aerospace Technology Center in São José dos Campos, the principal site of Brazil's rocket research and development program. Given the difficulties that other nations have experienced in pursuing laser enrichment, it is highly unlikely that Brazil is close to mastering the technology.
11. "EMFA Armed Forces General Staff Opposes Constitutional Restraint," *Fôlha de São Paulo*, June 22, 1985, translated in *JPRS-TND*, September 23, 1985, p. 51.

    For earlier indications of the military's interest in nuclear weapons, see "Production of Atomic Bomb Expected in 1990," *Fôlha de São Paulo*, April 28, 1985, translated in *FBIS-LAT*, May 1, 1985, p. D-1; "Editorial Questions . . ." *Fôlha de São Paulo*, October 14, 1984, translated in *JPRS-TND*, December 19, 1984, p. 36.

    In late 1983 and early 1984, when Argentina disclosed that it had built an unsafeguarded uranium enrichment plant in secret, a number of Brazilian military figures pointed openly to the weapons potential of the parallel program and declared that it would provide Brazil with the ability to produce nuclear arms by 1990. See Spector, *Nuclear Proliferation Today* (see note 1), pp. 259-262.
12. "Piva Confirms Country Able to Make Bomb in Five Years," *O Estado de São Paulo*, August 17, 1985, translated in *JPRS-TND*, October 21, 1985, p. 18.
13. "Army Minister Said to Favor Building Atomic Bomb," EFE, 1431 GMT, September 1, 1985, translated in *FBIS-LAT*, September 4, 1985, p. D-2; Richard Kessler, "Argentina Had No Official Comment on News Reports . . .," *Nucleonics Week*, September 5, 1985, p. 9.
14. "EMFA Chief Denies Interest in Building Atomic Bomb," *O Estado de São Paulo*, September 8, 1985, translated in *JPRS-TND*, December 13, 1985, p. 32; "Setabul Reassures Argentina on Atomic Bomb," *O Globo*, September 5, 1985, translated in *FBIS-LAT*, September 10, 1985, p. D-1. See also, "Army Minister Stresses Need to Complete Nuclear Fuel Cycle," *O Estado de São Paulo*, September 5, 1985, translated in *JPRS-TND*, September 23, 1985, p. 48.
15. "Navy Minister Said to Support Atomic Bomb Plan," EFE, 0135 GMT, September 14, 1985, translated in *FBIS-LAT*, September 17, 1985.
16. "Experts Discuss Development of Nuclear Bomb," *Fôlha de São Paulo*, March 22, 1987, translated in *FBIS-LAT*, March 25, 1987, p. D-1.
17. *Ibid.*
18. "Further Reportage of Alleged Nuclear Test Site," *Fôlha de São Paulo*, August

8, 1986, translated in *FBIS-LAT*, August 12, 1986, p. D-1; Richard House, "Brazil Steps Back from Race to Build Nuclear Weapons," *Washington Post*, August 28, 1986.

19. Richard Townley, "Brazil Rekindles Reactor Program and Denies Proliferation Plans," *Nucleonics Week*, August 14, 1986, p. 12.
20. *Ibid.*
21. According to one specialist, such a test would need only rudimentary instrumentation within the shaft itself; this would be augmented by seismic data from nearby pick-ups and after-the-fact analysis of radioactive debris in the shaft.

    The device tested by India in 1974, the first weapon designed by Sweden in the 1950s (but never built), and the bomb detonated at Hiroshima were all less than three feet in diameter. See Mitchell Reiss, *Without the Bomb* (New York: Columbia University Press, 1988), pp. 228-229, (India's device, including monitoring equipment and safety locks, was three feet in diameter; Indian test shaft was an "L-shaped trench built 107 meters (348 feet) below the surface"); Christer Larsson, "Build A Bomb!" *Ny Teknik*, April 25, 1985, (Swedish weapon was 1.9 feet in diameter). In addition, South Africa's Kalahari test site was discovered several years before that nation apparently possessed the material necessary for a nuclear device. See discussion in Chapter 15.

    Concerning U.S. peaceful nuclear explosive tests, see Edward Teller, *et al.*, *The Constructive Use of Nuclear Explosives* (New York: McGraw-Hill, 1968), pp. 129, 137 (cited in Antonio Rubens Britto de Castro, *et al.*, "Brazil's Nuclear Shakeup: Military Still in Control," *Bulletin of the Atomic Scientists* (May 1989): 22).
22. It is also possible that this faction hoped that by building the necessary infrastructure in advance, it could encourage a national decision to build nuclear arms, or perhaps it sought to ensure that it would "get a piece of the action" in the event that a decision to develop nuclear weapons was made at some later time.
23. "Former Nuclear Official Protests Military Role," *Fôlha de São Paulo*, November 11, 1986, translated in *JPRS-TND*, December 3, 1986.
24. "Parallel Program's Achievements, CNEN Secret Accounts Viewed," *Veja*, January 14, 1987, translated in *JPRS-TND*, March 4, 1987, p. 11.
25. "Joint Declaration on Nuclear Policy by the Governments of Brazil and Argentina," November 29, 1985. (Authors' files.)
26. Richard Kessler, "Argentina and Brazil Will Not Provide the IAEA with Information," *Nucleonics Week*, March 27, 1986, p. 9. By this point, support for the idea was growing among non-military elements of the government. This became clear in April, when a special inter-ministerial committee reviewing Brazil's troubled nuclear accord with West Germany released its final report. The committee, established in September 1985 and made up of industrialists and officials from a range of government departments, strongly endorsed the reciprocal inspection idea. See "Committee Reports on Talks with Argentina, Accord with FRG," *O Estado de São Paulo*, April 18, 1986, translated in *JPRS-TND*, June 13, 1986, p. 44; Richard Kessler, "Panel Favors Mutual Inspections by Argentina and Brazil," *Nucleonics Week*, May 1, 1986, p. 6.
27. "Nuclear Inspection Clause with Argentina Rejected," EFE, 1517 GMT, August 19, 1986, translated in *FBIS-LAT*, August 21, 1986, p. D-1. Nuclear cooperation was improved however, as each country agreed to notify the other in the case of nuclear accidents. See Richard Kessler, "Argentina-Brazil Protocol Promotes Nuclear Accident Accord," *Nucleonics Week*, July 31, 1986, p. 1.
28. Ivo Dawnay, "Brazil and Argentina to Set Seal on Collaboration," *Financial Times*, December 10, 1986; Richard Kessler, "Argentina-Brazil Protocols Detail Broad Nuclear Cooperation," *Nucleonics Week*, December 11, 1986, p. 9.

For additional details on cooperation between Brazilian and Argentine nuclear firms, see Richard Kessler, "Argentine-Brazilian Firms Pool Nuclear Design Know-How," *Nucleonics Week*, January 15, 1987.

29. Richard Kessler, "Sarney Visit to Pilcaniyeu Was Key to Reciprocal Inspections," *Nucleonics Week*, July 23, 1987, p. 11.
30. The facility, which has produced only low-enriched uranium but which is potentially capable of producing weapons-grade material, was secretly built by Argentina's military government between 1978 and 1983 and is not covered by IAEA inspections.
31. Richard House, "Brazil Says Uranium Enriched," *Washington Post*, September 10, 1987; Rik Turner and Michael Knapik, "Brazil's 'Parallel Program' Publicized by Sarney's Centrifuge Announcement," *Nuclear Fuel*, October 19, 1987, p. 3.
32. Also at this time, Brazil was completing the first module of a jet nozzle enrichment plant at Resende, which was subject to IAEA safeguards.
33. "Funding Needs, Suppliers of Parallel Program Discussed," *Exame*, October 14, 1987, translated in *JPRS-TND*, February 14, 1988, p. 21.
34. "Paper Discusses Plans for Nuclear Submarine," *O Estado de São Paulo*, April 12, 1987, translated in *FBIS-LAT*, April 22, 1987, p. D-2 (Navy's objective is to enrich uranium to a minimum of 70 percent, using 3,000 centrifuges); "Twenty Percent Enriched Uranium to be Produced by June," *Gazeta Mercantil*, February 5, 1988, translated in *JPRS-TND*, March 23, 1988 (20 percent-enriched uranium to be produced by June 1988; a "higher" level of enrichment to be achieved later).
35. Uranium enriched to 70 percent can itself be used for weapons, but it is not chosen for this purpose since physically smaller and more efficient devices can be made using material enriched to 90 percent or more.
36. "CNEN Official Speaks on Nuclear Program," *Fôlha de São Paulo*, September 6, 1987, translated in *FBIS-LAT*, September 10, 1987, p. 19.
37. Sarney has also had to play the role of broker during the country's transition to democracy between a military reluctant to cede power and a populace hungry for a greater say in the country's political affairs. One analysis of Sarney's announcement suggested that by extolling the military he was trying to blunt a controversial proposal by several opposition parties to prohibit the military under Brazil's new constitution from interfering in the country's internal affairs. See House, "Brazil Says Uranium Enriched" (see note 31).
38. Collor, who had been a dark horse candidate for much of 1989, came in first by a small plurality in the country's presidential balloting of November 15 and then won decisively in a December 17 run-off against socialist candidate Luis Inacio Lula da Silva.
39. The Resende operations were placed under a new state-owned corporation, Industriales Nucleares do Brazil (INB), but the CNEN controlled 51 percent of the stock of INB. The CNEN also obtained authority over Brazil's uranium production operations. The activities of the Brazilian Navy at Ipero and IPEN were not affected by the reorganization. See, "Navy Program Not Affected," Brasilia Domestic Service, 2200 GMT, September 1, 1988, translated in *JPRS-TND*, September 21, 1988, p. 22.
40. The reorganization and the role of the military are discussed in detail in Antonio Rubens Britto de Castro, Norberto Majlis, Luiz Pinguelli Rosa, and Fernando de Souza Barros, "Brazil's Nuclear Shakeup: Military Still in Control" (see note 21); "Text of INB Annual Report for 1988," *Correio Braziliense*, July 7, 1989, translated in *JPRS-TND*, September 18, 1989, p. 12. See also "Enrichment Process Changed," *O Estado de São Paulo*, September 2, 1988, translated in

FBIS-LAT, September 7, 1988, p. 33; "Major Changes for Brazilian Industry," *Nuclear Engineering International* (December 1988): 5; Rik Turner and Mark Hibbs, "Brazil's Nuclear Reorganization Limits Dual Program Development," *Nucleonics Week*, September 8, 1988, p. 1.

41. In late 1988, this facet of the reorganization triggered an embarrassing stand-off with the International Bank for Reconstruction and Development, the World Bank, one of the key supports for Brazil's shaky economy, where Electrobras was seeking a $500 million general purpose loan.

Historically, the bank had refused to make loans for nuclear energy projects, and, with the sudden inclusion of three such projects in Electrobras's portfolio, the bank balked at extending the $500 million line of credit. This led Brazil to try to establish the economic viability of the projects through a series of studies, but in March 1989, the bank declared that despite Brazil's analyses, it believed Angra III, then only ten percent complete, to be "non-viable." Brazil refused to abandon the project or shift it to another agency, creating an impasse that was not resolved as of mid-1990. See "Proposed World Bank Loan to Brazil for Electricity Prompts Opposition," *Nuclear Fuel*, December 26, 1988, p. 13; Ivo Dawnay, "Brazil Clashes with World Bank Over Nuclear Energy," *Financial Times*, December 1, 1988; "Government Decision 'Unjustified' " (José Goldemberg, "The Nuclear Agreement Mistakes"), *O Estado de São Paulo*, December 8, 1988, translated in *JPRS-TND*, February 6, 1989, p. 5; Rik Turner, "World Bank Rejects Angra-3 Viability," *Nucleonics Week*, March 3, 1989; Ivo Dawnay, "World Bank Strives to Mend Fences with Brazil," *Financial Times*, March 13, 1989; telephone interview with World Bank official.

One daily suggested in a series of editorials that the Brazilian military (through the CNEN) had a covert objective in defending Angra III to the World Bank. Completion of the facility was being pushed, the paper argued, in order to provide a legitimate "peaceful" rationale for enlarging Brazil's uranium enrichment capability; unless the third nuclear power plant were finished, the economies of scale necessary to justify the costs of such a project could not be realized. See "Paper Comments," *O Estado de São Paulo*, December 11, 1988, translated in *JPRS-TND*, January 25, 1989, p. 3; "Angra III Completion Sought," *O Estado de São Paulo*, January 26, 1989, translated in *JPRS-TND*, March 20, 1989, p. 12.

42. Constitution of Brazil (1988), Article 12.
43. Another factor that weakens the provision is that the constitution can be amended by simple majority vote of the Brazilian Congress, which could give the nuclear restrictions (and the remainder of the charter) little more permanence than ordinary legislation.
44. "Uranium Enrichment Process Perfected," *O Globo*, April 9, 1989, translated in *JPRS-TND*, April 25, 1989, p. 13.
45. "Uranium Enrichment Process Perfected," *O Globo* (see note 44); "20-Percent Uranium to be Produced," *Fôlha de São Paulo*, April 6, 1989, translated in *JPRS-TND*, May 23, 1989, p. 13.
46. "Nuclear Submarine Ready 'Within 5 or 6 Years,' " *Fôlha de São Paulo*, September 12, 1989, translated in *FBIS-LAT*, September 19, 1989; "Navy to Install PWR Nuclear Reactor in 1995," *Fôlha de São Paulo*, December 3, 1989, translated in *FBIS-LAT*, December 6, 1989; "Navy Official on Nuclear Submarine Program," Brasilia Radio Nacional da Amazonia Network, 1000 GMT, February 21, 1990, translated in *JPRS-TND*, March 2, 1990, p. 18.

In mid-1989, Pinheiro da Silva's goal had been to produce about 26 pounds (12 kilograms) of 20-percent enriched uranium by January 1990. See "Nuclear

Program Threatened," *O Globo*, July 11, 1989, translated in *FBIS-LAT*, July 13, 1989, p. 48.

47. In practice, the material is usually enriched to a level just below 20 percent, so that, technically speaking, it is still "low-enriched" uranium and thus cannot be used as the core of a nuclear device because it lacks the necessary physical properties.
48. For a similar assessment, see David Albright, "Bomb Potential for South America," *Bulletin of the Atomic Scientists* (May 1989): 16.

 According to a 1987 West German intelligence report, which was obtained by a parliamentary committee investigating illicit West German nuclear export activities, the original goal of the enrichment program was to produce a facility with 18,000 centrifuges, each capable of an output of three "separative work units" (SWUs) per year. This was scaled back to 1,000 centrifuges (3,000 SWUs), to be built by 1990. A plant of this size would permit an output of 33 pounds (15 kilograms) of weapons-grade material per year, possibly enough for a single nuclear device. See *ibid.*; Mark Hibbs, "Brazil Will Develop Domestic Fuel Cycle if Its Nuclear Power Capacity Warrants," *Nuclear Fuel*, September 18, 1989, p. 7.

 The 1987 report also stated that Brazil already had sufficient mastery of the uranium enrichment process to produce 70-percent enriched material, which approaches weapons-grade. See Mark Hibbs, "Germans Say Brazil Developing Two Production Reactors," *Nucleonics Week*, July 27, 1989.

49. "Nuclear Program Threatened," *O Globo*, July 11, 1989, translated in *FBIS-LAT*, July 13, 1989, p. 48. See Chapter 13.
50. "Nuclear Program, Enriched Uranium Plans Unveiled," *Jornal do Brasil*, May 22, 1988, translated in *JPRS-TND*, June 16, 1988, p. 1.
51. " 'Tricks' to Gain Nuclear Technology Disclosed," *O Globo*, April 9, 1989, translated in *JPRS-TND*, April 25, 1989, p. 11.
52. "Purchase of Nuclear Materials Reported," *Fôlha de São Paulo*, August 13, 1989, translated in *FBIS-LAT*, August 15, 1989, p. 28. A Gradel spokesman stated that the firm had not made any sales to the CNEN, however. Concerning earlier disclosures of secret CNEN accounts, see "Parallel Program's Achievements, CNEN Secret Accounts Viewed," *Veja*, January 14, 1987, translated in *JPRS-TND*, March 4, 1987, p. 11.
53. Hans Peter Schuetz, "There's Something Fishy Here," *Stern*, August 3, 1989, p. 110 (emphasis added).
54. "Nuclear Connection with Brazil Assessed," *Die Zeit*, July 28, 1989, translated in *JPRS-TND*, August 14, 1989, p. 37 (quoting Documents of the Federal Economics Ministry, Book Number 3/88, Volume 7a, p. 171).
55. Mark Hibbs, "Bonn: There is 'No Military Background' to Brazil's Unsafeguarded Program," *Nucleonics Week*, August 7, 1989, p. 13.
56. "Details of Secret Army Graphite Reactor Disclosed," *Fôlha de São Paulo*, October 24, 1989, translated in *JPRS-TND*, November 29, 1989, p. 14.
57. Hibbs, "Germans Say Brazil Developing Two Production Reactors" (see note 48). Plutonium extraction at this rate might allow the accumulation of 2.2 pounds of the material (1 kilogram) over a three-year period, the report stated. Since five to eight kilograms are needed for a nuclear device, it would be more likely that the IPEN unit would be used to permit Brazil to master this technology rather than as a source of plutonium for weapons, as such.

 The source of the plutonium-bearing spent fuel that might be processed at the IPEN unit was not specified in the report. If it were processing spent fuel from one of Brazil's IAEA-safeguarded reactors, then the IPEN unit, too, would have to be placed under the IAEA system, which it is not. Brazil is not known to possess a source of unsafeguarded spent fuel. (The "zero-power" MB-01 reactor, discussed below, cannot produce recoverable quantities of plutonium.)

A widely publicized press account in 1986 indicated that the facility was operating at that time. See "Brazil Says It Now Produces Small Amounts of Plutonium," *Washington Post*, December 18, 1986; Charles Thurston, "Brazil Denies Report of Reprocessing, Plutonium Production at IPEN," *Nuclear Fuel*, December 29, 1986, p. 9. The article may have been the result of confusion over a comment by CNEN officials, who stated that the organization had mastered the technology of plutonium extraction on a laboratory scale, but apparently meant only that the techniques involved had been successfully demonstrated using non-plutonium-bearing materials. Concerning allegations that the unit produced plutonium in 1983, see Spector, *Nuclear Proliferation Today* (see note 1), pp. 247-252.

58. Mark Hibbs, "Brazil Will Develop Domestic Fuel Cycle If Its Nuclear Power Capacity Warrants," *Nuclear Fuel*, September 18, 1989, p. 6.
59. Hibbs, "Brazil Will Develop Domestic Fuel Cycle" (see note 58).
60. Mark Hibbs, "Kohl Pressed By Opposition, IAEA Over Technology Sent to Brazil," *Nucleonics Week*, July 27, 1989, p. 5; Terence Roth, "Brazil Violating Nuclear Accord Files Indicate," *Wall Street Journal*, July 24, 1989. Under the 1975 agreement, *any* plutonium extraction plant using the standard "PUREX" process that is built by Brazil before 1995 is presumptively deemed to be based on German-supplied technology and is therefore to be subject to IAEA safeguards. The IPEN plutonium extraction laboratory would appear to fit this description, however, and is not under the IAEA system.
61. Hibbs, "Germans Say Brazil Developing Two Production Reactors" (see note 48). The report also noted that Brazil was processing and exporting beryllium, used as a neutron reflector in nuclear reactors and in nuclear weapons themselves.
62. Mark Hibbs, "West Germany to Investigate Charges of Brazilian Technology Diversions," *Nuclear Fuel*, September 4, 1989, p. 6.
63. The 1975 FRG-Brazil agreement provided that the IAEA was to be notified each time technology, in the form of blueprints or equipment, was transferred to Brazil so that the agency would be able to determine whether such equipment or replicas were being used improperly in an unsafeguarded installation. A January 1987 internal German Foreign Ministry report obtained by the parliamentary committee investigating the country's nuclear export activities stated that, in many instances, neither party had informed the IAEA of equipment transferred to the German-assisted Resende enrichment plant. As summarized by the trade journal *Nucleonics Week*:

    > According to a memorandum from the Foreign Office dated January, 27, 1987, IAEA safeguards officials informed Bonn that "a visit to the enrichment plant in Resende revealed that the IAEA had not been notified of all German equipment deliveries." The Brazilian record was worse, the IAEA reported, "On the Brazilian side, no single hardware delivery" had been reported to the agency after more than 10 years.

    Hibbs, "Kohl Pressed By Opposition, IAEA Over Technology Sent to Brazil" (see note 60).
64. *Ibid.*
65. In April 1990 the authors of this volume issued a report arguing that Brazil had engaged in clandestine transactions to acquire material and equipment for the coutry's unsafeguarded nuclear program. That report relied on many of the sources quoted above, especially the media accounts describing the internal

West German government documents discussed in the text. See Leonard S. Spector, with the assistance of Jacqueline R. Smith, *Nuclear Exports: The Challenge of Control* (Washington, D.C.: Carnegie Endowment for International Peace, April 1990), pp. 18-21.

In response to that study, Rex Nazareth and José Carvalho (the former and current directors, respectively, of the CNEN) issued blanket denials that Brazil had engaged in the illicit importation of any nuclear item. They did not, however, provide a direct rebuttal of the specific allegations in the report, which parallel those here. Nazereth, for example, offered no explanation of his contradictory comment, quoted above, that he would reveal the secret acquisition efforts of the parallel program "when I write my memoirs," nor did either official provide a rejoinder to Admiral Pinheiro da Silva's admission that he had personally obtained needed technology in West Germany under false pretenses. Both men were also silent regarding the transfer of West German-trained technicians and know-how from Brazil's safeguarded to its unsafeguarded program, which concerned the German Foreign Office. See "Charges of Smugggling Nuclear Equipment Refuted," *O Estado de São Paulo*, April 19, 1990, translated in *JPRS-TND*, June 12, 1990, p. 6.

The earlier study was also criticized by the Brazilian press for its reliance on press reports and its lack of first-hand information. See " 'Hypocrisy' Noted," *O Estado de São Paulo* (editorial), April 19, 1990, translated in *JPRS-TND*, June 12, 1990. Given the secrecy surrounding Brazil's unsafeguarded nuclear activities, such first-hand information is deliberately made scarce. The principal author of this volume, however, has seen copies of two of the internal German government documents discussed in the media reports cited above and has determined that they were accurately described in these accounts. There is no reason to believe that the other official documents have not been portrayed correctly, even though they have not been made available for verification.

66. "Brazilian Reactor Inaugurated," *Nuclear Engineering International* (February 1989): 5.
67. The Ipero facility was opened in April 1988. Since several months would have been needed to fabricate the fuel for the MB-01, it would have been necessary to obtain 4-percent enriched uranium in bulk form by mid-summer. When it began operating, however, Ipero was said to contain 50 to 75 centrifuges, and by the fall of 1988 had only 300. Albright, "Bomb Potential for South America" (see note 48). This appears to be insufficient to produce the 466 pounds of material required in the time available. Indeed, during all of 1989, the facility—whose capacity had presumably increased since the spring of 1988 with the installation of addition centrifuges—produced only 880 pounds of 5-percent enriched material. "Enriched Uranium Production Goals Perceived," *O Globo*, February 25, 1990, translated *JPRS-TND*, March 2, 1990, p. 18.
68. "Othon Reviews Successes of Nuclear Program," *O Estado de São Paulo*, March 13, 1990, translated in *JPRS-TND*, April 29, 1990, p. 12.

In the early to mid-1980s, through a West German middleman, China is believed to have sold unsafeguarded enriched uranium to such countries as South Africa and Argentina. Possibly Brazil also obtained some of this material from Beijing. In 1985, however, China and Brazil signed a nuclear cooperation agreement under which all transfers between them were to be placed under IAEA inspection, and by 1988, China had publicly pledged to subject all of its nuclear exports to this requirement. It therefore appears less likely that China would have made an unsafeguarded transfer of uranium to Brazil after 1985.

The 1987 West German intelligence agency report indicated that China supplied *un*enriched uranium to Brazil, apparently prior to 1985, a somewhat surprising transaction, inasmuch as Brazil produces uranium of its own. See Hibbs, "Bonn: There Is 'No Military Background' to Brazil's Unsafeguarded Program" (see note 55).

See also, "Physicists Call for Explanation of CNEN Debt," *Fôlha de São Paulo*, August 14, 1989, translated in *JPRS-TND*, October 6, 1989, p. 10 (suggesting that the fuel was obtained through the use of the secret CNEN account with the Banco di Roma in Luxembourg, noted in text, above).

69. Hibbs, "West Germany to Investigate Charges of Brazilian Technology Diversions" (see note 62); "Nuclear Cooperation Deals With Brazil Resumed," *Der Spiegel*, February 26, 1990, translated in *JPRS-TND*, March 2, 1990, p. 27.
70. Robert D. Shuey, *et al.*, *Survey of Emerging Missile Forces* (Washington, D.C.: Congressional Research Service, October 3, 1988) p. 92.
71. *Ibid*; "Sonda IV Rocket Successfully Launched 28 April," Brasilia Radio Nacional da Amazonia, 1000 GMT, April 29, 1989, translated in *FBIS-LAT*, October 25, 1989, p. 42.
72. Eustaquis de Freitas, "Tactical Missile Development Reported," *O Globo*, translated in *FBIS-LAT*, January 5, 1988, p.21.
73. *Ibid*; Christopher F. Foss, "New Family of Brazilian Missiles in Production," *Jane's Defence Weekly*, January 17, 1987.
74. "Warning on Missiles for Iraq," *Financial Times*, December 16, 1987; "Tactical Missile Development Reported," *O Globo*, January 3, 1988, translated in *FBIS-LAT*, January 5, 1988, p. 21; Alan Elsner, "Arms Expert Says Iraq Could Soon Have Ballistic Missiles," Reuters, AM cycle, December 15, 1987; Aaron Karp, "Ballistic Missile Development," *Journal of Defense and Diplomacy* (December 1987): 15; John Barham, "Brazil Ignores U.S. Protest Over Arms for Libya," *Sunday Times* (London), January 31, 1988.
75. "Brazil's Engesa May Be Following Avibras to Insolvency," *Defense and Foreign Affairs Weekly*, February 5-11, 1990, p. 2; "Engesa Asks Court Aid in Debt Liquidation," *O Estado de São Paulo*, March 22, 1990, translated in *FBIS-LAT*, April 17, 1990, p. 26.
76. "Brazil, Iraq to Develop Rockets Jointly," *O Globo*, March 19, 1989, translated in *FBIS-LAT*, April 6, 1989, p. 51; "Itamaraty Opposes Iraqi Satellite Construction, *Fôlha de São Paulo*, April 7, 1980, translated in *FBIS-LAT*, April 10, 1989, p. 35; "Government Offers to Sell Satellite to Iraq," *Fôlha de São Paulo*, June 19, 1989, translated in *FBIS-LAT*, June 22, 1989, p. 34.

It appears that Brazilian rocket scientists may have been recruited to assist Iraq's missile development program in 1988. See "Missile Technology Leak to Iraq Investigated," *O Globo*, May 15, 1990, translated in *FBIS-LAT*, May 17, 1990, p. 35.

77. Leonard S. Spector, *The Undeclared Bomb* (Cambridge, Mass.: Ballinger Publishing Company, 1988), p. 41; Shuey, *et al.*, "Missile Proliferation" (see note 70), p. 93.
78. Rik Turner, "Brazil Says Missile Technology Controls Hamper Launch Industry," *Defense News*, July 24, 1989; "Space and Missile Programs Discussed," *Correio Braziliense*, August 6, 1989, translated in *JPRS-TND*, October 26, 1989, p. 23.
79. David Silverberg, "French Proposal May Violate Pact on Proliferation," *Defense News*, July 17, 1989; "France and U.S. Discuss Brazilian Missile Deal," *Financial Times*, October 24, 1989; "Government Clarifies Position of Rocket Technology Transfer," *Le Quotidien de Paris*, October 7-8, 1989, translated in *JPRS-TND*, October 26, 1989, p. 45.

80. "O Globo: Nuclear Missiles Possible in Five Years," *O Globo*, January 13, 1986, translated in *FBIS-LAT*, January 16, 1986, p. D-1.
81. "Avibras Head Discusses PRC Joint Venture," *Fôlha de São Paulo*, June 16, 1989, translated in *FBIS-LAT*, June 19, 1989, p. 61; Radhakrishna Rao, "China's Launch Gets a NATO Boost," *South* (November 1989): 86.
82. "Brazil Pact Widened; Tlatelolco Considered," *Nuclear News* (October 1989): 59. The two countries would reduce tariff barriers in September, another indication of their cordial relations. See Ivo Dawnay, "Move on Brazil-Argentine Trade, *Financial Times*, September 13, 1989.

    The joint work on space-related projects was an apparent effort to offset the impact of restrictions on technology transfers imposed on both states under the Missile Technology Control Regime.
83. Mark Hibbs, "Argentine Officials Optimistic on Futures of Atucha-2, Exports," *Nucleonics Week*, October 12, 1989, p. 11.
84. The Brazilian Congress ratified Collor's proposals, reflecting his strong mandate on the issue. John Barham, "Brazil's Congress Agrees to Collor's Key Spending Cut," *Financial Times*, April 9, 1990.
85. See *e.g.*, "Government Decision 'Unjustified' " (see note 41).
86. "Santana [Carvalho] Replaces Nazareth as CNEN Chief," *O Estado de São Paulo*, March 22, 1990, translated in *JPRS-TND*, April 29, 1990, p. 18.
87. "Collor Selection of Military Ministers Viewed," *Istoe*, January 24, 1990, translated in *FBIS-LAT*, March 20, 1990, p. 53. Similarly, Collor sought to demilitarize the National Intelligence Service, purging numerous officers in favor of civilians. See John Barham, "Collor in Clash with Military," *Financial Times*, May 14, 1990; "Collor Demilitarization of Government Surveyed," *O Estado de São Paulo*, April 29, 1990, reprinted in *FBIS-LAT*, June 1, 1990, p. 39.
88. "Candidates on Nuclear Plan, Debt, Unemployment," *Fôlha de São Paulo*, September 10, 1989, translated in *FBIS-LAT*, October 6, 1989, p. 33.
89. "Santana [Carvalho] Replaces Nazareth as CNEN Chief" (see note 86).
90. "Secretary Denies Plans to Build Atomic Bomb," EFE, 2046 GMT, April 27, 1990, translated in *FBIS-LAT*, April 30, 1990, p. 42.
91. "Santana [Carvalho] Replaces Nazareth as CNEN Chief" (see note 86).
92. "Foreign Policy Changes Under Collor Previewed," *O Estado de São Paulo*, February 22, 1990, translated in *FBIS-LAT*, February 26, 1990.
93. "No Justification Seen in Not Signing the NPT," *O Estado de São Paulo*, February 25, 1990, translated in *JPRS-TND*, April 29, 1990, p. 20. The paper appeared to change positions abruptly in April, however. See " 'Hyprocrisy,' " *O Estado de São Paulo* (see note 65).
94. "Science Minister Discusses Nuclear Energy," *Jornal do Brasil*, April 8, 1990.
95. "Government Will Not Sign Treaty," *O Globo* May 25, 1990, translated in *FBIS-LAT*, May 29, 1990, p. 28.
96. Richard Kessler, "Panel Said to Endorse Integration of Brazilian/Argentine Programs," *Nuclear Fuel*, June 11, 1990, p. 5.
97. For a further discussion of this issue, see Chapter 13, "Prospects."

Part V: Africa - Introduction

1. "Nigerian FM Calls for Nuclear Weapons," *Defense and Foreign Affairs Daily,* August 26, 1987, p. 2.
2. "Babangida Remarks on Nuclear Policy," Agence France Presse, 1633 GMT, April 14, 1986, reprinted in *Foreign Broadcast Information Service-Near East and Africa,* April 15, 1986, p. T-7.
3. *Ibid.* On the possibility of black African proliferation, see Tunde Adeniran, "Nuclear Proliferation and Black Africa: The Coming Crisis of Choice," *Third World Quarterly* (October 1981): 673; Robert D'A. Henderson, "Nigeria: Future Nuclear Power?" *Orbis* (Summer 1981): 409; Ladi Adenrele, "Models of Nuclear Weapons Decision Strategy in Africa," *Arms Control* (September 1984): 148; Oye Ogunbadejo, "Africa's Nuclear Future," *Journal of Modern African Studies* (1984): 19.
4. But see, "Nigeria Offers Uranium in Exchange for Enrichment Technology," *O Estado de São Paulo,* September 11, 1987, translated in *Joint Publications Research Service-Nuclear Developments and Proliferation*, January 28, 1988, p.18 (alleging that Nigeria may be seeking to obtain uranium enrichment technology).

Chapter 15: South Africa

1. For a discussion of this period, see Leonard S. Spector, *Nuclear Proliferation Today* (New York: Vintage Books, 1984), pp. 297-282.
2. J.E. Spence, "The Republic of South Africa: Proliferation and the Politics of an 'Outward Movement,' " in Robert M. Lawrence and Joel Larus, eds., *Nuclear Proliferation Phase II* (Lawrence, Kans.: University Press of Kansas, 1974), p. 215. At the time China was aggressively cultivating ties with black African states and revolutionary groups. Chinese Premier Zhou Enlai toured Africa in 1964 and 1965.
3. *Ibid.*
4. Edouard Bustin, "South Africa's Foreign Policy Alternatives and Deterrence Needs," in Onkar Marwah and Ann Schulz, eds., *Nuclear Proliferation and the Near Nuclear Countries* (Cambridge, Mass.: Ballinger Publishing Company, 1975), p. 207.
5. Spence, "The Republic of South Africa" (see note 2), p. 215.
6. Zdnek Cervenka and Barbara Rogers, *The Nuclear Axis: Secret Collaboration Between West Germany and South Africa* (London: J. Friedman, 1978). See also, Murray Seeger, "Trial Testimony in S. Africa Indicates W. German A-Link," *International Herald Tribune*, November 26, 1975; *Facts on File* (1975), p. 789.

   In 1965, Great Britain is said to have provided Pretoria the technology for producing uranium hexafluoride, the form of uranium that is processed in an enrichment plant. See Spence, "Proliferation and Politics" pp. 217-218 (see note 2).

   The United States also authorized the sale of Foxboro computers, then the most advanced process control computers available, for use in the Valindaba pilot plant. See *Hearings on the Export Reorganization Act*, Committee on Government Operations, U.S. Senate, 94th Cong., 1st Sess., April 30 and May 1, 1975 (Washington, D.C.: U.S. Government Printing Office, 1975), pp. 68-69.
7. When the facility came on stream, Pretoria announced plans to build a commercial-scale enrichment plant, which was to produce significantly larger quantities of enriched uranium, both for export and for future domestic use as power reac-

tor fuel. (The project was later reduced to a "semi-commercial" scale facility, discussed below.)

8. This stance contrasted sharply with the aggressive efforts by the Nixon and Ford Administrations to curtail sensitive nuclear activities in Taiwan and South Korea and, before 1974, to warn India not to use plutonium made through the use of U.S. heavy water in the Cirus reactor for nuclear explosives.
9. "S. Africa Cites A-Bomb Capacity," *Washington Post*, July 12, 1984.
10. U.S. Central Intelligence Agency, "Prospects for Further Proliferation of Nuclear Weapons," DCI NIO 1945/74, September 4, 1974, p. 3. (Released to the Natural Resources Defense Council under the Freedom of Information Act.)
11. Thomas O'Toole, "French Get S. Africa A-Plant Sale," *Washington Post*, May 29, 1976. In addition to requiring that the Koeberg reactors be placed under IAEA safeguards, France gained South Africa's agreement that plutonium produced in the reactors' spent fuel would not be separated in South Africa, ensuring that the latter would not obtain access to this weapons-usable material.
12. International Institute for Strategic Studies, *The Military Balance*, 1971-2 through 1978-9, as cited in C. Raja Mohan, "Atomic Teeth for Apartheid: South Africa and Nuclear Weapons," in K. Subrahmanyam, ed., *Nuclear Myths and Realities: India's Dilemma* (New Delhi: ABC Press, 1981). The inauguration of the Carter Administration in January 1977, with its outspoken opposition to *apartheid* and support for majority rule in Namibia (the former German colony that South Africa had governed since World War I) meant that Pretoria would find little support in Washington.
13. "I Won't Call You a Liar," *Newsweek*, May 17, 1976, p. 53.
14. Confidential interview, spring 1985; Murray Marder and Don Oberdorfer, "How West, Soviets Moved to Head Off S. Africa A-Test," *Washington Post*, August 28, 1977.
15. Marder and Oberdorfer, "How West, Soviets Moved to Head Off S. Africa A-Test" (see note 14); Spector, *Nuclear Proliferation Today* (see note 1), pp. 291-293.
16. David Fishlock, "The South African Nuclear Weapons Scare," paper prepared for the Congressional Research Service, December 1977, p. CRS-5.
17. At the time the Kalahari test site was discovered in mid-1977, it may be noted, Washington reacted vigorously, given the clear evidence that a test site was being readied and the uncertainties that then prevailed as to Pretoria's enriched-uranium production capabilities.
18. Department of Political and Security Council Affairs, United Nations Centre for Disarmament, Report of the Secretary General, *South Africa's Plan and Capability in the Nuclear Field*, Report A/35/402 (New York: United Nations, 1980), pp. 30-31.
19. For a more detailed discussion, see Spector, *Nuclear Proliferation Today* (see note 1), pp. 299-300; "CBS Evening News with Walter Cronkite," February 21, 1980 (report by Dan Raviv); Benjamin Beit-Hallahmi, *The Israeli Connection* (New York: Pantheon, 1987), p. 134; but see James Adams, *The Unnatural Alliance* (New York: Quartet Books, 1984), p. 195 (source states that flash was not an Israeli test, but confirms nuclear collaboration between the two countries).
20. Caryle Murphy, "South Africa Powers Reactor with Uranium it Enriched," *Washington Post*, April 30, 1981. As noted earlier, the Ford Administration had suspended shipments of fuel for the U.S.-supplied SAFARI I reactor in 1975.
21. A widely cited 1980 U.N. Center for Disarmament study assumed South Africa was capable of producing 50 kilograms of highly enriched uranium annually,

beginning in March 1977. See *South Africa's Plan and Capability in the Nuclear Field* (see note 18). This estimate of the date of the first availability of South African highly enriched uranium now appears premature.

22. "Former IAEA Official on Bomb Production," *Star*, September 1986, reprinted in *Foreign Broadcast Information Service (FBIS)-Middle East and Africa*, September 17, 1985, p. U-16; *South Africa's Plan and Capability in the Nuclear Field* (see note 18).
23. Interviews with U.S. officials, Fall 1987. South Africa has argued that placing the pilot-scale enrichment plant under IAEA monitoring would divulge industrial secrets. However, it has never opened talks on the question with the IAEA to see if such disclosures could be avoided. See generally, summary of IAEA General Conference discussion of South African nuclear program, GC(XXVIII)/724 Annex 2 (Vienna: IAEA, 1984).

    One additional piece of circumstantial evidence that has subsequently come to light suggesting that Pretoria has indeed sought to build nuclear weapons is the report that in 1985 it obtained two 600-kilovolt flash X-ray machines from Sweden, through a clandestine purchasing network. The devices are used in nuclear weapon programs to test the non-nuclear "triggering package," a step that can increase confidence in the reliability of untested weapons. See Christer Larsson and Jan Melin, "Third World Countries Buy Swedish Nuclear Weapons Technology," *Ny Teknik*, May 2, 1986, p. 12, translated in *Joint Publications Research Service (JPRS)-Nuclear Development and Proliferation (TND)*, July 30, 1986, p. 1.
24. Nuclear Non-Proliferation Act of 1978, P.L. 95-242, amending the Atomic Energy Act of 1954, sections 127, 128 (1978).
25. Chester A. Crocker, "South Africa: A Strategy for Change," *Foreign Affairs* (Winter 1980/81).
26. In simplified terms, the Nuclear Non-Proliferation Act prohibited sales of nuclear fuel, reactors, and enrichment and plutonium extraction plants and know-how to South Africa unless it accepted safeguards on all of its nuclear activities. This left Pretoria still eligible to receive components for nuclear facilities; technology for facilities other than enrichment and plutonium plants; and dual-use equipment and technology (*i.e.*, those with both nuclear and non-nuclear applications) intended for nuclear purposes.
27. Ann MacLachlan, "U.S. Firms Play Role in South Africa Purchase of Uranium," *Energy Daily*, December 10, 1981; Thomas O'Toole, "U.S. Firms Help South Africa Get Uranium," *Washington Post*, April 13, 1982; Rob Laufer, "Kaiseraugst Purchase Seen As Freeing South Africa from U.S. Contract," *Nucleonics Week*, February 18, 1982, p. 1. The U.S. State Department learned of the U.S. uranium brokers' intentions before the sale to South Africa was arranged, but chose not to intervene. See *Hearings on Control of Exports to South Africa*, Subcommittee on Economic Policy and Trade, Committee of Foreign Affairs, U.S. House of Representatives, 97th Cong., 2nd Sess., February 9 and December 2, 1982 (Washington, D.C.: U.S. Government Printing Office, 1983), pp. 251-256.
28. See Spector, *Nuclear Proliferation Today* (see note 1), pp. 295-298; Rick Atkinson, "Reactor Operators Suspected of Working for South Africa," *Washington Post*, January 20, 1985.
29. Jack Anderson, "CIA Says China Has Sent A-Fuel to South Africa," *Washington Post*, July 23, 1981; "China Was Source of Atomic Fuel for South Africa, U.S. Believes," *Washington Post*, November 18, 1981; "China Says Study Shows It Wasn't Uranium Source," *Washington Post*, December 8, 1981; Judith

Miller, "U.S. Is Holding Up Peking Atom Talks," *New York Times*, September 19, 1982; Rob Laufer, "Interview With Malone," *Nucleonics Week*, August 19, 1982, p. 1; E.A. Wayne, "Secret Travels of Bombmaking Materials," *Christian Science Monitor*, July 8, 1988; "Ministerial Complicity in Illegal Material Shipments Charged," *Der Spiegel*, October 17, 1988, translated in *JPRS-TND*, December 23, 1988, p. 22.

There is still something of a mystery as to the source of the fuel used in the Koeberg units, which came on line in 1984 and 1985. Pretoria obtained 130 tons of low-enriched uranium from Belgian and Swiss sellers in 1981, with the aid of the U.S. uranium brokers, and France willingly agreed to fabricate the material into fuel rods. South Africa thus was able to bypass the U.S. fuel embargo. However, because each of the two Koeberg plants required 75 tons to start up and 30 tons for each annual refueling, Pretoria must have obtained additional supplies from some other outside source. China appears to be the source, although it denies having made the sale. See also, "China Has Been Selling Enriched Uranium Commercially Since 1981, Official Says," *Nuclear Fuel*, October 3, 1988, p. 11.

30. On a different front, Pretoria appeared to betray the Reagan Administration's conciliatory non-proliferation stand in 1983 and 1984 by quietly hiring twenty-five U.S. reactor operators and technicians to work at the Koeberg nuclear power station in disregard of U.S. regulations prohibiting such activities without special authorization. See Leonard S. Spector, *The New Nuclear Nations* (New York: Vintage Books, 1985), pp. 54-59. The Americans subsequently returned to the United States or were assigned to non-prohibited duties.

During 1984, Pretoria also reportedly sought to bypass supplier-country controls in the hopes of eventually obtaining a heavy water plant not subject to IAEA safeguards. Although no supplier nation would sell such a facility without IAEA controls, Pretoria was said to have entered into negotiations with the Swiss firm of Sulzer Brothers to obtain engineering services that would have enabled it to purchase unregulated components for such a plant and assemble them in South Africa. U.S. non-proliferation specialists feared that unsafeguarded heavy water might be used in a clandestine natural-uranium/heavy-water reactor to produce plutonium for nuclear arms. See Spector, *The New Nuclear Nations* (see above), pp. 220-222.

31. Preexisting contracts relating to the operation of the French-supplied Koeberg reactors were not to be affected, however. The July 1985 partial state of emergency was lifted on March 7, 1986, but a new, nationwide state of emergency was declared on June 12, 1986. See "Nuclear Hardware Not Affected," Umtata Capital Radio, 1100 GMT, July 25, 1985, reprinted in *FBIS-Middle East and Africa*, July 26, 1985, p. U-3.

32. Executive Order 12532. Exceptions were allowed to permit nuclear transfers for humanitarian reasons and in support of IAEA safeguards. These sanctions, originally imposed for a year, were extended in September 1986. See Bernard Weinraub, "Reagan Extending Limited Sanctions," *New York Times*, September 5, 1986. Concerning pre-existing U.S. nuclear restrictions, see note 26.

33. The change in the American stance reflected public and congressional concerns; the Reagan Administration had consistently opposed sanctions of any kind and implemented them only because of domestic political pressure. Interestingly, the most economically painful nuclear-related sanction—a prohibition on the *importation* of South African uranium—was not imposed at this time, and the United States and Western Europe continued to import significant quantities of the material; Canada, however, banned imports of uranium produced in South

African-administered Namibia. See James Branscome, "Canada Bans Future Conversion Contracts for Namibian Uranium at Eldorado Facility," *Nuclear Fuel*, July 29, 1985, p. 10; Dinah Wisenberg, "Antiapartheid Measure Poses Distant Threat to South African Uranium Imports," *Nuclear Fuel*, August 12, 1985, p. 3; James Branscome, "South Africa Will Offer SWU On World Market in 1988," *Nucleonics Week*, February 27, 1986, p. 1.

34. The six Commonwealth nations were Australia, the Bahamas, Canada, India, Zambia, and Zimbabwe; together with Britain they formed a seven-nation committee charged by the forty-nine member Commonwealth with developing a policy toward South Africa. Although the government of Prime Minister Margaret Thatcher in Great Britain rejected the sanctions package subsequently passed by the Commonwealth, it accepted a milder sanctions program proposed by the EC. The principal differences were that the Commonwealth package included a ban on extending landing rights to South African Airlines and a ban on new investment in South Africa. Mrs. Thatcher would not accept the former ban but did agree to "discourage" further British investment. Britain is one of South Africa's leading trading partners. Karen De Young, "Thatcher's Sanctions Stance Called 'Pathetic,' " *Washington Post*, August 6, 1986.

35. Anti-Apartheid Act of 1986, P.L. 99-440 (1986), section 307 (a).

The Anti-Apartheid Act prohibition on the importation of South African uranium into the United States (section 309) has been the subject of controversy, since it has been interpreted as applying only to uranium ore and oxide ("yellowcake") but not to more highly processed forms of uranium converted in third countries from South African uranium. Specifically, imports of uranium hexafluoride, a product made from yellowcake, are permitted under Treasury Department and Nuclear Regulatory Commission regulations if produced in third countries from South African uranium.

Uranium hexafluoride, a gas when slightly heated, is used as the feed for the uranium enrichment process. Thus the U.S. regulations permit uranium hexafluoride that has been produced abroad from South African uranium to enter the United States for enrichment in U.S. facilities. Thereafter the enriched uranium can be re-exported as fuel for foreign nuclear power plants or used as fuel in U.S. nuclear power reactors. The U.S. import ban thus affects only a portion of South Africa's uranium trade with the United States and does not affect U.S. uranium enrichment revenues. See Michael Knapik, "Treasury Decides All South African U_3O_8 [yellowcake] Is Banned from Import to U.S.," *Nuclear Fuel*, July 13, 1987, p. 12; "NRC Sets South African U. Import Ban," *Nuclear Fuel*, September 21, 1987, p. 13; "Congressmen Ask Court to Reverse NRC Policy on South African U. Import Policy," *Nuclear Fuel*, October 5, 1987, p. 12. This interpretation of the Anti-Apartheid Act was sustained after being challenged in federal court. See "Appeals Court Dismisses Challenge to NRC Rule on South African U Imports," *Nuclear Fuel*, December 26, 1988, p. 3.

36. Pretoria insisted on two provisos that the agency would not accept. First, it demanded the right to withdraw safeguarded enriched uranium produced in the facility for "non-explosive military purposes," *i.e.*, for nuclear submarine propulsion systems. Second, it insisted on the right to terminate the safeguards agreement in the event it decided that "extraordinary events related to the agreement have jeopardized its supreme interests" or in the event that its rights or privileges of membership in the IAEA were curtailed.

South Africa's demands were derived in part from stipulations found in IAEA safeguards agreements that are drawn up after a country has joined the NPT, and has therefore agreed to place all of its nuclear plants under IAEA inspection. These agreements allow the removal of safeguarded material for non-explosive military purposes, since this is permitted by the treaty itself, and they are subject to a country's right under Article X of the treaty to withdraw from the pact after ninety days if the country's supreme interests are jeopardized by actions relating to the subject of the accord. IAEA policy, however, has been to prohibit these reservations in safeguards agreements that cover only a single plant; such agreements come under safeguards rules that predate the NPT.

37. Ann MacLachlan and Gamini Seneviratne, "South Africa Hints NPT Signature, Averts IAEA Suspension Threat," *Nucleonics Week*, September 23, 1987.
38. MacLachlan and Seneviratne, "South Africa Hints NPT Signature . . ." (see note 37). For a procedural history of the events at the IAEA leading up to the General Conference vote, see "South Africa's Nuclear Capabilities: Report by the Board of Governors," IAEA Document GC(XXXI)/807 (August 24, 1987). The Soviet willingness to postpone sanctions—apparently out of a desire to avoid further politicization of the IAEA—was all the more surprising in view of the fact that the Soviet Union was fighting a proxy war with South Africa in Angola. Moscow had sent 1,000 advisors (and Cuba had sent 35,000 to 40,000 troops) to Angola to assist the Marxist government there to contain South African-backed forces of the National Union for the Total Independence of Angola (UNITA).
39. William Claiborne, "South Africa Acknowledges Fighting Soviets, Cubans in Angola," *Washington Post*, November 12, 1987.
40. William Claiborne, "S. African Military Says Intervention in Angola Staved Off Rebel Defeat," *Washington Post*, November 13, 1987.
41. "South African Leaders Visit Angola War Zone," *New York Times*, November 15, 1987. Heavy fighting continued in Angola into mid-1988.

    A second regional antagonist, Zimbabwe, was also in the process of significantly strengthening its military capabilities, and one report alleged that it was seeking to purchase the top-of-the-line MiG-29 from the Soviet Union. "Government Buying Arms from PRC, UK, USSR" (Clandestine) Radio Truth (Zimbabwe), 0430 GMT, March 9, 1988, reprinted in *FBIS-Near East and South Asia* (NES), March 11, 1988, p. 14.
42. These concerns were magnified by Pretoria's declining margin of military superiority vis-à-vis some of the front-line states in the region, a decline in large part attributable to the 1977 UN conventional arms embargo, which had prevented the country from obtaining advanced weaponry, especially aircraft, tanks, and naval equipment. Although the embargo had been far from airtight, and although South Africa had built up significant domestic conventional-arms production capabilities, some of its adversaries apparently obtained at least comparable weaponry from the Soviet Union, thereby slowly reducing Pretoria's equipment advantage.

    Angola, for example, is said to have MiG-21 and MiG-23 jet fighters in its arsenal. The Cheetah, an upgrade of South Africa's aging Mirage III/As, is said to be able to match the MiG-23, but Pretoria can produce the aircraft in only limited numbers. See "General Malan Comments," Johannesburg SAPA, 0025 GMT, July 16, 1986, reprinted in *FBIS-Middle East and Africa*, July 16, 1986, p. U-4; Glenn Frankel, "U.N. Arms Ban Proves Costly to South Africans," *Washington Post*, February 21, 1985; Christopher Coker, "South Africa: A New Military Role in Southern Africa," *Survival* (March/April 1983): 59; "T.V. Reviews Weapons Buildup in Southern Africa," Johannesburg Television Service, 1800 GMT, November 6, 1985, reprinted in *FBIS-Middle-East and Africa*, November 7, 1985, p. U-13.

For a detailed analysis reviewing South Africa's declining advantage in conventional armaments, see Christopher Coker, *South Africa's Security Dilemmas* (Washington, D.C.: Praeger/Center for Strategic and International Studies, 1987).

43. "Media Round-up," *Current News*, July 8, 1988, p. 18.
44. "Pretoria Says It Can Build A-Arms," *New York Times*, August 14, 1988.
45. Ann MacLachlan, "Israel, South Africa Remain in IAEA for Another Year," *Nuclear Fuel*, October 3, 1988, p. 10.
46. Ann MacLachlan, "Stumpf to Succeed De Villiers as Head of South African Atomic Energy Corp.," *Nuclear Fuel*, October 30, 1989, p. 5. Low-enriched uranium produced in the plant in excess of that required to fuel the Koeberg reactors was to be exported.
47. *Ibid.*; "Uranium Enrichment Plant at Valindaba to Start Commercial Production," *Financial Mail (Energy Supplement)*, July 24, 1987.
48. When operating at full capacity, the facility will have an annual output of 300,000 separative work units (the measure of enrichment plant capacity). Five thousand separative work units are sufficient to produce 55 pounds (25 kilograms) of uranium enriched to 90 percent or more. A nuclear device would require 33 to 55 pounds (15 to 25 kilograms) of the material, depending on its design.
49. Chester A. Crocker, "Southern Africa: Eight Years Later," *Foreign Affairs* (Fall 1989): 144. Although Pretoria had generally been successful in keeping its regional adversaries off balance, its hardline policies were increasingly costly, a factor that contributed to its desire for a settlement. As one analyst has stated:

    > South Africa was concerned about the cost of subsidizing the Namibian economy and security operations, and feared additional Western sanctions if it continued to refuse to grant Namibia independence. When its military became overextended at Cuito Cuanavale in southern Angola in early 1988 and suffered significant white casualties at the hands of newly reinforced Cuban troops, South Africa became even more interested in a settlement.

    See Gillian Gunn, "Unfulfilled Expectations in Angola," *Current History* (May, 1990): 213.
50. Christopher Wren, "Cutback in Draft for South Africa," *New York Times*, December 8, 1989; William Claiborne, "South Africa to Cut Military," *Washington Post*, December 8, 1989; Helmoed-Romer Heitman, "Major Cuts Will Go Ahead in S. Africa," *Jane's Defence Weekly*, January 27, 1990, p. 139.
51. Wren, "Cutback in Draft for South Africa" (see note 50); Patti Waldmeir, "Unthinkable Starts to Look Inevitable," *Financial Times*, December 28, 1989; Pauline Baker, "South Africa on the Move," *Current History* (May 1990): 197.
52. The state of emergency was retained in Natal Province because of continued fighting between rival black groups.
53. Ann MacLachlan, "Soviets Say They Have Got Assurances That South Africa Will Adhere to NPT," *Nucleonics Week*, October 2, 1989, p. 5.
54. "Notes 'Progress,' " Johannesburg Domestic Service, 0600 GMT, December 13, 1989, reprinted in *FBIS-Africa (AFR)*, December 13, 1989, p. 9.
55. "SA Willing to Sign No-Nukes Treaty," *SA Dialogue* (January 1990): 8; Paul Lewis, "South Africa Seen as Ready to Sign Nuclear Pact," *New York Times*, March 21, 1990.
56. Michael Knapik and Ann MacLachlan, "Closing of Pilot Enrichment Plant Seen as Sign that South Africa Will Sign NPT," *Nuclear Fuel*, May 14, 1990, p. 1; Mark Hibbs, "South Africa Says NPT Signature Is Near," *Nuclear Fuel*, May 28, 1990, p. 7.

57. Hibbs, "South Africa Says NPT Signature Is Near" (see note 56).
58. "Corporation to Close Uranium Enrichment Plant," Johannesburg Domestic Service, 0500 GMT, January 27, 1990, reprinted in *FBIS-AFR*, January 31, 1990, p. 5.
59. Confidential interview, Washington, D.C., June 1990; Michael Knapik and Ann MacLachlan, "Closing of Pilot Enrichment Plant Seen as Sign . . ." (see note 56).
60. Knapik and MacLachlan, "Closing of Enrichment Plant Seen as Sign . . ." (see note 56).
61. An additional question concerns the throughput of the facility. By processing higher-than-normal quantities of natural uranium and leaving a higher-than-normal quantity of uranium-235 in the plant's residue, or tailings, the amount of weapons-grade uranium could be increased over expected levels.
62. Judging from past international reactions to South African activities, foreign states would probably be more likely to accept Pretoria's *bona fides* on the matter if it continued to make progress on domestic political issues, as well.
63. "NBC Nightly News," National Broadcasting Company, October 25, 1989.
64. "NBC Sunrise," National Broadcasting Company, October 26, 1989; "Officials Continue to Deny South African Links," Jerusalem Domestic Service, 0500 GMT, October 30, 1989, reprinted in *FBIS-NES*, October 30, 1989, p. 36; "Pik Botha Denies Israeli Missile Links," SAPA, 1114 GMT, October 27, 1989, reprinted in *FBIS-AFR*, October 30, 1989, p. 18.

    Some Israeli denials contained an element of ambiguity. In a statement to NBC, for example, a spokesman for Israeli Prime Minister Yitzhak Shamir stated, "We absolutely deny that there is any link between Israel and South Africa in the nuclear field. And therefore there is no truth in this [NBC] story that was published yesterday night. Our cabinet decided in March 1987 that there is going to be no new contract between Israel and South Africa, and Israel scrupulously adheres and reenforces this decision of the cabinet." (The cabinet decision was taken in response to U.S. legislation that could lead to the termination of U.S. military aid to countries that engage in military sales to South Africa in violation of the 1977 United Nations arms embargo.) The Israeli spokesman's statement did not explicitly rule out, however, that the two countries were collaborating in the area of missile technology (as distinct from nuclear technology) under contracts signed prior to March 1987. See also, Michael Gordon, "U.S. Says Data Suggest Israel Aids South Africa on Missile," *New York Times*, October 27, 1989.
65. Michael Gordon, "U.S. Says Data Suggest . . ." (see note 64); David B. Ottaway and R. Jeffrey Smith, "U.S. Knew of 2 Nations' Missile Work," *Washington Post*, October 27, 1989.
66. See, *e.g.*, James Adams, *The Unnatural Alliance* (see note 19), p. 170.
67. It has also been alleged that the two countries were collaborating on the design of a nuclear-powered submarine. Charges that they were building a joint long-range-missile testing facility on Marion Island in the Antarctic and that South Africa was supplying Israel with enriched uranium for nuclear weapons date back to the mid-1980s. See Beit-Hallahmi, *The Israeli Connection* (see note 19) pp. 123, 129; "South Africa Reportedly May Test Atomic Weapons," United Press International, AM cycle, December 26, 1986. Charges of collaboration on a nuclear-powered submarine would be consistent with submarine acquisition activities in the two countries; Israel is acquiring two German conventional subs and South Africa has acquired the blueprints of the same design. This particular German submarine is being used by India and Brazil to develop the necessary

technology and know-how to advance their respective indigenous nuclear submarine programs. South Africa and Israel could harbor the same intentions in tandem. South Africa publicly claims to have dropped its conventional submarine plans, although reports have surfaced that the program is proceeding. The need to produce fuel for a nuclear-powered submarine, moreover, could establish a rationale for South Africa to accumulate highly enriched uranium. See Chapter 5.

68. Thomas L. Friedman, "Israel Approves Curbs on Pretoria," *New York Times,* September 17, 1987.
69. "Israel Offers Re-engined Kfir," *Flight International*, April 25, 1987, p. 10; Glenn Frankel, "Israeli Economy Depends on No-Questions-Asked Arms Sales," *Washington Post,* December 12, 1986; "Israel to Curb Arms for Pretoria," *New York Times,* March 19, 1987. The principal difference between the Kfir and the Cheetah is that the former uses the more powerful U.S. GE J79 engine.
70. See Hirsh Goodman, "RSA Seeking to Lure Lavi Engineers, Technicians," *Jerusalem Post,* November 9, 1987, reprinted in *FBIS-NES,* November 10, 1987, p. 44.
71. U.S. Department of State, "Report to Congress Pursuant to Section 508 of the Comprehensive Anti-Apartheid Act of 1986: Compliance with the U.N. Arms Embargo," April 2, 1987. See also, "Rabin Resisting 'Pressure' on S. African Contracts," *Jerusalem Post*, January 5, 1990, reprinted in *FBIS-NES*, January 10, 1990, p. 28; "Shamir: No New Agreements with South Africa," Jerusalem Domestic Service, 1800 GMT, February 11, 1990, reprinted in *FBIS-NES*, February 12, 1990, p. 29.
72. "Police Say Right Wing Planned Attack on de Klerk," SAPA, 1629 GMT, December 8, 1989, reprinted in *FBIS-AFR*, December 11, 1989; Christopher Wren, "South Africa's Infuriated Rightists Vow a Battle to Defend Apartheid," *New York Times*, February 6, 1990; David B. Ottaway, "Conservatives in S. Africa Want New Vote," *Washington Post*, February 8, 1990; David B. Ottaway, "Right-Wing Afrikaners Call for Early Elections," *Washington Post*, June 27, 1990.
73. Indeed, in 1988, when Pretoria was hinting that it might consider joining the pact, the Conservative Party strenuously objected. See "CP Highlights Disadvantages of Signing NPT," SAPA, 0902 GMT, August 16, 1988, reprinted in *FBIS-AFR*, August 16, 1988, p. 14.

Chapter 16: Controls and Safeguards

1. Article X.2 of the Non-Proliferation Treaty states that twenty-five years after the NPT enters into force, a conference shall be convened to decide on treaty extension. This meeting will take place in 1995.
2. For a detailed analysis previewing the 1990 NPT Review Conference, see Harald Müller and Richard Kokoski, *The Non-Proliferation Treaty: Political and Technological Dangers in 1990* (Stockholm: Stockholm International Peace Research Institute, April 1990). See also, Warren Donnelly, *The Nuclear Non-Proliferation Treaty: Upcoming International Review*, CRS Issue Brief (Washington, D.C.: Congressional Research Service, 1990).
3. In an initiative indirectly related to the NPT Review Conference, in April 1989, a group of forty neutral and non-aligned NNWS's, led by Mexico, Indonesia, Sri Lanka, Yugoslavia, Venezuela, and Peru, formally requested a conference to amend the Partial Test Ban Treaty in a manner that would transform it into a comprehensive ban on nuclear tests (CTB). The Partial Test Ban Treaty currently prohibits nuclear tests in the atmosphere, under water, and in outer space, but not underground. The accord provides, however, that it can be amended by a majority of its parties, together with all three of the nuclear-weapon state parties, the United States, the Soviet Union, and Great Britain; this arrangement gives each weapon-state a *de facto* veto over any proposed amendment.

   The initiators of the proposal had hoped to hold the Partial Test Ban Treaty Amendment Conference, and force a vote on the comprehensive test ban amendment, before the NPT Review Conference. They reasoned that the United States would be reluctant to veto the CTB amendment so soon before the NPT meeting, for fear that this would trigger a controversy there that might block agreement on a Final Document endorsing the non-proliferation pact.

   The nuclear-weapon-state parties to the Partial Test Ban Treaty successfully parried this gambit, however, by scheduling the amendment conference in January 1991, six months after the NPT meeting.
4. Fred Francis, "NBC Nightly News," National Broadcasting Company, October 28 and 29, 1989.
5. U.S. willingness to walk out of the 1982 IAEA General Conference after Israel's credentials were rejected (in response to its 1981 bombing of Iraq's Osiraq reactor) illustrated Washington's commitment to preventing Israel's international isolation. Subsequent Israeli assurances that it would not engage in future attacks on peaceful nuclear facilities eased this crisis at the IAEA. Nonetheless, at the 1985 NPT Review Conference, the Arab bloc pressed for the inclusion of language in the final statement of the session calling for sanctions against Israel and South Africa unless they signed the NPT. The proposed wording was subsequently weakened at U.S. insistence.
6. "IAEA Board of Governors Concludes," IAEA Press Release No. 89/15, June 16, 1989.
7. To compare this data with that from 1986, see Leonard S. Spector, *The Undeclared Bomb* (Cambridge, Mass.: Ballinger Publishing Company, 1988), pp. 312-313.
8. "IAEA To Supervise Nuclear Accord With PRC," Madrid EFE, 1532 GMT, January 16, 1988, translated in *Foreign Broadcast Information Service (FBIS)-Latin America (LAT)*, January 20, 1988, p. 35.
9. IAEA document GC(XXXIII)/873, July 1989, pp. 67-68.
10. IAEA document INFCIRC/369, October 1989, cited in *Programme for Promoting Nuclear Non-Proliferation Newsbrief* (Winter 1989/90): 5.

11. "Official Signs IAEA Safeguard Agreement," Hanoi VNA, 1445 GMT, November 2, 1989, reprinted in *Joint Publications Research Series (JPRS)-Nuclear Development and Proliferation (TND)*, November 29, 1989, p. 9.
12. Shahid-ur-Rehman Khan, "China Agrees to Supply 300-MW PWR to Pakistan," *Nucleonics Week*, November 23, 1989, pp. 1-2.
13. "Research Reactor Inaugurated," *Nucleonics Week*, December 14, 1989, p. 15.
14. West Germany also enacted legislation to tighten its nuclear export controls. See Chapter 4.
15. This relatively short distance was set as the criterion to ensure that the missiles usable in the Middle East theater would be covered. The payload standard is based on the assumption that emerging nuclear states will be unable to build nuclear warheads weighing less than 1,100 pounds.
16. "Joint Soviet-American Statement," TASS, February 10, 1990, reprinted by Federal News Service, SB-6-1, p. 3.
17. Don Oberdorfer and David Hoffman, "Scowcroft Warned China of New Hill Sanctions," *Washington Post*, December 15, 1989; "Bush Bid to Fix Beijing Ties Strains Those with Hill," *Congressional Quarterly*, December 16, 1989, pp. 3434-3436. But see Michael Gordon, "Beijing Avoids New Missile Sale Assurances," *New York Times*, March 30, 1990.
18. Victor Mallet, "Benelux Joins Missile Club Accord," *Financial Times*, May 17, 1990.
19. Richard Kessler, "Argentina, Brazil Agree to Mutual Inspection of Nuclear Facilities," *Nucleonics Week*, March 14, 1985, pp. 14-15.
20. The Ipero Declaration on Nuclear Policy pledged "[t]o improve existing mechanisms of political and technical cooperation by increasing the number of reciprocal visits and exchanges of information aimed at broadening the reciprocal acquaintance with the respective nuclear programs so as to optimize technological complementation and reciprocal trust." It was signed in Ipero, São Paulo State, Brazil, on 8 April 1988. See *Fôlha de São Paulo*, April 8, 1988, p. A-8, translated in *JPRS-TND*, May 5, 1988, p. 6.
21. In the latter statement, the two leaders pledged continued efforts to promote close contacts, joint projects, and exchanges of information, experience and technicians. (Ezeiza Declaration: Joint Declaration on Nuclear Policy, unofficial translation provided courtesy of the Embassy of Argentina.)
22. "Pakistan and India Sign Pact," *Washington Post*, January 1, 1989.
23. Both NFZs and NWFZs are directed primarily at deployed, or deployable, nuclear weapons. Nuclear free zones, however, can restrict, or attempt to restrict, broader nuclear activities such as nuclear waste disposal and nuclear testing.
24. Treaty on Principles Governing the Activities of States in the Exploration and Use of Outer Space, Including the Moon and Other Celestial Bodies.
25. Treaty on the Prohibition of the Emplacement of Nuclear Weapons and Other Weapons of Mass Destruction on the Seabed and the Ocean Floor and in the Subsoil Thereof.
26. U.S. Arms Control and Disarmament Agency (ACDA), *Arms Control and Disarmament Agreements* (Washington, D.C.: U.S. ACDA, 1982), p. 99.
27. Richard Kessler, "Argentine Position on NPT, Tlatelolco Unchanged, Says New Foreign Minister," *Nuclear Fuel*, September 4, 1989, p. 7.
28. The treaty also prohibits the disposal of nuclear waste within the boundaries of the zone.
29. See Richard H. Ullman, "The Covert French Connection," *Foreign Policy* (Summer 1989): 3.

Appendices

Appendix A: Producing Nuclear-Weapons Material – A Primer

It is generally accepted today that designing an atomic bomb—drawing the blueprint—is within the capabilities of most nations. Indeed, a number of American college students have come up with workable designs based on unclassified material.

The major technical barrier to making a nuclear device is obtaining the nuclear material for its core. Twenty-five kilograms (fifty-five pounds) of highly enriched uranium, or eight kilograms (about eighteen pounds) of plutonium are generally considered the necessary minimum, although in both cases more sophisticated designs relying on high compression of the core material, neutron reflecting "tampers," or both, enable a bomb to be built with considerably less material—perhaps fifteen kilograms of highly enriched uranium or five kilograms of plutonium, and even smaller amounts can apparently be used. Neither of these materials occurs in nature, however, and highly complex and expensive facilities must be built and operated in order to make them—an undertaking of considerable difficulty for developing nations without assistance from more advanced nuclear-supplier countries.

Highly enriched uranium. To make a weapon from uranium, the unstable "isotope" of uranium having a total of 235 protons and neutrons in its nucleus (U^{235}) is used. Since natural uranium consists of less than one percent U^{235}, while nuclear weapons use material that is made up of 90 percent or more U^{235}, natural uranium must be upgraded at an *enrichment plant* to achieve this concentration. Uranium enrichment is an extremely complex process and requires considerable investment. For this reason, the uranium enrichment route has been generally considered a less likely path to proliferation than the plutonium option. However, South Africa, Argentina, Brazil, India and (with extensive outside aid obtained mostly by clandestine means) Pakistan have all developed independent uranium enrichment capabilities, and Israel is known to be conducting research in the field.

Enriched uranium can also be used as a fuel in nuclear power or research reactors. The power reactors used in the United States and most other countries (called "light-water reactors") use *low-enriched uranium* fuel, i.e., uranium that has been enriched up to three percent U^{235}. Thus a country can have entirely legitimate, non-weapons-related reasons for developing uranium enrichment technology even though the same technology can be used to upgrade uranium to the high enrichment level useful for nuclear weapons. On the other hand, developing a sizable independent uranium enrichment capability is economically justifiable only for nations with large domestic nuclear power programs or significant potential export markets.

Because highly enriched uranium is sometimes used to fuel research reactors, a nation can have legitimate reasons for obtaining small quantities of this material, despite its usefulness in nuclear explosives. In recent years the United States and France have developed lower-enriched uranium fuels that can be used in lieu of highly enriched material in most of these reactors, however, considerably reducing the proliferation risks these research facilities pose.

Producing highly enriched uranium entails many steps apart from the enrichment process itself, and many other installations and capabilities are necessary. For nations wishing to obtain highly enriched uranium without international restrictions prohibiting its use for nuclear explosives, all of these would have to develop independently, or obtained illegally, since virtually all nuclear exporter states are unwilling to sell nuclear equipment and materials unless recipients pledge not to use them for nuclear explosives purposes and to place them under the inspection system of the International Atomic Energy Agency. (See Appendix C.) For illustrative purposes the basic nuclear resources and facilities that would be needed include:

- uranium deposits;
- a uranium mine;
- a uranium mill (for processing uranium ore containing less than one percent uranium into uranium oxide concentrate, or yellowcake);
- a conversion plant (for purifying yellowcake and converting it into uranium hexafluoride, the material processed in the enrichment plant);
- an enrichment plant (for enriching the uranium hexafluoride gas in the isotope U^{235}); and
- a capability for converting the enriched uranium hexafluoride gas into solid uranium oxide or metal (a capability usually associated with a yellowcake-to-hexafluoride conversion plant).

Plutonium. To obtain plutonium a country needs a *nuclear reactor*. This can be one designed specifically to maximize plutonium production (a "production reactor"), a large research reactor, or a power reactor for producing electricity. *Uranium fuel,* usually in the form of uranium-filled tubes (fuel rods) made of zirconium alloy (zircaloy) or aluminium, is placed in the reactor. For most production, power, and for a number of large research reactors, the fuel itself is either natural or low-enriched uranium, which is not usable for nuclear weapons at this point. As the reactor operates, the uranium fuel is partly transformed into plutonium. This is amalgamated in the fuel rods with unused uranium and highly radioactive waste products, however, and must then be extracted. To do this, "spent" fuel rods are taken to a *reprocessing plant* where they are dissolved in nitric acid and the plutonium is separated from the solution in a series of chemical processing steps. Since the spent fuel rods are highly radioactive, heavy lead casks must be used to transport them, and the rooms at the reprocessing plant where the chemical extraction of the plutonium occurs must have thick walls, lead shielding, and special ventilation to prevent radiation hazards.

Although detailed information about reprocessing was declassified by the United States and France in the 1950s and is generally available, it is still a complex procedure from an engineering point of view, and virtually every nation at the nuclear-weapons threshold that has attempted it—Argentina, Brazil, India, Iraq, Israel, and Pakistan—has sought outside help from the advanced nuclear-supplier countries.

Like enrichment facilities, however, reprocessing plants can also be used for legitimate civilian purposes, because plutonium can be used as fuel in nuclear-power reactors. Indeed, through the 1970s it was generally assumed that as the use of nuclear power grew, worldwide uranium resources would be depleted and plutonium extracted from spent fuel would have to be "recycled" as a substitute fuel in conventional power reactors.

In addition, research and development is under way in a number of nations on a new generation of reactors known as breeder reactors. These use plutonium as fuel surrounded with a "blanket" of natural uranium; as the reactor operates, slightly more plutonium is created in the core and the blanket together than is consumed in the core, thereby "breeding" new fuel.

Like plutonium recycling, the economic advantages of breeders depends on natural uranium becoming scarce and expensive. But over the past decade new uranium reserves have been discovered, nuclear power has reached only a fraction of its expected growth levels, and reprocessing spent fuel to extract plutonium (a critical step in both cases) has proven far more expensive and complex than anticipated. Moreover, concern over the proliferation risks of widescale use of plutonium as a fuel has grown. These factors have led one advanced nation, the United States, to abandon its plans to use plutonium fuel, although Japan, France, Britain, and West Germany are continuing to develop this technology actively. Nevertheless, the advanced nuclear-supplier countries are strongly discouraging plutonuim use in nations of proliferation concern.

The longstanding view that plutonium is a legitimate and anticipated part of civilian nuclear programs, however, has allowed Argentina, Brazil, India, and Pakistan to justify their reprocessing programs—even though they currently provide these nations with a nuclear-weapons capability, or may soon do so.

Like the production of enriched uranium, the production of plutonium entails many steps, and many installations and capabilities apart from the reactor and reprocessing plant are needed. Again for illustrative purposes, the following facilities and resources would be required for an independent plutonium-production capability assuming a heavy-water research or power reactor, employing natural uranium fuel, is used:

- uranium deposits;
- a uranium mine;
- a uranium mill (for processing uranium ore containing less than one percent uranium into uranium oxide concentrate, or yellowcake);
- a uranium purification plant (to further improve the yellowcake into reactor-grade uranium dioxide);
- a fuel fabrication plant (to manufacture the fuel elements placed in the reactor), including a capability to fabricate zircaloy or aluminium tubing;
- a heavy-water research or power reactor;
- a heavy-water production plant; and
- a reprocessing plant.

If the option of building a natural-uranium/graphite reactor is used, the needs would be the same although reactor-grade graphite would have to be produced instead of heavy water. A light-water reactor would necessitate use of low-enriched uranium, implying an enrichment capability was available; if so, highly enriched uranium could, in theory, be produced, obviating the need for plutonium as a weapons material.

Appendix B: Nuclear Weapons—A Primer

A nuclear weapon is a device in which most or all of the explosive energy is derived from either fission, fusion, or a combination of the two nuclear processes. *Nuclear fission* is the splitting of the nucleus of an atom into two (or more) parts. Highly enriched uranium and plutonium, when bombarded by neutrons, will release energy and emit additional neutrons while splitting into lighter atoms. In *nuclear fusion,* light isotopes of hydrogen, usually deuterium and tritium, join at high temperatures and similarly liberate energy and neutrons.

Fission Weapons. Many heavy atomic nuclei are capable of being fissioned; but only a fraction of these are *fissile,* which means fissionable by neutrons with a wide range of velocities. It is this property of fissile material, principally U^{235} and Pu^{239}, that allows a chain reaction to be achieved in weapons employing the fission process. In a *chain reaction,* fissile nuclei that have been bombarded by neutrons split and emit two or more neutrons, which in turn induce proximate nuclei to fission and sustain the process. With each successive fission "generation" additional energy is released, and, if the fission of one nucleus induces an average of more than one fission in the following generation, the energy yield of each generation is multiplied. A fission explosion in the range of 1 to 100 kilotons for example, would occur over a few microseconds and involve over fifty generations, with 99.9 percent of the energy released coming in the last seven. The minimum mass of material necessary to sustain a chain reaction is called the *critical mass.* This value may be lowered by increasing the material's density through compression or by surrounding it with "reflectors" to minimize the escape of neutrons; this makes it difficult to pin down the precise amount of uranium or plutonium required for a bomb.

Two basic nuclear-weapon-design approaches that are used to achieve a supercritical mass (i.e., exceeding the critical level) are the implosion technique and the gun assembly technique. In the *implosion technique,* a peripheral charge of chemical high explosive is uniformly detonated to compress a subcritical mass of plutonium or highly enriched uranium into a supercritical configuration. In the *gun assembly technique,* two (or more) subcritical masses of highly enriched uranium (plutonium cannot be used) are propelled together by a conventional explosion, resulting in a supercritical mass. In both cases, a *tamper* may be used to keep the material from exploding before enough generations of a chain reaction have occurred, and this tamper often doubles as a reflector to reduce the escape of electrons.

Fusion weapons. Fusion of light atomic nuclei requires a high density of fusion material and extraordinary heat, both of which are provided by a fission explosion in a "thermonuclear" or "hydrogen" bomb. Lithium-6 deuteride is the most widely used thermonuclear material, serving as a source of both deuterium and tritium, the atoms whose nuclei merge, in a fusion weapon.

In a *boosted weapon,* fusion material is introduced directly into (or next to) the core of fissile material, improving the efficiency of a fission weapon and thus increasing the yield of a given quantity of highly enriched uranium or plutonium. Although energy is released in the fusion reaction of a boosted weapon, the primary contribution of the fusion material to the explosion is that it provides additional neutrons for the fission process and therefore allows a more rapidly multiplying chain reaction to occur.

Other thermonuclear weapons are designed to capitalize on the energy released in a "secondary" fusion reaction triggered by a "primary" fission explosion. In such devices, fusion material is kept physically separate from a fissile or boosted fissile core that compresses and ignites it. Additional "stages" of fusion or fission material may be included to augment the weapon's yield, with each layer being triggered by ones closer to the core. For example, the *hydro-*

gen bomb includes a third stage or "blanket" of natural uranium, a widely available fissionable but not fissile material, that is fissioned by fast neutrons from the primary and secondary fission and fusion reactions. Hence, the energy released in the explosion of such a device stems from three sources—a fission chain reaction, the first stage; "burning" of the thermonuclear fuel, the second stage; and the fission of the U-238 blanket, the third stage—with, very roughly, half the total energy stemming from fission and the other half from fusion.

Source: Thomas B. Cochran, William M. Arkin, and Milton M. Hoenig, *U.S. Nuclear Forces and Capabilities* (Cambridge, Mass.: Ballinger Publishing Company, 1984), Chapter 2.

Appendix C: International Atomic Energy Agency (IAEA) Safeguards

The International Atomic Energy Agency, a Vienna-based U.N.-affiliated organization now having over 110 members, was founded in 1957. By the mid-1960s, it had established a program of on-site inspections, audits, and inventory controls known collectively as "safeguards." Today, the IAEA monitors some nine hundred installations in over fifty nations and the Agency is widely regarded as a principal bulwark against the spread of nuclear arms.

The basic purpose of IAEA safeguards is to deter the diversion of nuclear materials from peaceful uses to military purposes through the risk of timely detection. In simplified terms, the Agency monitors the flow of nuclear materials at nuclear installations by auditing plant records and conducting physical inventories. Seals and cameras are used to ensure materials are not diverted while IAEA inspectors are not present.

To date, the IAEA has never concluded that material under safeguards has been diverted. In September 1981, however, the Agency indicated that it was unable to determine whether material was diverted from Pakistan's KANUPP reactor or from a similarly designed reactor in India because the Agency had not been permitted to apply all of the monitoring apparatus needed at these installations. The Indian deficiency was quickly corrected, but it was nearly a year and a half before the safeguards at the KANUPP reactor were declared adequate. It is possible that during this period plutonium-bearing spent fuel was removed from the reactor without detection.

Apart from these unusual cases, the IAEA safeguards system has some well-recognized limitations. First, and most important, key installations in countries of proliferation concern including enrichment and reprocessing facilities, are not under the IAEA system. Thus, Argentina, Brazil, India, Israel, Pakistan, and South Africa all remain free to use unsafeguarded installations to manufacture material for nuclear weapons. (Even in each of these countries, however, some installations are subject to the IAEA system, and nuclear materials produced in them cannot, therefore, be used for this purpose.) Nations that have ratified the Non-Proliferation Treaty or for which the Treaty of Tlatelolco is in force have accepted safeguards on all their nuclear facilities.

Secondly, certain types of facilities, such as fuel fabrication, reprocessing, and enrichment installations, handle nuclear materials in bulk form, i.e., as powders, liquids, or gasses. Such materials are particularly difficult to safeguard since measurement techniques are not accurate enough to keep track of 100 percent of these substances as they move through the facilities processing them. This makes it theoretically possible to divert a certain small percentage of material for military purposes without detection since this could appear to be a normal operating discrepancy. The problem is especially dangerous at fuel-fabrication plants handling plutonium or highly enriched uranium in powdered form, at reprocessing plants where plutonium is dissolved in various liquids for processing, and at enrichment plants, which use uranium hexafluoride gas as feed.

Low IAEA budgets and manpower have also meant that far fewer inspections are conducted at safeguarded installations than are needed to meet the IAEA's safeguards objectives fully. Other problems include the fact that it is almost impossible for Agency inspectors to make unannounced visits to safeguarded installations. Nations subject to IAEA safeguards are also permitted to reject particular IAEA inspectors.

It must be stressed, however, that even if safeguards as applied are imperfect, their deterrent value remains strong since would-be diverters could not have confidence that their misuse of nuclear materials would go undetected.

Finally, even assuming that IAEA safeguards functioned perfectly, their usefulness may be limited when applied to highly enriched uranium and plutonium, materials directly usable for nuclear weapons. Here, even if the IAEA system reacted instantaneously to diversion, it might still be possible for the nation appropriating this material to manufacture nuclear weapons within a matter of weeks if all the non-nuclear components had been prepared in advance, presenting the world community with a *fait accompli.* In such a setting, safeguards cannot provide "timely warning" sufficient to allow the international community to react before the nation diverting the material has achieved its objective. For this reason, the United States has worked actively to curtail commerce with nations of proliferation concern involving plutonium, highly enriched uranium, or enrichment and reprocessing facilities—whether or not safeguards would be applied. Virtually all other nuclear supplier nations have adopted the cautious approach of the United States in transferring such items.

In the event of a safeguards violation, the Agency's Board of Governors has the authority to notify the United Nations Security Council, but not to impose sanctions of any kind.

Appendix D: Treaty on the Non-Proliferation of Nuclear Weapons

Signed at Washington, London, and Moscow, July 1, 1968
Entered into force March 5, 1970

The States concluding this Treaty, hereinafter referred to as "Parties to the Treaty",

Considering the devastation that would be visited upon all mankind by a nuclear war and the consequent need to make every effort to avert the danger of such a war and to take measures to safeguard the security of peoples,

Believing that the proliferation of nuclear weapons would seriously enhance the danger of nuclear war,

In conformity with resolutions of the United Nations General Assembly calling for the conclusion of an agreement on the prevention of wider dissemination of nuclear weapons,

Undertaking to cooperate in facilitating the application of International Atomic Energy Agency safeguards on peaceful nuclear activities,

Expressing their support for research, development and other efforts to further the application, within the framework of the International Atomic Energy Agency safeguards system, of the principle of safeguarding effectively the flow of source and special fissionable materials by use of instruments and other techniques at certain strategic points,

Affirming the principle that the benefits of peacful applications of nuclear technology, including any technological by-products which may be derived by nuclear-weapon States from the development of nuclear explosive devices, should be available for peaceful purposes to all Parties of the Treaty, whether nuclear-weapon or non-nuclear weapon States,

Convinced that, in furtherance of this principle, all Parties to the Treaty are entitled to participate in the fullest possible exchange of scientific information for, and to contribute alone or in cooperation with other States to, the further development of the applications of atomic energy for peaceful purposes,

Declaring their intention to achieve at the earliest possible date the cessation of the nuclear arms race and to undertake effective measures in the direction of nuclear disarmament,

Urging the cooperation of all States in the attainment of this objective,

Recalling the determination expressed by the Parties to the 1963 Treaty banning nuclear weapon tests in the atmosphere, in outer space and under water in its Preamble to seek to achieve the discontinuance of all test explosions of nuclear weapons for all time and to continue negotiations to this end,

Desiring to further the easing of international tension and the strengthening of trust between States in order to facilitate the cessation of the manufacture of nuclear weapons, the liquidation of all their existing stockpiles, and the elimination from national arsenals of nuclear weapons and the means of their delivery pursuant to a treaty on general and complete disarmament under strict and effective international control,

Recalling that, in accordance with the Charter of the United Nations, States must refrain in their international relations from the threat or use of force against the territorial integrity or political independence of any State, or in any other manner inconsistent with the Purposes of the United Nations, and that the establishment and maintenance of international peace and security are to be promoted with the least diversion for armaments of the world's human and economic resources,

Have agreed as follows:

Article I

Each nuclear-weapon State Party to the Treaty undertakes not to transfer to any recipient whatsoever nuclear weapons or other nuclear explosive devices or control over such weapons or explosive devices directly, or indirectly; and not in any way to assist, encourage, or induce any non-nuclear-weapon State to manufacture or otherwise acquire nuclear weapons or other nuclear explosive devices, or control over such weapons or explosive devices.

Article II

Each non-nuclear-weapon State Party to the Treaty undertakes not to receive the transfer from any transferor or whatsoever of nuclear weapons or other nuclear explosive devices or of control over such weapons or explosive devices directly, or indirectly; not to manufacture or otherwise acquire nuclear weapons or other nuclear explosive devices; and not to seek or receive any assistance in the manufacture of nuclear weapons or other nuclear explosive devices.

Article III

1. Each non-nuclear-weapon State Party to the Treaty undertakes to accept safeguards, as set forth in an agreement to be negotiated and concluded with the International Atomic Energy Agency in accordance with the Statute of the International Atomic Energy Agency and the Agency's safeguards system, for the exclusive purpose of verification of the fulfillment of its obligations assumed under this Treaty with a view to preventing diversion of nuclear energy from peaceful uses to nuclear weapons or other nuclear explosive devices. Procedures for the safeguards required by this article shall be followed with respect to source or special fissionable material whether it is being produced, processed or used in any principal nuclear facility or is outside any such facility. The safeguards required by this article shall be applied to all source or special fissionable material in all peaceful nuclear activities within the territory of such State, under its jurisdiction, or carried out under its control anywhere.

2. Each State Party to the Treaty undertakes not to provide: (a) source or special fissionable material, or (b) equipment or material especially designed or prepared for the processing, use or production of special fissionable material, to any non-nuclear-weapon State for peaceful purposes, unless the source or special fissionable material shall be subject to the safeguards required by this article.

3. The safeguards required by this article shall be implemented in a manner designed to comply with article IV of this Treaty, and to avoid hampering the economic or technological development of the Parties or international co-operation in the field of peaceful nuclear activities, including the international exchange of nuclear material and equipment for the processing, use or production of nuclear material for peaceful purposes in accordance with the provisions of this article and the principle of safeguarding set forth in the Preamble of the Treaty.

4. Non-nuclear-weapon States Party to the Treaty shall conclude agreements with the International Atomic Energy Agency to meet the requirements of this article either individually or together with other States in accordance with the Statute of the International Atomic Energy Agency. Negotiation of such agreements shall commence within 180 days from the original entry into force of this Treaty. For States depositing their instruments of ratification or accession after the 180-day period, negotiation of such agreements shall commence not later than the date of such deposit. Such agreements shall enter into force not later than eighteen months after the date of initiation of negotiations.

Article IV

1. Nothing in this Treaty shall be interpreted as affecting the inalienable right of all the Parties to the Treaty to develop research, production and use of nuclear energy for peaceful purposes without discrimination and in conformity with articles I and II of this Treaty.

2. All the Parties to the Treaty undertake to facilitate, and have the right to participate in, the fullest possible exchange of equipment, materials and scientific and technological information for the peaceful uses of nuclear energy. Parties to the Treaty in a position to do so shall also cooperate in contributing alone or together with other States or international organizations to the further development of the applications of nuclear energy for peaceful purposes, especially in the territories of non-nuclear-weapon States Party to the Treaty, with due consideration for the needs of the developing areas of the world.

Article V

Each party to the Treaty undertakes to take appropriate measures to ensure that, in accordance with this Treaty, under appropriate international observation and through appropriate international procedures, potential benefits from any peaceful applications of nuclear explosions will be made available to non-nuclear-weapon States Party to the Treaty on a nondiscriminatory basis and that the charge to such Parties for the explosive devices used will be as low as possible and exclude any charge for research and development. Non-nuclear-weapon States Party to the Treaty shall be able to obtain such benefits, pursuant to a special international agreement or agreements, through an appropriate international body with adequate representation of non-nuclear-weapon States. Negotiations on this subject shall commence as soon as possible after the Treaty enters into force. Non-nuclear-weapon States Party to the Treaty so desiring may also obtain such benefits pursuant to bilateral agreements.

Article VI

Each of the Parties to the Treaty undertakes to pursue negotiations in good faith on effective measures relating to cessation of the nuclear arms race at an early date and to nuclear disarmament, and on a treaty on general and complete disarmament under strict and effective international control.

Article VII

Nothing in this Treaty affects the right of any group of States to conclude regional treaties in order to assure the total absence of nuclear weapons in their respective territories.

Article VIII

1. Any Party to the Treaty may propose amendments to this Treaty. The text of any proposed amendment shall be submitted to the Depositary Governments which shall circulate it to all Parties to the Treaty. Thereupon, if requested to do so by one-third or more of the Parties to the Treaty, the Depositary Governments shall convene a conference, to which they shall invite all the Parties to the Treaty, to consider such an amendment.

2. Any amendment to this Treaty must be approved by a majority of the votes of all the Parties to the Treaty, including the votes of all nuclear-weapon States Party to the Treaty and all other Parties which, on the date the amendment is circulated, are members of the Board of Governors of the International Atomic Energy Agency. The amendment shall enter into force for each Party that deposits its instrument of ratification of the amendment upon the deposit of such instruments of ratification by a majority of all the Parties, including the instruments of ratification of all nuclear-weapon States Party to the Treaty and all other Parties which, on the date the amendment is circulated, are members of the Board of Governors of the International Atomic Energy Agency. Thereafter, it shall enter into force for any other Party upon the deposit of its instrument of ratification of the amendment.

3. Five years after the entry into force of this Treaty, a conference of Parties

to the Treaty shall be held in Geneva, Switzerland, in order to review the operation of this Treaty with a view to assuring that the purposes of the Preamble and the provisions of the Treaty are being realized. At intervals of five years thereafter, a majority of the Parties to the Treaty may obtain, by submitting a proposal to this effect to the Depositary Governments, the convening of further conferences with the same objective of reviewing the operation of the Treaty.

Article IX

1. This Treaty shall be open to all States for signature. Any State which does not sign the Treaty before its entry into force in accordance with paragraph 3 of this article may accede to it at any time.
2. This Treaty shall be subject to ratification by signatory States. Instruments of ratification and instruments of accession shall be deposited with the Governments of the United States of America, the United Kingdom of Great Britain and Northern Ireland and the Union of Soviet Socialist Republics, which are hereby designated the Depositary Governments.
3. This Treaty shall enter into force after its ratification by the States, the Governments of which are designated Depositaries of the Treaty, and forty other States signatory to this Treaty and the deposit of their instruments of ratification. For the purposes of this Treaty, a nuclear-weapon State is one which has manufactured and exploded a nuclear weapon or other nuclear explosive device prior to January 1, 1967.
4. For States whose instruments of ratification or accession are deposited subsequent to the entry into force of this Treaty, it shall enter into force on the date of the deposit of their instruments of ratification or accession.
5. The Depositary Governments shall promptly inform all signatory and acceding States of the date of each signature, the date of deposit of each instrument of ratification or of accession, the date of the entry into force of this Treaty, and the date of receipt of any requests for convening a conference or other notices.
6. This Treaty shall be registered by the Depositary Governments pursuant to article 102 of the Charter of the United Nations.

Article X

1. Each Party shall in exercising its national sovereignty have the right to withdraw from the Treaty if it decides that extraordinary events, related to the subject matter of is Treaty, have jeopardized the supreme interests of its country. It shall give notice of such withdrawal to all other Parties to the Treaty and to the United Nations Security Council three months in advance. Such notice shall include a statement of the extraordinary events it regards as having jeopardized its supreme interests.
2. Twenty-five years after the entry into force of the Treaty, a conference

shall be convened to decide whether the Treaty shall continue in force indefinitely, or shall be extended for an additional fixed period or periods. This decision shall be taken by a majority of the Parties to the Treaty.

Article XI

This Treaty, the English, Russian, French, Spanish and Chinese texts of which are equally authentic, shall be deposited in the archives of the Depositary Governments. Duly certified copies of this Treaty shall be transmitted by the Depositary Governments to the Governments of the signatory and acceding States.

IN WITNESS WHEREOF the undersigned, duly authorized, have signed this Treaty.

DONE in triplicate, at the cities of Washington, London and Moscow, this first day of July one thousand nine hundred sixty-eight.

Source: U.S. Arms Control and Disarmament Agency, *Arms Control and Disarmament Agreements: Text and Histories of Negotiations* (Washington, D.C.: U.S. ACDA, 1982).

A. Parties to the Non-Proliferation Treaty

Afghanistan	1970	Dominican Republic	1971
Antigua and Barbuda	1985	Ecuador	1969
Australia	1973	Egypt	1981
Austria	1969	El Salvador	1972
Bahamas, The	1976	Equatorial Guinea	1984
Bahrain	1988	Ethiopia	1970
Bangladesh	1979	Fiji	1972
Barbados	1980	Finland	1969
Belgium	1975	Gabon	1974
Belize	1985	Gambia, The	1975
Benin	1972	Germany (East)	1969
Bhutan	1985	Germany (West)	1975
Bolivia	1970	Ghana	1970
Botswana	1969	Greece	1970
Brunei	1985	Grenada	1975
Bulgaria	1969	Guatemala	1970
Burkina Faso	1970	Guinea	1985
Burundi	1971	Guinea-Bissau	1976
Cameroon	1969	Haiti	1970
Canada	1969	Holy See	1971
Cape Verde	1979	Honduras	1973
Central African Rep.	1970	Hungary	1969
Chad	1971	Iceland	1969
Colombia	1986	Indonesia	1979
Congo	1978	Iran	1970
Costa Rica	1970	Iraq	1969
Cyprus	1970	Ireland	1968
Czechoslovakia	1969	Italy	1975
Denmark	1969	Ivory Coast	1973
Dominica	1968	Jamaica	1970

Japan	1976
Jordan	1970
Kampuchea	1972
Kenya	1970
Kiribati	1985
Korea (North)	1985
Korea (South)	1975
Kuwait	1989
Laos	1970
Lebanon	1970
Lesotho	1970
Liberia	1970
Libya	1975
Liechtenstein	1978
Luxembourg	1975
Madagascar	1970
Malawi	1986
Malaysia	1970
Maldives	1970
Mali	1970
Malta	1970
Mauritius	1969
Mexico	1969
Mongolia	1969
Morocco	1970
Nauru	1982
Nepal	1970
Netherlands	1975
New Zealand	1969
Nicaragua	1973
Nigeria	1968
Norway	1969
Panama	1977
Papua New Guinea	1982
Paraguay	1970
Peru	1970
Philippines	1972
Poland	1969
Portugal	1977
Qatar	1989
Romania	1970
Rwanda	1975
San Marino	1970
Sao Tome & Principe	1983
Saudi Arabia	1988
St. Lucia	1979
St. Vincent & The Grenadines	1984
Senegal	1970
Seychelles	1985
Sierra Leone	1975
Singapore	1976
Solomon Islands	1981
Somalia	1970
Spain	1987
Sri Lanka	1979
Sudan	1973
Suriname	1976
Swaziland	1969
Sweden	1970
Switzerland	1977
Syrian Arab Republic	1969
[Taiwan	1970]
Thailand	1972
Togo	1970
Tonga	1971
Trinidad & Tobago	1986
Tunisia	1970
Turkey	1980
Tuvalu	1979
Uganda	1982
USSR*	1970
United Kingdom*	1968
United States*	1970
Uruguay	1970
Venezuela	1975
Vietnam	1982
Western Samoa	1975
Yemen (Aden)	1979
Yemen (Sana)	1986
Yugoslavia	1970
Zaire	1970

B. Countries that have neither signed nor ratified the Treaty

Albania
Algeria
Angola
Argentina
Brazil
Burma (Myanmar)
Chile
Comoros
China*
Cuba
Djibouti
France*
Guyana
India**
Israel
Mauritania
Monaco
Mozambique
Namibia
Niger
Oman
Pakistan
South Africa
Tanzania
United Arab Emirates
Vanuatu
Zambia
Zimbabwe

* Declared nuclear weapon state
** India has detonated a "peaceful nuclear device."

Appendix E: The Treaty of Tlatelolco

The Treaty on the Prohibition of Nuclear Weapons in Latin America (the Treaty of Tlatelolco) establishes a nuclear-weapons free zone in that region. Parties to the treaty agree not to manufacture, test, or acquire nuclear weapons or to accept weapons on their territory deployed by others. To verify that these pledges are kept, adherents agree to accept "full-scope" International Atomic Energy Agency (IAEA) safeguards (i.e., IAEA accounting and inspection measures on all of a nation's peaceful nuclear activities). In addition, the treaty establishes the Agency for the Prohibition of Nuclear Weapons in Latin America (OPANAL). OPANAL will undertake special inspections at the request of members who have reason to believe that another party is engaging in prohibited activity, a unique investigatory function not available under the IAEA system.

Under its entry-into-force provisions, the treaty becomes effective once it has been ratified by all eligible countries in the region. However, twenty-two nations have ratified the accord and waived this provision so that the treaty has become effective for these countries. Only four countries have yet to make the treaty operative. Brazil and Chile have ratified the treaty, but have not waived the entry-into-force requirement. Argentina has signed but not ratified, and Cuba has neither signed nor ratified the accord. None of these four states, therefore, is bound by the treaty's full-scope safeguards requirement. Because Cuba has made approval of the treaty contingent upon U.S. withdrawal from the Guantanamo naval base, progress toward full effectiveness has been stymied.

The treaty is supplemented by two protocols that apply to countries outside the region. Protocol I requires that outside nations with territories in Latin America respect the treaty's denuclearization requirements with respect to those territories. Protocol II prohibits nuclear-weapon states from using or threatening to use nuclear arms against treaty parties. All nations with territories in the region, the U.S., U.K., France, and the Netherlands have signed Protocol I, and all but France have ratified it. All nuclear-weapon states have ratified Protocol II.

One highly controversial issue arises from the treaty's definition in Article 5 of a nuclear weapon as a nuclear explosive "which has a group of characteristics that are appropriate for use for warlike purposes." When the United States and the Soviet Union ratified Protocol II they formally stated that in their view this phrase meant that the treaty's prohibitions applied to *all* nuclear explosives, including so-called "peaceful nuclear explosives" since there was no technological difference between them and nuclear weapons. Argentina and Brazil have objected to this interpretation and cited Article 18 of the treaty which states that parties "may carry out explosions for peaceful purposes—including explosions which involve devices similar to those used in nuclear weapons"—under IAEA supervision. Virtually all other treaty parties, however, have accepted the U.S.-Soviet view since they are also parties to the Non-Proliferation Treaty, which expressly prohibits the manufacture of any nuclear explosive.

*

Country	Year of Signature	Year of Ratification
Argentina	1967	—
Antigua and Barbuda	1983	1983
Bahamas, The	1967	1976
Barbados	1968	1969
Bolivia	1967	1969
Brazil	1967	1968*
Chile	1967	1974*
Colombia	1967	1972
Costa Rica	1967	1969
Dominican Republic	1967	1968
Ecuador	1967	1969
El Salvador	1967	1968
Grenada	1975	1975
Guatemala	1967	1970
Haiti	1967	1969
Honduras	1967	1968
Jamaica	1967	1969
Mexico	1967	1967
Nicaragua	1967	1968
Panama	1967	1971
Paraguay	1967	1969
Peru	1967	1969
Suriname	1976	1977
Trinidad and Tobago	1967	1975
Uruguay	1967	1968
Venezuela	1967	1970

* Not in force because entry-into-force provision not waived.

Protocol I to the Treaty

Country	Year of Signature	Year of Ratification
France	1979	—
Netherlands	1968	1971
United Kingdom	1967	1969
United States	1977	1981

Protocol II to the Treaty

Country	Year of Signature	Year of Ratification
China (PRC)	1973	1974
France	1973	1974
U.S.S.R.	1978	1979
United Kingdom	1967	1969
United States	1968	1971

Source: U.S. Arms Control and Disarmament Agency, *Arms Control and Disarmament Agreements: Texts and Histories of Negotiations* (Washington, D.C.: U.S. Arms Control and Disarmament Agency, 1982), and U.S. Department of State.

Appendix F: Nuclear Suppliers Organizations

Non-Proliferation Treaty Exporters Committee (Zangger Committee)

Shortly after the treaty came into force in 1970, a number of countries entered into consultations concerning the procedures and standards they would apply to nuclear fuel and equipment exports to non-nuclear-weapon states in order to implement the requirement in the pact that such exports and any enriched uranium or plutonium produced through their use be subject to IAEA safeguards. The countries engaged in those consultations, which were chaired by the Swiss expert, Claude Zangger, were parties to the Non-Proliferation Treaty (or have since become parties) and were also exporters or potential exporters of material and equipment for peaceful uses of nuclear energy.

In August 1974, the governments of Australia, Denmark, Canada, Finland, West Germany, the Netherlands, Norway, the Soviet Union, the United Kingdom, and the United States each informed the Director General of the IAEA, by individual letters, of their intentions to require IAEA safeguards on their nuclear exports in accordance with certain procedures described in memoranda enclosed with their letters. Those memoranda were identical in the case of each letter and included a "Trigger List" of materials and items of equipment which would be exported only under such safeguards. (The individual letters and the identical memoranda were published by the IAEA in September 1974 in document INFCIRC/209).

Subsequently, Austria, Czechoslovakia, East Germany, Ireland, Japan, Luxembourg, Poland, and Sweden sent individual letters to the Director General, referring to and enclosing memoranda identical to those transmitted by the initial groups of governments.

The agreed procedures and Trigger List represented the first major agreement on uniform regulation of nuclear exports by actual and potential nuclear suppliers. It had great significance for several reasons. It was an attempt to strictly and uniformly enforce the obligations of Article III, paragraph 2, of the Non-Proliferation Treaty requiring safeguards on nuclear exports. It was intended to reduce the likelihood that states would be tempted to cut corners on safeguards requirements, because of competition in the sale of nuclear equipment and fuel-cycle services. In addition, and very important in light of subsequent events, it established the principle that nuclear-supplier nations should consult and agree among themselves on procedures to regulate the international market for nuclear materials and equipment in the interest of non-proliferation. Notably absent from the list of actual participants or potential suppliers, as from the list of parties to the Non-Proliferation Treaty, were France, India, and the People's Republic of China.

Nuclear Suppliers Group

In November 1974, within a year of the delivery of these memoranda a second series of supplier negotiations were underway. This round, convened largely at the initiative of the United States, was a response to three developments: 1) the Indian nuclear test of May 1974, 2) mounting evidence that the pricing actions of the Organization of Oil Exporting Countries were stimulating Third World and other non-nuclear states to initiate or accelerate their nuclear power programs, and 3) recent contracts or continuing negotiations on the part of France and West Germany for the supply of enrichment or reprocessing facilities to Third World states.

The initial participants in these discussions, conducted in London were Canada, the Federal Republic of Germany, France, Japan, the Soviet Union, the United Kingdom, and the United States. One of the group's chief accomplishments was to induce France to join in such efforts, since France (which had not joined the Non-Proliferation Treaty or the Zangger Committee) could have undercut reforms of nuclear supply. The French, hesitant about becoming involved and uncertain as to where the effort might lead, insisted that any meetings be kept confidential—which was also the preference of some other participants. So the meetings in London were held in secret. But it soon became known that such meetings were taking place, and this led to suspicion and exaggerated fears of what they were about. The group was inaccurately referred to as a "cartel." Instead, one of its purposes was to foster genuine commercial competition based on quality and prices, untainted by bargaining away proliferation controls.

Two major issues were discussed in the series of meetings which led to a new agreement in late 1975. The first was if, and under what conditions, technology and equipment for enrichment and reprocessing, the most sensitive parts of the nuclear fuel cycle from a weapons proliferation perspective, should be transferred to non-nuclear states. The United States, with support from several other participants, was reported to argue in favor of both a prohibition on such transfer and a commitment to reprocessing in multinational facilities. France had already signed contracts to sell reprocessing plants to Pakistan and South Korea, and West Germany had agreed to sell technology and facilities for the full fuel cycle (including enrichment and reprocessing) to Brazil. They successfully resisted the prohibition proposed by others.

The second issue was whether transfers should be made to states unwilling to submit all non-military nuclear facilities to IAEA safeguards, or whether total industry (full-scope) safeguards should become a condition of sales. The Nuclear Suppliers Group came close to reaching consensus on requiring full-scope safeguards in recipient countries as a condition of future supply commitments, but was unable to persuade the French and the West Germans, though they did not rule out later reconsideration and possible changes by unanimous consent. The group did act to expand safeguards coverage by adopting a "Trigger List" of exports, similar to that of the Zangger Committee, which would be made only if covered by IAEA safeguards in the recipient state.

On January 27, 1976, the seven participants in the negotiations exchanged letters endorsing a uniform code for conducting international nuclear sales. The major provisions of the agreement require that before nuclear materials, equipment, or technology are transferred the recipient state must:

1. pledge not to use the transferred materials, equipment, or technology in the manufacture of nuclear explosives of any kind;
2. accept, with no provision for termination, international safeguards on all transferred materials and facilities employing transferred equipment or technology, including any enrichment, reprocessing, or heavy-water production facility that replicates or otherwise employs transferred technology;
3. provide adequate physical security for transferred nuclear facilities and materials to prevent theft and sabotage;
4. agree not to retransfer the materials, equipment, or technology to third countries unless they too accept the constraints on use, replication, security, and transfer, and unless the original supplier nation concurs in the transactions;
5. employ "restraint" regarding the possible export of "sensitive" items (relating to uranium enrichment, spent fuel reprocessing, and heavy-water production); and
6. encourage the concept of multilateral regional facilities for reprocessing and enrichment.

The guidelines have now been adopted by the United States, Great Britain, France, West Germany, Japan, Canada, the Soviet Union, Belgium, Italy, the Netherlands, Sweden, Switzerland, Czechoslovakia, East Germany, Poland, Australia, and Finland.

The Nuclear Suppliers' Guidelines extended the Zangger Committee's requirements in several respects. First, France (which had not participated in the Zangger group) agreed to key points adopted by that Committee, such as the requirement that nuclear export recipients pledge not to use transferred items for nuclear explosives of any kind and that safeguards on transferred items would continue indefinitely. Secondly, the Suppliers Group went beyond the Non-Proliferation Treaty and the Zangger Committee requirements, by imposing safeguards not only on the export of nuclear materials and equipment, but also on nuclear technology exports. India had demonstrated the existence of this serious loophole by building its own unsafeguarded replica of a safeguarded power reactor imported from Canada. The Suppliers Group was unable to reach agreement on the application of this reform to reactor technology, however, and so confined its recommended application to "sensitive" facilities—i.e., reprocessing, enrichment, and heavy-water production plants built with the use of exported technology. The group's acceptance of this limited reform was facilitated by the fact that such a condition was incorporated by West Germany in its safeguards agreements for sale of enrichment and reprocessing facilities to Brazil, and the French in their safeguards agreements covering their proposed sales of reprocessing plants to the Republic of Korea and Pakistan.

Third, the Suppliers' Guidelines, while not prohibiting the export of these sensitive facilities, do embody the participants' agreement to "exercise restraint" in transferring them, and where enrichment plants are involved, to seek recipient country commitments that such facilities will be designed and operated to produce only low-enriched uranium, not suitable for weapons.

Several supplier countries subsequently announced policies stricter than those in the guidelines. France, West Germany, and the United States all made separate public announcements that they did not, at least for the time being, contemplate any further new commitments to export reprocessing technology. In addition, the United States, Canada, Australia, and Sweden have all made recipient-country acceptance of full-scope safeguards a condition for nuclear transfer.

***

Excerpted and adapted from, Charles N. Van Doren, "Nuclear Supply and Non-Proliferation: The IAEA Committee on Assurances of Supply," A Report for the Congressional Research Service (Rep. No. 83-202-8) October 1983, pp. 61-64; U.S. Congress, Office of Technology Assessment, *Nuclear Proliferation and Safeguards* (Washington, D.C.: OTA, 1977), pp. 220-221; U.S. Department of State, "Report to the Congress Pursuant to Section 601 of the Nuclear Non-Proliferation Act of 1978" (January 1979), pp. 25-27.

Glossary

atomic bomb A bomb whose energy comes from the fission of uranium or plutonium.

beryllium A highly toxic steel-gray metal, possessing a low neutron absorption cross-section and high melting point, which can be used in nuclear reactors as a moderator, reflector, or cladding material. In nuclear weapons, beryllium surrounds the fissile material and reflects neutrons back into the nuclear reaction, considerably reducing the amount of fissile material required. Beryllium is also used in guidance systems and other parts for aircraft, missiles or space vehicles.

blanket A layer of fertile material, such as uranium-238 or thorium-232, placed around the core of a reactor. During operation of the reactor, additional fissionable material is produced in the blanket.

breeder reactor A nuclear reactor that produces somewhat more fissile material than it consumes. The fissile material is produced both in the reactor's core and when neutrons are captured in fertile material placed around the core (blanket). This process is known as breeding. Breeder reactors have not yet reached commercialization, although active research and development programs are being pursued by a number of countries.

CANDU (Canadian deuterium-uranium reactor.) The most widely used type of heavy-water reactor. The CANDU reactor uses natural uranium as a fuel and heavy water as a moderator and a coolant.

centrifuge See ultracentrifuge.

chain reaction The continuing process of nuclear fissioning in which the neutrons released from a fission trigger at least one other nuclear fission. In a nuclear weapon an extremely rapid, multiplying chain reaction causes the explosive release of energy. In a reactor, the pace of the chain reaction is controlled to produce heat (in a power reactor) or large quantities of neutrons (in a research or production reactor).

chemical processing Chemical treatment of materials to separate specific usable constituents.

coolant A substance circulated through a nuclear reactor to remove or transfer heat. The most common coolants are water and heavy water.

core The central portion of a nuclear reactor containing the fuel elements and, usually, the moderator. Also the central portion of a nuclear weapon containing highly enriched uranium or plutonium.

critical mass The minimum amount of fissionable material required to sustain a chain reaction. The exact mass varies with many factors such as the particular fissionable isotope present, its concentration and chemical form, the geometrical arrangement of the material, and its density. When fissionable materials are compressed by high explosives in implosion-type atomic weapons, the critical mass needed for a nuclear explosion is reduced.

depleted uranium Uranium having a smaller percentage of uranium-235 than the 0.7 percent found in natural uranium. It is a by-product of the uranium enrichment process, during which uranium-235 is culled from one batch of uranium, thereby depleting it, and then added to another batch to increase its concentration of uranium-235.

enrichment The process of increasing the concentration of one isotope of a given element (in the case of uranium, increasing the concentration of uranium-235).

feed stock Material introduced into a facility for processing.

fertile Material composed of atoms which readily absorb neutrons to produce fissionable materials. One such element is uranium-238, which becomes plutonium-239 after it absorbs a neutron. Fertile material alone cannot sustain a chain reaction.

fission The process by which a neutron strikes a nucleus and splits it into fragments. During the process of nuclear fission, several neutrons are emitted at high speed, and heat and radiation are released.

fissile material Material composed of atoms which readily fission when struck by a neutron. Uranium-235 and plutonium-239 are examples of fissile materials.

fusion The formation of a heavier nucleus from two lighter ones (such as hydrogen isotopes), with the attendant release of energy (as in a hydrogen bomb).

gas centrifuge process A method of isotope separation in which heavy gaseous atoms or molecules are separated from light ones by centrifugal force. See ultracentrifuge.

gaseous diffusion A method of isotope separation based on the fact that gas atoms or molecules with different masses will diffuse through a porous barrier (or membrane) at different rates. The method is used to separate uranium-235 from uranium-238. It requires large gaseous diffusion plants and significant amounts of electric power.

gas-graphite reactor A nuclear reactor in which a gas is the coolant and graphite is the moderator.

heavy water Water containing significantly more than the natural proportion (1 in 6,500) of heavy hydrogen (deuterium) atoms to ordinary hydrogen atoms. (Hydrogen atoms have one proton, deuterium atoms have one proton and one neutron.) Heavy water is used as a moderator in some reactors because it slows down neutrons effectively and does not absorb them (unlike light, or normal, water) making it possible to fission natural uranium and sustain a chain reaction.

heavy-water reactor A reactor that uses heavy water as its moderator and natural uranium as fuel. See CANDU.

highly enriched uranium Uranium in which the percentage of uranium-235 nuclei has been increased from the natural level of 0.7 percent to some level greater than 20 percent, usually around 90 percent.

hot cells Lead-shielded rooms with remote handling equipment for examining and processing radioactive materials. In particular, hot cells are used for reprocessing spent reactor fuel.

hydrogen bomb A nuclear weapon that derives its energy largely from fusion. Also known as a thermonuclear bomb.

irradiation Exposure to a radioactive source; usually in the case of fuel materials, being placed in an operating nuclear reactor.

isotopes Atoms having the same number of protons, but a different number of neutrons. Two isotopes of the same atom are very similar and difficult to separate by ordinary chemical means. Isotopes can have very different nuclear properties, however. For example, one isotope may fission readily, while another isotope of the same atom may not fission at all. An isotope is specified by its atomic mass number (the number of protons plus neutrons) following the symbol denoting the chemical element (e.g., U^{235} is an isotope of uranium).

jet-nozzle enrichment method A process of uranium enrichment that uses both uranium hexafluoride and a light gas flowing at high speed through a nozzle along curved walls.

kilogram A metric weight equivalent to 2.2 pounds.

kiloton The energy of a nuclear explosion that is equivalent to an explosion of 1,000 tons of TNT.

laser enrichment method A still experimental process of uranium enrichment in which a finely tuned, high-power carbon dioxide laser is used to differentially excite molecules of various atomic weights. This differential excitation makes it possible to separate uranium-235 from uranium-238.

light water Ordinary water (H_2O), as distinguished from heavy water (D_2O).

light-water reactor A reactor that uses ordinary water as moderator and coolant and low-enriched uranium as fuel.

low-enriched uranium Uranium in which the percentage of uranium-235 nuclei has been increased from the natural level of 0.7 percent to less than 20 percent, usually 3 to 6 percent. With the increased level of fissile material, low-enriched uranium can sustain a chain reaction when immersed in light-water and is used as fuel in light-water reactors.

medium-enriched uranium Uranium in which the percentage of uranium-235 nuclei has been increased from the natural level of 0.7 percent to between 20 and 50 percent. (Potentially usable for nuclear weapons, but very large quantities are needed.)

megawatt One million watts; used in reference to a nuclear power plant, one million watts of electricity.

metric ton One thousand kilograms. A metric weight equivalent to 2200 pounds or 1.1 tons.

milling A process in the uranium fuel cycle by which ore containing only a very small percentage of uranium oxide (U_3O_8) is converted into material containing a high percentage (80 percent) of U_3O_8, often referred to as yellowcake.

moderator A component (usually water, heavy water, or graphite) of some nuclear reactors that slows neutrons, thereby increasing their chances of fissioning fertile material.

natural uranium Uranium as found in nature, containing 0.7 percent of uranium-235, 99.3 percent of uranium-238, and a trace of uranium-234.

neutron An uncharged elementary particle, with a mass slightly greater than that of a proton, found in the nucleus of every atom heavier than hydrogen.

nuclear energy The energy liberated by a nuclear reaction (fission or fusion) or by spontaneous radioactivity.

nuclear fuel Basic chain-reacting material, including both fissile and fertile materials. Commonly used nuclear fuels are natural uranium, and low-enriched uranium; high-enriched uranium and plutonium are used in some reactors.

nuclear fuel cycle The set of chemical and physical operations needed to prepare nuclear material for use in reactors and to dispose of or recycle the material after its removal from the reactor. Existing fuel cycles begin with uranium as the natural resource and create plutonium as a by-product. Some future fuel cycles may rely on thorium and produce the fissionable isotope uranium-233.

nuclear fuel element A rod, tube, plate, or other mechanical shape or form into which nuclear fuel is fabricated for use in a reactor.

nuclear fuel fabrication plant A facility where the nuclear material (e.g., enriched or natural uranium) is fabricated into fuel elements to be inserted into a reactor.

nuclear power plant Any device or assembly that converts nuclear energy into useful power. In a nuclear electric power plant, heat produced by a reactor is used to produce steam to drive a turbine that in turn drives an electricity generator.

nuclear reactor A mechanism fueled by fissionable materials that give off neutrons, thereby inducing heat. Reactors are of three general types: power reactors, production reactors, and research reactors.

nuclear waste The radioactive by-products formed by fission and other nuclear processes in a reactor. Most nuclear waste is initially contained spent fuel. If this material is reprocessed, new categories of waste result.

nuclear weapons A collective term for atomic bombs and hydrogen bombs. Weapons based on a nuclear explosion. Generally used throughout the text to mean atomic bombs, only, unless used with reference to nuclear weapon states, (all five of which have both atomic and hydrogen weapons).

plutonium-239 A fissile isotope occurring naturally in only minute quantities, which is manufactured artificially when uranium-238, through irradiation, captures an extra neutron. It is one of the two materials that have been used for the core of nuclear weapons, the other being highly enriched uranium.

plutonium-240 A fissile isotope produced in reactors when a plutonium-239 atom absorbs a neutron instead of fissioning. Its presence complicates the construction of nuclear explosives because of its high rate of spontaneous fission.

power reactor A reactor designed to produce electricity as distinguished from reactors used primarily for research or for producing radiation or fissionable materials.

production reactor A reactor designed primarily for large-scale production of plutonium-239 by neutron irradiation of uranium-238.

radioactivity The spontaneous disintegration of an unstable atomic nucleus resulting in the emission of subatomic particles.

radioisotope A radioactive isotope.

recycle To reuse the remaining uranium and plutonium found in spent fuel after they have been separated at a reprocessing plant from unwanted radioactive waste products also in the spent fuel.

reprocessing Chemical treatment of spent reactor fuel to separate the plutonium and uranium from the unwanted radioactive waste by-products and (under present plans) from each other.

research reactor A reactor primarily designed to supply neutrons for experimental purposes. It may also be used for training, materials testing, and production of radioisotopes.

spent fuel Fuel elements that have been removed from the reactor after use because they contain too little fissile and fertile material and too high a concentration of unwanted radioactive by-products to sustain reactor operation. Spent fuel is both thermally and radioactively hot.

thermonuclear bomb A hydrogen bomb.

thorium-232 A fertile material.

tritium The heaviest hydrogen isotope, containing one proton and two neutrons in the nucleus, produced most effectively by bombarding lithium-6 with neutrons. In a fission weapon, tritium produces excess neutrons, which set off additional reactions in the weapon's fissile material. In this way, tritium can either reduce the amount of fissile material required, or multiply (i.e., boost) the weapon's destructive power as much as five times. In fusion reactions, tritium and deuterium, another hydrogen isotope, bond at very high temperatures, releasing approximately 14 million electron-volts of energy per set of neutrons.

ultracentrifuge A rotating vessel that can be used for enrichment of uranium. The heavier isotopes of uranium hexafluoride gas concentrate at the walls of the rotating centrifuge and are drawn off.

uranium A radioactive element with the atomic number 92 and, as found in natural ores, an average atomic weight of 238. The two principal natural isotopes are uranium-235 (0.7 percent of natural uranium), which is fissionable, and uranium-238 (99.3 percent of natural uranium), which is fertile.

uranium-233 (U^{233}) A fissionable isotope bred in fertile thorium-232. Like plutonium-239 it is theoretically an excellent material for nuclear weapons, but is not known to have been used for this purpose. Can be used as reactor fuel.

uranium-235 (U^{235}) The only naturally occurring fissionable isotope. Natural uranium contains 0.7 percent U^{235}; light-water reactors use about 3 percent and weapons grade, highly enriched uranium normally consists of 93 percent of this isotope.

uranium-238 A fertile material. Natural uranium is composed of approximately 99.3 percent U^{238}.

uranium dioxide (UO_2) Purified uranium. The form of natural uranium used in heavy water reactors. Also the form of uranium that remains after the fluorine is removed from enriched uranium hexafluoride (UF^6). Produced as a powder, uranium dioxide is, in turn, fabricated into fuel elements.

uranium oxide (U_3O_8) The most common oxide of uranium found in typical ores. U_3O_8 is extracted from the ore during the milling process. The ore typically contains only 0.1 percent U_3O_8; yellowcake, the product of the milling process, contains about 80 percent U_3O_8.

uranium hexafluoride (UF_6) A volatile compound of uranium and fluorine. UF_6 is a solid at atmospheric pressure and room temperature, but can be transformed into gas by heating. UF_6 gas (alone, or in combination with hydrogen or helium) is the feed stock in all uranium enrichment processes and is sometimes produced as an intermediate product in the process of purifying yellowcake to produce uranium oxide.

vessel The part of a reactor that contains the nuclear fuel.

weapons grade Nuclear material of the type most suitable for nuclear weapons, i.e., uranium enriched to 93 percent U^{235} or plutonium that is primarily Pu^{239}.

weapons-usable Fissionable material that is weapons-grade or, though less than ideal for weapons, can still be used to make a nuclear explosive.

yellowcake A concentrate produced during the milling process that contains about 80 percent uranium oxide (U_3O_8). In preparation for uranium enrichment, the yellowcake is converted to uranium hexafluoride gas (UF_6). In the preparation of natural uranium reactor fuel, yellowcake is processed into purified uranium dioxide. Sometimes uranium hexaflouride is produced as an intermediate step in the purification process.

yield The total energy released in a nuclear explosion. It is usually expressed in equivalent tons of TNT (the quantity of TNT required to produce a corresponding amount of energy).

zirconium A grayish-white lustrous metal which is commonly used in an alloy form (*i.e.*, zircaloy) to encase fuel rods in nuclear reactors.

Sources

1. Congressional Research Service, *Nuclear Proliferation Factbook* (Washington, DC: U.S. Govt. Printing Office, 1977).
2. Energy Research & Development Administration, *U.S. Nuclear Power Export Activities* (Springfield, VA: National Technical Information Service, 1976).
3. Nuclear Energy Policy Study Group, *Nuclear Power: Issues and Choices* (Cambridge, MA: Ballinger Publishing Co., 1977).
4. Office of Technology Assessment, *Nuclear Proliferation and Safeguards* (Washington, DC: Office of Technology Assessment, 1977); Nuclear Power in an Age of Uncertainty (Washington, D.C.: Office of Technology Assessment, 1984).
5. Wohlstetter, Albert, *Swords from Plowshares: The Military Potential of Civilian Nuclear Energy* (Chicago: The University of Chicago Press, 1977).
6. United Nations Association of the USA, *Nuclear Proliferation: A Citizen's Guide to Policy Choices* (N.Y.: UNA-USA, 1983).

Index

For Product Safety Concerns and Information please contact our EU representative GPSR@taylorandfrancis.com
Taylor & Francis Verlag GmbH, Kaufingerstraße 24, 80331 München, Germany

www.ingramcontent.com/pod-product-compliance
Ingram Content Group UK Ltd.
Pitfield, Milton Keynes, MK11 3LW, UK
UKHW022114190926
475702UK00013B/1428
* 9 7 8 0 3 6 7 1 6 5 6 3 5 *